Molecular and Cellular Biophysics

This book provides advanced undergraduate and beginning graduate students with a foundation in the basic concepts of molecular and cellular biophysics. Students who have taken physical chemistry and calculus courses will find this book an accessible and valuable aid in learning how these concepts can be used in biological research. The text provides a rigorous treatment of the fundamental theories in biophysics and illustrates their application with examples. Conformational transitions of proteins are studied first using thermodynamics, and subsequently with kinetics. Allosteric theory is developed as the synthesis of conformational transitions and association reactions. Basic ideas of thermodynamics and kinetics are applied to topics such as protein folding, enzyme catalysis and ion channel permeation. These concepts are then used as the building blocks in a treatment of membrane excitability. Through these examples, students will gain an understanding of the general importance and broad applicability of biophysical principles to biological problems.

Meyer B. Jackson is the Kenneth Cole Professor of Physiology at the University of Wisconsin Medical School. He has been teaching graduate level biophysics for nearly 25 years.

Molecular and Cellular Biophysics

Meyer B. Jackson
University of Wisconsin Medical School

CAMBRIDGE UNIVERSITY PRESS
Cambridge, New York, Melbourne, Madrid, Cape Town, Singapore,
São Paulo, Delhi, Dubai, Tokyo

Cambridge University Press
The Edinburgh Building, Cambridge CB2 8RU, UK

Published in the United States of America by Cambridge University Press, New York

www.cambridge.org
Information on this title: www.cambridge.org/9780521624701

First published 2006

A catalogue record for this publication is available from the British Library

ISBN 978-0-521-62441-1 Hardback
ISBN 978-0-521-62470-1 Paperback

Transferred to digital printing (with corrections) 2010

Contents

Preface

I have tried to present the subject of biophysics from a conceptual perspective. This needs to be stated because biophysics is too often defined as a collection of physical methods that can be used to study molecular and cellular biology. This technical emphasis often fosters narrowness, and in the worst cases leads to shallowness, where sophisticated measurements are interpreted with little consideration for the physical principles that govern the special complexities of the macromolecular world of biology.

The conceptual emphasis of this book has lead to a heavy dose of theory. Theoretical analysis is essential in a conceptual approach, but I must admit that the theoretical emphasis of this book also reflects my own personal fascination with the insights that can be gained by applying physical theory to biological questions. In developing theoretical topics I have tried to be practical. I have steered toward more basic forms of mathematics wherever possible. Much of the analysis is at the level of an introductory calculus course. Where more sophisticated mathematics is involved I have tried to teach the mathematics in parallel with the development of the subject at hand. Six mathematical appendices have been added to help the reader. These may be useful guides, but are certainly not rigorous or thorough. Readers who desire a better background in mathematics will have to find appropriate texts that treat subjects such as matrices and partial differential equations. The relevant chapters in a book on mathematical methods for physics or chemistry will probably fill the gap adequately.

The level of the mathematics is not the critical issue. The most essential pre-requisite here is physical chemistry. Everything has been written with the assumption that the reader has taken an undergraduate course that introduces thermodynamics, kinetics, and statistical mechanics. Some of the essentials are reviewed but my summaries cannot substitute for some intensive study focused on these topics. I also assume that the reader has had some exposure to biochemistry.

The concepts developed here are often quite general, and illustrations with specific examples are vital. Finding suitable examples has been a challenge. I have tried to avoid excessive reliance on examples from areas closer to my own research such as membranes and ion channels, but this has been hard to avoid. The concept teaches the example as often as the example teaches the concept. In order to make this book useful to an audience beyond those who share my particular research interests, I have attempted to cast a wide net and roam far and wide to present examples from the many different fields that biophysicists study.

Much of this book presents subjects that are fundamental but have not yet found their way into textbooks. Distilling such work and rendering it in an accessible form requires difficult decisions to be made about organization and topic selection. I can only hope that this has been successful. I am painfully aware of the many interesting and important aspects of biophysics that I have not written about. However, there is already more than enough here for a one semester course for advanced undergraduates and beginning graduate students. I can only hope that studying this book will bring the many omitted topics within reach of the initiated students.

The material covered in this book varies in difficulty. Sections that are more difficult and not essential for continuity are designated with a star (*).

Acknowledgements

I owe a very special thanks to two graduate students who worked in my laboratory while I was in the final stage of writing this book. I originally asked Payne Chang and Xue Han to read a few chapters, but in the end they read every page. They have done a remarkable job of finding errors and requesting greater clarity. They both followed the Chinese adage "I respect my professor but I respect the truth more," to the enormous benefit of this book.

I am also indebted to the following friends and colleagues for critical comments on one or more chapters: Ed Chapman, Claudio Grossmann, Enfu Hui, Matt Jones, Peter Jordan, Stuart Licht, Andrew Lokuta, Cathy Morris, Bob Pearce, Steve Redman, Kimberly Taylor, Jeff Walker, and Jim Weisshaar. A final thanks to Adam Van Wynsberghe for help with the cover picture.

Chapter 1

Global transitions in proteins

The relation between structure and function is central to molecular biology. But molecular structure can mean different things, especially when dealing with complex biological molecules. One can know the chemical structure of a molecule, how the atoms are connected by covalent bonds, but have no idea of the conformational state, how the atoms are arranged in space. The conformational state of a molecule has a profound impact on what it does, and much of the work in molecular biophysics deals with understanding molecular conformations, both what they are and how they perform biological functions.

The conformational state of a molecule can be studied at different levels. One might have a vague notion of its general shape, or one might have a detailed picture with the position of every atom specified. Thinking in terms of detailed structure is more difficult but more powerful. Some approaches to this problem will be taken up in later chapters. Here, we will start with something simple, introducing an approach to protein conformations that does not depend on all the structural details.

The approach of this chapter is based on the idea that proteins have "global" states. Global states are defined in terms of a protein's functional capability. We will assume that a protein has a few – perhaps just two – of these global states. Global states can interconvert, in what we will call global transitions. In terms of structure the global state is a black box. Without dealing explicitly with structure, the simplifying assumption of global states and transitions generates a robust quantitative framework for treating functional transitions of proteins.

These ideas can be applied to virtually any area of molecular biology. Conformational states of proteins are the basic building blocks in mechanistic models, and interconversions between these states are the basic molecular signaling events. Examples include the activation of membrane receptors, the regulation of gene expression, the control of cell division, the gating of ion channels, and the generation of mechanical force. This chapter focuses on two well-defined types of transitions, one induced by temperature and the other by voltage. These transitions are very different from an experimental point of view, and only rarely have they been studied in the same protein. However, from the theoretical point of view we can see striking parallels, and studying these two cases together provides a deeper understanding of the general nature of functional transitions in proteins.

1.1 | Defining a global state

The general strategy for now is to play down structural details, but we still need to define a global state rigorously. This will help make us aware of how the global state can at least in principle be related to a protein's structure. Global states are a coarse-grained view but they can be related to fine-grained views. The fine-grained view is based on what will be called microstates. A microstate has a conformation that is defined in great detail. We might know the positions of all of the atoms, or the dihedral angles of all of the rotating bonds (Chapter 3). A global state is envisioned as encompassing a large number of microstates. The microstates interconvert rapidly and the global state reflects the average behavior of these microstates. This view is taken directly from statistical mechanics. A collection of microstates forms an ensemble, and statistical mechanics provides the conceptual tools for understanding a global state in terms of its constituent microstates.

The free energy of a global state takes into account the internal potential energy of each of the microstates as well as the entropy arising from the conformational disorder of interconversions between microstates. We can express the free energy of a system containing N independent molecules as

$$G = -kT \ln (Q^N) \tag{1.1}$$

where Q is the partition function of the molecule and kT is Boltzmann's constant times temperature. The partition function is a sum over a set of microstates included in one global state

$$Q = \sum_{i}^{n_{GS}} e^{-E_i/kT} \tag{1.2}$$

Here E_i is the energy of the ith microstate and $e^{-E_i/kT}$ is its Boltzmann weight. When we focus on one particular global state, we limit this sum to a selected subset, or subensemble of microstates. This is indicated by the subscript GS; so n_{GS} is the total number of microstates comprised by a given global state. We can thus distinguish different global states formally by summing over different, non-overlapping subsets of microstates.

The Boltzmann distribution can be used to obtain the probability of finding a particular microstate, j, with energy E_j, among all of the possible microstates of a given global state

$$P(j) = \frac{e^{-E_j/kT}}{\sum_{i}^{n_{GS}} e^{-E_i/kT}} \tag{1.3}$$

The sum in the denominator is the partition function of Eq. (1.2). Here, it normalizes the probability function so that the probabilities all add up to one. In this context, as a sum over probabilities, we can see how the partition function embodies the notion of thermodynamic stability. A global state with a greater partition function has a higher probability of occurring, and so will have a lower free energy.

The definition of a microstate is very flexible and can be expanded to include many important features. For example, each microstate has an entropy resulting from the disordering effect of bond vibrations. To take this into account we can extend the sum in Eq. (1.2) to include vibrational energy levels. Microstates of a protein can be further distinguished by different positions and orientations of the surrounding water molecules, and possibly by ions in solution. If these contributions are included in the summation in Eq. (1.2), then the free energy in Eq. (1.1) will be more accurate. These are formal considerations that help us visualize in broad terms how different levels of detail can be incorporated into the picture. The free energy of a global state does indeed depend on all of these complex features. We will recognize these dependencies, but for now we will not deal with them explicitly.

An alternative way to think about the partition function of a global state is to view the discrete states in Eq. (1.2) as a continuum. The sum then becomes an integral over a specified range of all the internal coordinates of the molecule

$$Q = \int_{GS} e^{-E(\mathbf{r})/kT} d\mathbf{r} \qquad (1.4)$$

Here $\mathbf{r}$ is a vector containing the positions of all of the atoms, and $E(\mathbf{r})$ is the potential energy of the molecule as a function of these positions. Equation (1.4) is referred to as the classical configuration integral (McQuarrie, 1976). The range of integration, defined as GS, specifies a global state by limiting the region of coordinate space. This is analogous to limiting the total number of microstates to a subset of n_{GS} microstates in Eq. (1.2). Limiting the range of the integration of Eq. (1.4) and limiting the range of the summation in Eq. (1.2) are equivalent ways of dividing the vast state space of a protein into distinct regions corresponding to different global states.

It must be mentioned that Eqs. (1.2) and (1.4) leave out the contribution made by kinetic energy. This does not matter for our purposes because in classical physics (i.e. no quantum mechanical effects) every atom has an average kinetic energy of 3/2 kT. As long as classical physics is obeyed this will be the same for all microstates. For strong covalent bonds, quantum effects are important and the vibrational kinetic energy for an atom can be less than 3/2 kT. However, this contribution is not likely to change much during global transitions because the covalent bonds are not broken. These fortunate circumstances make Eqs. (1.2) and (1.4) especially well-suited for studying macromolecules in biophysical problems.

For now we have no intention of actually calculating the partition function of a protein. We will accept the existence of the free energy of a global state, defined in these terms. The free energy includes all of the energetic terms that make one global state more stable than another for a particular set of conditions. Without a complete accounting of all these terms we can still develop useful theories for what influences global transitions in proteins.

Models based on the idea of global states are very popular in molecular biology, but the conceptual basis and underlying assumptions often go unappreciated. In fact, we do not actually have a good reason to expect a priori that a protein will have just a few discrete global states, or that the behavior of a protein can be described well in these terms. One could imagine that functional changes in proteins involve shifts in the distribution of microstates. Each functional state could have access to all the same microstates, but would visit some more frequently than others. Thus, we have two extreme views of functional transitions in proteins. Global transitions at one end of the spectrum represent transitions between distinct nonoverlapping populations of microstates. The other extreme is a transition entailing a shift in the distribution within a single well-connected population of microstates. Deciding which of these extremes better approximates reality requires experiments, and some of these will be discussed as the theories are developed.

The variable E_i in the above expressions is the energy of a single molecule. However, we will often find it more convenient to use the energy of one mole of molecules. We can use these units provided that we replace Boltzmann's constant, k, with the gas constant, R. One often has to switch between molecular and molar energy units, and this is usually easy if we keep track of whether we are using k or R. To have some sense for energy magnitudes on the molecular versus the molar scale, consider the value of kT and RT for a physiological temperature of 300 K: $kT = 4.11 \times 10^{-21}$ J and $RT = 2.47 \times 10^3$ J (RT also equals 590 calories, and these units are more commonly used with moles). Energies are divided by these numbers when the Boltzmann distribution is used to determine the probability of a state, so they are well worth remembering. Therefore kT and RT are the fundamental reference points for energy. The energy of a state relative to these reference points determines how likely we are to encounter that state.

1.2 | Equilibrium between two global states

Consider an equilibrium between two global states, denoted as A and B, as shown in Scheme (1A) below

$$A \rightleftarrows B \tag{1A}$$

This is just like a chemical isomerization. Global state A has a molar free energy G_a and global state B has a molar free energy G_b.

The molar free energies of solutions with certain concentrations of A or B are then

$$G_a = G_a^o + RT \ln[A] \tag{1.5a}$$

$$G_b = G_b^o + RT \ln[B] \tag{1.5b}$$

where G_a^o and G_b^o are the molar free energies of the standard state (by convention a one molar solution). The free energy change for conversion from A to B is then

$$\begin{aligned} \Delta G &= G_b - G_a \\ &= G_b^o - G_a^o + RT \ln\frac{[B]}{[A]} \end{aligned} \tag{1.6}$$

At equilibrium $\Delta G = 0$. Taking the equilibrium concentrations as $[A]_{eq}$ and $[B]_{eq}$, we can relate these to the free energy difference between the two standard states as

$$\Delta G^o = -RT \ln\frac{[B]_{eq}}{[A]_{eq}} = -RT \ln K_{eq} \tag{1.7}$$

where $\Delta G^o = G_b^o - G_a^o$, and K_{eq} is the equilibrium constant for interconversions between A and B. Taking the exponential of this equation gives

$$e^{-\Delta G^o/RT} = \frac{[B]_{eq}}{[A]_{eq}} \tag{1.8}$$

We will drop the subscripts "eq" after this point because all concentrations are equilibrium values for the rest of the chapter.

Equation (1.8) looks very much like a Boltzmann distribution, which gives the ratio of the relative probabilities of two microstates as an exponential function of the difference in their energies. A corresponding relation for the relative probabilities of two microstates can be derived from Eq. (1.3). Likewise, Eq. (1.8) can be derived directly from Eq. (1.3) by summing over the microstates for each of the two global states and then taking the ratio of the two sums. This connection emphasizes the point that the equilibrium represented by Eq. (1.8) reflects the relative stability of two distinct collections of microstates. Equation (1.8) is a useful starting point for introducing assumptions about energies into models for global transitions. The general strategy in much of what follows is to make assumptions about ΔG^o and use Eq. (1.8) to explore the consequences.

1.3 | Global transitions induced by temperature

The thermal denaturation of a protein provides an excellent example of a global transition. At low temperatures a folded state is favored in which a relatively small number of microstates have a similar structure corresponding to a compact well-defined configuration (Fig. 1.1). This

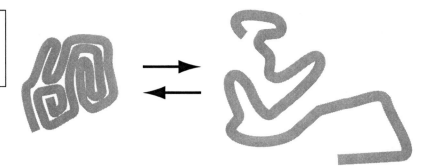

Fig. 1.1. In the thermal denaturation of a protein, a compactly folded native state undergoes a global transition to a randomly coiled denatured state.

configuration is maintained by a large number of specific contacts between the residues (Section 2.13). At high temperatures the protein unfolds (Fig. 1.1). This denatured state has a very large number of microstates corresponding to all the different configurations of a randomly coiled chain (Chapter 3). The transition is reversible, so that cooling restores the folded state.

It will be seen shortly that the difference in number of microstates in these two global states is what makes the transition temperature sensitive. In keeping with the spirit of this chapter in which structural details are played down, we simply accept the situation as given. Chapter 3 uses statistical mechanics to develop models for this process.

To understand how temperature influences thermal denaturation, we divide the free energy change into enthalpy and entropy

$$\Delta G^\circ = \Delta H^\circ - T\Delta S^\circ \tag{1.9}$$

The enthalpy can be loosely connected with the contacts that stabilize the native, folded state. The entropy represents the contribution arising from the greater disorder of the unfolded state. This is an oversimplification as the contact energies are not purely enthalpic. More will be said about this in Chapter 2, but the separation of enthalpy and entropy along these lines is useful and appropriate for the present discussion of thermal unfolding. It is also important to mention that contacts can stabilize the unfolded state. The unfolded protein has no real structure to stabilize, but stronger contacts between buried amino acids and water or another solvent in the unfolded state factor into ΔH°. This is the mechanism of denaturant action (Section 1.7).

Assuming that ΔH° and ΔS° themselves are independent of temperature allows us to focus on the term $T\Delta S^\circ$ as the drive for a temperature-induced shift in the equilibrium between A and B. There is a transition temperature, T^*, where the folded and unfolded states are present in equal concentrations. At that point $\Delta G^\circ = 0$, so by Eq. (1.9) $\Delta H^\circ = T^*\Delta S^\circ$. Below T^*, $\Delta H^\circ > T\Delta S^\circ$, and ΔG° is positive so that A is more stable. Above T^*, $\Delta H^\circ < T\Delta S^\circ$ and ΔG° is negative so B is more stable. This is readily seen by combining Eqs. (1.8) and (1.9).

$$\frac{[B]}{[A]} = e^{(-\Delta H^\circ + T\Delta S^\circ)/RT} \tag{1.10}$$

Practical applications often proceed by defining a variable x as the fraction of B or equivalently, the extent of the conversion from A to B.

$$x = \frac{[B]}{[B] + [A]} \qquad (1.11)$$

With the aid of Eq. (1.10), x can be expressed in terms of ΔG°. Factoring [B] from the numerator and denominator gives

$$x = \frac{1}{1 + ([A]/[B])} \qquad (1.12)$$

Substituting Eq. (1.10) then gives

$$x = 1/(1 + e^{\frac{\Delta H^\circ}{RT} - \frac{\Delta S^\circ}{R}}) \qquad (1.13)$$

At low temperatures the positive enthalpy term is larger, so the exponential is large, and x is close to zero. At high temperatures the negative entropy term dominates, so the exponential is small and x is close to one. Thus, Eq. (1.13) expresses the conversion from $x \sim 0$, where $[A] >> [B]$, to $x \sim 1$, where $[B] >> [A]$.

1.4 | Lysozyme unfolding

Thermal denaturation has been studied extensively in an effort to determine how different forces contribute to the stability of the native, folded state of a protein. These forces will be discussed in more detail in Chapter 2, but for the present purposes, these kinds of experiments provide a good illustration of how global transitions are modified by structural changes.

The enzyme lysozyme from T4 bacteriophage was subjected to a screen to select temperature-sensitive mutants. Lysozyme unfolds or denatures as the temperature is raised, and the unfolded state has no enzymatic activity. In temperature-sensitive mutants denaturation occurs at lower temperatures than in the wild type enzyme. Mutants with lower melting points identified residues that are critical to thermal stability. One of these was threonine 157. Using site-directed mutagenesis, this threonine was replaced with a number of different amino acids (13 in all) (Alber *et al.*, 1987).

In temperature-sensitive mutants the free energy difference between the folded and unfolded states is altered. This can occur through changes in the strength of the contacts of the amino acid side chain with its neighboring residues, leading to a change in enthalpy. Alternatively, the new amino acid side chain may have more or less conformational freedom, thus changing the entropy. It was found that T^* dropped for every amino acid replacement of threonine 157. When the mutations were studied by calorimetry (Connelly *et al.*, 1991) it was found that the destabilization reflected both reductions in ΔH° and increases in ΔS°.

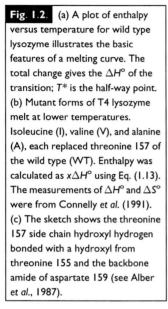

Fig. 1.2. (a) A plot of enthalpy versus temperature for wild type lysozyme illustrates the basic features of a melting curve. The total change gives the $\Delta H°$ of the transition; $T*$ is the half-way point. (b) Mutant forms of T4 lysozyme melt at lower temperatures. Isoleucine (I), valine (V), and alanine (A), each replaced threonine 157 of the wild type (WT). Enthalpy was calculated as $x\Delta H°$ using Eq. (1.13). The measurements of $\Delta H°$ and $\Delta S°$ were from Connelly et al. (1991). (c) The sketch shows the threonine 157 side chain hydroxyl hydrogen bonded with a hydroxyl from threonine 155 and the backbone amide of aspartate 159 (see Alber et al., 1987).

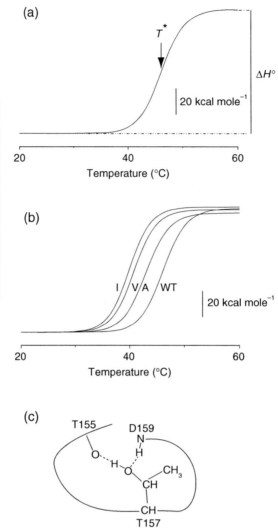

Figure 1.2 shows calculated melting curves for the wild type protein and a few mutants. As the temperature is raised the protein absorbs heat. This can be measured as enthalpy, using a calorimeter that scans temperature and records the heat absorption. These plots are based on Eq. (1.13), using experimentally measured values of $\Delta H°$ and $\Delta S°$. The variable x computed in this way from Eq. (1.13) is then multiplied by $\Delta H°$ to obtain the excess enthalpy, which is plotted. Baseline heat absorption by the solvent and by the global states of the protein unrelated to the transition are subtracted out to leave the heat of the transition. So the heat absorbed as a result of the unfolding transition is easily resolved.

The wild type melting curve (Fig. 1.2a) indicates that the total heat absorbed is the $\Delta H°$ for the protein. The half-way point is the melting temperature, $T*$, where $x = 1/2$, and the heat is $\Delta H°/2$. Once $\Delta H°$ and $T*$ have been read off such a plot, $\Delta S°$ is determined as $\Delta H°/T*$.

The mutant melting curves are shown in Fig. 1.2b, along with the wild type melting curve. Figure 1.2c shows a sketch of the structure around the mutated site (threonine 157). The alanine mutation shifts T^* down by 4 °C, and the high temperature plateau is reduced. Both of these changes in the melting behavior are consequences of the lower $\Delta H°$ of this mutant. The reduction in T^* follows from $T^*\Delta S° = \Delta H°$, for a mutation where $\Delta S°$ is the same as in the wild type protein. The reduction in $\Delta H°$ for this mutation reflects the loss of the two hydrogen bonds formed by the hydroxyl group of threonine 157 (Fig. 1.2c). Putting a valine in this position reduces T^* even more, but the final plateau is close to that of the wild type protein, so this destabilization is primarily caused by an increase in the $\Delta S°$ of the transition, rather than a decrease in $\Delta H°$. This may reflect a restriction in the motion of neighboring side chains caused by the larger valine side chain. Finally, with isoleucine $\Delta H°$ is higher than wild type. If $\Delta S°$ were the same then T^* would go up. But this mutation produces the greatest reduction in T^*, indicating that the increase in $\Delta H°$ is offset by a much larger increase in $\Delta S°$. The isoleucine side chain forms contacts with other parts of the protein in a structure stabilizing manner, but it is too big for the pocket occupied by threonine, so the motions of the side chains are more severely restricted. Furthermore, isoleucine can be more disordered in the unfolded state because of the flexibility of its large side chain, so $\Delta S°$ increases owing to changes on both sides of the transition.

1.5 | Steepness and enthalpy

A thermal transition reflects competition between enthalpy and entropy, so that changing one or the other moves the transition temperature up or down. Now we ask, what if $\Delta H°$ and $\Delta S°$ change in parallel? The value of T^* then stays the same but the plots change and this reveals another interesting aspect of thermal transitions. Equation (1.13) is plotted in Fig. 1.3 for transitions with the same $T^* = 300$ °K. Different values of $\Delta H°$ are used, and $\Delta S°$ was taken as $\Delta H°/T^*$ for each plot. Figure 1.3 shows how the transition becomes steeper as $\Delta H°$ and $\Delta S°$ increase together.

The increasing steepness with increasing $\Delta H°$ gives us a useful way of thinking about a thermal transition. The $\Delta H°$ (or equivalently, $\Delta S°$) dictates the sensitivity of the transition to temperature. If the $\Delta H°$ and $\Delta S°$ are larger, then the $\Delta G°$ changes more rapidly as T passes through T^*.

This relation between the temperature dependence of an equilibrium and its $\Delta H°$ and $\Delta S°$ is firmly rooted in classical thermodynamics. Start with the following basic relation

$$\frac{\partial G°}{\partial T} = -S°$$

(1.14)

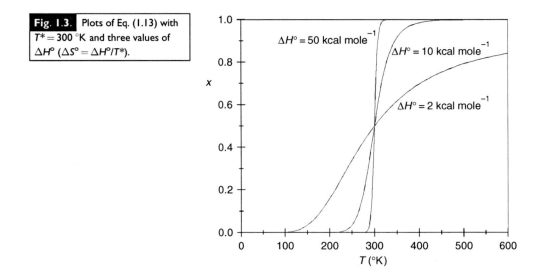

Fig. 1.3. Plots of Eq. (1.13) with $T^* = 300\ °K$ and three values of $\Delta H°$ ($\Delta S° = \Delta H°/T^*$).

This also applies to the changes in these quantities

$$\frac{\partial \Delta G°}{\partial T} = -\Delta S° \tag{1.15}$$

This amounts to a quantitative statement that $\Delta S°$ gives the temperature dependence of an equilibrium, and replacing $\Delta S°$ with $\Delta H°/T^*$ makes an analogous point for enthalpy.

Using Eq. (1.15) to substitute for $\Delta S°$ in Eq. (1.9) gives

$$\Delta G° = \Delta H° + T\frac{\partial \Delta G°}{\partial T} \tag{1.16}$$

Rearranging and dividing by T^2 leads to

$$\frac{1}{T}\frac{\partial \Delta G°}{\partial T} - \frac{\Delta G°}{T^2} = -\frac{\Delta H°}{T^2} \tag{1.17}$$

from which we obtain what is known as the Gibbs–Helmholtz equation

$$\frac{\partial(\Delta G°/T)}{\partial T} = -\frac{\Delta H°}{T^2} \tag{1.18}$$

This can be checked by evaluating the derivative on the left. Using Eq. (1.7) takes us one step further

$$\frac{\partial \ln([B]/[A])}{\partial T} = \frac{\Delta H°}{RT^2} \tag{1.19}$$

This expression is known as the van't Hoff equation, and provides an important way to interpret the temperature dependence of an equilibrium between two global states. The slope of a plot of $\ln([B]/[A])$ versus temperature can be used to determine $\Delta H°$. An enthalpy determined in this way is often referred to as a van't Hoff enthalpy, or $\Delta H_{vh}°$.

Note that $\Delta S°$ and $\Delta H°$ cannot be viewed in isolation from one another when thinking about temperature effects. For the sake of completeness we see that a parallel analysis can be made in terms of ΔS°. Combining Eqs. (1.7) and (1.15) gives

$$\frac{\partial RT \ln([B]/[A])}{\partial T} = \Delta S^\circ \tag{1.20}$$

If we chose to use this equation to analyze a thermal transition, we would obtain ΔS° from the slope. Because $\Delta H^\circ = T^* \Delta S^\circ$, the quantities T^*, ΔH°, and ΔS° do not vary independently; specifying any two determines the other.

1.6 | Cooperativity and thermal transitions

In general, steepness in a transition is associated with cooperativity. Cooperativity means that different parts of the system are tied together in some way so that the force on one part to undergo the transition is influenced by whether other parts have undergone the transition. Enthalpy and cooperativity do not appear to have much in common, but they both influence the steepness.

To see the connection, consider a protein with n identical subunits. If all n subunits undergo the transition in perfect unison, then we say that the transition is perfectly cooperative. A perfectly cooperative process can be treated as a single equilibrium (Scheme (1B))

$$A_n \rightleftharpoons B_n \tag{1B}$$

No mixed structures of the form A_iB_j (with $i+j=n$) are allowed, leaving a pure two-state, global transition between A_n and B_n. Because cooperativity forces the subunits to undergo the transition together, the enthalpy change per mole of n-mer is n times the single subunit ΔH°. This larger enthalpy is ΔH_{vh}°. This is the quantity that determines the steepness of the transition, and must be used in Eqs. (1.13) and (1.19) for the temperature dependence of the equilibrium. Thus, cooperativity in a multisubunit protein increases the effective enthalpy of the transition and therefore increases its steepness.

For perfect cooperativity, we have $\Delta H_{vh}^\circ = n\Delta H^\circ$. But if the cooperativity is less than perfect, the transition will be less steep and ΔH_{vh}° will be reduced. If the subunits undergo the transition completely independently of one another, then the steepness will be governed by ΔH° for a single subunit. Thus, we can formulate a measure of cooperativity in terms of the ratio of ΔH_{vh}° to ΔH°, where ΔH° for a single subunit is measured independently in a calorimeter. We call $\Delta H_{vh}^\circ/\Delta H^\circ$ the cooperative unit, and it can range from one to n. This number indicates how strongly subunits interact during a thermal transition.

These ideas have been used to test the critical assumption of whether thermal denaturation is a two-state global transition. Consider a protein made up of m small domains. Each domain undergoes a two-state thermal transition with an enthalpy change of Δh°. Note that ΔH_{vh}° would be Δh° if the domains were independent, and $m\Delta h^\circ$ if the domains interacted strongly to make the transition perfectly cooperative. The two-state hypothesis can thus be tested by comparing ΔH° measured calorimetrically with ΔH_{vh}° measured from the slope of the melting curve.

This comparison has been made with a number of proteins. For small globular proteins ($< 15\ 000$ kDa) ΔH_{vh}° and ΔH° are in almost perfect agreement (Privalov, 1979). This important result means that a protein does not unfold little by little as the temperature is raised. For temperatures near the transition, the two states coexist, but there are no significant amounts of intermediate states. Thus, thermal unfolding is a global transition of the entire protein, rather than a gradual loosening of the structure or many separate transitions in different structural domains. For larger proteins one often sees melting in a few distinct states as separate domains melt. The two-state model then applies to the domains (Privalov, 1982).

The cooperativity of thermal unfolding in globular proteins should not be taken for granted. It is not guaranteed that any sequence of amino acids will fold up into a well-defined compact structure, and many synthetic peptides do not. The idea that the good-folding sequences may represent a limited subset of all possible sequences will be discussed in Section 3.16. Some clues about the requirements for cooperative folding were obtained from calorimetric studies of lysozyme. Mutants have been described that reduce the cooperativity of thermal unfolding (Carra *et al.*, 1996). In these proteins, ΔH_{vh}° is about 80%–90% of the calorimetric ΔH°. The lysozyme crystal structure reveals that the N- and C-terminal domains are fairly well separated. They are linked by a single stretch of α-helix. The mutations in an α-helix in the N-terminal domain (alanine 42, serine 44, and lysine 48, quite far from threonine 157 of Fig. 1.2) disrupt the coupling between these two domains so that they melt on their own. Melting then proceeds as two separate processes, with each process showing its own cooperative two-state behavior.

1.7 | Transitions induced by other variables

The ideas developed for thermal transitions illustrate how assumptions concerning the energetics of a protein can be used to predict and interpret experimental behavior. We can generalize this approach to examine how other experimental variables affect a global transition. For example, when thinking about pressure, one must consider the change in volume associated with a global transition. The free energy change associated with a pressure-induced transition will then include a work term of the form $P\Delta V$.

Consider a global transition with a molar volume change, ΔV°. If each global state is incompressible, or equally compressible, then the only way that pressure can affect the relative stability of the two global states is through the work term. This allows us to divide the molar free energy into two parts

$$\Delta G^o = \Delta G_{pi}{}^o + P\Delta V^o \qquad (1.21)$$

where $\Delta G_{pi}{}^o$ is pressure independent. It is then straightforward to incorporate this into an expression for the ratio of concentrations of the two global states, and derive the appropriate exponential dependence on pressure. The result is Eq. (1.13), but with $\Delta G_{pi}{}^o + P\Delta V^o$ replacing $\Delta H^o - T\Delta S^o$. The steepness is then related to ΔV^o, just as the steepness of a thermal transition is related to ΔH^o. Continuing the analogy with thermal transitions, in a multisubunit protein, ΔV^o will be multiplied by a cooperative unit reflecting the degree to which the transitions in separate subunits are linked. The result is a higher effective ΔV^o, and a steeper pressure dependence if the transition is cooperative.

Another important example of a global transition is protein unfolding by denaturants (Tanford, 1968; Fersht, 1998). As substances such as urea, guanidinium, and many alcohols are added to a solution, a protein is converted to a state that is qualitatively similar to the unfolded state induced by heating (Fig. 1.1). They differ quantitatively in that a protein unfolded by heating has some residual structure; a protein unfolded by denaturants has almost none. Denaturants unfold proteins by interacting favorably with the protein interior. For example, urea can hydrogen bond with the backbone amides and carbonyls of the peptide chain. Once again, we defer the discussion of these forces to Chapter 2, but for now it suffices that buried residues of a protein have an unfavorable interaction with water. The exposure that results from unfolding makes ΔG^o positive. As the denaturant concentration increases, the unfavorable interaction with water is offset by an attractive interaction with the denaturant. This contribution is proportional to the denaturant concentration and the number of buried residues. For a denaturant D, we write

$$\Delta G^o = \Delta G_w{}^o + m_D[D] \qquad (1.22)$$

Here m_D is an empirical parameter that can be envisioned as a product of the number of buried residues times the average attractive interaction energy per buried residue.

Using this expression for ΔG^o again gives an equation like Eq. (1.13). Now we can measure and plot the fraction of unfolded protein as a function of the denaturant concentration [D] and measure m_D and $\Delta G_w{}^o$ as the best fitting parameters from the plot. Two-state behavior is the norm and studies of these kinds of plots have added a great deal to our understanding of protein folding.

There are many more examples. In each case we have a variable under experimental control, and a conjugate property of the system that can be used to express the sensitivity to that variable. For stretching an elastic molecule, the experimental variable is linear tension, ϑ, and the conjugate property is the length of the molecule, L. If a molecule under linear tension undergoes a global transition between two states of different length, we have a work term of the form $\vartheta \Delta L^{o}$, where ΔL^{o} is a change in length. For surface tension, Π, and surface area occupied by a membrane protein, we would have a work term of the form $\Pi \Delta A^{o}$, where ΔA^{o} is the change in area. A global transition can be induced by pH, and the drive is different numbers of protonated groups between the two global states. This can be generalized to ligand binding by using ligand concentration as the experimental variable and the number of ligand binding sites as the conjugate property. Yet another variable that can drive a global transition is voltage, and this is treated next.

1.8 | Transitions induced by voltage

When a protein resides within a membrane, a transmembrane voltage will exert a force on its charges. If the protein undergoes a global transition, then the voltage can favor one state over another depending on the charge distributions of the two global states. This is very important in the physiology of excitable membranes, and the ideas developed in this section will be used later, especially in Chapter 16. It should be noted that the concept of global transitions is especially clear for ion channels. Single-channel recording has revealed stepwise changes in membrane current, and this provides a direct view of global transitions at the level of individual molecules.

To incorporate voltage dependence into a global transition, we assume that the two global states A and B have different charge distributions. To illustrate this idea consider a single discrete charge of magnitude q moving between two locations (Fig. 1.4a). To simplify the analysis we assume that the voltage drop is linear across the membrane (Fig. 1.4b; see also Section 13.10). The electrostatic potential energy of a charge within the membrane is qxV, where x is the distance from the side of the membrane where the potential is

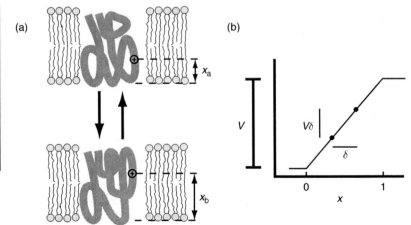

Fig. 1.4. (a) A protein embedded in a lipid bilayer, with the position of a single positive charge indicated. During a global transition the charge moves from x_a to x_b. (b) The charge has a potential energy that varies linearly with position as qVx. Movement of the charge from x_a to x_b changes the electrostatic energy by $qV(x_b - x_a) = qV\delta$ (with $\delta = x_b - x_a$).

taken as zero and V is the voltage drop across the whole membrane. The variable x is normalized to the thickness of the membrane and varies between 0 and 1, so the energy varies from 0 to qV as a charge moves from one side of the membrane to the other.

After the global transition the charge finds itself at a new location with a different electrostatic potential energy. If we take the displacement of the charge as $\delta = x_b - x_a$, then the change in electrostatic energy will be $qV\delta$ (Fig. 1.4b), and this constitutes the voltage dependent part of the free energy change of the global transition. Now that we know the contribution from one charge we can take a sum over all the charges in the protein and obtain the following expression for the total electrostatic free energy change

$$\Delta G_{es}{}^o = \sum_{i=1}^{m} q_i \delta_i V \tag{1.23}$$

Here m is the number of charges, and the index i is used to count them. It should be noted that if we want to relax the assumption that the voltage drops linearly through the membrane as depicted in Fig. 1.4b, we can rewrite Eq. (1.23) using δ_i as the fraction of the membrane potential. Then it would no longer represent the physical distance moved by the charge.

We can now separate the total free energy change of the global transition into voltage dependent and voltage independent parts

$$\Delta G^o = \Delta G_{vi}{}^o + \sum_{i=1}^{m} q_i \delta_i V$$
$$= \Delta G_{vi}{}^o + \alpha V \tag{1.24}$$

$\Delta G_{vi}{}^o$ is the voltage independent part, and α is introduced to represent the sum $\sum_{i=1}^{m} q_i \delta_i$ in the voltage dependent part. Note that the term αV is in a form that illustrates the idea of the preceding section about an external variable, in this case V, and a conjugate property that expresses the sensitivity to that variable, in this case α.

To illustrate how Eq. (1.24) can be used we take the global transition as the gating of an ion channel. The two global states of our protein are then the open (O) and closed (C) conformations of the channel as shown in Scheme (1C)

$$C \rightleftharpoons O \tag{1C}$$

Using Eqs. (1.8) and (1.24), the ratio of the concentrations of the two global states becomes

$$\frac{[O]}{[C]} = e^{-(\Delta G_{vi}{}^o + \alpha V)/RT} \tag{1.25}$$

We can then write the open probability in the same form as Eq. (1.11)

$$P_o = \frac{[O]}{[O] + [C]} \tag{1.26}$$

Combining with Eq. (1.25) and rearranging leads to

$$P_o = \frac{1}{1 + e^{(\Delta G_{vi}^{\circ} + \alpha V)/RT}} \tag{1.27}$$

Note the similarity with Eq. (1.13).

A more convenient form for this expression can be obtained by replacing RT/α by a new parameter, V_s, and $\Delta G_{vi}^{\circ}/RT$ by $-V_0/V_s$

$$P_o = \frac{1}{1 + e^{(V - V_0)/V_s}} \tag{1.28}$$

These new parameters are very useful in that they represent well defined properties of a voltage-induced transition: V_0 is the voltage at the midpoint of the transition (at $V = V_0$, $P_o = 1/2$), and V_s gives the voltage range over which the transition occurs; when it is small the transition is steeper.

Plots of Eq. (1.28) show how changing V_0 shifts the curve to the right or left (Fig. 1.5a), and changing V_s changes the steepness (Fig. 1.5b). Note that V_0 is analogous to T^* in a thermal transition. Indeed, one can call V_0 the "transition voltage," because when $V > V_0$ one state predominates and when $V < V_0$ the other state predominates. Likewise, V_s can be compared to ΔH°. Because it governs the steepness of the transition it is often referred to as the "steepness factor." The amount of charge movement was represented by the parameter α, and the movement of one unitary charge through the entire membrane potential corresponds to a steepness factor of 25 mV (for a discussion of units, see Section 13.1). This is because at physiological temperatures $RT/q \sim 25$ mV for a unitary charge. Moving a mole of monovalent ions up a 25 mV potential requires an amount of energy equal to RT.

Recall the relation between steepness and cooperativity in thermal transitions. We can make the same connection here if we realize that cooperativity is invoked when we say that all of the charges of the protein must move at once during the transition. The more charge that moves together, the larger α will be. Thus, increasing the number of charges that move during gating is analogous to increasing the size of the cooperative unit in a thermal transition.

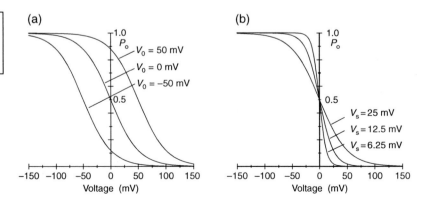

Fig. 1.5. Plots of Eq. (1.28) with various values of V_0 (a) and V_s (b). Note that $V_s = 25$ mV corresponds to $\alpha = 1$.

Membrane biophysicists refer to Eq. (1.28) as the Boltzmann equation (Hille, 1991). It is closely related to the Boltzmann distribution and contains the familiar exponential energy term. This equation is used very widely in investigations of voltage-gated channels, including those responsible for electrical impulses in neurons (Chapter 16). The following example illustrates how useful the Boltzmann equation is in interpreting channel data and identifying parts of a protein that are specifically dedicated to voltage-induced transitions.

1.9 | The voltage sensor of voltage-gated channels

The voltage-gated channels are a very large superfamily of proteins. They form channels that are closed at negative voltages and open at positive voltages. This superfamily includes channels that are selective for Na^+, K^+, and Ca^{2+}, and they all have some common structural features. One of these is a $\sim$25 amino acid segment known as S4 (the fourth of six putative membrane spanning segments) in which every third amino acid is a positively charged arginine or lysine and the other amino acids are hydrophobic. When the amino acid sequences of these proteins were first deduced, the S4 motif drew immediate attention as the part of the protein that could serve as the voltage sensor. This was confirmed with experiments using site-directed mutagenesis followed by biophysical analysis of channel function.

The biophysical analysis was carried out on proteins expressd in *Xenopus* oocytes. Channel-encoding mRNA was injected into these large cells. The oocytes then translate the RNA into protein, which finds its way to the cell membrane. The cell is voltage-clamped with microelectrodes, and voltage steps to various levels gate the channels and produce various amounts of current. The current is proportional to the fraction of channels that are open, so these experiments give plots of P_o versus V, which are interpreted with Eq. (1.28).

Some plots are shown in Fig. 1.6 for the first K^+ channel that was cloned, the *Shaker* protein from *Drosophila*. Some mutations that neutralized positive residues in S4 reduced the steepness of the voltage dependence, and the parameter V_s was increased. The R368Q mutation illustrates this with the replacement of a positively charged arginine by a neutral glutamine. This leaves less charge to move during the transition. Hence α goes down and V_s goes up. Similar experiments have been conducted with vertebrate Na^+ channels (Stuehmer *et al.*, 1989).

Mutations also shifted the voltage dependence in either the positive or negative direction, and such shifts arise from structural changes that alter the relative stability of the open and closed states. This is illustrated with the R371K mutation, in which one positively charged amino acid is replaced by another. Because this mutation

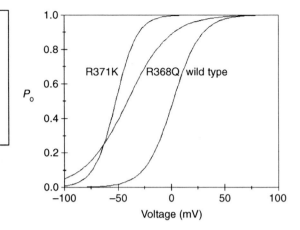

Fig. 1.6. Plots of Eq. (1.28) with different V_0 and V_s values give the open probability versus voltage. R368Q has a positively charged arginine replaced by a neutral glutamine. R371K has an arginine replaced by lysine so there is no change in charge (values of V_0 and V_s from Papazian et al., 1991).

leaves the charge in S4 unchanged, the steepness of the transition remains about the same. Both mutations shown in Fig. 1.6 shift the voltage dependence to the left, indicating stabilization of the open state of the channel.

Not all of the mutants of voltage-gated channels fall into this pattern, but by and large the results support the view that gating in these channels is triggered by the movement of positively charged residues in the S4 segment (Sigworth, 1994; Bezanilla, 2000). This view has been confirmed by a number of interesting experiments. Parts of S4 are accessible to chemical modification only from the inner surface of the membrane at negative voltages, and only from the outer surface at positive voltages (Ahern and Horn, 2004). The translocation of this highly charged segment through the membrane fits with the α values measured experimentally.

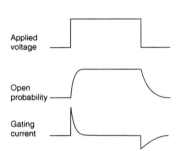

1.10 | Gating current

Charges moving within a membrane protein during a voltage-induced transition produce a transient current. This current is quite small, but it can be measured and is known as the gating current. The measurement of gating current provided an important early validation of the basic assumption that voltage-induced transitions are due to the movement of charges within a membrane protein. The gating current can be seen only while the conformational transition is actually occurring; once the transition is over, the channels may stay open and allow a steady ionic current to flow, but the gating current will have decayed to zero. The experimental appearance of a gating current is illustrated schematically in Fig. 1.7. Gating currents can be seen for virtually any voltage-gated channel, provided that it has a high density in the membrane and the ionic component can be blocked.

Fig. 1.7. A voltage step opens voltage-gated channels. The channels open with a characterisitic time course. If ion flow through the channels is blocked, the current due to channel gating can be seen. During the up-step, channel opening produces a positive gating current. During the down-step, channel closing produces a negative gating current. The different peaks and time courses reflect differences in the kinetics of opening and closing. The areas are equal because the total amount of charge for the forward and reverse transition is the same.

Each charge that moves during the transition makes a contribution to the gating current. Since current $= \mathrm{d}q/\mathrm{d}t$, the gating current can be integrated to obtain q. This quantity is an experimental measure of the charge that moves during gating (identical to α as defined in Eq. (1.24)). If there are N channels, each contributing an amount α to the total charge movement, then we have

$$\int I_g(t)dt = N\alpha \qquad (1.29)$$

where the time integral of the current, $I_g(t)$, is taken over the duration of the gating process. If the number of channels is known, then one can determine α from a measured gating current. This value can be compared with α computed from the steepness factor measured in a fit of Eq. (1.28).

This illustrates yet another parallel with thermal transitions. In both, the quantities characterizing the transition steepness can be measured directly. A calorimeter can be used to measure ΔH°, and gating current measurements provide a measure of α. In both kinds of transition a comparison of these measurements with a measurement derived from transition steepness provides some indication of the global extent of the transition. This relation is discussed in detail by Sigworth (1994) and Bezanilla (2000). For an excellent illustration of this kind of comparison in gating charge measurement see Schoppa et al. (1992).

1.11 | Cooperativity and voltage-induced transitions

Consider a voltage-induced transition of an n-subunit protein. As with thermal transitions, forcing the subunits to undergo a transition with perfect cooperativity leads to a multiplication in the steepness by n. This happens because α in Eq. (1.27) is replaced by $n\alpha$. Then V_s in Eq. (1.28) is divided by n so that the transition occurs in a narrower voltage range.

In a single subunit protein the idea of cooperativity is still relevant because the two-state model assumes that all of the different charges in the protein move concomitantly during the transition. One might also hope for a test of the two-state hypothesis with voltage-induced transitions in single-subunit membrane proteins, as has been made in thermal transitions. This has been difficult because the ion channels on which such experiments can be made are large proteins. Many are oligomers, and even those that are not have functionally distinct domains.

Perfect cooperativity between subunits is not very realistic. Voltage-induced transitions will now be used to illustrate a more plausible case of intermediate cooperativity. Consider a channel with two identical subunits, each of which undergoes a global transition characterized by the quantities in Eq. (1.24), ΔG_{vi}° and α. If both subunits undergo the transition together in a cooperative manner, then the observed steepness factor will be 2α. On the other hand, if each subunit can undergo the conformational transition independently, then we will have to consider three states of the dimer, namely AA, BB, and AB (where AB and BA are taken as identical by symmetry).

A transition in either subunit involves a change in free energy of $\Delta G_{vi}^{\circ} + \alpha V$, so

$$\frac{[AB]}{[AA]} = \frac{[BB]}{[AB]} = e^{-(\Delta G_{vi}^{\circ} + \alpha V)/RT} \tag{1.30}$$

If AA, AB, and BA are closed channels and BB is the only state in which the channel is open, then the open probability will be

$$P_o = \frac{[BB]}{[AA] + [AB] + [BA] + [BB]} \tag{1.31}$$

Factoring [BB] and substituting Eq. (1.30) gives

$$P_o = 1 \Big/ \left(\frac{[AA]}{[BB]} + \frac{[AB]}{[BB]} + \frac{[BA]}{[BB]} + 1 \right)$$
$$= \frac{1}{1 + 2e^{(\Delta G_{vi}^{\circ} + \alpha V)/RT} + e^{2(\Delta G_{vi}^{\circ} + \alpha V)/RT}} \tag{1.32}$$

Note that [BB]/[AA] = [AB]/[AA] times [BB]/[AB] from Eq. (1.30). The denominator of Eq. (1.32) is a complete square, so

$$P_o = \frac{1}{\left(1 + e^{(\Delta G_{vi}^{\circ} + \alpha V)/RT}\right)^2} = \frac{1}{\left(1 + e^{(V - V_0)/V_s}\right)^2} \tag{1.33}$$

The replacement of G_{vi}° and α by V_0 and V_s involves the same notation used to go from Eq. (1.27) to Eq. (1.28). Equation (1.33) is plotted together with Eq. (1.28) to illustrate how the interaction between the two subunits influences the shape of the transition (Fig. 1.8). (Equation (1.33) was offset to cross $P_o = 1/2$ at $V = 0$ to facilitate the comparison – see Problem 10).

Figure 1.8 shows that Eq. (1.33) is intermediate in steepness between simple two-state transitions with $\alpha = 1$ and 2. This is because the cooperativity is intermediate. Although the subunits undergo the transition independently, the restriction that both subunits must be in state B to have an open channel implies a

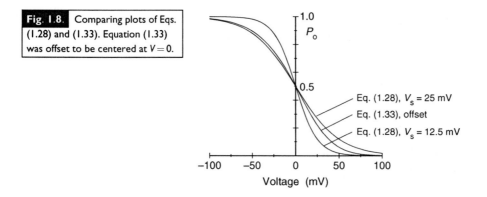

Fig. 1.8. Comparing plots of Eqs. (1.28) and (1.33). Equation (1.33) was offset to be centered at $V = 0$.

Eq. (1.28), $V_s = 25$ mV
Eq. (1.33), offset
Eq. (1.28), $V_s = 12.5$ mV

Voltage (mV)

more subtle form of interaction. If the subunits were truly independent then each one would have to conduct in the B state regardless of the state of the other subunit of the pair. Then the membrane conductance would exhibit a voltage dependence indistinguishable from that of a single-subunit channel. There would be no observable manifestation of the dimeric structure.

This example of the two-subunit channel can be extended to provide further insight into cooperativity. We introduce an interaction free energy for subunits in different states, ΔG_{int}°. This quantity reflects the unfavorable energy of contact between the two subunits when they are in different states (e.g. AB or BA). Incorporating this term into Eq. (1.30), and proceeding in the same way as in the derivation of Eq. (1.33) leads to

$$P_o = 1 \bigg/ \left(1 + 2e^{\frac{\Delta G_{vi}^{\circ} + \alpha V - \Delta G_{int}^{\circ}}{RT}} + e^{\frac{2\Delta G_{vi}^{\circ} + 2\alpha V}{RT}} \right) \qquad (1.34)$$

The presence of the term ΔG_{int}° prevents the simple factorization that allowed us to get Eq. (1.33) from Eq. (1.32). We can use this example to evaluate the extreme case of perfect cooperativity. This corresponds to a very large interaction energy, which would make the middle exponential term in the denominator go to zero. Equation (1.34) then becomes a simple Boltzmann equation with a gating charge of 2α. Thus, a high ΔG_{int}° makes the transition more cooperative by making mixed states unfavorable.

*1.12 | Compliance of a global state

Assuming that the voltage dependence of a conformational transition arises entirely from gating charge movement was used to develop a relatively simple theory. But one should wonder whether this assumption is physically realistic. For example, charges might move within a global state. If this happens then an electrical field could move the charges without a global transition. Then our simple electrostatic energy term of the form $qV\delta$ would be incomplete; δ would vary with voltage because as the voltage changed the charges would adjust their position within each global state. Real physical forces cannot keep a structure perfectly rigid. We will now consider the case of a protein with a charge held in place by an elastic force. The movement of the charge against this force reflects its compliance with the electric field.

When an elastic restoring force is added, we obtain a new expression for the potential energy of a charge as a function of position

$$U(x) = qVx + \tfrac{1}{2}\phi(x - x_0)^2 \qquad (1.35)$$

The first term is the familiar linear expression summed up in Eq. (1.23). The second term is the potential energy of a harmonic oscillator, with a force constant of ϕ. The charge feels a restoring force equal to $\phi(x - x_0)$

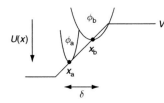

when it is displaced away from x_0, the position of the energy minimum in the absence of an electric field. The goal is now to incorporate this additional energy term into the derivation of the Boltzmann equation for a voltage-induced global transition. The global states differ here both in the position of the charge and the strength of the elastic force holding the charge in place (Fig. 1.9).

First we must see how the energy and position of the charge in each global state is influenced by the field. We replace $x - x_0$ in Eq. (1.35) by x', so that $x' = 0$ defines the center of one of the harmonic potential wells drawn in Fig. 1.9. Completing the square gives

$$U(x) = qVx_0 + \tfrac{1}{2}\phi(x' + qV/\phi)^2 - \tfrac{1}{2}(qV)^2/\phi \tag{1.36}$$

This shows that the voltage moves the position of minimum energy from zero to $-qV/\phi$. The last term of Eq. (1.36) tells us that the energy at this minimum is $1/2\,(qV)^2/\phi$ lower than it would be if there were no compliance and the charge were held in place. With the gating charge at its energy minimum, its energy is $qVx_0 - 1/2\,(qV)^2/\phi$. The difference between two energies of this form is then incorporated into the change in energy during the global transition.

$$\Delta G^\circ = \Delta G_{vi}{}^\circ + qV\delta - \tfrac{1}{2}(qV)^2\left(\frac{1}{\phi_b} - \frac{1}{\phi_a}\right) \tag{1.37}$$

Thus, we have Eq. (1.24) with a new quadratic term.

Now, instead of Eq. (1.28) we have a more complex Boltzmann equation.

$$P_o = 1 \left/ \left(1 + e^{\frac{\Delta G_{vi}{}^\circ + qV\delta - \frac{1}{2}(qV)^2\left(\frac{1}{\phi_b} - \frac{1}{\phi_a}\right)}{RT}} \right)\right. \tag{1.38}$$

This gives us an added exponential dependence on the square of the voltage. Allowing a multitude of gating charges to move in this way would mean using sums such as in Eq. (1.23), but the essential points can be understood by focusing on Eq. (1.38) for a single gating charge.

The key question is whether the V^2 term in the exponent is large enough to produce an observable effect. This would make the voltage dependence of P_o qualitatively different from that predicted by Eq. (1.28). If the force constants are the same in both conformations, then $\phi_a = \phi_b$ and this term will be zero. To see an effect of compliance on the voltage dependence of a membrane protein, the charge has to be more loosely tethered in one of the global states than in the other. Let's take an extreme case so that we have the best chance of observing an effect. We take one of the force constants as very high, so $1/\phi_a$ will be near zero. If the other force constant, ϕ_b, is small, then the factor multiplying V^2 will be large.

To estimate a reasonable value for ϕ_b, we think about how far the charge can move from its equilibrium position. A very loosely bound charge may move around 5 Å, which is about 10% of the thickness of the membrane (a typical lipid bilayer thickness is about 50 Å). This displacement is quite large compared to the displacements allowed by ordinary stretching of covalent bonds (see Section 2.12). It is also large compared to the scale of motions seen in proteins, so this value can be considered a generous maximum.

To convert this displacement to a force constant we note that any degree of freedom has a mean energy of $RT/2$ per mole (by the equipartition principle, see Section 12.4). The mean potential energy is then

$$\phi_b \overline{x'^2}/2 = RT/2 \tag{1.39}$$

With a root mean square displacement of 10% of the membrane thickness, we have $\overline{x'^2} = 0.01$. We can then solve for the force constant as

$$\phi_b = 100RT \tag{1.40}$$

Making this substitution, we can focus on the exponent in Eq. (1.38) and compare the linear and quadratic terms.

$$qV\delta - \frac{(qV)^2}{2\phi_b} = qV\delta - \frac{(qV)^2}{200RT} \tag{1.41}$$

Since $RT/q \sim 25$ mV at room temperature for a unitary charge, we can further simplify this expression.

$$qV\delta - \frac{q}{RT}\frac{qV^2}{200} = qV\delta - \frac{qV^2}{5 \text{ volts}} \tag{1.42}$$

Now factoring out qV gives

$$qV(\delta - V/(5 \text{ volts})) \tag{1.43}$$

When the dimensionless quantity $V/(5 \text{ volts})$ is comparable to δ (which is also dimensionless because it is normalized to the thickness of the membrane) the V^2 term will have an impact. If δ is small, say 0.1, then the quadratic term will become clearly visible when V gets up to around 0.5 V. This is very high for the voltage across a membrane. Biological membranes are usually destroyed at voltages less than half this large. In fact, if δ were zero we could focus all our attention on the term $V/(5 \text{ volts})$. This quantity changes so little within experimentally accessible membrane potentials (± 0.2 volts) that the protein would not appear to be voltage dependent. Thus, the compliance of charge in a membrane protein is unlikely to have an observable impact on voltage-induced transitions. This strengthens the justification for using the Boltzmann equation (Eq. (1.28)) in describing the effects of voltage on membrane proteins. In general, it is hard to find a property of a protein that exhibits this kind of quadratic-exponential dependence on an experimental variable. This analysis of compliance thus supports a focus on global transitions.

Problems for Chapter 1

1. With x defined as in Eq. (1.11), use the van't Hoff relation (Eq. (1.19)) to derive an expression for x as a function of T.

2. A protein undergoes an unfolding transition at 65 °C with a $\Delta H°$ of 80 kcal/mole. What is $\Delta G°$ at 25 °C? Repeat the calculation using the observation that the unfolded state has a heat capacity that is 9 cal/°C greater than the folded state.

3. How will a change in temperature alter the voltage dependence of a conformational transition as expressed in Eq. (1.28)? How would you interpret a change in V_0? Why is a substantial change in V_s less likely?

4. Use the Boltzmann equation to derive the probability of a freely rotating dipole in a membrane having an orientation θ with respect to an axis perpendicular to the plane of the membrane. First, consider the case where the dipole is stuck in one plane perpendicular to the membrane. Then consider the case where the dipole can rotate with complete freedom. For both cases the dipole should be taken as entirely within the membrane where the voltage varies linearly.

5. For the unfolding of a protein by denaturant, how would you expect m_D in Eq. (1.22) to vary with a protein's molecular weight for a series of spherical proteins.

6. For a two-state voltage transition with $V_0 = -20$ mV and $V_s = 10$ mV, calculate the $\Delta G°$ for this transition at $V = -70$ mV and $+30$ mV.

7. Derive the voltage dependence of channel opening for the model in Section 1.11, but with a trimer in which B_3 represents the only open state.

8. Derive the open probability for a gap junction channel between two cells as a function of the voltage difference between the two cell interiors. The gap junction channel is actually a pair of channels in series, each gating independently of the other, and each having the same V_0 and V_s. Assume each channel sees half the voltage difference.

9. Derive the expression corresponding to Eq. (1.34) for a thermal transition in a dimer.

10. Calculate the offset of Eq. (1.33) (V_0) that makes it cross 0 mV at $P_o = 0.5$ (as in Fig. 1.8). What is this offset for a trimer as in Problem 7?

11. Use the classical configuration integral (Eq. (1.4)) to calculate the partition function of a harmonic oscillator (the potential energy function is $U(x) = 1/2\phi(x - x_0)^2$). Write the free energy, entropy, and enthalpy (see Appendix 4).

Chapter 2

Molecular forces in biological structures

In Chapter 1 the global conformation of a protein was treated like a black box, without worrying about the internal machinations. This approach is useful in interpreting many kinds of experiments, but if we want to make use of detailed structural information about a macromolecule, we have to open up the black box and look inside. To do this we need to understand the molecular forces that act within a macromolecule. These forces govern how a protein folds, and which of its different conformations will predominate. Similar forces determine the structures of nucleic acids and lipid bilayers, and also drive the associations between macromolecules and ligands.

The forces at work in biological systems can be divided into various categories and examined in turn. They are generally well understood to the extent that good approximate mathematical expressions are available. Much is known about their relative strengths under various conditions. However, it must be emphasized that the structure and dynamics of biological macromolecules are determined by the interplay of many forces. This complexity makes it difficult to investigate these forces by studying the biological molecules directly. We therefore often turn to model systems that are very unbiological but nevertheless instructive. The results from model systems enable us to break these complicated problems down into simpler ones.

2.1 | The Coulomb potential

One of the fundamental tenets of electricity is that point charges interact with a potential energy that is inversely proportional to the distance of separation, r, and directly proportional to the product of the two charges, q_1 and q_2. This is Coulomb's law:

$$U = \frac{q_1 q_2}{\varepsilon r} \tag{2.1}$$

Here we will use cgs units, where charge is in esu (the charge of an electron is 4.8×10^{-10} esu), distance is in centimeters, and energy is in ergs; ε is the dielectric constant of the medium in which the charges are placed. This quantity characterizes the response of the surrounding medium to an electric field.

The dependencies on q and r are simple and easy to understand. The dielectric constant is a bit more subtle. It depends on how easily the molecules in the environment are polarized. A completely unresponsive medium, or a vacuum, will have $\varepsilon = 1$ because there are no molecules to polarize. A strongly responsive medium consisting of highly polar or polarizable molecules will have a large ε. The polarization of these molecules will counteract the electrical field and reduce the magnitude of the Coulomb potential energy. It is found that the value of ε varies from about 2 in nonpolar hydrocarbons to about 80 in water.

The high dielectric constant of water reflects the large dipole moment of the water molecule, and the ease with which these molecules can rotate in an electric field. This counteracts the effect of an electric field, making it quite small. At the distance of closest approach for Na^+ and Cl^- of 3 Å, Eq. (2.1) gives only about 1.3 kcal mole^{-1} or slightly more than $2RT$. This is why NaCl dissociates and dissolves in water. Because this interaction is weak, statistical mechanics must be used to describe the distribution of ions in solution (Chapter 11). On the other hand, the value of 2 for ε of hydrocarbons makes electrostatic interactions within the hydrophobic cores of proteins and membranes enormously strong. We will soon see that another aspect of electrostatics excludes charges from these places.

In a homogeneous dielectric it is easy to compute the electrostatic energy as a sum of Eq. (2.1) over all pairs of charges. However, biological systems are rarely homogeneous, and spatial variations in ε prevent the simple use of the Coulomb potential in most situations. The electric potential in a system with different regions having different dielectric constants can be worked out by solving the Poisson or closely related Laplace equations with boundary conditions given by the specifics of the geometry of a particular problem. These differential equations need not be written out here because we will not go into the details of their solution. For simple systems with planar, spherical, or cylindrical boundaries some standard mathematical methods can be employed. Some of these cases will be discussed below. For complex and irregular geometries it is usually necessary to use a computer to solve the equations numerically.

The simplicity of Coulomb's law conceals an interesting and often overlooked point regarding the thermodynamics of electrostatic interactions. When enthalpy is measured, the value is often erroneously attributed to electrostatic interactions (and hydrogen bonds, see Section 2.10). Likewise, large entropies are erroneously attributed to hydrophobic interactions (Section 2.8). The Coulomb potential energy gives the work done at constant temperature and pressure to bring the charges to their specified positions. This means that, in thermodynamic terms, it is a free energy rather than a simple potential energy to be equated with enthalpy. When the Coulomb potential is examined more carefully, one finds that in water this force is entropy driven.

To decompose the electrostatic potential into H and TS, take $S = -\partial G/\partial T$, where G is the potential energy in Eq. (2.1). Differentiation gives

$$S = -\frac{\partial}{\partial T}\left(\frac{q_1 q_2}{\varepsilon r}\right) = \frac{q_1 q_2}{\varepsilon^2 r}\frac{\partial \varepsilon}{\partial T} = G\frac{1}{\varepsilon}\frac{\partial \varepsilon}{\partial T} \qquad (2.2)$$

If the dielectric constant did not vary with temperature, then S would be zero, and the Coulomb potential would indeed be enthalpy. But the dielectric constant of water is strongly temperature dependent, and decreases by 0.46% per degree Kelvin near room temperature. That means that $(1/\varepsilon)(\partial \varepsilon/\partial T) = -0.0046$, giving $S = -0.0046G$. At $T = 300$ K, we get $TS = -1.38G$. Thus, the entropic contribution is greater than the free energy itself.

From this calculation we can conclude that the enthalpy opposes the attraction between oppositely charged ions in water, and it is the entropy that pulls them together. The molecular interpretation of this surprising result is that dissolved ions restrict the rotation (and other kinds of motion) of surrounding water molecules. The work done on these water molecules offsets the work done by the ions on one another, so that the Coulomb potential expresses a balance between ion–ion and ion–water interactions. The actual interactions are incredibly complicated in detail, but all of these details are absorbed into a single number, ε. What makes this simplified approach work is that the polarization of the solvent varies linearly with the applied electric field. This linearity is distilled into a single number, ε.

2.2 | Electrostatic self-energy

The preceding section made the point that an ion in water does a considerable amount of work on the surrounding water molecules by forcing them to rotate and orient their dipoles. This happens regardless of whether other ions are nearby, and this interaction between an ion and the surrounding medium plays an important role in the distribution of ions. The energy of placing an ion in a dielectric medium can be calculated from the Coulomb potential as follows. Consider the work done to bring a small increment of charge, $\delta q'$, to the surface of a sphere with radius r, already carrying a charge, q'

$$\delta G = \frac{q'\delta q'}{\varepsilon r} \qquad (2.3)$$

This is simply Eq. (2.1) with $q_1 = q'$ and $q_2 = \delta q'$. This charging process can be integrated to get the total work done, starting with zero charge and working our way up to the final value, q, as follows

$$G = \frac{1}{\varepsilon r}\int_0^q q'\,\mathrm{d}q' = \frac{q^2}{2\varepsilon r} \qquad (2.4)$$

This is called the self-energy of a charge. It is also referred to as the Born energy after the physicist Max Born who derived this result in 1920.

Equation (2.4) can be used to estimate the free energy of placing an ion in a solvent with a particular dielectric constant. In the case of water this is the hydration energy. Data on the enthalpy of hydration provide a test for the picture of hydration as a polarization of the surrounding water. To obtain the enthalpy from Eq. (2.4) one differentiates with respect to temperature to obtain $-S$, as illustrated in the preceding section, and obtains H as $G + TS$. The result is still an inverse dependence on r, and experimental data shown in Fig. 2.1 confirm this. However, some care is necessary in choosing r. It has been argued that r is the radius of a solvent cavity around the ion, and the good agreement with theory was obtained when the cavity radius was estimated from crystal structures of salts as the distance from the center of an ion to the outer electronic shell of the counterion in the crystal (Rashin and Honig, 1985).

It is remarkable that Eq. (2.4) works so well in estimating the energy of an ion's interaction with solvent. At atomic dimensions the details of the solvent structure should matter. One would expect these energies to depend on the specific form of the molecular interactions. Furthermore, the enormous strength of the electric field at the ion's surface should polarize the solvent to a maximal degree, producing what is known as dielectric saturation. Yet, Fig. 2.1 includes ions with valances ranging from one to four. If the dielectric were saturated, then the tetravalent ions would not fall on the same curve as the monovalent ions. It is fortunate that such a simple theory provides such an accurate quantitative description.

One can take the difference between the self-energy for two different values of ε and estimate the free energy of transfer of an ion between two mediums with different dielectric constants. For transfer of a sodium ion with $r = 0.95$ Å from water, where $\varepsilon = 80$, to a hydrocarbon medium where $\varepsilon = 2$, the potential energy difference is 0.15×10^{-12} erg. This works out to a very large energy of 85 kcal mole^{-1}. This is a reasonable result because inorganic ions

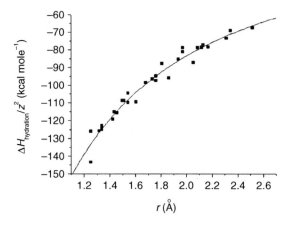

Fig. 2.1. Hydration enthalpy for different ions was divided by q^2 and plotted versus r. The inverse dependence on r arises from Eq. (2.4). The solid curve is proportional to $-1/r$ (data from Rashin and Honig, 1985).

are generally insoluble in organic solvents. Dielectric polarization keeps ions out of nonpolar environments. The partition coefficient for a sodium ion between oil and water is related to the free energy of transfer as $e^{-\Delta G/RT}$, which gives 10^{-63} when ΔG is taken as the self-energy. This shows why it is extraordinarily difficult to move an ion into the interior of a protein or lipid bilayer.

2.3 | Image forces

The strong dependence of the self-energy on the dielectric constant means that near a boundary between two mediums with different ε, a charge will be attracted toward the region where ε is higher. The ion will have a stronger polarizing action on the region with the higher ε, and effective dipoles created in this region will pull the ion toward the interface. Likewise, an ion in a high dielectric medium will be pushed away from a nearby low dielectric medium.

Consider a point charge near a planar boundary that separates two semi-infinite regions with dielectric constants ε_1 and ε_2 (Fig. 2.2). This example approximates the situation when an ion is near a lipid membrane.

The solution to the Laplace equation on the left side (where we have ε_1) turns out to be the sum of two terms that look like Eq. (2.1). One is the Coulomb potential of the charge q, and the other is the Coulomb potential of a fictitious charge referred to as an *image charge* (q' in Fig. 2.2). The image charge is located at a position the same distance on the opposite side of the boundary, as though it were an image in a mirror placed at the boundary. The force seen by a charge at a dielectric boundary is therefore called an *image force*.

The charge at the image point has a magnitude $q' = -q(\varepsilon_2 - \varepsilon_1)/(\varepsilon_2 + \varepsilon_1)$ for the case $\varepsilon_2 > \varepsilon_1$ (see Jackson, 1975). The potential energy is given by Eq. (2.1) for q and q' separated by a distance of $2r$

$$U = -\left(\frac{\varepsilon_2 - \varepsilon_1}{\varepsilon_2 + \varepsilon_1}\right)\frac{q^2}{2\varepsilon_1 r} \tag{2.5}$$

Thus, the force of a dielectric medium on a charge has the same form as the force of another charge.[1] The sign of Eq. (2.5) makes the force point toward the region with higher ε. With $\varepsilon_2 > \varepsilon_1$, the force is to the right.

To see how the image force influences energetics in some biological situations we can examine two interesting special cases. In one of these a charge is placed in the middle of a low dielectric slab of finite thickness, with a high dielectric medium extending to infinity on both

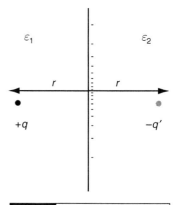

Fig. 2.2. A charge, q, at a distance r to the left of the boundary between ε_1 and ε_2 polarizes the surrounding mediums. The polarization of the distant medium (indicated by the dashes at the boundary) can be represented by an image charge, $-q'$, at a distance r on the other side of the boundary.

---

[1] Note that this expression is only valid for a charge that is small compared to the distance r. The singularity at $r = 0$ is unrealistic, and if the size of the charge is taken into account, the potential energy will change monotonically as the interface is crossed.

Fig. 2.3. (a) A charge with radius *a* is located in the center of a planar slab of thickness *b*. Within the slab the dielectric constant is ε_h (~2 for hydrocarbon), and outside the slab the dielectric constant is ε_w (~80 for water). Equation (2.6) gives the self-energy of this charge. (b) A charge with radius *a* lies in the center of a larger hydrocarbon sphere with radius *b*. Equation (2.7) gives the self-energy of this charge.

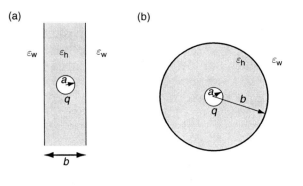

sides (Fig. 2.3a). In the other case a charge is placed in the center of a low dielectric sphere surrounded by a high dielectric medium (Fig. 2.3b).

For the first case of a dielectric slab, the free energy of putting a charge, *q*, at the very center has the following form (Parsegian, 1969)

$$G = \frac{q^2}{2\varepsilon_h a} - \frac{q^2}{\varepsilon_h b}\ln\left(\frac{2\varepsilon_w}{\varepsilon_w + \varepsilon_h}\right) \tag{2.6}$$

where ε_w is the dielectric constant of water and ε_h is the dielectric constant of hydrocarbon. The first term is the self-energy in an infinite medium (Eq. (2.4)). The second term tells how the finite thickness reduces that energy. For the thickness of a typical lipid bilayer of 50 Å, the second term is negligible. This means that putting an ion into the middle of a lipid bilayer is almost as hard as transferring it to a bulk hydrocarbon medium. The energy of placing an ion in a lipid bilayer is an important consideration in the study of ion permeation of membranes (Chapter 14).

The second case in Fig. 2.3 is a charge in the center of a dielectric sphere. This can be taken as an estimate for how hard it is to bury a charged residue in the core of a globular protein. This energy is evaluated by the same method used to derive Eq. (2.4). The movement of each element of charge $\delta q'$ to the surface at $r = a$ must be broken down into two steps. The first is a movement through the water to the surface at $r = b$, and the work is $q'\delta q'/\varepsilon_w b$. The second is a movement through the hydrocarbon to the surface at $r = a$, and the work is $q'\delta q'(1/\varepsilon_h)[(1/a) - (1/b)]$. Integrating over q' from zero to q gives

$$G = \frac{q^2}{2\varepsilon_h}\left(\frac{1}{a} - \frac{1}{b}\right) + \frac{q^2}{2\varepsilon_w b} \tag{2.7}$$

This is the energy of placing the charge in the center of the oily sphere in Fig. 2.3b. Note that as *b*, the radius of the outer hydrocarbon sphere, becomes large this expression becomes a difference between two self-energies of the form of Eq. (2.4). Since ε_w is much higher than ε_h, the last term in this expression can usually be ignored. We see that for a radius like that of a typical globular protein, say 20 Å, this energy is within 5% of the self energy in a bulk medium with dielectric constant ε_h (assuming that the charge has a radius of about 1 Å). The bottom line of these two results is that it is almost as difficult to place a charge in the middle of a lipid membrane or the middle of a globular protein as it is to place it in a bulk medium with a similar low dielectric constant.

Likewise, a small pocket of water can stabilize an ion within a nonpolarizable region. The structure of an ion channel revealed a water filled cavity with a radius of 5 Å. By stabilizing charges deep within the membrane, this cavity plays an important role in ion permeation (Section 14.10).

2.4 | Charge–dipole interactions

Charge is generally not distributed uniformly within a molecule, and this enables uncharged molecules to interact by electrical forces. One way to deal with these interactions is to introduce the idea of the electrical dipole, defined as two equal and opposite charges set a fixed distance apart. We envision a neutral molecule as in Fig. 2.4, in which positive and negative charges are effectively concentrated at two distinct foci, separated by a distance a. The molecule's dipole moment emerges from a calculation of the interaction between a charge, q, and each of these charged foci.

The interaction will be a sum of two Coulomb terms, one for the interaction between q and $-q'$, and the other between q and $+q'$, with distances as in Fig. 2.4

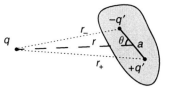

Fig. 2.4. A molecule with a dipole moment interacting with a point charge. The dipole is represented by two charges of opposite sign separated by a distance a.

$$
\begin{aligned}
U &= -\frac{q'q}{\varepsilon r_-} + \frac{q'q}{\varepsilon r_+} \\
&= -\frac{q'q}{\varepsilon(r - \frac{1}{2}a\cos\theta)} + \frac{q'q}{\varepsilon(r + \frac{1}{2}a\cos\theta)} \\
&= \frac{-q'qa\cos\theta}{\varepsilon(r^2 - (a^2/4)\cos^2\theta)}
\end{aligned} \tag{2.8}
$$

In the first step, r_- and r_+ are expressed as projections onto r, giving $r_+ = r + \frac{1}{2}a\cos\theta$ and $r_- = r - \frac{1}{2}a\cos\theta$. This is an approximation that is valid when a is small such that the lines are nearly parallel. The fractions are then simply added to obtain the final result in Eq. (2.8). When $r \gg a$ we can ignore the term $a^2\cos^2\theta/4$ in the denominator to obtain

$$
U = \frac{-qd\cos\theta}{\varepsilon r^2} \tag{2.9}
$$

where $d = aq'$ is defined here as the dipole moment.

The charge–dipole potential energy depends on a higher inverse power of distance $(1/r^2)$ compared with the Coulomb potential energy $(1/r)$. Thus, the interaction is shorter in range because the inverse square function decreases more rapidly. This illustrates a general trend that as higher order electrostatic interactions are considered, the inverse power of r increases, and the range becomes shorter. Thus, an interaction between two dipoles will decrease as

r^3 (and will depend on the orientation of both dipoles relative to the line between the centers, see Problem 3).

It is easy to forget that this derivation was based on $r \gg a$. Equation (2.9) is not valid at distances comparable to or smaller than the charge separation within the dipole. If we let the dipole of Fig. 2.4 point directly at the charge q, then $\theta = 0$ so we have

$$U = -\frac{q'q}{\varepsilon(r - \frac{1}{2}a)} + \frac{q'q}{\varepsilon(r + \frac{1}{2}a)} \tag{2.10}$$

As the dipole and charge get closer, r can fall below a so that only the first term matters. Then it is approximated by a Coulomb interaction with the nearer charge of the dipole.

The interactions just discussed are for dipoles with fixed orientations, as would be the case within a folded protein with a fairly rigid structure. For example, the carbonyl groups of polypeptides have substantial dipole moments and their electrostatic interactions make an important energetic contribution to the stability of an α-helix (Section 2.13). However, molecules interacting in solution can rotate, and the forces will vary with orientation. Then it becomes necessary to average over all orientations. The probability of a particular orientation is then given by the Boltzmann weight $e^{-U/kT}$. Allowing for rotations not only weakens the interaction by allowing both favorable and unfavorable interactions, it also shortens the range of the interaction. When a Boltzmann average is used to calculate the mean energy of interaction between two freely rotating dipoles, with dipole moments d_1 and d_2, one obtains the following expression (Setlow and Pollard, 1962)

$$U = -\frac{d_1^2 d_2^2}{3\varepsilon^2 r^6 kT} \tag{2.11}$$

It should be noted that the derivation of this expression depends on the assumption that the energy of changing orientations is small compared to kT. In fact calculations show that this is a valid assumption and that the dipole–dipole interaction does not restrict rotation very much. One very important exception is water, because water has a particularly large dipole moment. But for other molecules in solution this form of interaction is quite weak. As will be shown below, the closely related dispersion force is much stronger and plays a more important role in shaping biological structures.

2.5 | Induced dipoles

Even molecules with neither a net charge nor a permanent dipole moment can be influenced by electrical forces. Such molecules are polarized by an electrical field, resulting in an induced dipole moment. Molecules either positively charged, negatively charged, or polar will attract a neutral molecule as a result of this induction.

The induced dipole moment is given by the product of the field and the polarizability

$$d = \alpha E \tag{2.12}$$

where the polarizability, α, is a property of the particular molecule, and E is the field. Inserting this expression for d in Eq. (2.8), and taking $E = q/\varepsilon r^2$ gives the energy as a function of distance between a charge and a polarizable molecule.

$$U = -\frac{\alpha q^2}{\varepsilon r^4} \tag{2.13}$$

where $\cos\theta = 1$ because the induced dipole is assumed to point in the direction of the field.[2]

For the interaction between a permanent dipole and an induced dipole, the field of a dipole is used for E in Eq. (2.12). We then have

$$U = -\frac{4\alpha d^2}{\varepsilon r^6} \tag{2.14}$$

Again, the trend of increasing inverse powers in r can be seen as higher order electrostatic interactions are evaluated.

2.6 | Cation–π interactions

When a cation approaches an aromatic ring in the manner depicted in Fig. 2.5a, there is a strong attraction (Ma and Dougherty, 1997). In the gas phase benzene binds cations quite tightly. For Li^+ the energy is 38 kcal mole^{-1}; for Na^+ it is 28 kcal mole^{-1}; for K^+ it is 19 kcal mole^{-1}. The fact that smaller ions bind more tightly suggests that it is a simple electrostatic interaction, with smaller ions binding more tightly because they can approach more closely.

Part of this attraction is caused by the cation polarizing the aromatic ring and inducing a dipole, as in the preceding section. However, even without polarization an aromatic ring has a nonuniform charge distribution. The double-bonding π electrons form an electronegative region balanced by the partial positive charges in the plane of the carbon nuclei (Fig. 2.5b). This distribution of charge is

(a)   (b)   (c)

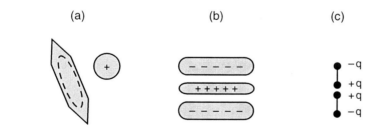
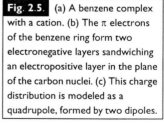

Fig. 2.5. (a) A benzene complex with a cation. (b) The π electrons of the benzene ring form two electronegative layers sandwiching an electropositive layer in the plane of the carbon nuclei. (c) This charge distribution is modeled as a quadrupole, formed by two dipoles.

---

[2] A molecule can have an anisotropic polarizability, meaning that a field of the same strength will polarize the molecule more in one direction than another. Then the polarization will not necessarily be in the same direction as the field.

termed a quadrupole, and can be represented by two dipoles (Fig. 2.5c). By extending the mathematics of Section 2.4 to this kind of charge distribution one obtains an energy that goes as $1/r^3$ (Problem 4). These charge–induced dipole and charge–quadrupole forces combine to varying degrees in different aromatic ring structures to generate a force that attracts cations (Cubero et al., 1998). The origin is rather complex, so this force is referred to as a cation–π interaction rather than with a specific electrostatic term such as quadrupole or induced dipole.

The amino acids phenylalanine, tyrosine, and tryptophan have aromatic side chains with the capacity for cation–π interactions. These amino acids often appear in proteins to interact with positively charged amino acid side chains, and they often appear in binding sites for cationic ligands and substrates. A theoretical calculation of the cation–π interaction energy between methylammonium and benzene gave -12.5 kcal mole^{-1} in a nonpolar solvent. This energy was reduced to -5.5 kcal mole^{-1} in water (Gallivan and Dougherty, 2000). By contrast, a Coulomb interaction between methylammonium and acetate was reduced from -53 kcal mole^{-1} to -2.2 kcal mole^{-1}. So according to this result, cation–π interactions are stronger in water than the electrostatic interactions between monovalent ions.

The neurotransmitter acetylcholine contains a positively charged quaternary amine, and two proteins that bind acetylcholine, acetylcholinesterase (Fig. 8.4) and the acetylcholine receptor, have aromatic amino acids in their binding sites. A study of the acetylcholine receptor has undertaken to evaluate the role of cation–π interactions in binding (Zhong et al., 1998). When one of the binding site tryptophans, at position 149, was replaced with a series of fluoro-tryptophans, the dose–response curve for acetylcholine was shifted. Fluorine atoms withdraw electrons from the indole ring system of tryptophan, and deplete the electronegative layer depicted in Fig. 2.5b. This weakens the cation–π interaction. A plot of the apparent binding energy versus the theoretically predicted strength of the cation–π interaction showed a good correlation, making a strong case for a substantial energetic contribution to acetylcholine binding (Fig. 2.6). It should be

Fig. 2.6. Plot of apparent binding energy computed from the EC$_{50}$ of the acetylcholine receptor versus the theoretically computed cation–π binding energy. The amino acid at position 149 (in the binding site) is indicated for each point (data from Zhong et al., 1998).

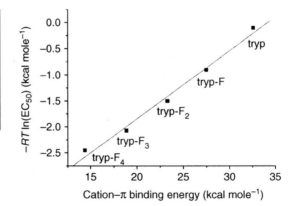

noted that the slope in this plot is less than one because the theoretical calculations are for the gas phase and for an inorganic ion and the experiments are for acetylcholine in water.

2.7 | Dispersion forces

The charge distribution of a molecule fluctuates rapidly with time. At any instant there will be a transient dipole moment. When two molecules are near one another the fluctuations in dipole moment will sometimes lead to an attraction and sometimes to a repulsion. In this sense the interaction is similar to that between freely rotating permanent dipoles orienting in either attractive or repulsive configurations. Since the attractive configurations have a lower potential energy than the repulsive configurations, they will have larger weights in a Boltzmann average, leading to a net attraction. There are many theoretical approaches to the calculation of this interaction energy, all of which are quite complicated. In general, the energy is proportional to $1/r^6$, as was the case for the interaction between rotating dipoles. The important physical factors are indicated in the following expression, which should be compared to Eq. (2.11) (Setlow and Pollard, 1962)

$$U = -\frac{1}{r^6}\frac{\alpha_1\alpha_2}{3n^4}\frac{I_1I_2}{I_1+I_2} \tag{2.15}$$

where α_1 and α_2 denote the polarizabilities of the two interacting molecules, I_1 and I_2 denote their ionization energies, and n denotes the refractive index of the medium.

Although both Eqs. (2.11) and (2.15) vary as $1/r^6$, the dispersion force differs from the force between freely rotating permanent dipoles in some important ways. In particular, the dielectric constant of the surrounding medium does not appear in Eq. (2.15). The strength of this interaction still depends on the medium, but the fluctuations in the electronic structure responsible for transient dipole moments are much faster than molecular rotations in a liquid. This makes the dielectric constant irrelevant. Instead, the refractive index of the medium appears in this expression, because it reflects the response of the medium to very rapid changes in electric fields (at frequencies like those of light). In water the refractive index is a little larger than one, and so it is much smaller than the dielectric constant. The important consequence is that the dispersion force is much stronger than interactions involving rotating permanent dipoles.

A more rigorous expression for this force would include polarizabilities as functions of frequency, as might be obtained from spectroscopic data. Quantities that vary with frequency are said to exhibit dispersion, and that is why this force is commonly called the dispersion force. It is also called the London force, after Fritz London, who developed the first satisfactory theory in 1930. An important consequence of dispersion is that two molecules may not attract one another strongly if the frequency ranges in which their polarizabilities are large do not overlap. For interactions between similar molecules the frequency ranges will be well matched, and the dispersion force will be stronger. This is the origin of the general rule in solution chemistry that "like dissolves like."

The $1/r^6$ dependence in Eq. (2.15) is used quite often in trying to quantify dispersion forces. The proportionality constant is usually obtained empirically rather than computed from molecular structure. Furthermore, it should be borne in mind that the $1/r^6$ dependence is valid for distances of separation that are large compared to the size of the interacting molecules. For short distances the dependence on distance is closer to $1/r$ (Parsegian, 1973).

Dispersion forces are quite general, acting between molecules with complex shapes, and between large macroscopic bodies. The theoretical analysis can get very complicated, and varies quite a bit for each case. There are only a limited number of results in the literature. One important case is parallel rod-shaped molecules, for which the dispersion energy varies as $1/r^5$. One can obtain this inverse fifth power by summing over interactions between linearly arranged centers each interacting as $1/r^6$. The attraction between parallel rods provides a cohesive force between hydrocarbon chains in a lipid bilayer (Section 2.16), and is relevant to the stability of assemblies of rod-shaped filamentous proteins. Planes or sheets separated by a medium with a different dielectric constant attract one another, and the leading term in the interaction varies as $1/r^2$ (Parsegian, 1973). This attraction, which draws two lipid bilayers together, is opposed by a repulsive hydration force (Section 2.9).

2.8 | Hydrophobic forces

One of the most important molecular forces in biology is the hydrophobic force. It is a familiar fact that oily substances will form a separate phase from water. Likewise, macromolecules will arrange into structures that minimize contact between their polar and non-polar domains. The attraction between the oily, hydrocarbon parts of molecules is then equated with an aversion or "phobia" of water, hence the term "hydrophobic." This term is highly appropriate because the hydrophobic force is driven primarily by an energy cost of creating hydrocarbon–water contact.

The dispersion force is the natural first choice in trying to explain the hydrophobic effect. But if the hydrophobic effect were caused by dispersion forces then one would expect to find a positive enthalpy change occurring when hydrophobic molecules come in contact with water. Mixing would then require strong water–water and hydrocarbon–hydrocarbon interactions to be broken and replaced by weak water–hydrocarbon interactions, leading to the absorption of heat. Water and hydrocarbons are weakly miscible, and calorimetric measurements with these mixtures showed that

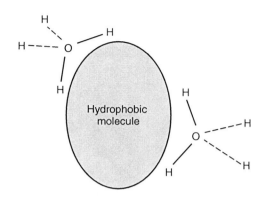

the enthalpy change is very small, and often of the wrong sign. This means that other physical processes must be considered to understand the hydrophobic effect.

With small positive enthalpy changes one must look for sources of entropy to drive the segregation of hydrocarbons and water. The primary source of entropy in the hydrophobic force is the ordering of water molecules at a hydrophobic surface. Computer simulations of vicinal water near a nonpolar surface have shown that there is a decrease in entropy because the water becomes structured, even ice-like. This ordering of vicinal water can be explained by noting that water cannot form strong hydrogen bonds (Section 2.10) with hydrocarbon. To maintain energetically favorable hydrogen bonding, water must straddle a hydrophobic surface (Fig. 2.7). This restricts the orientations of water molecules and lowers the entropy (Stillinger, 1980).

Another way of viewing the reduced entropy of vicinal water is to consider the number of ways a water molecule can form hydrogen bonds with its neighbors. If one pictures a tetrahedral cage of four water molecules hydrogen bonding a central water molecule, the central water can donate its hydrogen atoms in any combination of two of its four neighbors. This gives six ways to be fully hydrogen bonded. Replacing one water of the cage by a hydrophobic, nonhydrogen-bonding neighbor reduces the number of ways this can happen by a factor of about two.

The restriction in orientation of vicinal water varies with temperature. It becomes harder and harder to order molecules as the temperature is raised. As a result, hydrocarbon–water contacts have a very high heat capacity. Raising the temperature gradually melts the ice-like vicinal water. Interestingly, and somewhat paradoxically, as the temperature rises and the entropy goes up, the hydrophobic effect does not get weaker, but instead gets slightly stronger (Dill, 1990). This is because the dispersion force becomes stronger with increasing temperature, and this compensates for the loss of entropic drive. We must therefore view the hydrophobic force as entropy-driven at low temperatures (around room temperature) and enthalpy-driven at higher temperatures (near the boiling point of water). Extrapolations of thermodynamic quantities to temperatures

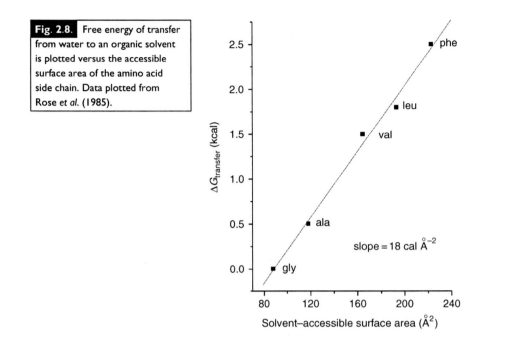

Fig. 2.8. Free energy of transfer from water to an organic solvent is plotted versus the accessible surface area of the amino acid side chain. Data plotted from Rose *et al.* (1985).

above the boiling point of water suggest that a maximum in the strength of the hydrophobic force will occur, and at this temperature the entropy change will be zero. Thus, the common label "entropy-driven" is only valid in a limited temperature range.

Although the complexity of the physical process that gives rise to the hydrophobic force has made it hard to develop a detailed quantitative theory, the idea that a hydrocarbon surface orders neighboring water molecules has motivated a very simple and useful approximate representation of the interaction energy in terms of the surface area of contact with water. Extensive measurements have been made of free energies of transfer of hydrocarbons of various shapes and sizes from oil to water. In general, this free energy scales with the surface area of the hydrocarbon. A plot for five amino acids is shown in Fig. 2.8. The slope of 18 cal Å^{-2} provides a useful scale factor for estimating the energetic cost of juxtaposing hydrophobic surfaces and water. Note that this plot was limited to amino acids with nonpolar side chains. If amino acids with polar side chains had been included, these points would have fallen below the line. The dependence of free energy on the surface area of contact is an empirical result, which can be very useful in estimating the energetics of hydrophobic interactions.

The hydrophobic effect becomes stronger as ions are added to a solution (Baldwin, 1996). Adding salt generally reduces the solubility of nonpolar molecules in water, and increases the slopes of plots such as Fig. 2.8. Ions attract and surround themselves with water because of its high dielectric constant. Nonpolar molecules are repelled owing to their much lower polarizabilities. This effect can be understood qualitatively in terms of the image forces of Section 2.3.

Extensive studies of different ions generated what are known as Hofmeister series in which the salting out actions of ions are ranked.

2.9 | Hydration forces

In contrast to hydrophobic surfaces, hydrophilic surfaces made up of charged or polar molecules interact favorably with water. It is hard to remove the water from these surfaces. Water removal is necessary to bring two hydrophilic surfaces close together, so there is an effective repulsion. This force is known as the hydration force. The hydration force has been studied most thoroughly in lipid bilayers, but it is thought to be quite general.

As noted above, lipid bilayers are attracted to one another by dispersion forces between the two hydrocarbon interiors. The hydration force opposes this attraction and prevents them from getting closer than about 20–40 Å. This force is studied not by directly pushing two bilayers together, but by osmotically withdrawing water. In this way one observes how the distance of separation varies as a function of the chemical potential of water. From this information one can determine the force as a function of distance. These experiments showed that the interbilayer repulsive force increases exponentially as the distance is reduced. The distance over which the force changes by a factor of e is quite short, roughly 1–3 Å. Hydration forces thus rise very steeply, and the distances at which they become significant vary depending on the chemical nature of the phospholipid headgroup. Lipid bilayers will aggregate under some conditions, and the distance of separation represents an equilibrium between attractive dispersion forces and repulsive hydration forces (Rand, 1981).

Hydrophilic surfaces are often charged so one would expect electrostatic forces to play a greater role. For lipid bilayers, electrostatic interactions dominate only for bilayer separations of greater than roughly 25 Å. At closer distances ionic strength does not have much of an effect, indicating that the hydration force has assumed a dominant role. In thinking about cellular processes such as membrane fusion, where two opposing lipid bilayers join together, it is important to remember that the hydration force presents a formidable obstacle. This force can be overcome by divalent cations, which can bridge some of the polar head groups found on phospholipids. Alternatively, proteins associated with vesicle and target membranes can catalyze membrane fusion with speed and specificity.

2.10 | Hydrogen bonds

Hydrogen atoms have a unique chemical property of being able to form bridges between pairs of electronegative atoms (typically oxygen or nitrogen). The standard theory of chemical bonding allows the hydrogen atom to form only a single covalent bond, but in a hydrogen bond a hydrogen atom shares its electron with two bonding atoms. The electronegative atom serving as the hydrogen acceptor also violates the standard rules of bond number. For example, a carbonyl oxygen atom already has two covalent bonds so it forms a third bond when it hydrogen bonds. To address the problem of maintaining correct bond numbers, it has been proposed that the hydrogen bond is electrostatic rather than covalent, resulting from the dipole moments of the interacting groups. This is supported by the fact that hydrogen bonds are much weaker than typical covalent bonds. However, quantum mechanical calculations have shown that the hydrogen bond has a strong covalent character, and that the electrostatic picture is an oversimplification (Weinhold, 1997).

Hydrogen bond lengths are generally in the range 2.5–3 Å (for the distance between the two electronegative atoms). The three participating atoms preferentially orient along a single line. When the hydrogen acceptor is an oxygen or nitrogen atom with single rather than double bonds, the hydrogen bond will be incorporated into the tetrahedral symmetry of the p orbitals of that atom. However, the orientation effects are weak so that the energy cost of bending a hydrogen bond is low compared to similar distortions of covalent bond angles.

Hydrogen bonds vary quite a bit in strength. Among the weakest are bonds involving aliphatic hydrocarbons, and these can have energies of less than 1 kcal mole^{-1}. On the other hand, very strong hydrogen bonds have energies exceeding 20 kcal mole^{-1}. These constitute a special class of very strong hydrogen bonds, which can drive catalysis in some enzymes (Section 10.14).

Hydrogen bonds are stronger when more electronegative atoms are involved. For example, bonds with oxygen atoms are usually stronger than bonds with nitrogen atoms. Some energies are listed in Table 2.1. Their magnitudes are such that hydrogen bonds can be expected to play a role in biomolecular structure. But unlike

Table 2.1. *Energies of hydrogen bond formation*

Bond	ΔH (kcal mole^{-1})
H$_2$O - - - HOH	
gas	−5.4
liquid	−3.4
ice	−3.0 to −7.7
C – OH - - - O = C acetic acid (gas)	3.7
C – OH - - - OH ethanol (gas)	−3.4 to −4.0
NH - - - O = C formamide (theory)	−7.9

From Jeffrey and Saenger (1991), p. 27.

stronger covalent bonds, they break easily so that a structure with hydrogen bonds will exhibit some flexibility.

One of the most important roles of hydrogen bonds is endowing water with unusual properties such as a negative change in volume with melting, a high dielectric constant, a high melting point, a high freezing point, and a high heat capacity (Stillinger, 1980). Recall that the hydrogen bonding capacity of water was part of the explanation of the hydrophobic effect (Section 2.8).

Water readily forms hydrogen bonds with dissolved substances and this is an important factor in evaluating the tendency of groups to hydrogen bond. An intramolecular hydrogen bond between two groups within a protein forms at the expense of two hydrogen bonds with water. Likewise, the formation of an intermolecular hydrogen bond between a protein and ligand also involves breaking hydrogen bonds with water. This diminishes the energy that hydrogen bonds can contribute to structural stability.

To account for the role of water, one must then think of hydrogen bonding as an exchange reaction rather than simply bond formation. A full accounting of hydrogen bonding between a donor AH and an acceptor B gives Scheme (2A) as follows

$$\mathrm{AH \cdots OH_2 + B \cdots HOH \rightleftarrows AH \cdots B + HOH \cdots OH_2} \qquad (2A)$$

Each side has the same number of hydrogen bonds (two). On the left-hand side, AH and B each hydrogen bond with water. On the right-hand side AH hydrogen bonds with B and two water molecules hydrogen bond with each other. Because there is no change in net bond number, from a simple bond-inventory perspective there should be no change in enthalpy. However, as Table 2.1 shows, the energies of hydrogen bond formation vary with bond type and environment. Furthermore, although the ΔH for a water–water hydrogen bond in the liquid state is not particularly large, this value is for an isolated pair of water molecules. In a cluster the hydrogen bonds are stronger owing to a cooperativity effect to be discussed further a little later. Finally, water that is not bound to a protein has a higher entropy than bound water. All these factors tend to drive the equilibrium of the above reaction to the right, but the precise free energy contributed by a hydrogen bond is difficult to quantify.

The free energies of hydrogen bond exchange have been estimated in a variety of ways, and some of these results are summarized in Table 2.2. The estimate for the NH - - - OC hydrogen bond within the backbone of a polypeptide chain was made by a series of measurements for model compounds.

Free energies for enzyme–substrate hydrogen bonds were derived from measurements employing site-directed mutagenesis to replace a hydrogen bonding residue with a nonhydrogen bonding residue. This is illustrated in Fig. 2.9, with the enzyme tyrosyl–tRNA synthase. The substrate tyrosine is bound to the enzyme by several contacts, including hydrogen bonds with aspartate 176 to the substrate tyrosyl hydrogen, and with tyrosine 34 to the substrate

Table 2.2. *Energies of hydrogen bond exchange with water*

Bond	Free energy (kcal mole^{-1})
NH - - - O = C peptide[a]	−1.4
enzyme–substrate with neutral partner[b]	−0.8 to −1.5
enzyme–substrate with charged partner[b]	−3 to −6

[a] Jeffrey and Saenger (1991).
[b] Fersht (1987).

Fig. 2.9. Hydrogen bonds formed by the enzyme tyrosyl-tRNA synthase with water (a and c) or substrate tyrosine (b and d). In the wild type enzyme hydrogen bonds form with tyrosine 34 (a and b). In a mutant with phenylalanine at this location this hydrogen bond cannot form (c and d) (see Fersht *et al.*, 1985).

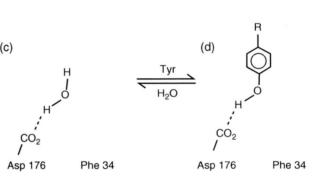

tyrosyl oxygen (Fig. 2.9b). In the absence of the substrate a water molecule occupies this site and is held in place by hydrogen bonds with the same residues (Fig. 2.9a). In a mutant enzyme with phenylalanine in place of tyrosine, a hydrogen bond is subtracted both from the enzyme–water complex and from the enzyme–substrate complex (Fig. 2.9c and d), and the binding is weakened by 0.5 kcal mole^{-1}. This then constitutes a measurement of the energy of the exchange reaction depicted in Scheme (2A). These kinds of experiments provided the estimates for the enzyme substrate values in Table 2.2.

The importance of the environment has already been mentioned for hydrogen bonds, and polarity is a particularly critical factor. Amides and carbonyls are very polar, but when they form a hydrogen bond their polarity is reduced. As a result, it is easier to transfer a donor–acceptor pair to a nonpolar environment if they are joined by a hydrogen bond. This will also increase the energetic drive for hydrogen bond formation in a nonpolar environment. Thus, hydrogen bonds are generally stronger in the nonpolar interiors of proteins and lipid bilayers (Roseman, 1988).

When an atom participates in more than one hydrogen bond, there will be cooperativity between those bonds. This happens because, when an atom serves as a hydrogen donor, its charge distribution changes, increasing its electronegativity and making it a better hydrogen bond acceptor. This promotes the formation of chains, rings, and other networks of hydrogen bonds in which the stabilizing energies are considerably greater than one would estimate by pairwise addition (Stillinger, 1980; Weinhold, 1997). This cooperativity can strengthen hydrogen bonds in liquid water. When a network of hydrogen bonds forms within a protein this network will have a stability that exceeds the sum of the individual hydrogen bond energies.

2.11 | Steric repulsions

Two objects cannot occupy the same space at the same time, and the result is steric repulsions. The origin of this force is the Pauli exclusion principle, which states that two electrons cannot have the same quantum state. This leads to a very steep rise in energy when the electron shell of one atom starts to penetrate the electron shell of another. This force has been modeled as a steep exponential function or as a high inverse power of distance. It is common to represent the steric repulsion as $1/r^{12}$ because when this is added to the $1/r^6$ attractive dispersion force (see, for example, Eq. (2.15)) the mathematics simplifies in a convenient way. The potential energy as a function of distance then takes the form

$$U(r) = 4\varepsilon \left(\left(\frac{r_0}{r} \right)^{12} - \left(\frac{r_0}{r} \right)^{6} \right) \tag{2.16}$$

This function has a minimum just above $r = r_0$ with a depth of ε. Equation (2.16) is known as the *Lennard–Jones potential*. It has the generic features widely found in intermolecular potentials. An attractive part draws the molecules together and a steep repulsion at short range prevents them from penetrating one another (Fig. 2.10). It has been widely used in the theory of liquids, and can be used whenever an attractive dispersion force must be balanced by steric repulsion.

For virtually all practical purposes, the precise mathematical shape of this repulsive force does not matter. As long as the energy rises very steeply when atoms get close together, then each atom has a virtually impenetrable space that can be treated as a volume from which other atoms are excluded. The hard surfaces defined by steric repulsive forces are of vital importance as the basis for the lock-and-key stereospecific interactions that dictate specificity in molecular biology.

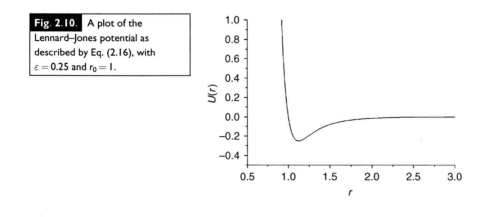

Fig. 2.10. A plot of the Lennard–Jones potential as described by Eq. (2.16), with $\varepsilon = 0.25$ and $r_0 = 1$.

2.12 | Bond flexing and harmonic potentials

One might think that steric forces would make a biological macro-molecule behave almost like a hard ceramic object. This would then define stereospecificity as an absolute process, with no possibility of squeezing or stretching to permit recognition between imperfectly matching molecules. However, the atoms in a macromolecule are not fixed in place. They can move as the bonds that hold them together strain and twist. Each of these motions can be thought of as the stretching of a spring. The displacement is opposed by a force, F, proportional to the displacement, x. Thus, we have $F(x) = -\phi(x - x_0)$, where ϕ is a force constant and x_0 is the equilibrium position where the force is zero. The potential energy is then

$$U(x) = \tfrac{1}{2}\phi(x - x_0)^2 \tag{2.17}$$

When this expression is inserted into Newton's equation of motion, we have a classical harmonic oscillator. It is easy to show that the solution gives a periodic sine or cosine function with a frequency of $\sqrt{(\phi/m)}$. This is the vibrational frequency of the bond, and so vibrational spectroscopy provides a direct route to the determination of force constants (Wilson et al., 1955). The determination of force constants can be made more comprehensive by including in addition to vibrational frequencies other forms of information such as equilibrium conformations and enthalpies of fusion (Lifson and Warshel, 1968).

For the carbon–carbon bond the stretching force constant is 5×10^5 dyne cm^{-1}, or 660 kcal Å^{-2} mole^{-1}. This means that a displacement of only 0.1 Å will exceed thermal energy at room temperature. Many covalent bonds have stretching force constants that are within an order of magnitude of the carbon–carbon bond force constant, which means that bond stretching is not going to soften things up very much. Bond bending force constants are generally about ten-fold weaker, and deformations of about 0.2 Å can be expected for thermal energies. Finally, a typical force constant for stretching a hydrogen bond is about 30 kcal Å^{-2} mole^{-1}. Although any one of these degrees of freedom does not have much compliance, the effects of many such motions in a large macromolecule can accumulate to allow considerable flexibility.

The harmonic potential arises quite naturally from a formal representation of the potential energy of a macromolecule as a general function of all the positions of its atoms. The internal coordinates are denoted as a vector $\mathbf{x} = (x_1, x_2, x_3 \ldots x_N)$, where the xs cover the x, y, and z coordinates of all the atoms. We further specify the coordinate system to be centered at a potential energy minimum, U_0, so we do not have to deal with the x_0 in Eq. (2.17). Near this minimum, the potential energy is

$$U(\mathbf{x}) = U_0 + \sum_{ij} a_{ij} x_i x_j \tag{2.18}$$

This is a multidimensional parabola, and any nonzero set of x_i will increase U above U_0. The quantities a_{ij} can ultimately be related to the force constants of stretching and bending the bonds.

We can rewrite Eq. (2.18) in matrix-vector notation (Appendix 2) as

$$U(\mathbf{x}) = U_0 + \mathbf{x}\mathbf{A}\mathbf{x}^t \tag{2.19}$$

The vector $\mathbf{x}$ is a row vector on the left and $\mathbf{x}^t$, the transpose of $\mathbf{x}$, is a column vector on the right. The matrix $\mathbf{A}$ has the property of being *positive semidefinite*, which guarantees that all displacements away from $\mathbf{x} = 0$ leave U unchanged or increased. In this form the potential energy can be transformed with the aid of powerful methods from matrix algebra (Appendix 2). A positive semidefinite matrix can be diagonalized by a similarity transform $\mathbf{T}\mathbf{A}\mathbf{T}^{-1} = \Gamma$, where Γ is a matrix in which the only nonzero elements fall on the diagonal. These diagonal elements, γ_{ii}, are referred to as the eigenvalues of the matrix $\mathbf{A}$. How to find the matrix $\mathbf{T}$ is another problem that need not concern us here. It is enough to know that it exists to rewrite Eq. (2.19) as

$$\begin{aligned} U(\mathbf{x}) &= U_0 + \mathbf{x}\mathbf{T}^{-1}\mathbf{T}\mathbf{A}\mathbf{T}^{-1}\mathbf{T}\mathbf{x}^t \\ &= U_0 \varphi \Gamma \varphi^t \end{aligned} \tag{2.20}$$

in which $\varphi = \mathbf{x}\mathbf{T}^{-1}$ and $\varphi^t = \mathbf{T}\mathbf{x}^t$. Thus, each element of φ is a linear combination of the x_i.

Casting off the matrix-vector notation allows us to express Eq. (2.20) as

$$U(\phi) = U_0 + \sum_i \gamma_i \phi_i^2 \tag{2.21}$$

The potential energy has now become a sum of terms like Eq. (2.17). It is an important general property of any arbitrary potential energy function that near the minimum it can be rewritten in the form of a sum over terms identical to that of a harmonic oscillator. If this representation of the potential energy is incorporated into an equation for the atomic motions, a similar mathematical analysis yields the modes of vibration of a complex molecule. These are referred to as the *normal modes* of vibration of a molecule (Wilson *et al.*, 1955). In fact, in a polyatomic molecule it is actually these that are measured by vibrational spectroscopy rather than the frequencies of individual bonds. A potential energy function also has normal modes, and it often helps to think about molecular flexibility in terms of a sum of deformations of the form in Eq. (2.21).

Since $\mathbf{x}$ in Eq. (2.19) is composed of the x, y, and z coordinates for each atom, a molecule with N atoms is represented by a vector with $3N$ dimensions. The matrix $\mathbf{A}$ must therefore have $3N$ eigenvalues (Section A2.3). However, not all of these eigenvalues will reflect deformations of the molecule. There are three modes corresponding to movement of the whole molecule along the x, y, and z axes and three modes corresponding to rotations around the x, y, and z axes. There is no potential energy change associated with such displacements, and the eigenvalues corresponding to these modes of motion will reflect this by assuming values of zero. This leaves $3N - 6$ nonzero eigenvalues of $\mathbf{A}$, and this is how many normal modes of deformation the molecule will have. These nonzero eigenvalues can span a considerable range. Smaller eigenvalues correspond to "soft" modes of motion for which deformation is easy. These are often the most interesting because they reflect global changes in structure (Levitt *et al.*, 1985).

Statistical mechanics is especially easy with the harmonic potential. With Eq. (2.17) the classical configuration integral (Eq. (1.4)) becomes

$$Q = \int_{-\infty}^{\infty} e^{-\phi(x - x_0)^2 / 2kT} d\mathbf{r} = \sqrt{\frac{2kT\pi}{\phi}} \tag{2.22}$$

and the free energy is

$$G = -kT \ln Q = \frac{1}{2} kT \ln\left(\frac{\phi}{2kT\pi}\right) \tag{2.23}$$

Thus, increasing the force constant raises the free energy. This reflects the decrease in entropy resulting from confining x to a smaller region. In general, a strong force constant that might exist in a tight complex between two molecules has this hidden free energy cost, which must be considered in performing a complete free energy tally (Section 4.8). The extension of this analysis to the multidimensional harmonic potential leads to a remarkably simple form (Problem 11).

2.13 | Stabilizing forces in proteins

To function properly a protein must fold up into its native state. The native state is a global state comprising a minute subset of all the possible conformations available to a protein (Sections 1.1 and 3.11). The molecular forces between the different amino acid side chains and backbone groups must pull the protein together into its compact native state. A great deal of effort has gone into assessing the energetic contributions made by the various forces to the stability of a protein's native state. Studies have focused on the relative contributions made by hydrogen bonds, salt bridges (contacts between oppositely charged side chains), and hydrophobic contacts. Indeed, one should also ask whether general guiding principles of protein folding exist, or whether each individual protein has its own unique energetic mix of forces that maintain its native structure.

Some of the earliest work on this subject attempted to assess electrostatic contributions. This was based on the idea that salt bridges form between positively and negatively charged amino acid side chains. However, because of the large self-energies associated with burying charges, even as ion pairs, such interactions would be restricted to the surfaces of proteins exposed to water. Crystallographic studies reveal very few buried ion pairs in proteins. There are more on the surface, but the high dielectric constant of water reduces the energetic contribution of an ion pair to 1–2 RT (as noted in Section 2.1). Ion pairing at protein surfaces therefore does not play a major stabilizing role. In general, ion pairs are not highly conserved in evolution, and when charged amino acids are replaced the effect on the stability of the native state is usually rather small. Experimentally, electrostatic interactions can be manipulated by varying the pH or salt concentration. The results of these investigations are somewhat equivocal, but overall it appears that ion pairs are of secondary importance in stabilizing the native state of a protein (Dill, 1990).

There is growing evidence that cation–π interactions (Section 2.6) play a role in protein stability (Gallivan and Dougherty, 1999). The energy is greater than that of a salt bridge and the aromatic side chain could lie slightly below the surface, within range of a positively charged side chain at the surface. Tryptophans in particular are frequently positioned with the right orientation near arginines and lysines. Experiments have not yet assessed the energetic contribution of these contacts to protein stability, but work in model peptides indicates that properly oriented pairs of tryptophan and arginine at the appropriate spacing can stabilize α-helices, presumably through a cation–π interaction (Shi et al., 2002b). However, the magnitude of this effect is below theoretical estimates of the strength of the cation–π interaction.

A large body of research indicates that hydrophobic interactions do most of the work in protein folding (Dill, 1990). Structural studies clearly show that hydrophobic groups are buried in the interior (Fig. 2.11). Analysis of protein structures and interpretations with the aid of solvent transfer data (Fig. 2.8 and Section 2.8) provide a quantitative assessment of the hydrophobic contribution to protein stability (Spolar et al., 1992). As solvents are made more hydrophobic, the native states of proteins are destabilized and proteins

Fig. 2.11. The folding of a protein internalizes residues with hydrophobic side chains (gray), allowing polar groups (black) to coat the surface. Charged groups form occasional salt bridges at the surface.

are denatured. For example, alcohols are more hydrophobic than water, and they are very effective denaturants. Propanol works better than ethanol, reflecting its greater hydrophobicity. The denaturing action of alcohols and other hydrophobic solvents reflects their ability to compete with the hydrophobic contacts between residues in the protein interior. Additional evidence for the importance of hydrophobic interactions in protein folding comes from the analysis of mutants and their effect on stability (Section 1.4). This work indicates that each buried CH_2 group contributes about $-2 RT$ to the free energy of the native state (Pace, 1992).

As noted in Section 2.8, the hydrophobic force is temperature dependent. Cooling weakens the hydrophobic force to the point where it is no longer strong enough to stabilize the native state of a protein. At very low temperatures (usually below 0 °C) there is enough order in bulk water so that the presence of a hydrophobic surface makes relatively little difference. This weakening of the hydrophobic force denatures proteins. Thus, both low temperatures and high temperatures can denature proteins, although to see cold denaturation requires supercooling, or adding urea to shift the cold denaturation temperature to an accessible range (Griko and Privalov, 1992).

Nearly 90% of all the groups in a protein that can form a hydrogen bond do so, but the abundance of hydrogen bonds in proteins is not sufficient to conclude that they are energetically important in protein folding. When viewed as an exchange process (Scheme (2A) and Section 2.10), there is no guarantee that hydrogen bonding will favor the folded state. How can hydrogen bonds stabilize the folded state when they can form just as easily with water? One explanation for the abundance of hydrogen bonds in proteins is that they form after hydrophobic interactions take the lead in pulling parts of a protein together to form a compact state (Section 3.16). Once hydrogen-bonding groups find themselves sequestered away from water they hydrogen bond with one another, driving the formation of α-helices and β-sheets. According to this view, the hydrophobic effect indirectly drives the formation of secondary structure.

As noted above, experiments with solvents such as alcohols suggest that hydrogen bonds are less important to protein stability than hydrophobicity. However, experiments on the stability of mutant proteins suggest that hydrogen bonds are doing more than merely following the lead of hydrophobic interactions (Myers and Pace, 1996). Proteins are in fact destabilized by mutations of hydrogen bonding residues. Although any mutation has multiple energetic effects, after carefully assessing the various contributions, losing a hydrogen bond was found to destabilize the native state by as much as 4 RT.

An interesting approach to the role of hydrogen bonds has been developed using hydrogen–deuterium exchange (Shi *et al.*, 2002a). Deuterium partitions selectively into weaker hydrogen bonds. Stronger hydrogen bonds prefer protons. The deuterium/hydrogen ratio (measured by NMR) can thus be used to estimate the energy of individual hydrogen bonds in a protein. Furthermore, weakening hydrogen bonds by isotope exchange shifts the equilibrium between the native and denatured states. These experiments lead to an estimate of 0.7 kcal mole^{-1} (about 1 RT) for the contribution of a hydrogen bond in an α-helix to the stability of the folded state. The stabilizing effect varies among different proteins, and in proteins with less α-helix and more β-sheet, the stabilizing effect of hydrogen bonds is no longer evident. Thus, the variable strengths of hydrogen bonds between different atoms, in different geometries, and in different environments can produce a significant drive for protein folding in some but not all proteins. Water–water hydrogen bonds appear to be stronger than water–protein and intramolecular protein hydrogen bonds. This drives the exchange equilibrium (Scheme (2A)) to the right, allowing hydrogen bond formation to contribute to protein stability.

In an α-helix the amide dipoles and carbonyl dipoles are fixed in orientation and position by the structure of the helix (Fig. 2.12). The dipole–dipole interaction energy varies as $1/r^3$ and has a complex dependence on orientation (Section 2.4). Adjacent bonds are nearly side-by-side so their interaction is repulsive (Fig. 2.12b). But for more

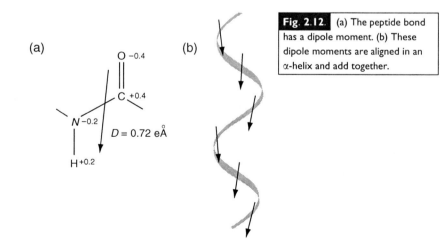

Fig. 2.12. (a) The peptide bond has a dipole moment. (b) These dipole moments are aligned in an α-helix and add together.

distant pairs of bonds the orientations are nearly end-to-end so the interactions become attractive. Despite the short range, when all these interactions are summed the attractive terms win out to give a net stabilizing energy of about 1–2 kcal mole^{-1} per residue. This stabilizing effect is greater for longer helices.

An important consequence of the alignment of dipole moments is a substantial net dipole moment of the entire α-helix, as though there were half an elementary charge at each end. This dipole creates a field that can influence ligand binding and enzyme catalysis (Hol *et al.*, 1978). The dipole moments of α-helices draw cations into the aqueous cavity of an ion channel selective for K^{+} (Section 14.10). The dipole moments of transmembrane helical segments allow the membrane potential to influence orientation, providing a potential driving force for voltage-induced transitions (Section 1.8).

2.14 | Protein force fields

The quantitative expressions for the various forces developed above should in principle allow one to calculate the total potential energy of a protein in a particular configuration. In fact, a major effort has been made to find good quantitative representations for the potential energy of a protein as a function of the positions of all the atoms. These *protein force fields* are generally written as a sum of many terms of the form (Brooks *et al.*, 1983; Weiner *et al.*, 1986; Levitt *et al.*, 1995)

$$
\begin{aligned}
U = {} & \sum_{\text{bonds}} a_\alpha (x_i - x_{i0})^2 \\
& + \sum_{\text{bond angles}} b_\beta (\theta_i - \theta_{i0})^2 \\
& + \sum_{\text{dihedral angles}} c_\gamma (1 - \cos(n_i(\phi_i - \phi_{i0})))^2 \\
& + \sum_{\text{charges}} \frac{q_i q_j}{\varepsilon(r) r_{ij}} \\
& + \sum_{\text{neutral atoms}} 4 d_\delta \left(\left(\frac{r_{ij0}}{r_{ij}} \right)^{12} - \left(\frac{r_{ij0}}{r_{ij}} \right)^6 \right)
\end{aligned}
\tag{2.24}
$$

The first sum represents the stretching energy of all the covalent bonds in the protein. The a_α are the force constants for the bonds (Section 2.12), with the subscript α indicating the particular type of bond, and x_{i0} is the length of the bond for which the stretching energy is at its minimum. The second sum represents the bending energy of all the bonds, with the b_β and θ_{i0} taking on meanings analogous to the corresponding parameters in the first sum. The third term represents rotational potentials of the dihedral angles, with the parameters defined as in the first two sums. The fourth sum is for electrostatic interactions between charges, with a dielectric constant that depends on the position within the protein. The fifth term represents nonbonded interactions represented by a Lennard–Jones potential (Eq. (2.16)). A computer program can take the coordinates of all the atoms of a protein (typically a "pdb" file) as input, and use Eq. (2.24) to calculate U.

Detailed force fields have a wide range of applications. One of the most important is simulating dynamic fluctuations in structure. Here one starts with Newton's law of motion, $\mathbf{F} = m\mathbf{a}$, and uses Eq. (2.24) to calculate $\mathbf{F}$ as the gradient of U. It is thus a vector with three components for each atom in the protein. The acceleration, $\mathbf{a}$, is also a vector with three components per atom. One then takes an initial state with thermally randomized velocities, incorporates a random element to reflect thermal effects, and uses a computer to integrate the equation numerically. These kinds of computer simulations have been widely used to study rapid processes occurring on timescales of up to nanoseconds (Karplus, 2002).

It is tempting to try to use the protein force field to predict the structure of a protein. Presumably, a structure obtained by minimizing the potential energy in Eq. (2.24) would be the native state. The energy difference between two different local minima would then give the relative stability of two global states. However, some serious problems arise when such calculations are attempted. First of all, the computer time needed to find the true minimum increases exponentially with the size of the molecule. For even small proteins the problem is impossible for present day computers.

Another problem is that each term in the potential energy function is of limited accuracy. It is very difficult to calculate the electrostatic terms accurately because the dielectric constant is position dependent and is rarely known to better than a factor of about 2. The environment dependences of other interactions are poorly understood as well. Aside from these uncertainties, one must bear in mind that the total energy is a sum of thousands of such terms. The statistics in this kind of situation dictates that the error in the sum will be roughly equal to the average error per term times the square root of the number of terms (see Chapter 12). The error in this sum turns out to be much larger than the energy differences between many of the local minima in this potential energy function.

To illustrate this point, consider an average energy per term of 10 ± 1 kcal mole^{-1}. With N such terms the total energy will be $10N \pm 1\sqrt{N}$. If $N = 10\,000$ then we have $100\,000 \pm 100$ kcal mole^{-1}. The error in this quantity of 100 kcal mole^{-1} is a relatively small fraction of the total energy (0.1%), but 100 kcal mole^{-1} is very large compared to the free energy changes occurring during typical conformational transitions (for example thermal unfolding, see Chapter 1). This problem cannot be overcome by better computers. It depends not only on having a better understanding of molecular forces, but also on an accuracy in the potential energy function that is probably unattainable.

Monumental efforts have been made to overcome these difficulties. Year by year methods improve in the use of force fields to calculate the native structure of a protein, and these methods are starting to become useful. The determination of a protein's structure from its amino acid sequence has spawned a large scale semi-annual competition known as CASP (critical assessment of protein structure prediction). Much more goes into these efforts than just protein force fields, but the CASP competitions provide an excellent venue for testing and evaluating developments in this area (Lesk *et al.*, 2001).

2.15 | Stabilizing forces in nucleic acids

Molecular forces determine the conformation and behavior of oligo-nucleotides, single-stranded polynucleotides (e.g. RNA), and double-stranded polynucleotides (e.g. DNA) (Bloomfield *et al.*, 1974; Record *et al.*, 1981). The three most important interactions in nucleic acids are (1) stacking interactions between the aromatic bases, (2) electrostatic interactions between the charged phosphates of the polymer backbone, and (3) hydrogen bonds in Watson–Crick base pairs.

The stacking interaction between base pairs is quite strong and makes the flat aromatic bases stack together like plates. It causes monomeric nucleotides to associate weakly in solution, and more importantly, provides the major driving force in the formation of the DNA double helix. Because this force has a relatively small entropic contribution, it is not the same as the hydrophobic interaction. The stacking of nucleic acid bases depends on dispersion forces, which are especially strong because of the very similar electronic structures of the different bases. The bases have large overlaps in the frequency dependence of their polarizabilities, and as emphasized above, this is a key element in generating strong dispersion forces (Section 2.7).

Hydrogen bonds form between bases of the different strands in a DNA double helix, and this forms the basis for complimentarity in Watson-Crick base pairs. The energy contributed by a single hydrogen bond in a DNA duplex has been estimated to fall in the range 0.8–1.6 kcal mole^{-1} (Fersht, 1987).

Guanine–cytosine base pairs have three hydrogen bonds and adenine-thymine base pairs have two, so one would naturally think that this accounts for why a double helix rich in guanine-cytosine base pairs is more stable than a double helix rich in adenine-thymine base pairs. This hypothesis has been difficult to test because of the thermodynamic complexity of double helix melting. Although the specific energetic contributions have been difficult to quantify, it is a remarkable result that the temperature, ΔH, and ΔS of melting all are simple linear functions of the fraction of guanine–cytosine base pairs. The linear dependence is significant because it indicates that, overall, the base composition of DNA is far more important to the stability of a double helix than the actual sequence.

The interplay between hydrogen bonding and stacking makes double helix formation a highly cooperative process. When the first base pair forms between two single-stranded molecules, there are only hydrogen bonds to hold it together, so the complex is not very stable. After the first base pair forms, the immediately adjacent bases are aligned in a pairing configuration by the stacking force. This gives the double helix a rapid rate of growth following the nucleation process of forming the first basepair. This nearest neighbor interaction is an excellent example of the kind of cooperativity used to develop theories for helix–coil transitions (Section 3.12).

Nucleic acids have strongly acidic phosphate groups, a large fraction of which are ionized at neutral pH. The repulsive force between these charges tends to straighten the molecule out, extending single-stranded polymers and increasing their persistence length (Section 3.4). The repulsion between phosphates on the two chains also opposes strand association and double helix formation. Increasing the salt concentration reduces this repulsion and stabilizes a double helix. There is a surprisingly simple linear relation between melting temperature and logarithm of the salt concentration. The effect can be accurately explained with a theory based on counterion binding (Section 11.11).

2.16 | Lipid bilayers and membrane proteins

The lipid bilayer provides a perfect solution to the puzzle of how to arrange amphipathic molecules in such a way as to minimize contact between hydrocarbon and water. The approximately parallel alignment of the phospholipid hydrocarbon chains excludes water. This allows the polar headgroups to form a flat surface that is solvated by water, while the hydrocarbon tails are exposed to one another and to the hydrocarbons of the opposite leaflet of the bilayer. Integral membrane proteins accommodate this scheme with hydrophobic membrane-spanning segments (see Fig. 2.13, and compare it with Fig. 2.11). The interior of a membrane thus maintains its hydrophobic character as one passes from lipid to protein and back to lipid. Segments of a membrane protein with polar side chains are immersed in water and contact the polar phospholipid headgroups.

The principle of hydrophobic matching is particularly useful in membrane proteins. The membrane-spanning segments of a protein vary quite a bit in their content of hydrophobic amino acids. The segments in contact with surrounding lipid are more hydrophobic than the segments that are buried in the protein core (Rees et al., 1989). Indices of hydrophobicity were calculated for the α-helical membrane-spanning segments of photosynthetic reaction centers, which are composed of several membrane proteins. The mean hydrophobicity of residues buried within the interior of these membrane proteins is similar to that for residues buried inside globular proteins. So the factors that influence the internalization

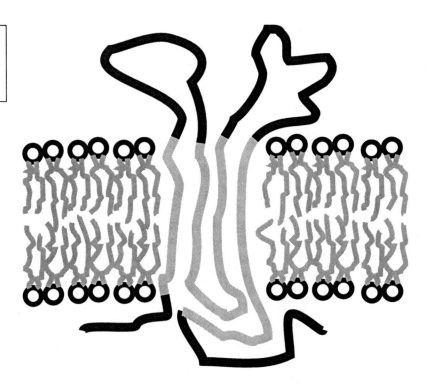

Fig. 2.13. A lipid bilayer with an integral membrane protein. Hydrophobic (gray) elements are buried. Polar elements (black) line the surface exposed to water.

of amino acid side chains appear to be generally applicable to both classes of protein. The peripheral domains of membrane proteins that come in contact with the lipid hydrocarbon tails are much more hydrophobic than the buried domains, and this can be seen as a matching of the protein and lipid tail hydrophobicities.

The parallel hydrocarbon chains in a lipid bilayer attract one another by dispersion forces, which vary as $1/r^5$ (Section 2.7). Without headgroups, solid paraffin has an average area per hydrocarbon chain of about 18–19 $Å^2$, and this reflects the minimum potential energy for the combination of attractive dispersion forces and steric repulsion. In the ordered gel phase of lipid bilayers, the chains are straight and parallel with an average area of about 21 $Å^2$. This is greater than in paraffin because of the polar headgroups, which have greater packing areas than the two appended hydrocarbon chains. The headgroups repel one another, and this spreads the hydrocarbon chains out.

Raising the temperature causes a melting transition as the hydrocarbon chains of a lipid bilayer become disordered. In this fluid state the chains become disordered, assuming kinked and bent conformations. The hydrocarbons spread out further, reaching an area of nearly 30 $Å^2$. The packing density is then very similar to that of liquid paraffin, and the increased disorder of the hydrocarbon chains decreases the thickness of the bilayer (Weiner and White, 1992). The lipids in biological membranes tend to be fluid at physiological temperatures, allowing membrane proteins greater flexibility and freer diffusion.

Problems for Chapter 2

1. Derive an expression for the enthalpy of hydration from Eq. (2.4). The solid curve in Fig. 2.1 has the form C/r. Calculate a numerical value for C from your result using the temperature dependence of ε_w mentioned below Eq. (2.2) (Rashin and Honig, 1985).

2. The energy of an arrangement of charges is given by the integral over all space of the quantity $\varepsilon E2/(8\pi)$, where E is the field (Jackson, 1975). Use this to derive Eqs. (2.4) and (2.7).

3. Derive the potential energy of interaction between two dipoles lying in the same plane as a function of distance and orientation.

4. Derive the potential energy of interaction between the quadrupole of Fig. 2.5c and a cation on the axis of the quadrupole.

5. Calculate the difference in Born energy for transferring tetraphenylphosphonium from oil to water. Treat the molecule as a sphere and use your own estimate of its size based on the chemical structure.

6. The amino acid lysine has a primary amino group at the end of its side chain. Take the radius of this amino group as 1.7 Å, and calculate the energy of transfer from water to oil. Take the pK of the primary amino group as 10 and calculate the free energy of deprotonation of the acid form (RNH_3^+) at a physiological pH of 7.3. Based on a comparison of these energies would you expect the protonated or deprotonated form of lysine to be inserted into a lipid bilayer?

7. Calculate the hydrophobic free energy for transferring tetraphenylphosphonium from oil to water. As in Problem 5, base your calculation on your own estimate of area from the chemical structure. Compare the magnitude of this with that from Problem 5.

8. Estimate the difference between the free energy of transferring tyrosine and phenylalanine from oil to water.

9. Derive the position of minimum potential energy for the Lennard–Jones potential (Eq. (2.16)) and the depth of that minimum.

10. Calculate the force constant at the energy minimum of the Lennard–Jones potential in terms of ε and r_0. What is the force constant when $\varepsilon = 10\,\text{kcal mole}^{-1}$ and $r_0 = 2.8$ Å. Compare the value with the force constant given for the stretching C–C bond in Section 2.12.

11. Evaluate the partition function and free energy of a multi-dimensional harmonic potential in Eq. (2.19) using the classical configuration integral (Eq. (1.4)). Express the result in terms of the matrix **A**.

Chapter 3

Conformations of macromolecules

The idea of a global state was introduced in Chapter 1 as a way of looking at the behavior of a large molecule without getting bogged down in details. Here we will face the details head on, examining the vast variations of microstates, and in the process gaining some insight into the nature of certain types of global states. With help from the ideas about molecular forces in Chapter 2, we will develop more detailed descriptions of macromolecules. We will rely heavily on simple polymers as model systems for the more complex biological macromolecules.

3.1 | n-Butane

To get a feel for the more complicated models, we start off with a simple model based on the small molecule, n-butane. This molecule has three favored conformations, defined by rotating the central C–C bond to give different dihedral angles. These conformations are *trans, gauche-left,* and *gauche-right* (Fig. 3.1). The electronic structure of the C–C bond restricts rotations around the central bond,[1] so that the molecule will spend most of its time in these three configurations.

Trans and *gauche* conformations differ in energy by about $1/2$ kcal mole^{-1} (Rosenthal *et al.*, 1982). With the aid of the Boltzmann distribution, this information can be used to calculate the relative probabilities of an n-butane molecule being *trans* or *gauche*. If the *trans* conformation has an energy E_t and a gauche conformation has an energy E_g, then the Boltzmann distribution gives the relative abundance of the two states in terms of the energy difference $\Delta E = E_g - E_t$ as

---

[1] The traditional explanation for the rotational preferences of n-butane and other straight-chain aliphatics is steric repulsion. However, rigorous analysis suggests that the repulsion is negligible, and that the rotational preferences are dictated by bond orbital interactions (Reed and Weinhold, 1991).

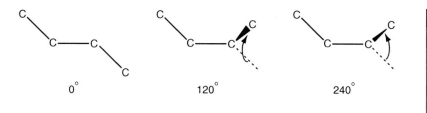

Fig. 3.1. Three rotational isomers of *n*-butane. In the *trans* conformation the carbon atoms all lie on a plane, and the dihedral angle formed by the three C–C bonds is assigned the value zero. Rotation of this angle to 120° and 240° gives the *gauche-right* and *gauche-left* conformations.

$$\frac{P_{g\text{-}r}}{P_t} = \frac{P_{g\text{-}l}}{P_t} = e^{-\Delta E/RT} \tag{3.1}$$

Here we specify *trans, gauche-right,* and *gauche-left* with the subscripts t, g-r, and g-l, respectively. Energy is in units per mole so the gas constant R can be used (Section 1.1).

It is convenient to use the Boltzmann weight (Section 1.1), which for state i is

$$w_i = e^{-E_i/RT} \tag{3.2}$$

In the context of *n*-butane, the subscript i could be t, g-r, or g-l. With $E_t = 0$, $E_{g\text{-}r} = E_{g\text{-}l} = 0.5$, and $RT \sim 0.6\,\text{kcal mole}^{-1}$ (at room temperature), the Boltzmann weights are

$$w_t = e^{-0/0.6} = 1 \tag{3.3a}$$

$$w_g = e^{-0.5/0.6} = 0.43 \tag{3.3b}$$

where w_g is used for both the *gauche-left* and *gauche-right* configurations. The probabilities of *n*-butane being *trans* or *gauche* can be expressed in terms of these weights

$$P_t = \frac{w_t}{w_t + 2w_g} = \frac{1}{1 + 2 \times 0.43} = 0.54 \tag{3.4a}$$

and

$$P_g = P_{g\text{-}r} + P_{g\text{-}l} = \frac{2\,w_g}{w_t + 2\,w_g} = \frac{2 \times 0.43}{1 + 2 \times 0.43} = 0.46 \tag{3.4b}$$

Note that $P_t + P_g = 1$, because a sum over all probabilities must add up to one.

The factor 2 multiplying w_g in Eqs. (3.4a) and (3.4b) enters because we stopped worrying about whether we had a *right-* or *left-gauche* bond and lumped the two configurations together. This factor of 2 is referred to as a degeneracy, the term used for a count of the energetically equivalent variants of a state. We can multiply the Boltzmann weight and degeneracy together and call the product an *elementary weight* ($2e^{-\Delta E/RT}$ for the *gauche* state). These terms often facilitate the calculation of the probabilities.

Equations (3.4a) and (3.4b) show how the energies and degeneracies influence the relative abundance of each configuration. As E_g gets larger, w_g gets smaller, P_g approaches zero, and P_t approaches one. When RT is small compared to the energy differences, a molecule spends most of its time in the lowest energy state. As the temperature increases, the effect of the energy difference becomes smaller as the Boltzmann weights approach one. The states then become probable in proportion to their degeneracies.

To illustrate how these statistical ideas come into play, the mean end-to-end distance of n-butane will be calculated. Each of the two conformations discussed above has a length denoted as l_t and l_g. If all of the molecules were *trans* then the average would be l_t, and if they were all *gauche* it would be l_g. But since there is a mixture in the proportion determined by P_g and P_t (from Eqs. (3.4a) and (3.4b)), we take the average of the two lengths, weighted by their relative abundance.

$$\bar{l} = l_t P_t + l_g P_g \tag{3.5}$$

It is common to take the end-to-end length as a vector, but then the average over all orientations gives zero. For this reason the root-mean-square (rms) length is widely used as a basic quantitative index for the size of a molecule.

$$\sqrt{\bar{l^2}} = \sqrt{l_t^2 P_t + l_g^2 P_g} \tag{3.6}$$

The rms length is a very important property of flexible macro-molecules, and will be calculated later in this chapter for longer molecules using the same basic method of averaging over all conformations.

3.2 | Configurational partition functions and polymer chains

The denominator in Eqs. (3.4a) and (3.4b) is the sum of the Boltzmann weights of all the configurations of n-butane. Recall that this is how the partition function is defined in statistical mechanics (Section 1.1). More generally, for a molecule with n states the partition function is

$$Q = \sum_{i=1}^{n} e^{-E_i/RT} \tag{3.7}$$

One can also express the partition function using energy as the summation index and take the sum over elementary weights

$$Q = \sum_{E} \omega_E e^{-E/RT} \tag{3.8}$$

This version is more convenient for many problems. The variable ω_E is the degeneracy of the states with energy E. When several of the states indexed with i in Eq. (3.7) have the same energy, then that energy level is degenerate, so we count it ω_E times in Eq. (3.8).

To illustrate the use of the partition function, take the average energy of a molecule

$$\overline{E} = \frac{\sum\limits_{i=1}^{n} E_i e^{-E_i/RT}}{\sum\limits_{i=1}^{n} e^{-E_i/RT}} \tag{3.9}$$

Differentiating Eq. (3.7) with respect to temperature gives

$$\frac{\partial Q}{\partial T} = \frac{1}{RT^2} \sum\limits_{i=1}^{n} E_i e^{-E_i/RT} \tag{3.10}$$

Dividing through by Eq. (3.7), and using Eq. (3.9) for $\overline{E}$ gives the mean energy as

$$\overline{E} = RT^2 \frac{\partial \ln Q}{\partial T} \tag{3.11}$$

In general, once a convenient form has been found for the partition function, many important properties can be calculated.

The partition function can be used to illustrate an important relation between a polymer and its constituent monomers. Take a chain of N monomers. Each monomer can have a different energy and we denote the energy of the nth monomer as E_n. The value of n ranges from 1 to N, and each E_n can assume the values defined by a set of rotational states. If the monomers are independent, the elementary weight for a particular configuration of the whole molecule is then a product over all the individual monomer elementary weights: $\prod\limits_{E_n} \omega E_n e^{-E_n/kT}$. The partition function is then the sum over all these products

$$Q_N = \sum\limits_{E_1, E_2, \ldots E_N} \prod\limits_{n=1}^{N} \omega_{E_n} e^{-E_n/kT} \tag{3.12}$$

We can reverse the order of evaluating sums and products, and then write out the product explicitly as

$$Q_N = \left(\sum\limits_{E_1} \omega_{E_1} e^{-E_1/kT} \right) \left(\sum\limits_{E_2} \omega_{E_2} e^{-E_2/kT} \right) \cdots \left(\sum\limits_{E_N} \omega_{E_N} e^{-E_N/kT} \right) \tag{3.13}$$

Each sum in this product is the partition function of a monomer, and can be denoted as Q_1. The result is

$$Q_N = Q_1{}^N \tag{3.14}$$

So if the monomers are independent, then the partition function of the polymer is just the partition function of an isolated monomer raised to the power N.

Take the polymer polyethylene as an example. This is a chain of CH_2 groups. If we think of the rotations around each C–C bond as being like the rotations in n-butane (Fig. 3.1), then Q_1 is the partition function of n-butane, and the partition function of polyethylene is this to the Nth power. (This ignores the fact that the end segments are different. For N carbon atoms we should really take Q_1^{N-3} because the total number of n-butane-like bonds will be $N-3$.)

This illustrates how easy things are when you assume that each of the component parts is independent. Trying to take interactions between adjacent bonds into account makes things more difficult. This will be apparent in the analysis of random coils and the helix–coil transition below, where limited forms of interaction between adjacent chain segments are considered.

3.3 | Statistics of random coils

Flexible polymer chains such as polyethylene tend to bend and coil randomly. This is a common property of polymers, including many of biological interest, and statistical models for this random behavior are important tools in understanding their physical properties. These models are generally referred to as random coils. There are a number of versions and they fall into a hierarchy with increasing constraints on the bonds between segments. The least constrained model is the freely jointed chain, where each segment has complete freedom to rotate, and any angle is equally probable. The freely rotating chain model has fixed bond angles, but the dihedral angles can assume any value. Finally, the rotational isomer model has bonds like n-butane, for which dihedral angles have a few favored values.

We will first look at the freely jointed chain, and calculate the mean square end-to-end distance. This is illustrated in Fig. 3.2. For clarity the picture shows all of the bonds in one plane, but the treatment here will deal with a three-dimensional chain of N segments.

The mathematics of the freely jointed chain is essentially identical to that of a random walk, a model for diffusion that will be studied in Chapter 6. The analysis of the freely jointed chain will be developed here in a way that helps us understand the physics of polymers. In a chain of N segments, we can define the vector from the beginning of the first segment to the end of the Nth segment as $\mathbf{R}_N$ (Fig. 3.2). This end-to-end vector is a sum over the N segment vectors.

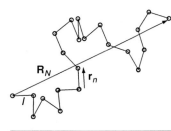

Fig. 3.2. A freely jointed chain contains N segments connected by bonds of length l. The end-to-end vector, $\mathbf{R}_N$, is a vector sum of the N segment vectors, r_1 through r_N.

$$\mathbf{R}_N = \sum_{n=1}^{N} \mathbf{r}_n \tag{3.15}$$

The average over all configurations is $\overline{\mathbf{R}_N} = 0$, because for every $\mathbf{R}_N$ there is a vector pointing in the exact opposite direction that will cancel (as noted above for n-butane). Instead, we average the square of this length

$$\overline{\mathbf{R}_N^2} = \overline{\left(\sum_{n=1}^{N} \mathbf{r}_n \right)^2} = \overline{\sum_{n=1}^{N} \sum_{m=1}^{N} \mathbf{r}_n \cdot \mathbf{r}_m} \tag{3.16}$$

where the vector dot product is indicated. We can break the double sum up into terms according to whether $n = m$ or $n \neq m$

$$\overline{\mathbf{R}_N^2} = \sum_{n=1}^{N} \overline{\mathbf{r}_n^2} + \sum_{n \neq m}^{N} \sum^{N} \overline{\mathbf{r}_n \cdot \mathbf{r}_m} \qquad (3.17)$$

For the freely jointed chain all the terms with $n \neq m$ are zero because with completely free joints the orientations of any two different segments are completely uncorrelated. Only terms of the form $\overline{r_n^2}$ remain, and these are all equal to l^2. With N such terms in the sum, we end up with a mean square end-to-end length of

$$\overline{\mathbf{R}_N^2} = Nl^2 \qquad (3.18)$$

So the rms end-to-end-length increases as $\sqrt{N}$. This is a fundamental property of random chain molecules. This proportionality appears again and again not only in chain molecules but in a large number of statistical problems. The key to obtaining this simple result is the independence of the segments.

When neighboring segments interact, the proportionality between $\overline{R_N^2}$ and N still holds as long as the interactions do not extend indefinitely up and down the chain. This is illustrated with the freely rotating chain, where the angle formed by two connected bonds is fixed to a value, θ. Now, the segments are correlated. The terms in Eq. (3.17) of the form $\mathbf{r}_n \cdot \mathbf{r}_{n+1}$ are now $l^2 \cos\theta$. This is a straightforward trigonometric result. The angle between adjacent segments is always θ, so averaging over θ is not necessary. To evaluate $\overline{\mathbf{r}_n \cdot \mathbf{r}_{n+2}}$, we divide $\mathbf{r}_{n+2}$ into components parallel and perpendicular to $\mathbf{r}_n + \mathbf{r}_{n+1}$. The perpendicular component contributes zero to the average and the parallel component gives $l^2 \cos^2\theta$. This reveals the trend that for any pair of segments with $n \neq m$: $\overline{\mathbf{r}_n \cdot \mathbf{r}_m} = l^2 \cos^{|n-m|}\theta$. For the lower values of $n-m$ each sum is taken nearly $2N$ times in Eq. (3.17). Returning to Eq. (3.17), the terms with $n \neq m$ become a geometric series in $2N\cos\theta$. When all the terms are added together with the aid of the sum of a geometric series (Appendix 1), the mean square end-to-end length is

$$\overline{\mathbf{R}_N^2} = \frac{1 + \cos\theta}{1 - \cos\theta} l^2 N \qquad (3.19)$$

Note that except for the angular factor in front, this expression is the same as Eq. (3.18) for a freely jointed chain. The result becomes nonsensical as θ goes to $0°$ or $180°$. The approximations in this derivation loose their validity because the sums no longer converge, but these angles are not relevant to flexible polymers.

For the rotational isomer model rotations around bonds are restricted, and this makes the mathematics more complicated. The solution will simply be stated as

$$\overline{\mathbf{R}_N^2} = \left(\frac{1 + \cos\theta}{1 - \cos\theta}\right)\left(\frac{1 + \overline{\cos\phi}}{1 - \overline{\cos\phi}}\right) l^2 N \qquad (3.20)$$

And we still have $\overline{\mathbf{R}_N^2} \propto l^2 N$. The angle ϕ is the dihedral angle formed by three successive bonds. For more thorough treatments of these models see Flory (1969) and Cantor and Schimmel (1980).

The rotational isomer model is very much like stringing the monomers along the points of a lattice. If the bonds were constrained to 90° and the dihedrals to multiples of 90°, then we have a cubic lattice. There is no such molecule but for polyethylene the angles are quite close to 120°, so a hexagonal lattice is not a bad approximation. The lattice approach gives us a helpful visual image of macromolecule statistics, and will appear in a number of guises later in this chapter.

3.4 | Effective segment length

These random chain models lead to increasingly complicated solutions as more constraints are added, but the complexity is all contained in the factor multiplying $l^2 N$. The proportionality with N remains. This suggests that the results can be generalized if we absorb the various model-specific factors into an effective segment length, l_{eff}. We can then generalize Eqs. (3.19) and (3.20) to

$$\overline{\mathbf{R}_N^2} = l_{eff}^2 N \tag{3.21}$$

In this form any of the models will look like a freely jointed chain, but the segments are not the true segments. Instead, they are effective segments with an *effective* length l_{eff}. This provides a useful measure of the stiffness of a polymer. If l_{eff} is only slightly longer than l, that means the molecule is very flexible; $l_{eff} >> l$ means that the molecule is stiff.

Another index widely used for conformational flexibility is the *characteristic ratio*, defined as

$$C = \frac{\overline{\mathbf{R}_N^2}}{l^2 N} = \frac{l_{eff}\, z}{l} \tag{3.22}$$

where z is a geometric factor used to account for the fact that a fully extended chain may have a length less than Nl (for example, in fully extended polyethylene the bond angles are 112°, so the length Nl must be reduced by a geometric factor). The characteristic ratio tells us how much longer the mean square end-to-end distance is than it would be if it were a freely jointed chain. Values for l, l_{eff}, z, and C are given in Table 3.1 for a few common polymers. Values for l_{eff} and C are quite large for DNA because it is stiff. Flexible molecules such as polyethylene and poly-L-alanine have smaller values. It is important to appreciate that even a very stiff molecule such as double-stranded DNA will coil randomly if it is long enough. In this case, long enough means several times longer than l_{eff}. Looking at these numbers for DNA we can expect random-coil behavior for a 3000 Å molecule containing on the order of 1000 base pairs.

Table 3.1. *Flexibility parameters for chain molecules*

	l (Å)	l_{eff} (Å)	z	C
Polyethylene[a]	1.54	12.4	0.83	6.7
Poly-L-alanine[b]	3.8	36	0.95	9.0
Double-stranded DNA[b] (in 0.2 M salt)	3.5	300	1	86

[a] Cantor and Schimmel (1980).
[b] Record *et al.* (1981).

For the sake of completeness we should mention one other widely used measure of polymer stiffness, persistence length. This quantity, denoted as l_p, is defined as the sum of the projections of all segments on one bond

$$l_p = \lim_{N \to \infty} \overline{\frac{\mathbf{r}_j}{l} \cdot \sum_{n=1}^{N} \mathbf{r}_n} \tag{3.23}$$

The average is taken over all conformations. If a chain is stiff, segments separated by significant distances will still be correlated and contribute to this sum. Thus, a stiffer chain will have a longer persistence length. The persistence length can be used to replace many of the terms in the double sum of Eq. (3.17) that gives the mean square end-to-end distance. After suitable rearranging, one can recover the following relationship between persistence length and characteristic ratio (Cantor and Schimmel, 1980).

$$l_p = \tfrac{1}{2}(C + 1)l \tag{3.24}$$

For a stiff chain, with $z = 1$, we can combine this with Eq. (3.22) to obtain the simple approximate result that $l_p = l_{eff}/2$.

3.5 | Nonideal polymer chains and theta solvents

In the above analysis of random chains, the only intersegment interactions that were considered were constraints on bond angles. Two additional kinds of interaction are also important. They strongly influence the conformations available to a molecule, but the mathematical treatment is much more difficult. The first of these is attractive interactions such as hydrogen bonds and hydrophobic forces (Chapter 2), which pull distant segments together and tend to make a chain more compact. The second is steric repulsion (also discussed in Chapter 2), which prevents two segments from occupying the same space at the same time. This is referred to as the excluded volume effect, and tends to work against the attractive forces to spread a chain out. Exact mathematical solutions to models with these interactions have not been found. This makes the random-coil model especially important as a reference point. From this perspective random-coil models are referred to as *ideal chains*, and the more realistic but approximate models that incorporate attractive and excluded volume interactions are referred to as *nonideal chains*.

For the excluded volume effect a number of approximate treatments indicate that the rms end-to-end distance is proportional to a fractional power of the number of segments.

$$\sqrt{\overline{\mathbf{R}_N^2}} \propto N^\nu \tag{3.25}$$

For an ideal chain $\nu = 1/2$ (Section 3.3). The theories for the excluded volume effect give values somewhat greater than $1/2$; $\nu = 3/5$ is typical. A theory by De Gennes (1972) gave $\nu = 0.5975$. These theoretical results make the reasonable point that when a polymer is not allowed to overlap onto itself it will spread out further; the ends will on average be separated by a greater distance. Thus, the excluded volume effect tends to increase the effective size of the molecule relative to that expected for an ideal chain.

The set of all configurations available to a chain subject to excluded volume is necessarily a subset of the configurations available to the corresponding ideal chain. Excluded volume reduces the total number of configurations, and this means that the random-coil model overestimates the configurational entropy of a polymer. If we consider an ideal chain strung out randomly on a cubic lattice (recall the final comments about the rotational isomer model in Section 3.3), the number of possible configurations would be 6^N. This follows because each site has six neighboring sites, so that each new segment can be added in six ways.

Computers have been used to count the number of non-overlapping configurations of a chain on a cubic lattice, and the result was empirically fitted by the expression $N^{1/6}4.68^N$ (Chan and Dill, 1991; Camacho and Thirumalai, 1993). Dividing 6^N by this expression indicates that the ideal chain model on a cubic lattice overestimates the true number of configurations by a factor of $1.28^N/N^{1/6}$. We could easily improve the random-coil model by realizing that the value six, the number of ways to add a new segment to a lattice, must include at least one site already occupied. Subtracting this one site is clearly an improvement, and this leaves five ways to add a new segment. There are then 5^N configurations, and this simple improvement of the random-coil model overcounts by a factor of $1.07^N/N^{1/6}$.

Attractive interactions between segments have the opposite effect. They tend to draw the segments together and make the molecule more compact. They thus oppose the spreading out effect of excluded volume interactions. This depends strongly on the choice of solvent. If the polymer segments are weakly soluble, then the segments will clump together to reduce the unfavorable interactions with the solvent. On the other hand, a good solvent will make the chain more extended.

A solvent can be chosen to make intersegment attractions counteract the excluded volume repulsions. If the right balance is found, then the rms length will scale according to ideal chain theory. The solvent is then called a *theta* solvent. Because a solution of polymers then conforms to theories for ideal chain molecules, the theta solvent creates unique conditions for evaluating the basic properties of chain molecules. However, one should bear in mind that although the rms length scales as $N^{1/2}$, in other respects the behavior is different from that of a true random coil. Excluded volume tends to prevent more of the compact configurations from forming, and this is compensated by attractive forces that make the more extended conformations less likely. The result is that the probability distribution for the end-to-end distance of a polymer in a theta solvent is sharper than that of an ideal chain (Sanchez, 1979).

3.6 | Probability distributions

The mean square end-to-end distance is only one of many properties of a flexible polymer. A more complete description requires knowledge of the probability distribution of the end-to-end distance, $P(R)$. In an ideal chain the effective segments are uncorrelated, and this information is sufficient to permit the use of a very powerful general result from probability theory called the central-limit theorem. This theorem states that any sum of independent random quantities with the same mean will be distributed as a Gaussian function. Thus, from a fundamental perspective the answer is already in hand. Recall that in the calculation of the rms end-to-end distance above, the assumption of no correlations between segments was also the key to obtaining a simple solution.

To see how the Gaussian distribution arises in polymers, we start with a simplified model of a random chain in one dimension. Each segment can point in only two possible directions, right or left. If we see that a chain of N segments ends at position x, then we know that the chain of $N-1$ segments leading up to this point must have terminated at either $x+l$ or $x-l$. We can therefore relate the probability $P_N(x)$ to the two probabilities $P_{N-1}(x-l)$ and $P_{N-1}(x+l)$ as follows

$$P_N(x) = \tfrac{1}{2}(P_{N-1}(x-l) + P_{N-1}(x+l)) \tag{3.26}$$

where the factor of 1/2 appears because for either of the two possible locations of segment $N-1$, the next step can be either toward or away from x. Expanding each of the terms on the right as a Taylor series around x gives

$$P_N(x) = \tfrac{1}{2}\left(P_{N-1}(x) - l\frac{\partial P}{\partial x} + \tfrac{1}{2}l^2\frac{\partial^2 P}{\partial x^2} + P_{N-1}(x) + l\frac{\partial P}{\partial x} + \tfrac{1}{2}l^2\frac{\partial^2 P}{\partial x^2}\right) \tag{3.27}$$

$$= P_{N-1}(x) + \tfrac{1}{2}l^2\frac{\partial^2 P}{\partial x^2} \tag{3.28}$$

Bringing $P_{N-1}(x)$ to the left-hand side, we see the difference $P_N(x) - P_{N-1}(x)$ can be taken as $\Delta P(x)$. Since $\Delta N = 1$, this then can be approximated as a derivative, leading to a differential equation in P

$$\frac{\partial P}{\partial N} = \tfrac{1}{2}l^2 \frac{\partial^2 P}{\partial x^2} \tag{3.29}$$

This is the diffusion equation (to be studied further in Chapter 6), and it has a Gaussian function as a solution.[2] This takes the form

$$P_N(x)dx = \frac{1}{\sqrt{2\pi N l^2}} e^{-\frac{x^2}{2Nl^2}} dx \tag{3.30}$$

as can be verified by substitution back into Eq. (3.29).

This result was for an artificial one-dimensional chain. A more realistic three-dimensional chain is a simple extension. The probability of finding the end of the chain at a position (x, y, z) is taken as the product of three Gaussians of the form in Eq. (3.30)

$$P_N(x,y,z)dxdydz = \frac{1}{(2\pi N l^2)^{3/2}} e^{-\frac{x^2+y^2+z^2}{2Nl^2}} dxdydz \tag{3.31}$$

To go from this form to a function of the end-to-end distance means transforming from Cartesian to spherical coordinates, where $x^2 + y^2 + z^2 = R^2$ and $dxdydz = 4\pi R^2 dR$

$$P_N(R)dR = \frac{4\pi R^2}{(2\pi N l^2)^{3/2}} e^{-\frac{R^2}{2Nl^2}} dR \tag{3.32}$$

This actually represents a random chain that follows the points on a cubic lattice (mentioned at the end of Section 3.3). For other forms and models of random chain molecules we would expect that Eq. (3.32) should still apply but with l_{eff} replacing l.

3.7 | Loop formation

How easy is it to bring the two ends of a random coil together? Equation (3.31) can be used to calculate the probability of loop formation. The molecule will form a loop when the end has a position near $x = y = z = 0$ as

$$P_N(0)dxdydz = (2\pi N l^2)^{-3/2} dxdydz \tag{3.33}$$

Of course, the actual probability of two ends being within a certain distance will depend on the size of an arbitrary volume element, $\delta x \delta y \delta z$. But for a given value of this volume element, the probability will vary as $(2\pi N l^2)^{-3/2}$. Thus, the probability of loop formation in an ideal chain decreases as the number of segments or the segment length increases. Equation (3.33) provides a good estimate of the

---

[2] This solution also satisfies the appropriate boundary conditions that $P \to 0$ as $x \to \pm\infty$, and $P(x) \to \delta(x)$ as $N \to 0$.

rate of formation of circular DNA from linear DNA (Record *et al.*, 1981; Crothers *et al.*, 1992). For example, the enzyme DNA ligase will join the ends of a molecule of linear DNA together if they are "sticky," i.e. have short single-stranded ends that are complementary. The rate of this cyclization reaction effectively measures the probability of loop formation. With very short pieces of DNA Eq. (3.33) fails because the stiffness of the molecule makes it harder for the ends to join than expected for a random-coil. The stiffness of short chains is more effectively treated with the "worm-like chain" model (see chapter 19–7 of Cantor and Schimmel, 1980).

3.8 | Stretching a random coil

If we hold the ends of a random coil at fixed positions, Eq. (3.31) tells us that the molecule will have more configurations when the ends are closer together than when they are far apart. As a consequence, the molecule will exert a force to pull its ends together. This force is entropic, resulting from the thermal motions of the chain. It can be calculated by looking at how the number of configurations changes during an extension. For a given value of x the number of configurations available to the chain molecule will be

$$\Omega P(x)dx = \frac{\Omega}{\sqrt{2\pi Nl^2}} e^{-\frac{x^2}{2Nl^2}} dx \tag{3.34}$$

where Ω represents the total number of configurations of the chain. The chain entropy as a function of length will then be given by Boltzmann's expression for the entropy as $k \ln$ (number of states)

$$S(x) = k \ln \left(\Omega P(x)dx \right)$$

$$= k \ln \left(\frac{\Omega}{\sqrt{2\pi Nl^2}} e^{-\frac{x^2}{2Nl^2}} dx \right)$$

$$= k \ln \left(\frac{\Omega dx}{\sqrt{2\pi Nl^2}} \right) + k \ln \left(e^{-\frac{x^2}{2Nl^2}} \right) \tag{3.35}$$

The left-hand term is independent of x, so it can be ignored. In ideal chains there are no other physical interactions, so the free energy arises entirely from entropy

$$G(x) = -TS$$

$$= -kT \ln \left(e^{-\frac{x^2}{2Nl^2}} \right)$$

$$= \frac{kTx^2}{2Nl^2} \tag{3.36}$$

The force pulling the ends together is the derivative with respect to x

$$F(x) = \frac{kTx}{Nl^2} \tag{3.37}$$

This is identical to Hook's law for a spring; the force increases linearly with distension. This means that a random coil behaves like a simple elastic spring. A longer random coil (larger N) will be looser and easier to stretch. The factor $1/N$ makes the force smaller as the number of segments increases. It is notable that the restoring force increases with temperature. This reflects the fact that the force arises from the thermal tendency toward disorder.

Equation (3.37) is an old result first derived from the theory of rubber elasticity. This spring-like elasticity is a basic property of materials made up of chain molecules, and rubber is a good example. Recent experiments with microscopic probes tethered to the ends of chain molecules have measured the force of extension of individual molecules. For short extensions one does see a linear force, but for longer extensions, and especially for stiff molecules like DNA, the random-coil model does not work well. Instead, as with loop formation just discussed, the worm-like chain model is preferred because it treats the stiffness property more effectively (Kellermayer *et al.*, 1997).

3.9 | When do molecules act like random coils?

With the exact results derived above for random-coil models, it is important to know when this behavior is observed. Poly-amino acids and long linear DNA often behave like random coils. For a denatured protein the random-coil model may seem like a natural choice, but denatured proteins can be quite compact even though they are very disordered.

In its native state, a protein has a well-defined structure, which is anything but random. However, even a well-folded protein has some regions that fail to adopt the standard secondary structures of α-helix and β-sheet (Section 3.10). In some of these cases parts of the crystal structure look blurred, indicating some disorder. These may represent stretches of random coil, so even native proteins have occasional small segments for which some form of random-coil model may apply.

From the perspective of a rotational isomer model, a well-defined native state of a protein can be thought of as one particular rotational isomer among a very large number of possibilities. How one particular configuration comes about is a challenging question which is discussed at the end of this chapter (Section 3.16).

3.10 | Backbone rotations in proteins: secondary structure

The polypeptide backbone has three kinds of chemical bonds, the bond between the α and carboxyl carbons (the C^α–C' bond), the bond between the amide nitrogen and α carbon (the N–C^α bond), and the

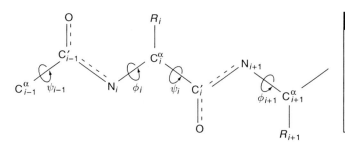

Fig. 3.3. Rotations of the backbone bonds of a peptide chain. The partial double bond character of the peptide bond prevents rotation. Rotations about the N–C^α bond, ϕ, and about the C^α–C′ bond, ψ, define the backbone conformation of a protein.

bond between the carboxyl carbon and amide nitrogen (the C′–N peptide bond) (Fig. 3.3). The C^α–C′ and N–C^α bonds can rotate, but the partial double bond character of the peptide bond, due to resonance with the C–O double bond, forces the carbonyl group and adjacent C^α and N atoms to lie in one plane. With this bond angle fixed, the problem of specifying the conformation reduces to specifying the dihedral angles of the C^α–C′ and N–C^α bonds. By convention we denote the C^α–C′ bond dihedral angle as ψ and the N–C^α bond dihedral angle as ϕ. A set of ψ and ϕ then specifies the conformation to a considerable extent, leaving only the side chains to be defined.

As with n-butane, ψ and ϕ for the backbone rotations have three energy minima. However, the positions of these minima have not been unambiguously located and the energies are not known. Steric repulsions can restrict the allowed bond angles of a polypeptide chain considerably. For a given value of ψ there will be a range of ϕ for which adjacent backbone and side chain atoms overlap. Such values are energetically prohibited. A useful way to evaluate this effect is by plotting energy contours as a two-dimensional function of ψ and ϕ, and highlighting regions not excluded by steric repulsion. Such plots are referred to as Ramachandran plots, after the originator of this approach (Ramachandran and Sasisekharan, 1968). An example is shown in Fig. 3.4 for a peptide containing alanine. The contours enclose regions where ψ and ϕ are allowed, i.e. not excluded by steric repulsions.

In the case of glycyl–glycine about 50% of the area in the ϕ–ψ plot is energetically excluded. Adding a side chain restricts the available area much more. Alanine, with its methyl side chain, has only about 25% of the area accessible (regions within the dashed contours of Fig. 3.4).[3] Adding a second methyl group to the α-carbon produces a nonstandard amino acid called α-aminoisobutyric acid. This further constrains rotations, restricting access in a ϕ–ψ plot to only a small percentage of the total area.

One especially useful feature of the ϕ–ψ plot is that certain values of the angles correspond to the common secondary structure

---

[3] The precise shapes of the contours as well as their areas depend on the choice of the potential energy functions for bond rotations and steric repulsion.

Fig. 3.4. A contour plot of potential energy versus ψ and ϕ for an alanine dipeptide (Ramachandran plot). The dashed contours represent zero and each solid contour marks a 1 kcal mole^{-1} decline. Points corresponding to α-helix, β-sheets, and polyproline II (P II) are indicated (after Ramachandran et al., 1966).

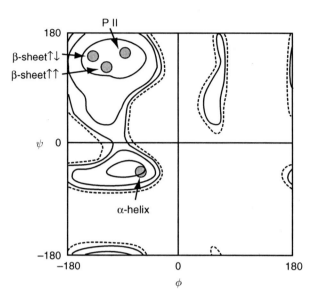

motifs of proteins. For an α-helix we have $\phi = -57°$ and $\psi = -47°$. This point is indicated in Fig. 3.4. The other important forms of secondary structure, parallel and antiparallel β-sheets, are also indicated. It is interesting that although much of the space in a ϕ–ψ plot is prohibited by excluded volume interactions, the α-helix region and β-sheet regions are usually allowed. This helps explain the high incidence of α-helix and β-sheet in proteins. Even in peptides formed from the highly constrained amino acid α-aminoisobutyrate, the region around $(-57°, -47°)$ is still energetically accessible. Incorporating this amino acid into a polypeptide virtually forces the formation of an α-helix.

There is growing experimental and theoretical evidence that the minimum potential energy conformation of simple hydrated polypeptides is a somewhat extended, left-handed helix in which $\phi = -75°$ and $\psi = 145°$ (Han et al., 1998; Shi et al., 2002c). This point is also plotted in Fig. 3.4 and it falls in an allowed region of the ϕ–ψ space. This conformation is referred to as polyproline II, one of the two conformations allowed to a polypeptide formed from proline. Polyproline I and II refer to the N–C$^\alpha$ bond being either cis or trans, respectively. The five-membered pyrrolidine ring of proline restricts rotation around the N–C$^\alpha$ bond (ϕ) and makes polyproline much less flexible than peptides formed from other amino acids. This prevents it from forming α-helices.

Figure 3.5 shows ϕ–ψ plots based on the structures of four well-studied proteins. Each residue provides a point on the plot. The points generally are clustered in the α-helix and β-sheet regions. Hemoglobin has a large proportion of its points in the helical zone, reflecting its high helical content. Green fluorescent protein (GFP) has a large barrel formed from β-sheet, and this is reflected in the high density of points in this zone. Lysozyme has both structural

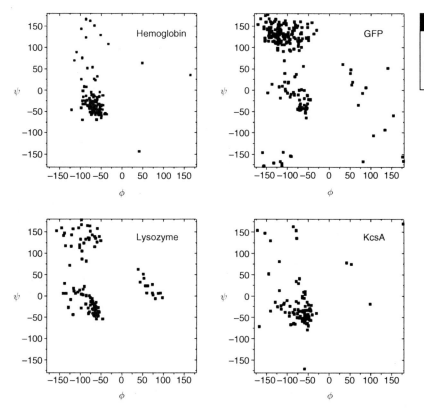

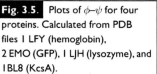

Fig. 3.5. Plots of ϕ–ψ for four proteins. Calculated from PDB files 1 LFY (hemoglobin), 2 EMO (GFP), 1 LJH (lysozyme), and 1 BL8 (KcsA).

motifs. The microbial K^+ channel (KcsA) contains two membrane spanning α-helices, and little else in the way of secondary structure. All of the proteins have some points scattered through the plot, indicating that they have some residues that do not participate in the more common forms of secondary structure.

3.11 | The entropy of protein denaturation

The native state of a protein depends on many internal contacts between the constituent amino acids. There are two general kinds, local contacts such as those between neighboring residues in an α-helix, and distant contacts between residues separated by tens or hundreds of amino acids. Both kinds of contacts can have an impact on the overall stability of the native state. With such a complex web of interconnections it is not possible to break one or a few contacts without simultaneously breaking many others. This makes the formation of the native structure a highly cooperative process. A protein cannot be partially folded, at least not for any significant amount of time; it must be either completely folded or highly disordered. Experimental studies support the view that thermal unfolding is a two-state transition between a unique native state and a denatured state consisting of a large ensemble of configurations (Section 1.6).

The origins of cooperative protein folding will be discussed further in Section 3.16, but here we will accept the cooperativity as a fact and try to use the random-coil model to estimate the unfolding entropy. The native state is taken as a single rotational configuration, with a configurational entropy of $k \ln 1 = 0$. The denatured state is taken as a random coil where each bond can assume three different rotational energy minima. With two rotating bonds per residue (ψ and ϕ mentioned above) each residue has $2^3 = 8$ rotational states in the random coil. The side chains have some conformational flexibility as well. The number of additional rotating bonds varies quite a bit between different amino acids, but on average there are 3. This gives an additional factor of $3^3 = 27$. The number of configurations for a chain with N residues is then $(8 \times 27)^N = 216^N$ (the resemblance to Eq. (3.14) is not a coincidence). This gives a configurational entropy of $k \ln 216$ per residue per molecule, or $R \ln 216 = 10.6$ cal K^{-1} per residue per mole of protein.

If this seems too simplistic there are ways to improve it. First, allow the native state to have more configurations. Let 10% of the polypeptide backbone be disordered in the native state. This effect is small, and only increases the entropy of the native state by $0.1 \times R \ln 8 = 0.4$ cal K^{-1}. A more important source of entropy in the native state is the side chains, especially on the protein surface. Allowing half of these bonds to rotate increases the entropy by $R \ln (3^{1.5}) = 3.3$ cal K^{-1}. These two considerations reduce our estimate of the entropy of denaturation to 6.9 cal K^{-1}. Recall that the random-coil model overestimates the entropy because excluded volume reduces the number of configurations. Using the factor of $1.28^N/N^{1/6}$ mentioned above (Section 3.5) for nonideal chains, we find that for a 100 residue chain there will be a correction of 2.4×10^{10}. This reduces the entropy by only 0.47 cal K^{-1} to 6.4 cal K^{-1} per residue.

An extensive experimental analysis of thermal denaturation yielded an estimate of 4.2 cal K^{-1} per residue per mole (Privalov, 1979). This estimate was made at 110 °C, a temperature where the hydrophobic interaction is driven mostly by enthalpy rather than entropy (see Section 2.8). Our rough estimate of the configurational entropy of the polypeptide and side chains is about 50% higher than this experimental value. A very likely explanation for this discrepancy is that the denatured state of a protein does not have as much freedom as a truly random coil. The polyproline II conformation (Section 3.10) is probably the lowest energy conformation once a polypeptide has unfolded, so rather than rotating freely, many bonds have angles indicated for this conformation in Fig. 3.4. Thus, using eight rotational states per residue is probably an over-estimate, and computer modeling studies of unfolded proteins gave estimates of 1.5 to 3 (Dinner and Karplus, 2001).

The large entropy decrease that occurs when a protein folds into its native state raises an interesting question about how much time folding is expected to take. With an entropy of 420 cal K^{-1} mole^{-1} for a 100 residue chain we have $e^{212} = 10^{92}$ unfolded configurations.

It was pointed out by Levinthal (1968) that if an unfolded protein sampled configurations at a rate of 1 per nanosecond (a reasonable value based on the known rate of rotation of individual bonds), then this sampling process could go on for 10^{75} years before chancing upon the correct native state. This is a hopelessly long time and this conundrum is commonly referred to as Levinthal's paradox. The fact that protein folding requires only seconds to minutes rather than 10^{75} years indicates that the sampling of configurations cannot be random. Without some form of guidance the protein will never fold. Indeed, almost any model that incorporates some energetic bias toward the correctly folded state can give realistic folding times. Models have been proposed in which certain parts of a protein form secondary structure and this nucleates the spread of structure throughout the protein. Other models have incorporated a small bias toward the native state at every residue. Distinguishing between these and other models will require experiments to follow the actual pathway that a protein takes through configuration space as it folds (Section 7.10).

3.12 | The helix–coil transition

The α-helix is one particular rotational configuration of a polypeptide chain, characterized by $\psi = -47°$ and $\phi = -57°$ (Fig. 3.4). This configuration is stabilized by hydrogen bonds between backbone carbonyl oxygen atoms and amide nitrogen atoms three and four residues apart. However, restricting the rotations around the backbone bonds reduces the entropy. These two forces oppose one another, and depending on factors such as temperature, solvent, and amino acid side chain, either the ordering effect of hydrogen bonds or the disordering effect of chain entropy will prevail.

The relative stability of the α-helix versus the random coil is the subject of an important theory. Helix–coil theory provides a conceptual basis for many efforts to predict the conformational state of a polypeptide, and illustrates some of the basic ideas of cooperative transitions in macromolecules (Zimm and Bragg, 1959; Poland and Scheraga, 1970; Cantor and Schimmel, 1980).

The essence of the helix–coil transition is recognizing that the tendency of a residue to assume a helical conformation is very sensitive to the state of its immediate neighbors. In the above discussion of the entropy of protein unfolding it was argued that it is hard to break a few contacts in the native state of a protein without breaking many others. In the α-helix we make a less restrictive assumption about the interdependence of different contacts. We assume that it is harder for one residue to break its own helix stabilizing hydrogen bonds if its nearest neighbor is also helical. If one residue is already helical, it will be easier for a neighbor to fall into step. Thus, the helical configuration of one residue restricts the

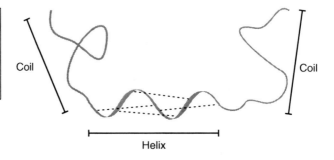

Coil   Coil

Helix

ψ, ϕ angles of its neighbors, and as a result these neighbors do not loose any more rotational entropy when they hydrogen bond with their neighbors. Some of these points are illustrated in Fig. 3.6, where overlapping hydrogen bonds reinforce one another.

Another important point seen in Fig. 3.6 is that part of the molecule is helical and part is coiled. Thus, the helix–coil transition is not global. However, the interactions between neighboring residues tends to force helical sections to join together into longer contiguous stretches. Although only one helical stretch is shown in Fig. 3.6, multiple stretches of helix are certainly possible, and the average over all configurations, worked out in the following section, takes this into account.

*3.13 | Mathematical analysis of the helix–coil transition

The theory of the helix–coil transition starts with a derivation of the partition function for a polypeptide chain. Recall that the partition function for n-butane was a simple sum of the Boltzmann weights for *trans* and *gauche* molecules. For polyethylene, a partition function was computed as the n-butane partition function raised to the Nth power (Eq. (3.14)). One might think that this example can be applied to a polypeptide. Just determine the partition function for the backbone bond rotations of each amino acid, and then multiply them together. (That was the main idea in Sections 3.2 and 3.11.) However, this ignores the interactions between residues, and these interactions are the essential feature of the helix–coil transition. The Boltzmann weights have to include these interactions, so we will have to estimate the interaction energy for each configuration.

In an α-helix, a residue interacts with residues three and four positions away. To make the mathematics easier we will use a simpler model with interactions just between adjacent residues. A rigorous analysis of the more realistic model shows that this simplification leaves intact the important features of the helix–coil transition (Lifson and Roig, 1963). As already noted, a long chain will have parts that are helical and parts that are coiled (Fig. 3.6). To represent this essential feature symbolically, we denote helical segments with the letter h and coiled segments with the letter c. We might have a segment of a chain with the following arrangement of helix and coil

cccccchhhhhccchccccccchhhhhhhhhhhccccccccc

This is just one of the 2^{40} possibilities for this 40-residue chain. This particular configuration has three helical stretches of 5, 1, and 10 residues.

To calculate the elementary weight for a configuration such as this we define three elementary weights for the states available to individual residues as products of a state degeneracy and Boltzmann weight, $we^{-E/kT}$, as in Eq. (3.8). The need for elementary weights for the h and c states is clear, but we also need to take into account the interaction between nearest neighbors. This requires an additional elementary weight for a helical residue at a helix–coil boundary. We can then combine these three elementary weights for single residues into a product (Eq. (3.12)) to obtain the elementary weight of any configuration of the entire chain. These elementary weights can be calculated directly if we know the relevant degeneracy and energy. They are more often left as free parameters to be determined from fits to helix–coil transition data.

The coil is taken as the ground state with zero energy. If we then normalize all degeneracies to that of the coil then its elementary weight is 1. The elementary weight for a helical residue is denoted as s, and represents only the favorable contribution from forming the hydrogen bonds. This applies to the continuation of an already started helical stretch; s is often referred to as the helix continuation parameter. For a helical residue at a helix–coil boundary, the elementary weight must also include the unfavorable entropy term of restricting backbone bond rotations. If we denote this entropy term as σ, the elementary weight for a boundary residue becomes the product σs. The term σ is often referred to as the helix initiation parameter. Note that for a stretch of helix we have two boundaries with coil on each side. Each helical stretch should have only one helix initiation factor, so we will include a factor of σ only for c–h boundaries and not the h–c boundaries.

Now we can calculate the elementary weight of a particular configuration as a product of the three segment elementary statistical weights, 1, s, and σs. For the configuration of the 40-residue chain above we see one 5-residue stretch of helix, which gives $s^5 \sigma$. The isolated h gives $s\sigma$, and the remaining 10-residue stretch of helix gives $s^{10} \sigma$. Thus, we have $s^5 \sigma s \sigma s^{10} \sigma = s^{16} \sigma^3$. The partition function is the sum of such products of weights over all possible combinations of c and h as follows

$$Q = \sum_{ij} \omega_{ij} s^i \sigma^j \qquad (3.38)$$

where $\omega_{i,j}$ is the degeneracy for the number of ways that a chain can have i helical residues arranged in j helical stretches.

Recall that the partition function can be used to calculate the average energy of a molecule (Eq. (3.11)). Similarly, the mean number of helical residues is the average of i

$$\bar{i} = \frac{\sum\limits_{i,j} i\omega_{i,j}s^i\sigma^j}{\sum\limits_{i,j} \omega_{i,j}s^i\sigma^j} \tag{3.39}$$

Differentiating Eq. (3.38), dividing, and comparison with Eq. (3.39) (in the same manner used to obtain Eq. (3.11)) provides a useful expression for the mean number of helical residues

$$\bar{i} = \frac{s}{Q}\frac{\partial Q}{\partial s} = s\frac{\partial \ln Q}{\partial s} \tag{3.40}$$

The mean number of helical residues is clearly a useful index for the extent of the transition, and knowing the partition function will allow us to calculate it.

Now we will derive the partition function for an N-residue chain. First, divide it into two parts. The first part is the partition function of the molecule in which the right-most segment is coiled (Q_{Nc}) and the second part is the partition function of the molecule in which the right-most segment is helix (Q_{Nh}). Then $Q_N = Q_{Nc} + Q_{Nh}$. If we add another segment to the chain and try to calculate Q_{N+1}, we see that for a chain with a c at the end, adding another c multiplies Q_{Nc} by 1 (the elementary weight for a coiled residue), and adding an h multiplies Q_{Nc} by σs (the elementary weight for a helical residue at the boundary). Likewise, if the chain has an h at the end, adding a c multiplies Q_{Nh} by 1 and adding an h multiples it by s. This gives

$$Q_{N+1} = (1 + \sigma s)Q_{Nc} + (1 + s)Q_{Nh} \tag{3.41}$$

If we separate out the parts according to their right-most segment, we have two equations

$$Q_{(N+1)c} = Q_{Nc} + Q_{Nh} \tag{3.42a}$$

$$Q_{(N+1)h} = \sigma s Q_{Nc} + s Q_{Nh} \tag{3.42b}$$

These two equations can be expressed in matrix-vector form (Appendix 2). We designate the two parts of the partition function as components of a vector (Q_{Nc}, Q_{Nh}), and Eqs. (3.42a) and (3.42b) become

$$\begin{pmatrix} Q_{(N+1)c} \\ Q_{(N+1)h} \end{pmatrix} = \begin{pmatrix} 1 & 1 \\ \sigma s & s \end{pmatrix} \begin{pmatrix} Q_{Nc} \\ Q_{Nh} \end{pmatrix} \tag{3.43}$$

If our first segment is coiled then its state vector is (1, 0). Adding N segments means multiplying by the 2×2 matrix in Eq. (3.43) (denoted as **M**) a total of N times. This gives

$$\begin{pmatrix} Q_{Nc} \\ Q_{Nh} \end{pmatrix} = \mathbf{M}^N \begin{pmatrix} 1 \\ 0 \end{pmatrix} \tag{3.44}$$

Multiplying from the left by the row vector $(1, 1)$ performs the task of adding the two components to get the partition function for an N-residue polypeptide.

$$Q_N = (1, 1)\mathbf{M}^N \begin{pmatrix} 1 \\ 0 \end{pmatrix} \tag{3.45}$$

Putting the partition function into matrix form allows us to use the powerful mathematical method of matrix diagonalization. It is possible to find a matrix $\mathbf{T}$ and its inverse $\mathbf{T}^{-1}$ that diagonalize $\mathbf{M}$ such that

$$\mathbf{T}^{-1}\mathbf{MT} = \Lambda = \begin{pmatrix} \lambda_1 & 0 \\ 0 & \lambda_2 \end{pmatrix} \tag{3.46}$$

where λ_1 and λ_2 are the eigenvalues of $\mathbf{M}$. Since $\mathbf{TT}^{-1}$ gives the identity matrix, Eq. (3.45) can be written as

$$Q_N = (1, 1)\,\mathbf{T}\mathbf{T}^{-1}\,\mathbf{M}\mathbf{T}\mathbf{T}^{-1}\,\mathbf{M}\mathbf{T}\dots\mathbf{T}^{-1}\,\mathbf{M}\mathbf{T}\mathbf{T}^{-1} \begin{pmatrix} 1 \\ 0 \end{pmatrix}$$

$$= (1, 1)\mathbf{T}\Lambda^N\mathbf{T}^{-1} \begin{pmatrix} 1 \\ 0 \end{pmatrix} \tag{3.47}$$

This simplifies things quite a bit because raising a diagonal matrix to the Nth power gives a diagonal matrix with elements $\lambda_1{}^N$ and $\lambda_2{}^N$. The eigenvalues of $\mathbf{M}$ are[4]

$$\lambda_1 = \frac{1 + s + \sqrt{(1-s)^2 + 4\sigma s}}{2} \tag{3.48a}$$

$$\lambda_2 = \frac{1 + s - \sqrt{(1-s)^2 + 4\sigma s}}{2} \tag{3.48b}$$

For $\mathbf{T}$ and $\mathbf{T}^{-1}$ we have

$$\mathbf{T} = \begin{pmatrix} \frac{\lambda_1 - s}{\sigma s} & \frac{\lambda_2 - s}{\sigma s} \\ 1 & 1 \end{pmatrix} \tag{3.49a}$$

and

$$\mathbf{T}^{-1} = \frac{1}{\lambda_1 - \lambda_2} \begin{pmatrix} \sigma s & s - \lambda_2 \\ -\sigma s & \lambda_{1-s} \end{pmatrix} \tag{3.49b}$$

One can verify these expressions for $\lambda_1, \lambda_2, \mathbf{T}$ and $\mathbf{T}^{-1}$ by substitution back into Eq. (3.46). Putting these expressions for $\mathbf{T}$ and $\mathbf{T}^{-1}$ into Eq. (3.47) and performing the matrix and vector multiplications gives us the following expression for the partition function

---

[4] This result is obtained by first expanding the determinant of the matrix obtained by using $\mathbf{M}$ in Eq. (A2.17). The resulting quadratic equation has these two solutions.

$$Q_N = \frac{\lambda_1^{N+1}(1 - \lambda_2) - \lambda_2^{N+1}(1 - \lambda_1)}{\lambda_1 - \lambda_2} \tag{3.50}$$

If $\lambda_1 > \lambda_2$, then with large N the first term in the numerator dominates, leaving

$$Q_N \sim \lambda_1^N \tag{3.51}$$

From this we can use Eq. (3.40) to obtain the average fraction of helix as

$$\frac{\bar{i}}{N} = \frac{s}{N} \frac{\partial \ln Q}{\partial s} = s \left(\frac{1 + \frac{s - 1 + 2\sigma}{\sqrt{(1-s)^2 + 4\sigma s}}}{1 + s + \sqrt{(1-s)^2 + 4\sigma s}} \right) \tag{3.52}$$

This gives us an expression for the extent of the transition as a function of the basic parameters s and σ.

3.14 | Results of helix–coil theory

Equation (3.52) tells us what fraction of a polypeptide will be helical as a function of s. Remember that s has the form $w e^{-E/kT}$, where E is the energy of forming a hydrogen bond within a helical segment. Thus, if conditions such as solvent or temperature change so that helix is favored, s will become larger. For small s, $\bar{i}/N$ is near zero, and for large s, $\bar{i}/N$ approaches 1; $s = 1$ defines the midpoint of the transition where $\bar{i}/N = 1/2$ (Problem 11).

To see how the fraction of helix varies with a condition that affects helical stability, we plot $\bar{i}/N$ versus s for various values of the cooperativity parameter, σ (Fig. 3.7). The fraction of helix increases with s, as expected. The shape of the transition varies in a striking way with different choices of σ. The reason for this is that σ is a cooperativity parameter reflecting the ease of forming a boundary between a helical stretch and a coiled stretch. When σ is small, then it is hard to form a boundary, and large stretches are forced to undergo the transition together. The transition is thus more

Fig. 3.7 Plots of Eq. (3.52) give the fraction of helix, $\bar{i}/N$, versus s, the helix continuation parameter. Decreasing the helix initiation parameter, σ, makes the transition steeper.

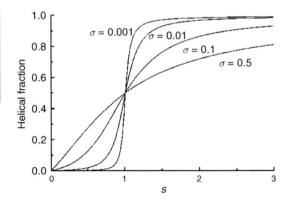

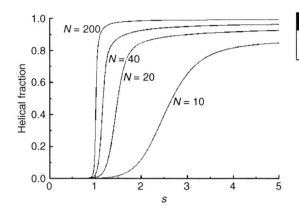

cooperative. On the other hand, when $\sigma = 1$ the state of one residue is completely insensitive to the state of its neighbor. It can be shown that Eq. (3.52) then reduces to $s/(1+s)$ (Problem 10). This is how a collection of noninteracting segments should behave. Small values of σ can be thought of as increasing the size of a cooperative unit of the transition, providing an interesting parallel with the idea of size and cooperativity in global transitions discussed in Chapter 1.

For shorter chains cooperativity becomes less of an issue. Equation (3.52) cannot show this because it is a limit based on large N. To see how chain length matters we must go back to Eq. (3.50) and differentiate $\ln Q$ with respect to s. The result is quite a bit more complicated, so it will not be written out. Plots for a few values of N are shown in Fig. 3.8. As the chain gets shorter we see a characteristic end effect, i.e. the transition shifts and gets broader. The shift of the transition point to higher s reflects the fact that end segments cannot be flanked by helical segments. For a long chain this does not make much of a difference, but for a short chain the end segments contribute significantly to the total chain free energy.

Finally, we note that near the transition point, $s = 1$, one sees the greatest number of separate helical and coil stretches. To see this we differentiate the partition function, Eq. (3.38), with respect to σ rather than s. This gives the average of j, the number of boundaries, or equivalently, the number of helical stretches.

$$\bar{j} = \frac{\sigma}{Q}\frac{\partial Q}{\partial \sigma} = \frac{\sigma \partial \ln Q}{\partial \sigma} \tag{3.53}$$

A plot of the resulting function is shown in Fig. 3.9.

In Fig. 3.9 we see that, at the transition midpoint, a 1000-segment chain has about 16 separate helical stretches. The average size will by $500/16 = 31.25$ segments. We might think of this as a cooperative unit for the transition, by analogy with the concept of cooperative unit from Section 1.4. Global transitions get sharper as the size of the cooperative unit increases. However, in the helix–coil transition the cooperative unit size is not fixed and the transition is definitely not two-state.

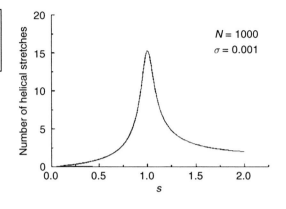

Fig. 3.9 Plot of the number of helical stretches versus s, calculated with the aid of Eqs. (3.50) and Eq. (3.53).

$N = 1000$
$\sigma = 0.001$

3.15 | Helical propensities

The equation for the helical fraction, $\bar{i}/N$, from helix–coil theory has been fitted to a large body of experimental data, generally with good results. The parameters obtained from these fits vary for peptides composed of different amino acids. This reflects the variable tendencies of different amino acids to form α-helices. In proteins, some amino acids are found within α-helices far more frequently than others. This indicates that the parameters σ and s might be useful in the prediction of secondary structure. These parameters can thus be viewed as helical propensities, and a great deal of work has gone into measuring and interpreting them.

Experiments with homopolymers are impossible for many of the amino acids because of solubility problems. Instead, a "host–guest" strategy has been developed. One makes a series of peptides using a host sequence that has one guest location where different amino acids are inserted. The helix–coil transition will then be shifted to different temperatures according to the helical propensity of the guest amino acid. From the analysis of these kinds of data the initiation parameter, σ, was found to be small, $\sim 10^{-4}$, as expected. The continuation parameter, s, is what changes as a peptide goes through the transition, but the value can be determined for a standard set of conditions, and used as an index for comparing the tendencies of different amino acids. Table 3.2 shows a set of $\Delta\Delta G_\alpha$ values determined by measuring the shifts in a dimerization equilibrium in which the peptides are helices as dimers and coils as monomers. The numbers reflect the preference of each amino acid for residence in an α-helix. These values show a rough agreement with a variety of other indices of helical preference (O'Neil and DeGrado, 1990)

The values of $\Delta\Delta G_\alpha$ give some clues about why amino acids differ in their individual tendencies to form α-helices. One of the most important factors already mentioned (Section 3.10) is steric hindrance caused by the side chain. The group attached to the α-carbon restricts the range of ϕ–ψ angles quite a bit, but usually

Table 3.2.	*Helix formation energies*		
Amino acid	$\Delta\Delta G_\alpha$	Amino acid	$\Delta\Delta G_\alpha$
ala	− 0.77	cys	− 0.23
arg	− 0.68	ile	− 0.23
lys	− 0.65	tyr	− 0.17
leu	− 0.62	asp	− 0.15
met	− 0.50	val	− 0.14
trp	− 0.45	thr	− 0.11
phe	− 0.41	asn	− 0.07
ser	− 0.35	his	− 0.06
gln	− 0.33	gly	0
glu	− 0.27	pro	3

Source: from O'Neil and DeGrado (1990).

the values corresponding to an α-helix are still allowed. Thus, glycine is one of the weakest helix formers. With no side chain to restrict rotation, the entropy cost for α-helix formation is high. By restricting these rotations, the α-methyl carbon of alanine makes it one of the best helix formers. Another factor is the conformational entropy of the side chain. Side chains are tightly packed in an α-helix, which means that large flexible side chains will lose rotational freedom. This explains why, for example, alanine has a higher helical propensity than valine (Aurora *et al.*, 1997). The alanine side chain has no conformational entropy to lose, but the valine side chain does.

There are a number of other factors relevant to the stability of an α-helix. For example, end effects are quite important, as reflected by the low value of σ. Serine and threonine have side chains that can form hydrogen bonds with the backbone. This makes it easier to form a junction between helical and coiled stretches. This effect is called "capping" and the strategic location of capping residues can make a big difference in the ease of helix formation. On the other hand, the side chain of asparagine interacts with the backbone in a manner that destabilizes the α-helix. Proline cannot form an α-helix because its ring structure constrains the rotations of its backbone bonds; ϕ and ψ cannot assume the correct values for an α-helix so proline has the lowest s value by far of all the amino acids (and highest $\Delta\Delta G_\alpha$ value, see Table 3.2).

Despite understanding much about the stability of α-helices, it is hard to predict where one will find helices in the amino acid sequence of a protein. In general, predictions based on helical propensities have a success rate of about 60%–70%. Furthermore, many helical stretches in proteins are short, and this is hard to reconcile with the unfavorable boundary energy, which should favor longer helices (Dill, 1990). Capping cannot account for the large number of short helices observed.

One reason that it is so hard to predict which parts of a protein will form α-helices is that helix stability is sensitive to the environment. The polarity of the environment influences the energetics of hydrogen bonds (Section 2.10). In water, hydrogen bonding is an exchange process in which contacts with water are replaced by intramolecular contacts; the free energy change associated with hydrogen bond exchange is small. When a hydrogen bond forms between two groups confined within a protein interior, water has already been excluded. Hydrogen bonding is then no longer an exchange process, so that the full bond energy is available to drive secondary structure formation.

From this perspective, the side chain of an amino acid is less relevant. If nonlocal forces pull an amino acid into the interior of a protein, it is much more likely to form highly hydrogen bonded secondary structures, regardless of its side chain. This means we have to figure out which stretches will fold up inside the protein and which will remain on the surface. Of course, this is a hard problem. But if we can make some headway towards its solution, we may be able to improve our ability to predict secondary structure as well.

3.16 | Protein folding

Models examined so far were mostly based on homopolymers such as poly-L-alanine. These models were good at explaining homopolymer behavior. They coil randomly and undergo a helix–coil transition. But some key properties elude them. They cannot fold up into a highly structured native state, and they cannot undergo global transitions. Proteins are special in these regards. The closest a homopolymer can come to a native state is a compact "molten-globule," which is compact but still very random and fluid-like. The molten-globule state behaves very much like a nonideal random coil with very strong attractive interactions between the monomers.

The formation of a molten-globule from a random coil is accompanied by a drastic reduction in the number of conformations. A homopolymer with 100 segments can be arranged in about 10^{67} ways on a cubic lattice without having any segments overlap (Camacho and Thirumalai, 1993). Compacting this chain to its maximum density reduces the number of configurations to about 10^{17}, and this is a rough estimate of the number of possible compact configurations of a 100-residue protein. The big question is what makes a protein prefer only one of these 10^{17} possibilities.

Of course, describing the native state as having only one conformation is an oversimplification. It is more realistic to think of the native state as a cluster of structurally similar conformations, or as a global state comprising a small but significant number of microstates. Even from this perspective, the native state of a protein is vastly more restricted than the set of all compact states, and we still have a very challenging problem of figuring out what determines the folding of a protein to its native state.

The crucial property of a protein that enables it to fold is its sequence of amino acids. The molecular forces discussed in Chapter 2 – electrostatic interactions, hydrogen bonds, and hydrophobic interactions – can be incorporated into a full potential energy function of a protein that depends on thousands of these interactions. Efforts to use these protein force fields to solve the protein folding problem were discussed in Section 2.14, and we pointed out that they have yet to succeed.

Some basic principles of protein folding have been elucidated by models based on potential energy functions that are much simpler than the full protein force fields. One example is the H–P lattice model, in which two amino acid types, hydrophobic (H) and polar (P) occupy sites on a lattice. Taking the view that the hydrophobic interaction is the predominant force responsible for the stabilization of the native state (Section 2.13), the H–P model focuses exclusively on these interactions. If we think of the standard amino acids as a 20-letter alphabet, the H–P model can be thought of as a reduction to an alphabet with only two letters.

The H–P model simplifies things not only by ignoring nonhydrophobic interactions, but also by lumping all of the hydrophobic interactions together. The hydrophobicities of the hydrophobic amino acids vary quite a bit (Section 2.8), but the H–P model ignores this, employing a single energy parameter, ε, for any H–H contact. The value is typically ~ -2 kcal mole^{-1}, but it can be varied to see how sensitive predictions are to the strength of this interaction. Most H–P models use a lattice representation, so that a configuration is defined as a sequence of connected lattice points (Fig. 3.10). This makes it easy to identify contacts. To calculate the energy for any particular configuration, we count the total number of H-residues that are in contact with other H-residues. The number of such contacts times the basic energy parameter gives the energy of that configuration.

Most of the analysis of H–P models was carried out with a computer. This was the only practical way to examine each sequence and enumerate all of its configurations. A thorough analysis turned up an interesting result for short sequences on a two-dimensional lattice. Although the number of configurations was enormous, some

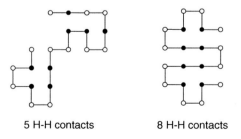

5 H-H contacts   8 H-H contacts

Fig. 3.10 A 20-residue H–P chain with the sequence PHPPHPPHHPPHHPPHPPHP (H-filled, P-open). The left configuration has five H–H contacts and the right has eight. Three other configurations of this sequence have five H–H contacts, but no other configuration has eight. Thus, this sequence has a unique energy minimum (Dill et al., 1995).

sequences had very few configurations with the minimum energy. To illustrate this point, the 20-segment H–P chain shown in Fig. 3.10 has a minimum energy state (right) with eight H–H contacts, giving an energy of 8ε. This is the only configuration of this particular sequence with this many H–H contacts; all the others have less. Another configuration of this sequence has five H–H contacts (Fig. 3.10, left). Counting up all the configurations of this sequence revealed a total of four configurations with this number of H–H contacts.

Relatively few H–P sequences are like the one in Fig. 3.10, with only one minimum energy configuration. Other sequences had many, making their energy minima degenerate. For chain lengths ranging from 11 to 18, about 2.5% of the sequences had only a single configuration with the minimum energy, and this configuration was generally highly compact. The low degeneracy energy minimum of some of the sequences implied that if you find the right sequence, an H–P chain can simulate the special ability of proteins to fold up into a unique configuration. From this perspective, the mystery of protein folding is explained as the sequestering of hydrophobic residues into the protein core. Since the hydrophobic effect is generally viewed as not very specific, it comes as a surprise that a model based solely on hydrophobic interactions gives a unique minimum energy configuration. The key is the sequence. Its specificity allows a nonspecific force to generate a uniquely folded structure.

H–P sequences vary greatly in how many of their configurations share the energy minimum. Those with just a few minimum energy configurations can be considered "good-folding" sequences and come closer to capturing the protein-like property of a native state. The odds of finding a good-folding sequence are better among those with a ratio of H to P residues of 1:1. Higher ratios of H-residues favor collapse into compact structures, but there are then many such compact states with the same minimum energy. On the other hand, at high ratios of P-residues sequences will not fold into a compact configuration, but will assume extended configurations to solvate the P-residues. The optimum ratio of 1:1 is close to the ratio of hydrophobic to hydrophilic amino acids found in globular proteins (Camacho and Thirumalai, 1993). The small number of good-folding H–P sequences is probably relevant to the similar small number of folding patterns exhibited by proteins. Protein sequences of unknown structure can often be threaded into known structures of proteins with very different amino acid sequences. We do not yet know how hard it is to find this property (good-folding) among randomly generated sequences composed of the 20 amino acid building blocks of proteins.

When the chains get longer and a three-dimensional lattice is employed to make things more realistic, H–P chains no longer fold uniquely. Sequences with only one minimum energy configuration were not found, but some sequences still had low degeneracy energy minima. In a limited study of 88-residue H–P chains on a three-dimensional lattice, it was found that some had fewer than five configurations with the minimum free energy (Dill *et al.*, 1995). Another study of 48-segment sequences on a three-dimensional lattice found 10^3–10^6 minimum energy configurations for all of the 10 sequences examined (Yue *et al.*, 1995). Thus, although optimizing H–H contacts narrows down the search for the native state considerably, this factor alone does not get us all the way to a unique endpoint.

Clearly other interactions are necessary to achieve a unique native configuration, and more sophisticated potential energy functions have more success in predicting the formation of unique native states (Kolinski *et al.*, 1993). This amounts to adding letters to the alphabet. Thus, side-chain packing can be incorporated into the potential energy function as well as cooperativity parameters for hydrogen bonding such as the parameter σ introduced above in helix–coil theory.

A parallel experimental approach to this issue involves investigating the association of synthetic peptides (Raleigh and DeGrado, 1992; Betz *et al.*, 1995; Schafmeister *et al.*, 1997). A prototypical H–P peptide was made with leucines as the Hs and glutamates as the Ps. The leucines and glutamates were spaced so that when the peptide adopted an α-helical conformation it had distinct polar and hydrophobic surfaces. Four such helical segments bundled together into a tetramer with a hydrophobic leucine core (Fig. 3.11a). Spectroscopic analysis indicated that this structure was quite disordered. The side chains exhibited fluid-like behavior, indicating that there is a lot of "wiggle-room" in the hydrophobic core.

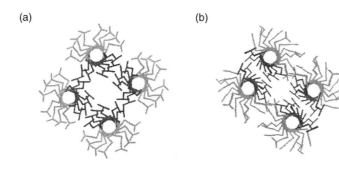

(a) GLEELLEELEELLEEG

(b) EELLKQALQQAQQLLQQAQELAKK

Fig. 3.11. (a) An assembly of helices with an H–P-like sequence (sequence below in one-letter code) forms a bundle as viewed along the helical axes. The hydrophobic groups (black) form a surface along the edge of each helix. These buried side chains are disordered and the bundle does not resemble the native state of a protein (Raleigh and DeGrado, (1992). (b) A more complicated sequence with strategic alternation of leucines and alanines gives a roughly flat hydrophobic surface. The complementary surfaces pack very well so that this bundle is structured and mimics the native state of a protein (Schafmeister *et al.*, 1997). Abbreviations: G = gly, L = leu, E = glu, A = ala, K = lys, Q = gln.

Mixing up the leucines with other amino acids which are also hydrophobic but have different sizes freezes the four-helix bundle into a highly structured state. Alternate leucines and alanines form a flat interior surface that packs more tightly. The lower side chain mobility of this structured bundle comes closer to mimicking the native state of a protein (Fig. 3.11b). According to this picture, a protein interior is like a jigsaw puzzle in which the pieces interlock to form a tight structure with little or no room for motion. By contrast, the interior of the bundle in Fig. 3.11a is more like an oil droplet.

The oil droplet and the jigsaw puzzle represent two extreme views of a protein interior (Kallenbach, 2001). H–P sequences are good at forming oil-droplet interiors but not jigsaw-puzzle interiors. How closely do the interiors of proteins resemble oil droplets versus jigsaw puzzles? Protein crystal structures generally show a high density in the interior, suggesting that packing is tight. A protein with a jigsaw-puzzle interior should be far less tolerant of mutations than a protein with an oil-droplet interior. Large scale mutagenesis studies have been undertaken, replacing core interior hydrophobic residues with other hydrophobic residues and testing catalytic activity and stability. Two of these studies gave quite different answers, suggesting that protein interiors do not adhere to a unique model. The ribonuclease barnase was found to tolerate hydrophobic substitutions at many sites (Axe *et al.*, 1996). By contrast, in triosephosphate isomerase mutation of most residues in the protein core substantially reduced catalytic activity (Silverman *et al.*, 2001). For triosephosphate isomerase it was estimated that only one in 10^{10} variants of the core sequence preserves activity. By contrast, one fourth of the variants of the barnase core sequence are functional. These two examples indicate that proteins vary in how tightly their amino acid side chains pack in the folded structure. Both oil droplets and jigsaw puzzles can be found in nature.

3.17 | Cooperativity in protein folding

One particularly significant failure in the theoretical performance of H–P sequences is that they can be disordered in small increments by small increases in energy. They melt gradually, not cooperatively. This is counter to the cooperative nature of protein folding. In Section 1.6 protein unfolding was seen to be a global transition. Adding energy (actually, increasing the temperature) induces a cooperative transition from a highly structured compact state to a random mixture of many looser states. There are relatively few stable intermediate states. The fact that one can slightly expand a good-folding H–P chain to realize small increases in energy and entropy means that the H–P chain misses an essential aspect of protein behavior.

Simply adding letters to the alphabet does not overcome this problem (Kolinski *et al.*, 1996; Kaya and Chan, 2000; Pokarowski *et al.*, 2003). The general problem is that treating the energy as a simple sum over pairwise contacts does not produce a molecule with a two-state unfolding transition, even when there are many different types of contacts. In order to generate cooperative folding, models must have some form of multibody (nonpairwise) interaction. The simplest form of interaction that satisfies this requirement is the cooperative interaction introduced in helix–coil theory (Section 3.12). The parameter σ of this theory reflected the dependence of contact formation (in this case a hydrogen bond) on the state of a neighbor. If the potential energy function includes this kind of cooperativity term (which was intuitively very reasonable in the context of helix–coil theory), cooperative folding can be produced.

Modeling studies of cooperative folding took as their benchmark the calorimetric criterion (Section 1.6) that the van't Hoff enthalpy of the transition equals the calorimetric enthalpy (Kaya and Chan, 2000; Pokarowski *et al.*, 2003). A distribution of conformations could be evaluated for any temperature by using a computer to generate large numbers of configurations and calculating the energy for each. These distributions had fairly large gaps between the native and denatured states, and increasing the temperature shifted the distribution from one group of configurations to the other. From the slope of this shift versus temperature one obtains a theoretical van't Hoff enthalpy. From the difference between the mean energies of these two populations one obtains the calorimetric enthalpy. The models with cooperative terms in the potential energy function satisfied the two-state condition: the calorimetric enthalpy equaled the van't Hoff enthalpy. These models thus mimicked the cooperative nature of protein folding.

In summary, simple pairwise contact energies for two residue types, hydrophobic and polar, are sufficient to account for the compactness of proteins in their native state. However, additional packing interactions are necessary to define a unique native state of a protein. Cooperative transitions between folded and unfolded states cannot be explained by pairwise contacts between residues. This property requires a multibody term in the potential energy function to reflect the dependence of contact energies on the states of neighboring residues.

Problems for Chapter 3

1. Show that Eqs. (3.4a) and (3.4b) are consistent with Eq. (3.1).
2. Calculate the average energy of butane as a function of temperature. Evaluate this quantity at $T = 0$ K, $T = 298$ K, and in the limit $T \to \infty$.
3. Calculate the probability of finding a chain of *n*-pentane with no *gauche* bonds, and with only one *gauche* bond.
4. Use Eq. (3.19) to determine an explicit value for l_{eff} of polyethylene assuming that it is a freely rotating chain.

5. Use Eq. (3.20) to determine l_{eff} for polyethylene, using the probabilities for different dihedral angles derived from *n*-butane.

6. Use the data in Table 3.1 to calculate the rms end-to-end distance of poly-L-alanine and DNA. Use lengths of 100 and 1000 (residues and base pairs, respectively) for both.

7. Calculate the concentration of one end of a polymer at the other end for poly-L-alanine and DNA using the lengths from Problem 6.

8. Use Eq. (3.23) to calculate l_p for a freely rotating chain.

9. A protein called "modulin" has a melting temperature of 350 K and a $\Delta H°$ of 50 kcal mole^{-1}. It is treated with a chemical reagent that selectively modifies valine residues so that their rotational freedom is reduced. Prior to modification the valine side chain has three rotational configurations, and after treatment there is only one. Modulin has five valines, all in the interior. Estimate the changes caused by the chemical modification in entropy, enthalpy, and temperature for thermal unfolding.

10. Show how Eq. (3.52) reduces to a simple form when $\sigma = 1$.

11. Evaluate Eq. (3.52) for $s = 1$. What is the significance of the absence of σ from the answer?

Chapter 4

Molecular associations

The preceding chapters treated molecules as isolated entities. Now we will look at how molecules interact with one another. In biological systems molecules are continually binding together and coming apart. Molecular associations are the first step in most forms of biological signaling, as well as in enzyme catalysis. Hormones, neurotransmitters, second messengers, and metabolites bind to proteins to regulate their activity. Pharmacology is rooted in the molecular association between drugs and receptors. On a larger scale, associations between macromolecules direct the assembly of organelles. Here, we will examine the thermodynamic and statistical mechanical principals underlying chemical association processes. These concepts will serve as a useful prelude to the theory of allosteric interactions in the following chapter.

There are two guiding principles in understanding association processes in molecular biology. (1) The forces that control associations are usually noncovalent. These kinds of forces were covered in Chapter 2. Here we will discuss how noncovalent interactions such as electrostatic forces, hydrogen bonds, and hydrophobic interactions combine in various ways to stabilize molecular complexes. (2) Associations are stereospecific, and depend on a precise spatial arrangement of the interacting groups. A binding site within a protein is viewed as a lock, and a ligand that fits into this binding site is a key. As a result biological associations are highly specific; molecules can recognize one another and distinguish subtle variations in structure.

4.1 | Association equilibrium in solution

If two molecules A and B in solution associate to form a complex C, then we have a chemical equilibrium, as shown in Scheme (4A):

$$A + B \rightleftarrows C \tag{4A}$$

A will usually be taken as a large protein molecule, and B will be taken as a small ligand molecule. Alternatively, both A and B can be macromolecules. For either case, the complex of the two is denoted as C. For now, this doesn't matter; the formulation is general. At equilibrium, the concentrations of these three molecules will be related by the basic equilibrium condition

$$\frac{[C]}{[A][B]} = \frac{1}{K_{dis}} \tag{4.1}$$

where K_{dis} is the dissociation constant, and has units of concentration. The reciprocal of the dissociation constant is also often used, and is referred to as an association constant. A higher association constant means tighter binding so this quantity can be taken as a measure of ligand affinity or the strength of the interaction. The advantage of using the dissociation constant is that with units of concentration, it can more readily be used to assess concentrations. Overall, binding will be favored when the concentrations are above K_{dis}.

With B as a small ligand and A as a protein with a binding site for B, we can use the equilibrium relation above to calculate the fraction of binding sites occupied as a function of ligand concentration. The total amount of protein is a constant, so that the concentrations of free and bound protein must add up to the total $[P_{tot}]$

$$[P_{tot}] = [P_f] + [P_b] \tag{4.2}$$

Here P_f, the free protein, takes the place of A in Scheme (4A), and P_b, the protein with a ligand bound to it, takes the place of C. We can now rewrite Eq. (4.1) in terms of $[P_b]$, replacing $[P_f]$ by $[P_{tot}] - [P_b]$

$$\frac{[P_b]}{[L]([P_{tot}] - [P_b])} = \frac{1}{K_{dis}} \tag{4.3}$$

where L denotes ligand, and takes the place of B. A simple rearrangement gives us f, the fraction of protein binding sites occupied by ligand, as

$$f = \frac{[P_b]}{[P_{tot}]} = \frac{[L]}{[L] + K_{dis}} \tag{4.4}$$

Plotting this fraction versus [L] shows the basic saturation behavior characteristic of a protein–ligand binding interaction (Fig. 4.1).

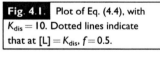 **Fig. 4.1.** Plot of Eq. (4.4), with $K_{dis} = 10$. Dotted lines indicate that at $[L] = K_{dis}$, $f = 0.5$.

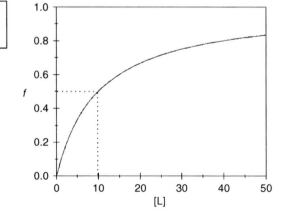

Figure 4.1 shows that f increases smoothly with increasing [L], starting from 0 and approaching 1 asymptotically at high [L]. The point where half of the sites are occupied is $[L] = K_{dis}$. The behavior illustrated in Fig. 4.1 is very common. It is seen whenever a protein has a single binding site, and some measurable form of activity depends on the fraction of binding sites occupied. A familiar example is the Michaelis–Menten equation for enzyme activity (see Chapter 10).

4.2 | Cooperativity

There are many situations where ligands bind to proteins in a cooperative fashion, and when this happens there is a clear departure from the behavior depicted in Fig. 4.1. Cooperativity is seen when a macromolecule has more than one ligand binding site, and the binding sites influence one another. There are many ways to picture interactions that make binding cooperative. A few examples will be examined here.

4.2.1 Concerted binding
Let a protein have n binding sites. We assume that some interaction through the whole protein forces all the binding sites to be either simultaneously occupied or simultaneously empty. This is called concerted binding, because the binding sites all fill up at once rather than in sequence. We represent this process with Scheme (4B) as follows

$$P_0 + nL \rightleftarrows P_n \tag{4B}$$

where P_0 denotes a protein with empty binding sites, and P_n denotes a protein with all sites occupied. There are no intermediate partially occupied states. This situation is reminiscent of global transitions in a multisubunit protein where all of the subunits undergo a transition concomitantly (Section 1.6).

The absolute exclusion of intermediate states follows from the assumption of perfect cooperativity. For ligand binding this assumption is especially artificial because there is no reason to believe that the physical forces holding different ligands in their binding sites should be so strongly interdependent. We will have to wait for the treatment of allosteric processes in Chapter 5 to see a reasonable model for cooperative binding, and then we will see that the cooperativity of binding is less than perfect. The present example may not be very realistic, but it is instructive, and very widely used.

Scheme (4B) specifies an equilibrium condition for the concentrations

$$\frac{[P_n]}{[L]^n [P_0]} = \frac{1}{K_n} \tag{4.5}$$

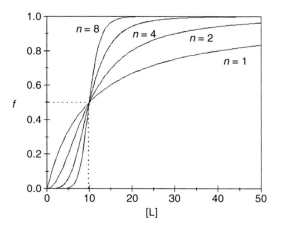

Fig. 4.2. Plots of Eq. (4.7), with n as indicated and with $K_n = 10^n$. (This assures that at $[A]^n = K_n$, $f = 0.5$, as indicated by the dotted lines.) The curve for $n = 1$ is identical to that plotted in Fig. 4.1.

where K_n is a compound dissociation constant reflecting the concerted binding of all n ligands. From here we proceed as before. Just as with Eq. (4.2), $[P_0]$ and $[P_n]$ must add up to $[P_{tot}]$. So we can eliminate P_0.

$$\frac{[P_n]}{[L]^n ([P_{tot}] - [P_n])} = \frac{1}{K_n} \tag{4.6}$$

Rearranging as in the derivation of Eq. (4.4), we obtain

$$f = \frac{[P_n]}{[P_{tot}]} = \frac{[L]^n}{[L]^n + K_n} \tag{4.7}$$

This is often referred to as the Hill equation. A plot of Eq. (4.7) has a sigmoidal character when $n > 1$ (Fig. 4.2). This means that the curve is steeper in the middle than at the start. Cooperativity makes the binding curve rise more steeply because of the $[L]^n$ term in the numerator of the right-hand side of Eq. (4.7). Recall that the same increase in steepness was observed when global transitions became more cooperative (Section 1.4).

Equation 4.7 can be rearranged into the following form

$$\ln\left(\frac{1-f}{f}\right) = n \ln ([L]) - \ln (K_n) \tag{4.8}$$

This form suggests that a plot of $\ln((1-f)/f)$ versus $\ln([L])$ should be linear, with a slope of n. This kind of plot, often referred to as a Hill plot, is widely used to evaluate cooperativity. The value of n obtained as the slope of a Hill plot is referred to as the Hill coefficient. Plots of f versus L can also be referred to as Hill plots. It is common to fit Eq. (4.7) directly to such a plot of f versus $[L]$ in order to estimate n.

One of the most important examples of a cooperative association is O_2 binding to hemoglobin. Each of the four subunits of hemoglobin binds O_2 at an iron–heme binding site. Plots of binding site occupancy versus oxygen tension are sigmoidal; Hill plots give a value of $n = 2.7$. However, hemoglobin has four subunits, each with an oxygen binding site. The reason n is not 4 is that binding is not perfectly cooperative. There is some sequential binding so that partially saturated intermediates can exist. This makes binding less cooperative, so the observed Hill coefficient is less than the number of binding sites. The cooperativity of oxygen binding by hemoglobin is examined with allosteric theory in Section 5.8.

In general, one never finds a Hill coefficient exactly equal to the number of binding sites in a multisubunit protein because cooperativity cannot be absolute. The general rule is that an experimentally determined value of n, obtained from a Hill plot or a curve fit of Eq. (4.7), is a lower bound to the number of binding sites that the protein can have. If one sees $n = 2.5$ in an experiment, then there must be at least three binding sites, and there could in principle be many more. This point can be seen more clearly with sequential occupancy models, as will be discussed next.

4.2.2 Sequential binding

To see why Hill coefficients are less than the actual number of binding sites, we consider a more general situation in which binding to the different sites on a protein is not concerted, but sequential (Tanford, 1961; Cantor and Schimmel, 1980)

$$P_0 \underset{\longleftarrow}{\overset{L}{\rightleftharpoons}} P_1 \underset{\longleftarrow}{\overset{L}{\rightleftharpoons}} P_2 \underset{\longleftarrow}{\overset{L}{\rightleftharpoons}} \cdots \cdots \cdots P_{n-1} \underset{\longleftarrow}{\overset{L}{\rightleftharpoons}} P_n \qquad (4C)$$

For each binding step in Scheme (4C) we have an equilibrium condition of the form

$$\frac{[P_i]}{[L][P_{i-1}]} = \frac{1}{K_i} \qquad (4.9)$$

This forms a system of equations relating the concentrations of the various states of occupancy. There are $n[P_{tot}]$ total binding sites, and the fraction that are occupied is

$$f = \frac{[P_1] + 2[P_2] + 3[P_3] + \cdots + n[P_n]}{n[P_{tot}]} \qquad (4.10)$$

With the aid of Eq. (4.9) we can express the concentration of each partially occupied state, $[P_i]$, in terms of $[P_0]$ as

$$[P_i] = \frac{[L]^i [P_0]}{K_1 K_2 K_3 \ldots K_i} \qquad (4.11)$$

This result is obtained by deriving the result first for $[P_1]$, then $[P_2]$, etc. We use this result for each of the $[P_i]$ in Eq. (4.10), and then factor out $[P_0]$ to obtain

$$f = \frac{\frac{[L]}{K_1} + \frac{2[L]^2}{K_1 K_2} + \cdots + \frac{n[L]^n}{K_1 K_2 \ldots K_n}}{n \left(1 + \frac{[L]}{K_1} + \frac{[L]^2}{K_1 K_2} + \cdots + \frac{[L]^n}{K_1 K_2 \ldots K_n} \right)} \qquad (4.12)$$

Equation (4.12) cannot rise as steeply as $[L]^n$ because of all the terms in the numerator with lower exponents. Thus, intermediate

Fig. 4.3. A long molecule has many binding sites arranged in a row. A ligand can interact with another ligand in one of the adjacent sites, and this can stabilize the association.

states of receptor occupancy reduce the sigmoidicity of the saturation curve, and this in turn lowers the value of n that would be determined by fitting Eq. (4.7). With only n binding sites, there is no way that f can increase with a higher power than $[L]^n$. So as a general rule, an experimentally determined Hill coefficient must be less than or equal to the number of binding sites.

4.2.3 Nearest neighbor interactions

Another way to incorporate cooperativity into a binding process is to assume that binding depends on the state of occupancy of neighboring binding sites. This is depicted in Fig. 4.3 for a linear array of binding sites on a long molecule. This model provides a good description of nucleotide binding to a long single strand of DNA or RNA, and is also useful in understanding ligand binding to filamentous proteins.

To understand this form of cooperative binding we need two equilibrium constants, one for the binding of an isolated ligand to a site, K_0, and another for the binding of a ligand adjacent to an occupied site, K_1. At this point we can see a strong resemblance to the helix–coil transition (Section 3.12), and in fact the mathematical treatment of this binding problem is essentially the same. The helix continuation parameter corresponds to K_1, and the helix initiation parameter corresponds to K_0. Because of this close correspondence, the general observations regarding how the shape of the helix–coil transition varies with the different parameters carries over to the present example (Figs. 3.6–3.8). The binding curve will be more sigmoidal if the ligand binds more tightly in the presence of a neighboring ligand than in its absence. When this multisite molecule is short, then the steepness will be limited by its length. Lengthening the molecule will increase the number of sites, and increase the sigmoidicity of the binding curve. However, this effect reaches a limit determined by the strengths of the interaction between sites. Adding more binding sites by lengthening the molecule will only go so far to increase the sigmoidicity of binding. As with the helix–coil transition, the cooperativity can be viewed in terms of the average length of a contiguous stretch of occupied sites. Increasing the length of the molecule beyond this point will not make binding significantly more cooperative.

4.3 | Thermodynamics of associations

So far, the phenomenon of molecular association has been discussed without looking into the specific interactions that hold molecules together. Now we will try to relate dissociation constants to energetics, and evaluate the strengths of the noncovalent contacts that stabilize a complex. We take the molar free energy for a solute in an ideal solution

$$G = G^{\circ} + RT \ln [A] \tag{4.13}$$

where G° is the molar free energy of the standard state.

Equation 4.13 is now used to write down an expression for each of the three species. Subtracting the free energy of the reactants from the free energy of the products gives ΔG, which is taken as equal to zero when the associating molecules are at equilibrium. This provides a relation between ΔG° and K_{dis} (Eq. (4.5)).

$$\Delta G^{\circ} = -RT \ln \frac{[P_b]}{[L][P_f]} = RT \ln K_{\text{dis}} \tag{4.14}$$

This gives the molar free energy change for the standard state, nearly always taken as the state where all species are present at concentrations of 1 M. Thus, ΔG° is the free energy change that occurs when 1 M protein and 1 M ligand associate to form 1 M complex. We will soon see that the choice of standard state is critical in determining the magnitude of one of the contributions, the translational contribution, to the binding free energy.

This is the starting point in attempting to understand associations in terms of molecular interactions. We will now assess ΔG° in terms of the relevant physical forces.

4.4 | Contact formation

The first thing one thinks about when trying to visualize what drives an association process is energetically favorable contacts between the two molecules. A complex is viewed as being held together by contacts such as hydrogen bonds, salt bridges, and juxtaposed hydrophobic surfaces. If we know the structure of the complex then we might try to figure out which interactions hold it together, and then estimate the energies of those contacts with some of the formulas from Chapter 2. We could then add them all together to get the total contribution made by these contacts to ΔG° in Eq. (4.14). If the ith contact contributes an energy ε_i, then the contribution of contacts to the binding free energy could be approximated by the sum

$$\Delta G_{\text{ct}}{}^{\circ} = \sum_i \varepsilon_i \tag{4.15}$$

For hydrogen bonds, ε might be taken from Tables 2.1 and 2.2. If the contacts obeyed the Lennard–Jones potential (Eq. (2.13)), then the contact energy would be ε from that equation. For the attractive forces that stabilize a complex, these terms will be negative, and they will help drive the association. There might also be positive, repulsive terms if the fit between the two molecules is less than perfect, so that some nonattracting groups are forced together. If charges of the same sign are near one another, or if groups are pushed together against their steric repulsions, or if hydrophobic groups are in contact with hydrophilic groups, these contacts will oppose the association.

This sum (Eq. (4.15)) assumes that the contacts are independent. For example, it ignores possible distortions of the structures of the two molecules brought about by their association. There could be small stretching and twisting distortions of the bonds within each molecule, or larger scale conformational changes. The conformational changes will be discussed further below. In general, distortion effects must raise the free energy of the complex, and thus oppose the association process.

We could generalize Eq. (4.15) to express the energy with the aid of an all-atom force field between the two molecules (Section 2.14), so that all of the possible intermolecular interactions are included. However, no matter how carefully we try to evaluate this energy in terms of molecular forces, the value computed will fail to give the actual free energy of association. The reason for this is that there are other important effects that have to be considered. When two molecules associate there is a large decrease in entropy due to restricted motion of the two molecules with respect to one another. These contributions are extremely important, and so we will now evaluate them.

4.5 | Statistical mechanics of association

We take the partition function of a molecule, denoted by Q (Section 1.1), and break it down into various terms

$$Q = q_{ct}q_tq_rq_vq_{cf}q_s \tag{4.16}$$

where q_{ct} denotes the contributions of intermolecular contacts just discussed, q_t denotes the translational contribution, q_r denotes the rotational contribution, q_v denotes the vibrational contribution, q_{cf} denotes the contribution of conformational flexibility, and q_s denotes solvation effects. Electronic states are also discussed in standard texts, but they do not matter much in association processes, so they will be ignored here.

Equation (4.16) is our starting point. It helps that this partition function is a simple product of terms representing the various contributions. The partition function factors in this way because of an assumption that each of the different forms of energy is independent and additive. Because the Boltzmann weights are exponentials of energy, additivity of energies allows the Boltzmann terms to be factored into the product on the right-hand side of Eq. (4.16).

One first takes the ensemble partition function as the product over all N molecules, Q^N, divided by a combinatoric or "counting" factor, $N!$, due to the indistinguishibility of states resulting from interchanging identical molecules. The free energy is then given in terms of the partition function for an ensemble formed from N molecules of the same species.[1] This is written as

$$G = -kT \ln (Q^N/N!) \qquad (4.17)$$

To treat the association of molecules A and B into the complex C, we extend the partition function to a mixture of these three species. We assume that the molecules in a solution do not interact with one another, except through the association process we are trying to understand. This independence makes the partition function of the mixture of A, B, and C a product of the partition functions of each of the three species. The free energies are therefore additive, so we can express the free energy of the mixture of A, B, and C by adding expressions of the form in Eq. (4.17) as follows

$$G = -kT \ln \left(\frac{Q_A^{N_A}}{N_A!} \right) - kT \ln \left(\frac{Q_B^{N_B}}{N_B!} \right) - kT \ln \left(\frac{Q_C^{N_C}}{N_C!} \right)$$

$$= -kT \ln \left(\frac{Q_A^{N_A} Q_B^{N_B} Q_C^{N_C}}{N_A! N_B! N_C!} \right) \qquad (4.18)$$

If the association process has reached equilibrium, then the free energy is at its minimum. Converting an infinitesimal amount of A and B into C will then not change the value of the free energy. We apply this principle by combining just one molecule of A and one molecule of B into one molecule of C and calculating G. Therefore N_A and N_B will each decrease by one and N_C will increase by one. Since this small conversion leaves the free energy unchanged, we have

$$kT \ln \left(\frac{Q_A^{N_A} Q_B^{N_B} Q_C^{N_C}}{N_A! N_B! N_C!} \right) = kT \ln \left(\frac{Q_A^{N_A-1} Q_B^{N_B-1} Q_C^{N_C+1}}{(N_A - 1)!(N_B - 1)!(N_C + 1)!} \right) \qquad (4.19)$$

Equating the terms in parentheses we can easily derive the following equilibrium condition.[2]

$$\frac{Q_A Q_B}{Q_C} = \frac{N_A N_B}{N_C} \qquad (4.20)$$

Since concentration is the number of moles per unit volume, we can convert the number of molecules to concentration by multiplying and dividing by volume, V. With concentrations, we can then obtain an expression for the dissociation constant, K_{dis}, of Eqs. (4.1) and (4.14)

---

[1] See texts such as MacQuarrie (1976), Hill (1960) or Moore (1972) for a derivation of the partition function of a system of N molecules, and for further discussion of the free energy contributions of translation, rotation, and vibration.

[2] The method used in standard texts is to differentiate with respect to N_A, N_B, and N_C, set the derivative equal to zero, and solve for the Ns using Stirling's formula to approximate the factorial.

$$\frac{Q_C V\tilde{A}}{Q_B Q_A} = \frac{[C]}{[A][B]} = \frac{1}{K_{dis}} \tag{4.21}$$

where $\tilde{A}$ is Avogadro's number, which entered because [A] in moles per unit volume is equal to $N_A/V\tilde{A}$.

Now we can go on to dissect the various contributions made by contacts, translation, rotation, etc. Replacing each Q with the appropriate product from Eq. (4.16) gives an expression for the dissociation constant in terms of factors reflecting each contribution

$$\frac{1}{K_{dis}} = \left(\frac{q_{Cct}}{q_{Act}q_{Bct}}\right) \left(\frac{q_{Ct}V\tilde{A}}{q_{At}q_{Bt}}\right) \left(\frac{q_{Cr}}{q_{Ar}q_{Br}}\right) \left(\frac{q_{Cv}}{q_{Av}q_{Bv}}\right) \left(\frac{q_{Ccf}}{q_{Acf}q_{Bcf}}\right) \tag{4.22}$$

The reason for placing $V\tilde{A}$ together with the translational term will become clear immediately in the discussion of translational free energy.

If we take the logarithm of Eq. (4.22) and multiply by RT, we can return to an expression for $\Delta G°$. We can now view $\Delta G°$ as a sum of contributions reflecting each of the factors in Eq. (4.22). It was noted at the beginning of this section that the additivity of each form of energy in a single molecule allows us to factor the partition function into terms as in Eq. (4.16). So now when we have reached the level of molar free energy by taking $RT \ln K_{dis}$, we see that we have recovered this additivity property. Equation (4.22) allows us to decompose the association free energy into its separate parts.

In evaluating the translational, rotational, and vibrational contributions, it helps to think in terms of numbers of degrees of freedom. All molecules have three translational degrees of freedom, corresponding to movement in the x, y, and z directions. Except for certain symmetrical molecules, they also have three rotational degrees of freedom, corresponding to rotations about the x, y, and z axes. Two molecules have a total of 12 translational and rotational degrees of freedom and when they associate, the resulting complex has only three of each for a total of six. Six degrees of freedom appear to be lost when the two molecules combine. Actually, they are converted in the complex to six new internal vibrational and librational (rocking-rotational vibrations) degrees of freedom (to be analyzed in Section 4.8). The important point is that the total number of degrees of freedom stays the same. The complete picture emerges by sequentially examining translation, rotation, and vibration, and then putting them all together.

4.6 | Translational free energy

The translational contribution to the free energy arises from the freedom that molecules have to roam within a given volume, V. This freedom can be quantified as entropy by counting the number of positional states. We divide the space available to a molecule into many small cells or cubicles. The molecule will have the same potential energy in any one of these cells (unless a force is applied), so the partition function will just be a count of the total number of cells. This makes the translational partition function proportional to the volume available to the molecule. The volume of one cell or cubicle is derived from quantum mechanics using the wave function of a particle in a box. The result is $(h^2/(2\pi mkT))^{3/2}$ (McQuarrie, 1976; Hill, 1960; Moore, 1972), where m is mass and h is Planck's constant. This gives us the translational partition function for a molecule in a volume V, as

$$q_t = V\left(\frac{2\pi mkT}{h^2}\right)^{3/2} = \frac{V}{\Lambda^3} \tag{4.23}$$

Denoting the volume of the quantum mechanical cell as Λ^3 gives the result a simple form, which makes clear how q_t is the number of such cells in a given volume. This result is generally quoted for an ideal gas, but it can be extended to solutions (Steinberg and Scheraga, 1963; Tidor and Karplus, 1994; Gilson et al., 1997). The variable Λ has units of length, and is called the thermal de Broglie wavelength. Because Λ^3 is a volume, the units in q_t cancel to give a dimensionless count of the number of translational states. Applying this expression for q_t to each species, A, B, and C, we can calculate the translational factor in Eq. (4.22) as

$$\frac{q_{Ct}V\tilde{A}}{q_{At}q_{Bt}} = \frac{\Lambda_A^3\Lambda_B^3\tilde{A}}{\Lambda_C^3} \tag{4.24}$$

This represents the translational contribution to the association free energy.

The only molecule-specific property that appears in this expression is the mass, so we can make a reasonably general estimate of the translational contribution. If we consider a protein, A, binding a small ligand, B, then Λ_A and Λ_C will be nearly the same and cancel, leaving $\tilde{A}\Lambda_B^3$ as a good approximation for Eq. (4.24). For a molecular weight of 100, and at a temperature of 300 K, $\Lambda_B = 10^{-9}$ cm. The standard state translational free energy is simply $-RT \ln (1\,\text{M})$, reflecting the choice of 1 molar for all of the relevant species. Converting Λ_B^3 to liters then lets us calculate the translation contribution to $\Delta G°$ as follows

$$\Delta G_t° = -RT \ln (\tilde{A}\Lambda_B^3/1\,\text{M}) = 0.588 \ln (6.02 \times 10^{-7})$$
$$= 8.4 \,\text{kcal mole}^{-1} \tag{4.25}$$

This is a fairly general result, with the only ligand-specific part being the mass that appears in Λ_B. A ten-fold increase in the mass of molecule B increases the free energy in Eq. (4.25) by 2 kcal mole^{-1}. Thus, the result is relatively insensitive to the specifics of the ligand. Furthermore, when we take the vibrational contribution into account, we will see that this mass dependent part of the translational free energy cancels out, leaving a binding free energy with essentially no dependence on mass (Gilson et al., 1997).

To get an intuitive feel for the translational contribution, we can think of binding as reducing the volume available to a molecule from 1 liter per $\tilde{A}$ molecules in a $1\,M$ solution (1.7×10^{-24} liter molecule^{-1}) to Λ^3 per one molecule (10^{-30} liter molecule^{-1}). The translational partition function is proportional to the volume per molecule (Eq. (4.23)), so the change in translational free energy upon association is the logarithm of the ratio of these two volumes.

It is important to emphasize that ΔG_t° depends on the choice of standard state. In the above derivation, the standard state was $1\,M$, and a different choice would change ΔG_t°. In fact, the only part of the free energy that depends on concentration is the translational contribution. Thus, an association equilibrium should be thought of as a balance between the translational part and all the other parts. That is perhaps the most fundamental explanation for why changing the concentration changes the degree of association.

The term "cratic" is widely used for the translational contribution to the free energy of an association process (Gurney, 1953). This is a useful idea because it distinguishes the translational part, which depends on standard state rather than molecular properties, from intrinsic contributions, which depend intimately on molecular properties and on the detailed nature of the interactions. Early efforts to estimate the cratic contribution used the volume per molecule of water at its concentration of $55\,M$ to calculate the entropy of the bound state. It is now generally recognized that this seriously underestimates the degree to which association restricts translation.

The translational contribution is mostly, but not entirely, entropy. To see this, take the derivative of Eq. (4.25) with respect to temperature. (Note that Λ depends on temperature so the derivative is not simply $R \ln \tilde{A}\Lambda_B^3$.)

$$\Delta S_t^{\circ} = R \ln \tilde{A}\Lambda_B^3 - \tfrac{3}{2}R \qquad (4.26)$$

Subtracting $T\Delta S_t^{\circ}$ from ΔG_t° in Eq. (4.25) gives

$$\Delta H_t^{\circ} = \tfrac{3}{2}RT \qquad (4.27)$$

This result has a natural interpretation in terms of the number of degrees of freedom. Association removes three translational degrees of freedom, each of which has $1/2\,RT$ of kinetic energy per mole. The value of ΔH_t° is only about $0.9\,kcal\ mole^{-1}$ at room temperature, and it is generally recovered as kinetic energy in new vibrational degrees of freedom within the complex (to be discussed in Section 4.8). The only situation where this fails to occur is when the association is so tight that quantum mechanical effects come into play to reduce the kinetic energy of a harmonic oscillator below $1/2\,RT$. Since the enthalpic contribution to the translational free energy is small, and is conserved during association, we can think of ΔG_t° as essentially purely entropic in origin.

An alternative approach to estimating the translational contribution can be developed by looking at the change in translational entropy as the volume per molecule in the standard state (1 liter/Å) divided by a volume defined in terms of the restricted movement that a bound molecule experiences (Finkelstein and Janin, 1989; Gilson et al., 1997). This restricted volume takes the form $(\delta x)^3$, where δx is defined as the freedom within the complex permitted by vibrations. The change in translational free energy then reflects the effective free volume available to the bound ligand within the confines of its binding site.

$$\Delta G_t^\circ = -RT \ln((\delta x)^3 \tilde{A}/1 \, \text{M}))$$ (4.28)

This equation is the same as Eq. (4.25), but with δx replacing Λ. The motivation for doing it this way is that within the complex, translational freedom is not really as restricted as implied by the very short length, Λ. However, the relative motion of the two molecules within the complex, reflected in the value of δx, is due to the new vibrational modes, and this contribution will be estimated separately below. The advantage of keeping them separate is that the translational free energy change is virtually independent of the specific properties of the two interacting molecules, whereas the vibrational contribution is not.

4.7 | Rotational free energy

A molecule in solution can rotate about its axes; typical asymmetrical biological molecules have three orthogonal axes. The number of rotational states for each axis might be visualized as proportional to the circumference of a circle of molecular radius, r; for example $\propto 2\pi r$. The actual number can be calculated explicitly for a rigid molecule using quantum mechanics to obtain the rotational partition function. This follows the same strategy used for the translational contribution, and gives an expression for the rotational factor in Eq. (4.16) (McQuarrie, 1976; Hill, 1960; Moore, 1972)

$$q_r = \pi^{1/2}\left(\frac{8\pi I_x kT}{h^2}\right)^{1/2}\left(\frac{8\pi I_y kT}{h^2}\right)^{1/2}\left(\frac{8\pi I_z kT}{h^2}\right)^{1/2}$$ (4.29)

where I_x, I_y, and I_z represent the principal moments of inertia for rotation about the three orthogonal axes. Each has the form of a sum $\Sigma(mr^2)$ over all atoms, where m is the mass of an atom and r is the distance to the axis of rotation. (Symmetry factors included in most standard texts are not considered here because they are largely irrelevant in biological associations.)

We can use this expression to calculate the rotational factor $q_{Cr}/q_{Ar}q_{Br}$ in Eq. (4.22). Again, if we take molecule A as the protein, B as the ligand, and C as the complex, the rotational factors for A and C will nearly cancel because $I_{Ax} \sim I_{Cx}$, etc. So we are left with $1/q_{Br}$. For a reasonable value of the moment inertia of a small molecule $(2 \times 10^{-37}$ g.cm^2), we obtain $q_{Br} = 10^6$ from Eq. (4.29). Thus, the loss of three rotational degrees of freedom makes the following contribution to the free energy of association

$$\Delta G_r{}^\circ = -RT \ln \left(1/q_{Br} \right) = 8.2 \, \text{kcal mole}^{-1} \qquad (4.30)$$

This is similar in magnitude to the translational contribution (Eq. (4.25)).

As with the translational contribution, the entropy can be obtained by differentiation with respect to temperature. Most of $\Delta G_r{}^\circ$ is entropy, but as with the translational contribution, at room temperature one obtains $\Delta H_r{}^\circ = 3/2RT \sim 0.9 \, \text{kcal mole}^{-1}$, reflecting rotational kinetic energy around the three principal axes. Again, as with translational kinetic energy, the lost rotational kinetic energy will reappear in the new vibrational modes of the complex. So the rotational contribution to ΔG°, like the translational contribution, can be viewed as purely entropic.

This calculation was based on the assumption that the two molecules do not rotate relative to one another within the complex. If they were attached by only one contact, then there would be an internal rotation that would have to be considered, and this would reduce the change in rotational free energy. However, most non-covalent complexes encountered in biology are stabilized by multiple contacts, and that will prevent free rotation.

4.8 | Vibrational free energy

The translational and rotational free energy contributions just evaluated were necessarily positive. Because association removes these modes, the resulting loss of freedom opposes association. Since these modes are converted into vibrational modes, we would expect vibrations to make a compensatory negative contribution to the free energy of association. The complex formed by the association of two molecules has six new vibrational degrees of freedom that were not present when the molecules were separate. These new modes of vibration arise from stretching and distorting the contacts that hold the molecules together. The fact that there are six modes is fundamental to a binary association and is completely independent of the actual number of contacts (as long as there are more than one). We will now evaluate the free energy associated with these vibrations.

The factor in Eq. (4.22) for the vibrational contribution was $q_{Cv}/q_{Av}q_{Bv}$. To evaluate this expression we exploit the fact that the many vibrations of a complex molecule can be decomposed into normal modes (Section 2.12). A normal mode is a collective vibration of many atoms that has the important mathematical character of behaving like an isolated harmonic oscillator. With the energy as a sum of separate terms representing each of the normal modes (Eq. (2.21)), the vibrational partition function becomes a product of

terms for each normal mode. A harmonic oscillator has the following partition function (McQuarrie, 1976; Hill, 1960; Moore, 1972)

$$q_{vi} = \frac{e^{-h\nu_i/2kT}}{1 - e^{-h\nu_i/kT}} \tag{4.31}$$

where $\nu_i = (1/2\pi)\sqrt{(\phi_i/\mu_i)}$ is the vibrational frequency, with μ as the
reduced mass, and ϕ as the force constant. The index i specifies
a particular normal mode. The complete vibrational partition function for a molecule is then the product of terms of the form of
Eq. (4.31) over all of the normal modes.

We first simplify Eq. (4.31) by noting that the noncovalent contacts
that hold the two molecules together are much weaker than covalent
bonds. This will lead to weak force constants compared to those for the
covalent bonds in each molecule. The vibrational frequencies for these
weak noncovalent contacts will be low, so $h\nu << kT$. We can thus
expand the exponential as $e^{-h\nu/kT} \sim 1 - h\nu/kT$. The numerator of
Eq. (4.31) is then one and the denominator is $h\nu/kT$, so we have

$$q_{vi} = \frac{kT}{h\nu_i} = \frac{kT}{2\pi h}\sqrt{\frac{\mu_i}{\phi_i}} \tag{4.32}$$

using the expression immediately following Eq. (4.31) for ν_i. This is
the classical limit for a harmonic oscillator in which the average
kinetic energy is $\frac{1}{2}RT$ per mole. Thus, the lost translational kinetic
energy is completely recovered in vibrations as long as the contacts
are weak enough to make the vibrations classical.

In the formation of a complex C from A and B, the normal modes
all change, because they are functions of the complete potential
energy function of a molecule. Thus, the vibrational contribution
depends on the details of the molecule much more strongly than the
translational and rotational contributions. Tidor and Karplus (1994)
carried out a detailed computer analysis of changes in vibrational
energy in the dimerization of insulin. They showed that in addition
to adding the six new vibrational modes, association alters many of
the preexisting internal modes of the monomers. Their analysis
yielded a vibrational entropy contribution of 23 cal K^{-1} mole^{-1} for
dimerization, to give -7.2 kcal mole^{-1} of free energy contributed
by vibrations.

Although a quantitative analysis of vibrations requires structural information and a detailed potential energy function, we can
get a good qualitative understanding of the situation by taking a
simplified approach. We assume that the internal vibrations of
A and B do not change much when the complex forms. This allows
us to approximate q_{Cv} as the product of $q_{Av}q_{Bv}q_v'$, where q_v' represents
the six new vibrational modes due to stretching and flexing the
contacts that hold the two molecules together. Now when we insert
this expression for q_{Cv} into the vibrational factor $q_{Cv}/q_{Av}q_{Bv}$ from
Eq. (4.22), we see that $q_{Av}q_{Bv}$ cancels out, leaving just q_v'. So the
problem is reduced to estimating this quantity.

We start by making reasonable estimates of ϕ and μ. By equipartitioning of energy (Section 12.4), we know that the mean potential energy of harmonic motion $\phi(\delta x)^2/2 = RT/2$ (in the classical limit). If we take δx from typical rms motions observed in proteins as ranging from 0.15 to 0.25 Å, we can then solve for $\phi = RT/\delta x^2$. We take the molecular weight of a typical ligand as 100 Da and use this value for μ. Inserting these values for μ and ϕ into Eq. (4.32) tells us that q_v for a single mode can range from 3.7 to 6.3. For six such modes we have

$$\Delta G_v{}^\circ = -RT \ln (q_v'^6) = -4.6 \text{ to } -6.5 \text{ kcal mole}^{-1} \qquad (4.33)$$

With this number in hand we can estimate the total entropic barrier to association due to restricted motion. Summing the translational and rotational terms from Eqs. (4.25) and (4.30), and subtracting Eq. (4.33) gives 10.0–12.2 kcal mole^{-1}. The favorable free energy from contact formation must exceed this value to drive an association process.

We can also make an estimate for the case of insulin dimerization just discussed (Tidor and Karplus, 1994). We use the molecular weight of insulin (5700 Da), and estimate the force constant as above for an rms variation in position of 0.15 Å. For a single mode $q_v = kT/h\nu = 29$ (Eq. (4.32)). This gives 2 kcal mole^{-1} per mode or 12 kcal mole^{-1} when all six new vibrational modes are counted. To compute the entropy we note again that a classical harmonic oscillator has a total internal energy per mole equal to RT ($\frac{1}{2} RT$ kinetic energy and $\frac{1}{2} RT$ potential energy). Subtracting $6RT$ leaves -8.4 kcal mole^{-1} as the entropic contribution, which is not far from the result of -7.2 kcal mole^{-1} obtained by Tidor and Karplus.

Now that we have worked out expressions for the translational, rotational, and vibrational contributions, we can combine these results to make the important point that the equilibrium constant for molecular association has essentially no dependence on mass. Extracting the mass from the translational factor ($\Lambda_B{}^3$ in Eq. (4.25)) gives $m_B{}^{-3/2}$. For the rotational factor the mass dependence is contained in the principal moments of inertia, so from Eq. (4.30) we extract $(I_{Bx}I_{By}I_{Bz})^{-1/2}$. The vibrational factor reduces approximately to q_v' for the modes of stretching and twisting the contacts between the molecules, and will be proportional to the product of the square roots of the six reduced masses of the new vibrational modes. If we consider a large protein binding to a small ligand, then the reduced mass for three of the vibrational modes will be close to the mass of the small molecule, giving a factor of $m_B{}^{3/2}$. The other three modes are librations, and their reduced masses will be the principle moments of inertia of the ligand, giving $(I_{Bx}I_{By}I_{Bz})^{1/2}$. When the translational, rotational, and vibrational factors are multiplied together, these mass dependent terms completely cancel out. The masses in the translational factor cancel the masses in the vibrational factor. Likewise, the masses in the rotational factor cancel the masses in the librational terms. The bottom line is that equilibrium constants for association will have very little dependence on mass (Gilson et al., 1997).

4.9 | Solvation effects

Since associations of biological interest occur in an aqueous solution, we have to consider the effect of water at the surfaces of the associating molecules. When two molecules associate, water will be displaced (unless these waters remain behind to help hold the complex together by forming bridges). The interaction between hydrophobic surfaces and water is unfavorable and the displacement of solvent is the basis for the attractive hydrophobic interaction (Section 2.8). For polar groups on a ligand or protein surface, water will be attracted rather than repelled. X-ray diffraction and NMR studies of protein-bound water support the general conclusion that a large amount of water is bound weakly and nonspecifically. However, most protein molecules have a few sites that can bind water rather tightly (Bryant, 1996). When association displaces these bound water molecules, there will be a significant free energy cost.

For a tightly bound water molecule the strengths of the contacts holding the water in place might be roughly similar to the contacts formed by the same site when binding to a ligand. For example, a hydroxyl group on a ligand might form a hydrogen bond to replace one formed with a bound water molecule. If the contact strengths are similar, then contact formation will not provide the impetus for a protein to bind a ligand rather than water. On the other hand, the loss of translational and rotational entropy will be quite different for the water and the ligand. If one large ligand displaces several water molecules, then there should be a large net increase in translational and rotational entropy. However, the tally of entropy for water is complicated. The high concentration of water lowers its molar entropy. The statistical mechanical formulas used above to calculate these contributions do not work well for water because water molecules are densely packed and highly correlated. This makes the entropy lower than that predicted above on the basis of free translation and rotation.

Instead of using the statistical mechanical formulas above, we take advantage of thermodynamic measurements (Dunitz, 1994). The entropy of liquid water is 16.7 cal mole^{-1} K^{-1} at 298 K. For water bound in crystals, entropy measurements give values of about 10 cal mol^{-1} K^{-1}. This value reflects the vibration of a water molecule within a crystal. The difference between these two measurements can be used as an upper bound to the entropy cost of binding water to the surface of a protein. With a difference of 6.7 cal mole^{-1} K^{-1} between the entropy of water in the liquid and bound states, we see that at 298 K the free energy cost for displacing the most tightly bound water molecule is about 2 kcal mole^{-1}, and of course it will be less for the more typical water molecules that are only loosely tethered to the surface of a protein.

This free energy cost of 2 kcal mole^{-1} for immobilizing water at the surface of a protein is much less than that for immobilizing a ligand at the standard-state concentration of 1 M. Recall that summing Eqs. (4.25), (4.30), and (4.33) gives a total of more than 10 kcal mole^{-1} that opposes association. Now, imagine that an association between two molecules is driven by contacts similar in strength to those that bind water. The much lower entropy cost of immobilizing water would strongly favor the binding of water. So a ligand that displaces only one water molecule would not be competitive in binding. (This point is actually trivial as it also reflects the greater availability of water in an aqueous solution.) However, if a ligand is large and displaces several water molecules, the tables start to turn. A ligand that displaces 10 water molecules might be expected to form as many contacts with the protein as those 10 displaced water molecules. If the contact energies balance out, ΔH will be zero. However, freeing 10 water molecules could increase the entropic free energy by $10 \times 2 = 20$ kcal mole^{-1}, which is roughly twice that lost by binding this one large ligand, so the ligand will bind tightly. Entropy thus favors the binding of large ligands.

4.10 | Configurational free energy

Associations can produce large conformational changes. When small molecules bind to proteins, structural changes in the small molecule are usually not considered because small molecules do not have much flexibility. On the other hand, proteins sometimes undergo large conformational changes when they bind small ligands, and this is the basis of allosteric theory (Chapter 5).

The conformational change induced by association is often viewed as a restriction in the number of configurations. Thus, a portion of a molecule may behave like a random coil (Chapter 3) prior to association, and be forced to assume a single conformation after association (Fig. 4.4). For these molecules association and folding are coupled. We can look at thermal unfolding of proteins for a comparison. The native state is taken as one unique configuration and the denatured state is taken as a random coil. An analysis of this model provided an estimate of the entropy change per residue during protein folding (Section 3.11). If we knew how many residues were ordered by the association process, we could use the per residue entropy change in thermal denaturation to estimate the expected configurational entropy change for ordering of a bound peptide.

Thermodynamic and structural data have been examined in an effort to test these ideas (Spolar and Record, 1994). The translational and rotational contributions were each estimated with the aid of formulas similar to Eqs. (4.25) and (4.30). A hydrophobic entropy term was estimated from the hydrocarbon areas coming into contact (Section 2.8). These terms were then subtracted from the

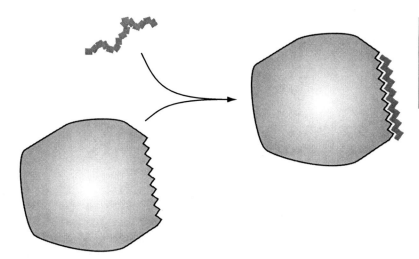

Fig. 4.4. The binding of a randomly coiled peptide to a protein binding site forces the peptide to assume a rigid conformation. This will result in an unfavorable entropy decrease.

measured ΔS° of association. In some association processes there was nothing left, and these were interpreted as rigid-body associations (subtilisin binding its inhibitor was an example). Structural data confirmed that there were hardly any conformational changes in these cases. However, in other associations there was a large excess entropy, which was attributed to a change in configurational freedom. This excess entropy was divided by a per residue value to estimate the number of residues that were folded during association.[3] These numbers compared well with estimates of the number of folded residues from crystal and NMR data.

Data from the enzyme ribonuclease A illustrate this analysis well. This protein can be cleaved into two fragments, one of which is a 20-amino acid peptide, the S-peptide. These pieces reassociate, and the entropy change can be measured. The translational, rotational, vibrational, and hydrophobic contributions were estimated and subtracted out, and the remainder was taken as configurational. Dividing this configurational entropy by the per residue value indicated that 14 of the 20 amino acids became ordered during association. Analysis of structural data identified 17 of the original 20 amino acids of the S-peptide as being ordered in the complex. Thus, this association gives an entropy change near that expected for conversion of the S-peptide from a random coil to a fixed conformation, as depicted in Fig. 4.4.

4.11 | Protein association in membranes – reduction of dimensionality

When a protein is in a membrane, its translation is restricted to two dimensions within a plane (Fig. 4.5). Membrane-bound proteins generally do not tumble in and out of the lipid hydrocarbon core.

---

[3] Spolar and Record (1994) used 5.6 cal K^{-1} for the entropy of folding per residue, based on their own analysis of thermal unfolding of a large number of proteins.

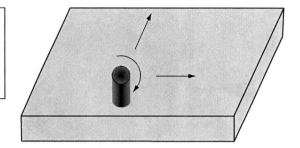

Fig. 4.5. The translational movement of a protein in a membrane is restricted to two dimensions. Rotation is restricted to one axis normal to the membrane plane.

Their orientation is fixed by membrane spanning segments, so they can only rotate around a single axis perpendicular to the membrane plane. Attaching molecules to a membrane thus constrains them, and this reduces the entropic barrier for subsequent associations.

Association of two proteins already bound to a membrane will remove only two translational degrees of freedom instead of three, so instead of 8.4 kcal mole^{-1} in Eq. (4.25) we have 5.6 kcal mole^{-1}. Association will remove one rotational degree of freedom, so instead of 8.2 kcal mole^{-1} in Eq. (4.30), we have 2.7 kcal mole^{-1}. The complex will have three new vibrational modes instead of six, so this contribution will range from -2.3 to -3.2 kcal mole^{-1} instead of the range given in Eq. (4.33). This adds up to an overall free energy of 5–6.1 kcal mole^{-1} that has to be overcome by energetically favorable contacts. This is considerably less than the entropic barriers to association estimated above for two molecules in solution. This makes it easier for two molecules to associate with one another when they are in a membrane. Sequestration of molecules into a membrane thus will facilitate their association. This effect has been referred to as dimensionality reduction. It can facilitate association processes, making membranes better places for some forms of signaling processes to occur.

Reduction of dimensionality not only makes association easier; it can also make it quite a bit faster. Breaking the process down into two steps can reduce how long it takes two freely diffusing molecules to encounter one another for the first time. First the molecules bind to a membrane. Then they search for one another within the membrane. Both the membrane-binding step and the two-dimensional diffusion-controlled association within the membrane can be faster than diffusion-controlled association in three dimensions. The kinetic implications are examined in Sections 8.8 and 8.9.

4.12 | Binding to membranes

The restricted translation and rotation of a molecule in a membrane will influence how that molecule partitions between the water and the membrane. This has to be considered when trying to understand the binding of proteins and peptides to a lipid bilayer. A membrane-

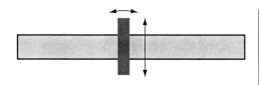

Fig. 4.6. A protein in a membrane will have one vibrational mode perpendicular to the membrane (straight double arrow), and two librational modes, one in the plane of the drawing (curved double arrow) and the other perpendicular to the plane of the drawing (not shown).

bound protein is free to move in two dimensions, compared to three in solution. So it loses one translational degree of freedom when it binds. The loss of one translational degree of freedom amounts to one third of the value in Eq. (4.25), or 2.8 kcal mole^{-1}. A membrane protein can rotate around one axis in a membrane, versus three in solution. The loss of two rotational degrees of freedom gives two thirds of the value in Eq. (4.30), or 5.5 kcal mole^{-1}. A total of 8.3 kcal mole^{-1} then opposes the binding of a molecule to a membrane.

Once bound, there will be some wobbling along the axis normal to the plane of the membrane, as well as angular wobbling of the axis (Fig. 4.6). These contributions to the binding energy have been estimated by Ben-Shaul *et al.* (1996), and they add up to about 3.7 kcal mole^{-1}. For a peptide with many hydrophobic residues the energy released by immersion in the lipid hydrocarbon chains can easily overwhelm this opposition and lead to a strong association with a membrane. Nevertheless, efforts to estimate the binding energy quantitatively must take the immobilization factor into account.

Problems for Chapter 4

1. Derive Eq. (4.11) from Eq. (4.9).

2. Take the limit of Eq. (4.12) for $K_n \gg K_1 \ldots K_{n-1}$.

3. Estimate the change in rotational free energy for binding of O_2 to a site in myoglobin or hemoglobin. Note that rotation around the axis of the O–O bond is the same for bound and free O_2.

4. Estimate the change in translational free energy for a binding process in which a ligand is confined within a 1 Å cubic cavity at 298 K.

5. Assume all the contacts with a ligand are hydrogen bonds with energies of -2.5 kcal mole^{-1}. Approximately how many hydrogen bonds are necessary to give a dissociation constant of 1 μM, or 1 nM? Assume a value of $T = 298$ K. Ignore water displacement.

6. Assume the contacts are hydrophobic. Approximately how large does the binding surface have to be to obtain a dissociation constant of 1 μM, or 1 nM? Assume that water displacement effects are already included in the parameter that relates free energy to the area of hydrophobic water contact (see Section 2.8).

7. Consider an ideal chain molecule with a known l_{eff} and with ends that can bind one another. A ligand is present at 1 mM, and this ligand is chemically equivalent to one of the polymer ends. How long does the chain have to be before the free ligand can displace half of the polymer ends (see Section 3.7)?

8. Consider an ideal chain with known l_{eff} and two ends that can bind to two distinct sites on a protein separated by a distance r. How does the binding free energy vary with chain length (see Section 3.8)?

9. Calculate the vibrational free energy at 298 K for a 16 Da atom held in place by a Lennard–Jones potential with $\varepsilon = 2\,\text{kcal mole}^{-1}$ and $r_0 = 1\,\text{Å}$ (hint: use the second derivative evaluated at the minimum for the force constant). Compare the vibrational free energy arising from this entropy with the depth of the potential energy well (ε).

Chapter 5

Allosteric interactions

The importance of molecular associations in biological signaling processes was mentioned in the preceding chapter. That chapter concentrated on the physical aspects of the association process and paid little attention to the signaling events that are initiated by ligand binding. This chapter will accept the binding event as given, and go on to look at what consequences this has on the biological function of a protein.

Powerful theories to explain this kind of signaling can be developed by combining the concepts of molecular associations from Chapter 4 with the concepts of global states and transitions from Chapter 1. In putting these two ideas together, a key point to remember is that both processes are governed primarily by the kinds of noncovalent forces covered in Chapter 2. As a result the energies for global transitions and binding events are often in the same range. This enables an association reaction to trigger a conformational transition in a protein, and this is what makes allosteric interactions possible. Here, we will develop this theory, known as allosteric theory, and illustrate its use with examples.

The word allosteric is quite popular in molecular biology. The word was introduced as a combination of the Greek words *allo* and *steric* to mean *other-site*. A classical usage in this sense is when a ligand binds to a regulatory site of an enzyme and alters the enzyme's effectiveness as a catalyst. The regulatory site of the enzyme can be altogether different from the catalytic site where the chemistry takes place. With no structural overlap whatsoever, the mechanism by which these sites influence each other is an especially intriguing question. The seminal paper by Monod *et al.* (1965) offered a theoretical solution to this problem, and in doing so extended the meaning of the word allosteric. Their theory has come to be known as allosteric theory, and they used the term allosteric transition to describe a special kind of a global transition that makes allosteric interactions possible.

5.1 | The allosteric transition

We start with a protein that has two functionally distinct global states. Monod *et al.* proposed that one of these states is "tense," using the letter T, and the other is "relaxed," using the letter R. These states interconvert, as in Scheme (5A), so there is an equilibrium

$$T \rightleftharpoons R \tag{5A}$$

If this protein were an enzyme then the T state might have low catalytic activity and the R state might have high catalytic activity. If the protein were an ion channel, then the T state could be closed and the R state open. If the protein were the repressor of a gene, then the T state might be able to bind to a specific sequence of DNA and block transcription; in the R state the protein would fall off the DNA and allow transcription to proceed. The idea is very general and has many applications.

The allosteric transition is a special case of a global transition. Both global states must be folded and functional for a transition to be allosteric. Thus, thermal unfolding is global, but it is not allosteric because there is no function associated with the unfolded state. It is essential, as will become clear below, that both states in an allosteric transition are able to bind ligand. Thus, an allosteric state corresponds to a well-folded, well-defined energy minimum on a protein's potential energy surface. The transition from one allosteric state to another starts as a brief excursion into higher energy configurations, but the intramolecular forces within the protein pull it back into one of the well-folded low-energy states in which the binding specificity and functional activity are precisely defined.

5.2 | The simplest case: one binding site and one allosteric transition

To illustrate the principles of allosteric theory, we will start with the simplest possible example: a protein with a single ligand binding site undergoes an allosteric transition between two allosteric states. The allosteric states have different affinities for ligand, so we have two separate binding equilibria (Schemes (5Bi) and (5Bii))

$$T_0 + L \rightleftharpoons T_1 \tag{5Bi}$$

$$R_0 + L \rightleftharpoons R_1 \tag{5Bii}$$

where L denotes the ligand. The subscripts of R and T denote the number of ligands bound, which in this case is either zero or one.

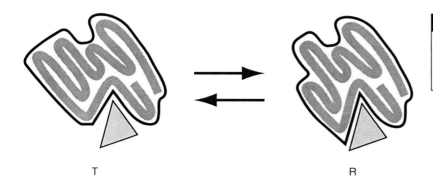

Fig. 5.1. An allosteric transition between two allosteric states, T and R. The stronger binding by the R state is indicated by a better fit between the binding site and ligand.

The change in ligand binding affinity accompanying the allosteric transition suggests a rearrangement of the polypeptide chain so that either the number of contacts with the ligand increases, or the existing contacts get stronger (Fig. 5.1). An essential element of this theory is that the transition is concerted, with the whole protein changing concomitantly. The protein cannot have R-like function and T-like binding. What this assumption means in terms of the molecular physics of a protein is that the potential energy minimum corresponding to each allosteric state specifies a structure of the entire protein. The R and T states correspond to well-separated minima in this multi-dimensional potential energy function, and each of these minima has its own characteristic binding energy and functional activity.

The protein structure does not change at the instant of ligand binding. The structure of T_0 is essentially the same as T_1 and the structure of R_0 is essentially the same as R_1. With the two binding equilibria and the allosteric transitions of the liganded and unliganded protein, we have the following complete scheme (Scheme (5C))

$$
\begin{array}{ccc}
 & K_T & \\
T_0 + L & \rightleftharpoons & T_1 \\
Y_0 \updownarrow & & \updownarrow Y_1 \\
R_0 + L & \rightleftharpoons & R_1 \\
 & K_R &
\end{array}
\qquad (5C)
$$

where K_T and K_R denote the two binding equilibrium constants, and Y_0 and Y_1 denote the equilibrium constants of the allosteric transition for the unliganded and liganded protein, respectively.

To derive the biological activity of the protein as a function of ligand concentration, we assume that it is proportional to the fraction of the total protein in the R state. Denoting biological activity as A, we have

$$
A = \frac{[\text{Total R}]}{[\text{Total Protein}]} = \frac{[R_0] + [R_1]}{[R_0] + [R_1] + [T_0] + [T_1]}
\qquad (5.1)
$$

We then write the four equilibrium conditions for Scheme (5C) above.

$$K_T = \frac{[T_1]}{[T_0][L]} \tag{5.2a}$$

$$K_R = \frac{[R_1]}{[R_0][L]} \tag{5.2b}$$

$$Y_0 = \frac{[R_0]}{[T_0]} \tag{5.2c}$$

$$Y_1 = \frac{[R_1]}{[T_1]} \tag{5.2d}$$

These equilibrium conditions can be used to express R_1, T_0, and T_1 in terms of R_0. For example, $[T_0] = [R_0]/Y_0$ (Eq. (5.2c)). These are then substituted into Eq. (5.1)

$$A = \frac{[R_0] + [L]K_R[R_0]}{[R_0] + [L]K_R[R_0] + [R_0]/Y_0 + [L]K_R[R_0]/Y_1} \tag{5.3}$$

Factoring out R_0 gives the response as a function of ligand concentration.

$$A = \frac{1 + [L]K_R}{1 + 1/Y_0 + [L]K_R(1 + 1/Y_1)} \tag{5.4}$$

This expression tells us that at $[L] = 0$, $A = Y_0/(1 + Y_0)$, which equals $[R_0]/([R_0] + [T_0])$. This is how it would be with only the transition between R_0 and T_0 to consider (Eq. (5.2c)). For very high $[L]$, Eq. (5.4) goes to $A = Y_1/(1 + Y_1)$, which is what we would get if we only considered the transition between R_1 and T_1 (Eq. (5.2d)).

Remember that the protein is designed to respond to the ligand, and have minimal activity in its absence. This means that we can assume Y_0 is very small, so the value of A at $[L] = 0$ will be close to zero; Y_1 is very large, so the value of A at high $[L]$ approaches one. These points lead to a useful simplification of Eq. (5.4). First, multiply the numerator and denominator by Y_0

$$A = \frac{Y_0 + [L]K_RY_0}{1 + Y_0 + [L]K_RY_0(1 + 1/Y_1)} \tag{5.5}$$

Since Y_0 is small compared to the other terms in both the numerator and denominator, we can ignore it (except where it is part of a product). Since Y_1 is large, $1/Y_1$ can also be ignored. This leads to a simple approximate expression

$$A \sim \frac{[L]K_RY_0}{1 + [L]K_RY_0} \tag{5.6}$$

Equation (5.6) tells us that the activity function is a rectangular hyperbola, and resembles a simple binding process. The behavior is indistinguishable from binding without an allosteric transition (Fig. (4.1) and Eq. (4.4)). This is an important result because one might expect two allosteric states with different binding affinities to give rise to qualitatively different behavior. An allosteric protein responds to ligand as though it has a single binding site with an affinity that is intermediate between that of the T and R states. When $[L] = 1/(K_R Y_0)$, $A = 1/2$, and the response is half maximal. Judging from the response function, an allosteric protein behaves as though it has a single binding site with an apparent affinity $K_{app} = K_R Y_0$.

5.3 | Binding and response

We call the quantity $K_R Y_0$ the "apparent" affinity because it is not a true binding equilibrium constant. The variable K_R is a true binding affinity, but it cannot be determined solely from a dose–response experiment. In fact, it can also not be determined from a plot of binding site occupancy as a function of $[L]$. To see this we denote the fraction of binding sites on the protein that are occupied as B.

$$B = \frac{[\text{Occupied sites}]}{[\text{Total Protein}]} = \frac{[R_1] + [T_1]}{[R_0] + [R_1] + [T_0] + [T_1]} \tag{5.7}$$

Once again, we use the equilibrium conditions (Eqs. (5.2a)–(5.2d)) to express everything in terms of R_0

$$B = \frac{[L]K_R[R_0] + [L]K_R[R_0]/Y_1}{[R_0] + [L]K_R[R_0] + [R_0]/Y_0 + [L]K_R[R_0]/Y_1} \tag{5.8}$$

We factor out R_0 and multiply through by Y_0 to give

$$B = \frac{[L]Y_0 K_R(1 + 1/Y_1)}{1 + Y_0 + [L]Y_0 K_R(1 + 1/Y_1)} \tag{5.9}$$

Again noting that Y_0 is small and Y_1 is large, we obtain the approximation

$$B \sim \frac{[L]K_R Y_0}{1 + [L]K_R Y_0} \tag{5.10}$$

This is identical to the dose–response equation (Eq. (5.6)). Thus, a binding measurement would tell us the same thing as a dose–response plot. It would give us the apparent affinity, $K_{app} = K_R Y_0$, without helping us separate out the true binding equilibrium constant.[1]

These results show how the conformational transition is invisible to many forms of experiments. An allosteric protein behaves like a simple ligand-binding protein without a conformational transition. The activation scheme (Scheme (5D))

$$P + L \rightleftharpoons P^* \tag{5D}$$

---

[1] One important case where the binding step can be distinguished from the allosteric transition is when the allosteric transition is relatively slow. If one has a technique for rapid measurement then one can capture the binding of ligand to the T state prior to the transition to the R state.

in which P is a protein and P* is the protein activated by the ligand, would fit a binding curve or dose–response plot just as well as the simplest allosteric model.

One of the most profound conclusions drawn from this analysis is that if a protein is manipulated in some way so that the conformational equilibrium is altered, the apparent binding affinity will also change. This is because the equilibrium constant, Y_0, is part of the apparent affinity. The allosteric transition is global, so a change in the structure far from the binding site can alter K_{app} through Y_0. This is what makes allosteric interactions possible. A binding curve can now change without any structural change in the binding site itself. This will be illustrated with examples below.

5.4 | Energy balance in the one-site model

Pushing a protein from the T state to the R state requires energy. The simple explanation is that this comes from ligand binding, but it is actually the difference between the energy of ligand binding to the two states that counts. We can get an appreciation of this distinction by looking carefully at the free energy changes associated with each step.

In Scheme (5C) there are two pathways by which we can get from T_0 to R_1. One pathway is via T_1; ligand binds first and then the allosteric transition occurs. The other pathway is via R_0; the allosteric transition occurs first and then the ligand binds. Regardless of which one of these pathways is more often traveled in reality, the fact that they both exist has important implications for the energetics. The reason for this is that if each of the steps is reversible, then the net free energy change for either pathway will be the same. This allows us to write down an energy balance equation

$$RT \ln K_T + RT \ln Y_1 = RT \ln Y_0 + RT \ln K_R \tag{5.11}$$

The left-hand side is for the pathway through T_1 and the right-hand side is for the pathway through R_0. This immediately leads to the following.[2]

$$K_T Y_1 = K_R Y_0 \tag{5.12}$$

Thus, $K_{app} = K_R Y_0$ could just as easily be expressed as $K_T Y_1$. This result can also be obtained from the condition of detailed balance, a concept from kinetic theory where forward and reverse rates must

---

[2] Equation (5.12) could also be derived without thinking explicitly about free energy. The four equilibrium expressions, Eqs. (5.2a)–(5.2d), give two ways to compute the ratio of concentrations of any two states in the scheme. This allows the ratio to be eliminated, leaving Eq. (5.12) as a relation between the equilibrium constants (Problem 1). This is a general result for a model with a loop such as Scheme (5C). Whenever a model has loops, the equilibrium constants are no longer independent of one another.

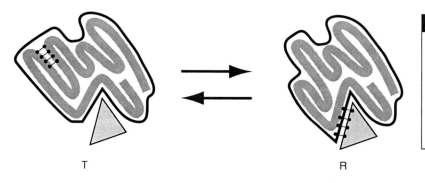

Fig. 5.2. The allosteric transition of a ligand–protein complex is shown again (see Fig. 5.1), but this time key contacts (●—●) are highlighted. The internal contacts in the T state are lost in the transition to the R state. In the R state new contacts form between the protein and ligand.

balance out if a system is at equilibrium such that concentrations do not change with time (see Section 7.3).

If we arrange Eq. (5.12) to give

$$\frac{Y_1}{Y_0} = \frac{K_R}{K_T} \tag{5.13}$$

we see that the factor by which Y changes during binding is equal to the factor by which the binding equilibrium constant changes during the allosteric transition. The expression RT times the logarithm of the right-hand side of Eq. (5.13) gives the difference between the binding free energy of the two allosteric states. Thus, the *improvement* in binding energy is what drives the protein into the R state, rather than the actual binding energy itself. If the ligand bound equally well to both states, there would be no drive for the allosteric transition, no matter how strong the binding.

We can pursue this point a bit further and try to visualize the process in more detail with a simple sketch (Fig. 5.2). Here we see that in the T state there are more contacts within the protein, and fewer contacts between the protein and ligand. In the R state the situation is reversed. There are more protein–ligand contacts and fewer contacts within the protein. Equation (5.13) expresses the balance between the improvement in ligand–protein contacts in the ratio K_T/K_R, and the different relative stabilities of the two conformations in the ratio Y_0/Y_1. Thus, for the ligand to be an effective signal, the strengthening of these added contacts with the binding site has to win out against the contacts within the protein that stabilize the T conformation. It is easiest to picture this process as forming and breaking contacts, but contacts could also be stretched or displaced from their positions of minimum potential energy. The idea encompasses virtually all forms of energy relevant to protein stability.

5.5 | G-protein coupled receptors

The G-protein coupled receptors make up a very large family of proteins that mediate responses to chemical signals and light. All members of this protein family contain seven hydrophobic segments that span the plasma membrane and anchor the protein at the cell surface. When activated by the binding of ligand, these receptors initiate a cascade of events by interacting with GTP-binding proteins, hence the name G-protein coupled receptors.

Most G-protein coupled receptors exist as monomers and have a single ligand binding site. Plots of binding and functional response for these receptors generally show simple saturation behavior, according to Eqs. (5.6) and (5.10). But direct comparisons between the two are often complicated by the G-protein signaling cascade, which places several additional steps between ligand binding and the biological response. A great deal of work has gone into elucidating the detailed mechanisms of activation of G-protein coupled receptors.

One approach is to use site-directed mutagenesis to search for the ligand binding site. The receptor is mutated, and a function such as binding or response is evaluated. However, as is clear from Eqs. (5.6) and (5.10), the concentration for half saturation of either response or binding gives an apparent binding constant, K_{app}, equal to the product $K_R Y_0$. This suggests that there are two ways that the saturation behavior of a receptor can be altered: (1) a mutation in the binding site will change K_R; (2) a mutation elsewhere in the protein alters the equilibrium between the two allosteric states, leading to a change in Y_0.

Figure 5.3a shows a sketch of one of the G-protein coupled receptors, the muscarinic acetylcholine receptor, which can be activated by two closely related ligands, acetylcholine (a neurotransmitter) and carbamylcholine (a drug that mimics acetylcholine). In an effort to locate the binding site, the residues indicated in Fig. 5.3a were mutated, and experiments showed that all of these mutations altered acetylcholine binding to variable degrees (Fig. 5.3b1). Carbamylcholine binding showed mostly parallel variations (Fig. 5.3b2). Looking at how these sites are distributed through the protein in Fig. 5.3a, it is hard to imagine how molecules as small as acetylcholine and carbamylcholine could interact with all of these residues.

Equation (5.10) suggests a way to make sense of these results. Note that K_{app}, which equals $K_R Y_0$, depends on both Y_0 and K_R. Clearly, mutating a residue directly involved in binding will change K_R, and thus K_{app}. But a mutation that alters Y_0 will also change K_{app}. A change in Y_0 will change K_{app} by the same factor for different ligands, so the ratio of K_{app} for two ligands will not change. The situation is different for mutations that alter residues that interact with a bound ligand. Then K_R is likely to change by different amounts for different ligands.

The ratio of K_{app} for acetylcholine and carbamylcholine is plotted in Fig. 5.3b3. This shows that in spite of the shifts in K_{app} for acetylcholine and carbamylcholine, the ratios are quite similar ($\sim$3) for wild type and most mutants. This suggests that many of the residues that

(a)

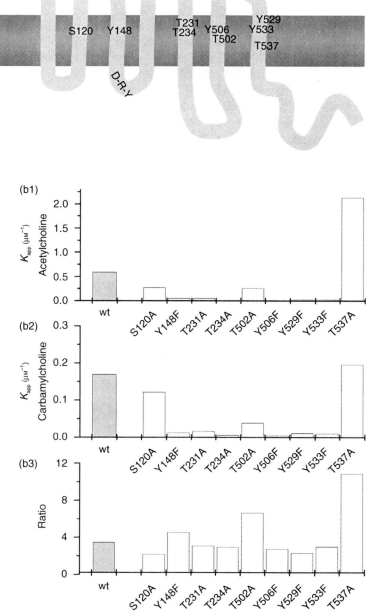

Fig. 5.3. (a) A sketch of the muscarinic acetylcholine receptor shows the characteristic seven membrane-spanning segments. The residues indicated were mutated in a study of ligand binding. D-R-Y (aspartate–arginine–tyrosine) is a conserved triad that influences Y_0.

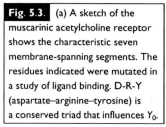

Fig. 5.3. (b1) Plot showing K_{app} for acetylcholine in wild type and mutant receptors. (b2) Plot of K_{app} for carbamylcholine. (b3) The ratio of K_{app} for acetylcholine and carbamylcholine. Data from Wess et al., 1990. S = serine, A = alanine, T = threonine, F = phenylalanine, Y = tyrosine.

influence binding are not part of the actual binding site. Rather, they exert their influence through the allosteric transition. Only two of the mutants, at positions 502 and 537, change the ratio significantly. The unequal shifts in the binding of acetylcholine and carbamylcholine in these mutants cannot be explained solely by a change in Y_0. These residues, which are quite close together in the sketch of Fig. 5.3a, therefore interact directly with the ligand. This is a more plausible conclusion than direct binding interactions with residues distributed throughout the protein. In general, looking at ratios of K_{app} for a series of similar ligands is a very powerful method for finding residues involved in ligand binding (Jackson, 1993b).

Another interesting way to evaluate allosteric aspects of G-protein coupled receptors is to find mutations that alter Y_0 so that there is higher basal activity. Mutations that elevate ligand-independent activity still respond to a ligand by increasing activity, but $K_{app} = Y_0 K_R$ is higher because Y_0 is higher.

The D-R-Y motif shown in the sketch in Fig. 5.3a is highly conserved and thought to play an important role in stabilizing the T conformation of the protein. Replacing the aspartate residue of this triad (the D) in the α_{1B} adrenergic receptor (a receptor that mediates responses to epinephrine and norepinephrine) with any other amino acid increases the basal level of activity (Fig. 5.4; in this case activity refers to the enzyme phospholipase C, the target of the G-protein activated by the α_{1B} receptor).

Figure 5.4 shows the qualitative trend expected: the greater the basal activity, the greater K_{app}. But we can go further by remembering that Eqs. (5.6) and (5.10) were derived from more complicated expressions by assuming that Y_0 was very small. Since we are now looking at mutant receptors with unusually high levels of basal activity, we should be concerned about this approximation. So we return to the exact expression (Eq. (5.9)), and estimate K_{app} by setting $B = 1/2$ and solving for $1/[L]$. The new apparent affinity is

$$K_{app} = \frac{Y_0 K_R (1 + 1/Y_1)}{1 + Y_0} \tag{5.14}$$

This equation fits the data in Fig. 5.4 well. The curve has a limiting value for K_{app} at large Y_0 of about 3 μM^{-1}, which according to Eq. (5.14) is $K_R(1 + 1/Y_1)$. If $Y_1 >> 1$, then this limiting value becomes K_R, the binding constant of the R conformation. This makes sense because as Y_0 gets large the receptor spends most of its time in the

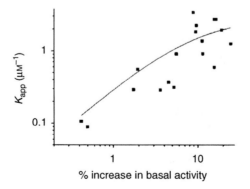

Fig. 5.4. Plot of K_{app} versus basal activity. Each point represents a pair of measurements from one molecular variant of the α_{1B} adrenergic receptor. The fit of Eq. (5.14) is shown. (Data from Scheer et al., (1997). The arginine mutant was excluded because it gave anomalous results, possibly due to defective G-protein coupling.)

R state regardless of whether ligand is bound or not. The T state then becomes irrelevant.

Although G-protein coupled receptors provide a number of illustrations of the simple one-site, two-state model developed here, it should be noted that this model cannot explain experimental results from some of these proteins. In these cases models with a third allosteric state have been proposed (Tucek, 1997).

5.6 | Binding site interactions

A major goal of allosteric theory is to provide a mechanism for how the occupation of a protein binding site can influence distant parts of the protein. To illustrate this, consider a single-subunit protein that undergoes an allosteric transition and binds two distinct ligands at different sites (Fig. 5.5).

The complete model for binding of ligands A and B to this protein is a combination of two schemes similar to Scheme (5C) above. This gives a total of eight states (T_0, T_A, T_B, T_{AB}, R_0, R_A, R_B, and R_{AB}), so the scheme is more complicated. But we can get some understanding of how the binding sites interact without the full treatment. To simplify things, we look at the case where the binding of either ligand alone does not provide enough drive to convert the protein from the T state to the R state (i.e. the equilibrium constants, Y_A and Y_B, are $<< 1$), but binding both ligands does (i.e. $Y_{AB} >> 1$). In the presence of an excess of ligand B, its site will essentially always be occupied, but without A the T state will prevail. We can describe the binding of A in the presence of excess B with an adaptation of Scheme (5C) as follows (Scheme (5E))

$$
\begin{array}{ccc}
& K_{TA} & \\
T_B + A & \rightleftharpoons & T_{AB} \\
Y_B \updownarrow & & \updownarrow Y_{AB} \\
R_B + A & \rightleftharpoons & R_{AB} \\
& K_{RA} &
\end{array}
\qquad (5E)
$$

With $Y_B << 1$ the protein will not switch from T_B to R_B (the same is true for Y_A and transitions from T_A to R_A). Because Y_{AB} is greater than one, binding the second ligand will bring about the switch.

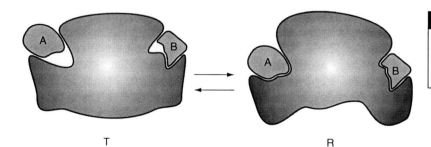

T   R

Fig. 5.5. The allosteric transition of this protein is accompanied by structural changes that increase the affinity of its two binding sites on opposite sides of the protein.

With reference to Scheme (5C), we can use Eq. (5.10) to get the apparent affinity of the protein for A in the presence of excess B; $K_{app} = K_{RA}Y_B$. Likewise, in the presence of excess A, the apparent affinity for B will be $K_{app} = K_{RB}Y_A$. Exploiting the energy balance condition (Eq. (5.12)), we see that K_{app} can also be expressed as $K_{TB}Y_{AB}$ and $K_{TA}Y_{AB}$, for A and B, respectively.

It is interesting to compare these values with the apparent affinities in the absence of the other ligand. For the binding of the first ligand, the allosteric transition is irrelevant. Binding is then described by the affinities of the T state, K_{TA} and K_{TB} (see Problem 2). Comparing with $K_{TB}Y_{AB}$ and $K_{TA}Y_{AB}$ for the apparent affinities in the presence of an excess of the other ligand, we see that each ligand increases the affinity of the protein for the other by the same factor, Y_{AB}.

Now look at another case, where binding only one ligand is sufficient to switch the protein. There is a striking parallel with the foregoing analysis. The approximations that lead to Eq. (5.10) for Scheme (5C) apply here, because each ligand can drive the transition on its own. This makes $K_{app} = K_{RA}Y_0$ and $K_{RB}Y_0$ for A and B, respectively, in the absence of the other ligand. With either one of these ligands bound, the protein will be in the R state. The allosteric transition is then irrelevant for further binding, and the affinities for the second ligand are simply K_{RA} and K_{RB}. We again see that one ligand increases the affinity of the protein for the other, and in this case the factor is $1/Y_0$ for both ligands.

For both of the above cases one ligand can influence the binding of another at a remote site on the protein. The key is that an allosteric transition forces both sites to change concomitantly. Interactions such as this exemplify the general concept of *linkage*. The coupling of one equilibrium to another can be studied by a thermodynamic analysis of the cycles to derive a variety of formal relations between the various parameters (Cantor and Schimmel, 1980).

These examples dealt with the binding of one ligand when the other ligand is either absent or saturating. However, when the other ligand is present at intermediate levels we see qualitatively different behavior. If A is present at intermediate concentrations and B is varied, A will be bound to a fraction of the receptors, and those receptors will have an apparent affinity of K_{RA}. But as B occupies its site, the remaining sites for B on receptors having no A will have an affinity of K_{TA}. The saturation behavior will now be biphasic. Such an experiment thus reveals the two allosteric states. A variant of this model was used in the analysis of G-protein coupled receptors to explain the biphasic behavior in adrenergic receptors, as well as the change in ligand binding affinity induced by guanine nucleotides (De Lean et al., 1980). This was actually a two-protein system consisting of a G-protein and G-protein-coupled receptor. These two proteins associate to form a complex in which the receptor binds its ligand and the G-protein binds GTP. The binding of GTP to the G-protein converts the associated receptor to a form with high affinity for its ligand. Likewise, the binding of the ligand to the receptor converts the G-protein to a form with a high affinity for GTP.

5.7 | The Monod–Wyman–Changeux (MWC) model

The focus until now has been on proteins with a single subunit. However, the early efforts in allosteric theory were motivated by a desire to understand cooperative processes in multisubunit proteins. The idea that an allosteric transition in a multisubunit protein can lead to cooperativity was introduced by Monod *et al.* (1965) to explain a large body of results from many proteins. This idea formed the basis for the MWC model.

The model starts with a protein composed of n identical subunits, each with a single binding site. This protein undergoes an allosteric transition so that the binding affinities change. The transition is a concerted change in the tertiary structure of all the subunits and a change in quaternary structure as the subunits rearrange with respect to one another. The entire process of tertiary and quaternary structural change is assumed to occur concomitantly. This assumption is a defining feature, and the MWC model is often referred to as the "concerted transition" theory. Separate tertiary transitions by individual subunits are forbidden. We will see later in this chapter that this assumption can be relaxed to develop other classes of models (Koshland–Nemethy–Filmer and Szabo–Karplus). For now we recall the analysis of global transitions in multisubunit proteins (Section 1.6) and assume that the cooperative unit is equal to the number of subunits.

Consider a sequence of binding steps for the T state. As long as the protein is in this state, all binding events will be governed by the same equilibrium constant, K_T. Likewise, we can envision a sequence of binding steps for the R state, with K_R governing all of these. Now we allow any state of occupancy to undergo the allosteric transition from T to R. The complete scheme can be depicted as (Scheme (5F))

$$
\begin{array}{ccccccc}
& nK_T & & \frac{1}{2}(n-1)K_T & & & K_T/n \\
T_0+L \rightleftharpoons & T_1+L & \rightleftharpoons & T_2 & \cdots\cdots & T_{n-1}+L & \rightleftharpoons T_n \\
Y_0 \updownarrow\updownarrow & \updownarrow\updownarrow\,Y_1 & & \updownarrow\updownarrow\,Y_2 & & \updownarrow\updownarrow\,Y_{n-1} & \updownarrow\updownarrow\,Y_n \\
R_0+L \rightleftharpoons & R_1+L & \rightleftharpoons & R_2 & \cdots\cdots & R_{n-1}+L & \rightleftharpoons R_n \\
& nK_R & & \frac{1}{2}(n-1)K_R & & & K_R/n
\end{array}
\qquad (5F)
$$

The subscript denoting the number of molecules of bound ligand ranges from 0 to n. It is important to realize that K_T and K_R describe the binding to individual sites. Using these binding constants for the binding steps in the model requires accounting for the multiple sites and multiple occupancy states. This means multiplying K_T and K_R by the number of sites available to accept a new ligand during a move to the right, and dividing by the number of occupied sites on the state to the right that could be vacated during a move to the left. This is why K_T and K_R in Scheme (5F) are multiplied by n, $(n-1)/2$, etc., to obtain the equilibrium constants for the relevant steps.

The equilibrium conditions corresponding to Eqs. (5.2a) and (5.2b) have to be modified by including these factors

$$K_T = \frac{(i+1)[T_{i+1}]}{(n-i)[T_i][L]} \tag{5.15a}$$

$$K_R = \frac{(i+1)[R_{i+1}]}{(n-i)[R_i][L]} \tag{5.15b}$$

For each state of occupancy we have an equilibrium constant for the allosteric transition

$$Y_i = \frac{[R_i]}{[T_i]} \tag{5.16}$$

With these relations we can derive the response generated by the protein, or if we wish, the fraction of binding sites occupied. We will illustrate the derivation of occupied binding sites here and leave the response function as an exercise (Problem 4).

The fraction of occupied binding sites can be written as (see Eq. (5.7))

$$B = \frac{[R_1] + 2[R_2] + \cdots + n[R_n] + [T_1] + 2[T_2] + \cdots + n[T_n]}{n([R_0] + [R_1] + [R_2] + \cdots + [R_n] + [T_0] + [T_1] + [T_2] + \cdots + [T_n])}$$

$$= \frac{1}{n} \frac{\sum\limits_{i=0}^{n} i[R_i] + \sum\limits_{i=0}^{n} i[T_i]}{\sum\limits_{i=0}^{n} [R_i] + \sum\limits_{i=0}^{n} [T_i]} \tag{5.17}$$

Equations (5.15a) and (5.15b) are then used to derive expressions for $[R_i]$ and $[T_i]$ in terms of $[R_0]$ and $[T_0]$

$$[T_1] = n[T_0]K_T[L] \tag{5.18a}$$

$$[T_2] = \frac{n-1}{2}[T_1]K_T[L]$$

$$= \frac{n(n-1)}{2}[T_0](K_T[L])^2 \tag{5.18b}$$

Recognizing the trend leads to a general expression for $[T_i]$

$$[T_i] = \frac{n!}{(n-i)!i!}[T_0](K_T[L])^i \tag{5.19a}$$

and likewise for $[R_i]$

$$[R_i] = \frac{n!}{(n-i)!i!}[R_0](K_R[L])^i \tag{5.19b}$$

Since $n!/(n-i)!i!$ is the expression for the coefficients of the binomial expansion (Eq. (A1.7)), the sums in the denominator of Eq. (5.17) can be evaluated

$$\sum_{i=0}^{n} [T_i] = [T_0] \sum_{i=0}^{n} \frac{n!}{(n-i)!i!} (K_T[L])^i = [T_0](1 + K_T[L])^n \qquad (5.20a)$$

$$\sum_{i=0}^{n} [R_i] = [R_0] \sum_{i=0}^{n} \frac{n!}{(n-i)!i!} (K_R[L])^i = [R_0](1 + K_R[L])^n \qquad (5.20b)$$

The sums in the numerator of Eq. (5.17) can be evaluated by taking the derivative of Eqs. (5.19a) and (5.19b) with respect to [L], and then multiplying by $K_T[L]$ or $K_R[L]$ as follows

$$\sum_{i=0}^{n} i[T_i] = [T_0] \sum_{i=0}^{n} i \frac{n!}{(n-i)!i!} (K_T[L])^i = n[T_0]K_T[L](1 + K_T[L])^{n-1}$$

$$(5.21a)$$

$$\sum_{i=0}^{n} i[R_i] = [R_0] \sum_{i=0}^{n} i \frac{n!}{(n-i)!i!} (K_R[L])^i = n[R_0]K_R[L](1 + K_R[L])^{n-1}$$

$$(5.21b)$$

Substituting these sums into Eq. (5.17) and taking $[T_0] = [R_0]/Y_0$ (Eq. (5.16)) gives the fraction of occupied sites versus ligand concentration

$$B = \frac{K_R[L](1 + K_R[L])^{n-1} + (K_T[L]/Y_0)(1 + K_T[L])^{n-1}}{(1 + K_R[L])^n + (1/Y_0)(1 + K_T[L])^n} \qquad (5.22)$$

Equation (5.22) gives sigmoidal binding for $n > 1$. This may seem surprising because no explicit assumption was made that binding itself was cooperative. Each binding step was treated as a simple association equilibrium. The cooperativity of the MWC model arises from the assumption that the transition is concerted: all of the subunits of the protein make the transition together. So when the protein is in the T state and a few of the binding sites are occupied, the protein will switch more readily to R. The remaining empty binding sites will then have a higher affinity and the curve will rise more steeply. The important result here is that even when cooperative binding is not explicitly assumed, a cooperative global transition will still lead to cooperative binding. If biological activity is measured as the fraction of protein in the R state, then biological activity will have the same sigmoidal concentration dependence (Problem 4).

The quantity n in Eq. (5.22) is much more meaningful than the Hill coefficient, n, in Eq. (4.7). Because the assumption of concerted binding is unrealistic, the Hill coefficient cannot be equated with the actual number of binding sites. The MWC model is more realistic because it allows different numbers of binding sites to be occupied. Thus n in this model is a real estimate of the number of subunits. It should therefore take on an integral value. In the case of hemoglobin it was noted that the Hill coefficient of 2.7 was less than the number of subunits, 4 (Section 4.2.1). We will now see that

Eq. (5.22) with $n = 4$ replicates the binding behavior of hemoglobin very well. The n in the MWC model is thus more than just a qualitative index of the degree of cooperativity.

5.8 | Hemoglobin

Although Monod *et al.* (1965) developed their model for cooperative enzymes, they considered hemoglobin as well, calling it an "honorary enzyme." Hemoglobin provided the earliest example of sigmoidal binding, and this protein has received a great deal of attention in efforts to understand the physical basis of cooperativity. The concerted binding model (Section 4.2.1) could account for this qualitative behavior with the artificial assumption of simultaneous binding, but when we use the knowledge that hemoglobin has four subunits and set $n = 4$, the Hill equation does a poor job of fitting the O_2 binding data (Fig. 5.6). By contrast, the MWC model with $n = 4$ fits the data very well.

Aside from the good fit to the binding data, more detailed study of hemoglobin supports many of the principles developed for the MWC model (Perutz *et al.*, 1998). For example, crystals of the T form bind O_2 with a low affinity, and because the crystal environment suppresses large-scale quaternary structural transitions, the binding curve is a simple rectangular hyperbola with no cooperativity. Likewise, crystals of the R form bind noncooperatively, but with high affinity. Thus, without the allosteric transition, there is no cooperativity, and this is exactly what MWC theory predicts.

The T form really does appear to have tension. The conformation is held in a state of high potential energy by interactions between the subunits. When the subunits are dissociated the tension is lost and the isolated polypeptides bind O_2 noncooperatively with a high affinity characteristic of the R state.

Fig. 5.6. Plot of O_2 binding data for hemoglobin (points) fitted to Eq. (5.22) for the MWC model (solid curve) and Eq. (4.7) (the Hill equation, broken curve). In both models, n was set to 4 (see Szabo and Karplus, 1972, for tabulated data). Note that the convention for hemoglobin is to plot the partial pressure of O_2; P_{O_2} is proportional to $[O_2]$).

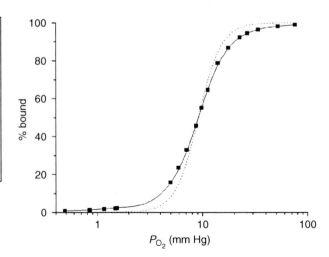

Finally, hemoglobin has a number of allosteric effectors. Diphosphoglycerate binds to a single site at a subunit interface and alters the affinity. Note that pH also shifts the binding curve in what is known as the Bohr effect. These results are explained within the MWC framework by allowing diphosphoglycerate and protons to bind preferentially to different conformations. They shift the binding curve without directly interacting with the binding sites by altering the preference for the T state versus the R state. See Section 5.14 for a discussion of the regulation of hemoglobin at a finer level of detail.

5.9 | Energetics of the MWC model

With the multisubunit MWC model we can expand upon the point made in Section 5.4 about how the improvement in binding drives the transition. Consider a diagram of an allosteric transition in a tetramer with two occupied binding sites (Fig. 5.7). As in Figs. 5.1 and 5.2, the binding site matches the shape of the ligand better in the R conformation.

When the allosteric transition takes place binding is strengthened by

$$\Delta\Delta G_b^\circ = -RT \ln (K_R/K_T) \tag{5.23}$$

This increase in binding energy reflects the strengthening of the contacts between the protein and ligand and this is energy that can drive the allosteric transition. With two ligands bound (as in Fig. 5.7) the energy is twice that given in Eq. (5.23). If the unliganded receptor has a ΔG° for the conformational transition of $-RT \ln Y_0$, then each occupied binding site will change the free energy of the transition by the amount specified in Eq. (5.23). This leads to the following expression for the free energy change of the allosteric transition as a function of the number of occupied binding sites

$$\Delta G_a^\circ(i) = -RT \ln Y_i = -RT \ln Y_0 - iRT \ln (K_R/K_T) \tag{5.24}$$

This amounts to a reformulation of the MWC model. The additivity assumption could have been used as the starting point of the theory to derive the equilibrium conditions leading ultimately to

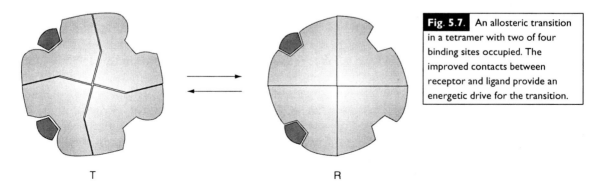

T   R

Fig. 5.7. An allosteric transition in a tetramer with two of four binding sites occupied. The improved contacts between receptor and ligand provide an energetic drive for the transition.

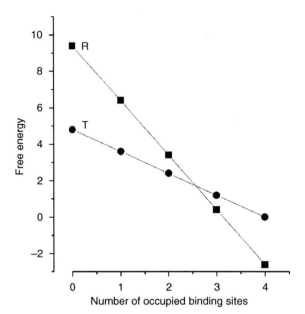

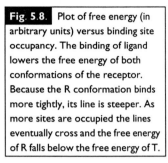

Fig. 5.8. Plot of free energy (in arbitrary units) versus binding site occupancy. The binding of ligand lowers the free energy of both conformations of the receptor. Because the R conformation binds more tightly, its line is steeper. As more sites are occupied the lines eventually cross and the free energy of R falls below the free energy of T.

Eq. (5.22). Equation (5.24) indicates that there are constant increments in the energy of the allosteric transition for each binding site occupied. This is illustrated graphically in Fig. 5.8.

This plot shows linear decreases in the free energy of each conformation as more binding sites are occupied. Because the R conformation binds more tightly, the decreases are larger for this conformation than for the T conformation. In the completely unliganded receptor, the R conformation has a much higher free energy, given by $-RT\ln Y_0$. As the binding sites are filled, this gap narrows, until the two lines cross and the R conformation is favored. This cross-over marks the point where the energy arising from the improvement in binding overcomes the unfavorable energy of the allosteric transition.

---

*5.10 | Macroscopic and microscopic additivity

The preceding section made the point that the MWC model depends on the assumption that the $\Delta\Delta G_b{}^\circ$ of binding and the ΔG° of the allosteric transition are additive. We can refer to this form of additivity as macroscopic, because the microscopic details are ignored. Now we will consider the microscopic details and try to get some insight into what macroscopic additivity means. In particular, we will consider how this can be related to additivity of energies for specific contacts. Examining the assumption of additivity in more detail helps us understand the relation between the various interactions within the protein and between the protein and ligand.

Additivity should not be taken for granted. One could imagine that a ligand interacts with a protein in such a way that internal contacts within the protein near the binding site are perturbed. If the protein is "soft" then these perturbations could spread and if the regions of distortion from different binding sites overlap then the two binding energies might no longer be additive.

First, we picture the stability of two allosteric states as dependent on two different sets of internal contacts within the protein. One set, with energies ε_T, stabilizes the T conformation, and the other set, with energies ε_R stabilizes the R conformation. The allosteric transition is accompanied by breaking one set of contacts and forming the other. Thus, we can view the free energy change of the allosteric transition as the difference in energy of these sets of contacts

$$\Delta G_a{}^\circ = \sum \varepsilon_R - \sum \varepsilon_T \qquad (5.25)$$

Ligand binding can also be viewed in terms of contacts. We define a set of contacts between the protein and ligand with energies ε_L. Contact energies are only part of the binding free energy. As discussed in Chapter 4, there are also translational, rotational, vibrational, and configurational contributions. The translational and rotational contributions will be essentially the same for binding to different allosteric states, and since we are trying to evaluate the difference, $\Delta\Delta G^\circ$, we do not have to worry about them. We also do not have to worry about configurational factors as long as the ligand assumes the same configuration when bound to T or R. The protein configurational change is already taken into account in the allosteric transition, and we will assume that without the allosteric transition there is no differential configurational contribution to the energetics of ligand binding. The vibrational term will depend on the specific nature of the contacts that stabilize the complex, so this contribution can be lumped together with the contribution made by ligand–protein contacts. The change in binding energy associated with the allosteric transition can then be expressed as a sum over changes in the energies of contacts between the ligand and protein

$$\Delta\Delta G_b{}^\circ = \sum \delta\varepsilon_L \qquad (5.26)$$

If the contacts that stabilize the receptor–ligand complex are independent of the internal protein contacts that stabilize the two allosteric states, then Eq. (5.26) can be added to Eq. (5.25) for each binding site occupied. This gives us the free energy change of the allosteric transition as a function of the number of binding sites occupied

$$\Delta G_a{}^\circ(i) = \sum \varepsilon_R - \sum \varepsilon_T + i \sum \delta\varepsilon_L \qquad (5.27)$$

This equation is equivalent to Eq. (5.24) in showing a linear dependence on i.

Thus, we see that macroscopic additivity follows from a microscopic picture if we assume that all the relevant contacts obey microscopic additivity. The question of whether the microscopic forces involved are additive brings up some complicated issues (Jackson, 1997a). Many of the forces discussed in Chapter 2 are not additive. But in general, greater distances between binding sites favor additivity. Thus, if the binding sites are distant from one another, their energies are more likely to be additive. Some of the contacts that stabilize the allosteric states could be near a binding site, and this would lead to nonadditivity between binding energies and allosteric transition energies. However, as long as this effect is local the dependence on i in Eq. (5.26) will remain linear.

This analysis indicates that the assumption of macroscopic additivity embedded in the MWC model is plausible if the binding sites are far apart. These ideas have not been subjected to rigorous testing, but at this stage we can conjecture that microscopic additivity implies macroscopic additivity, and that the assumptions underlying the MWC model are more likely to be valid when binding sites are separated by greater distances.

5.11 | Phosphofructokinase

The primary motivation for developing the MWC model was to explain the cooperativity and regulation of allosteric enzymes. These enzymes have multiple substrate binding sites, as well as regulatory binding sites where other molecules can control activity. The enzyme phosphofructokinase is an excellent example. The rate of catalysis shows a cooperative, sigmoidal dependence on substrate concentration, and various regulatory molecules shift the concentration dependence in a manner that fits the MWC theory very well (Blangy et al., 1968).

Phosphofructokinase catalyzes the phosphorylation of fructose-6-phosphate. ATP is hydrolyzed to yield fructose 1,6-diphosphate and ADP. This enzyme is at a pivotal step of the glycolytic pathway, and is thus an important regulatory point in controlling the cellular energy supply. ADP as a product might be expected to inhibit catalysis by the common mechanism of feedback inhibition. Instead the opposite occurs. A rise in ADP is a signal that carbohydrate oxidation is not keeping up with energy needs. Phosphofructokinase meets this need by accelerating its catalytic activity after binding ADP at a regulatory site. ADP thus acts as an allosteric enhancer of enzyme activity. The regulatory site is distinct from the site that binds ATP as a substrate (Schirmer and Evans, 1990). Conversely, ATP, citrate, and phosphoenolpyruvate all indicate that energy levels in a cell are high, and when any of these metabolites occupies the regulatory site catalysis is inhibited.

Phosphofructokinase is a tetramer, so to illustrate the allosteric mechanism we take a sketch like Fig. 5.7 and add a regulatory site

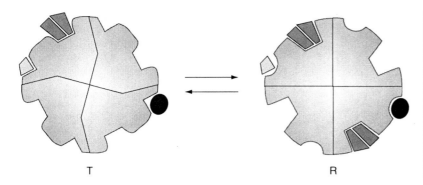

T   R

(Fig. 5.9). Substrates bind to the catalytic sites, and the allosteric transition converts these binding sites to high affinity. If the transition occurs in an enzyme with less than full occupancy, the empty sites can now bind substrate more readily and the speed of fructose-6-phosphate phosphorylation goes up. Monod *et al.* (1965) called this kind of allosteric protein a "K system", in which the binding affinity changes but V_{max}, the maximum rate of phosphorylation, is the same for both conformations (Section 10.5). For a K system, enhancement of catalysis is brought about by better binding.

The theoretical analysis of the MWC model with a regulatory site is an extension of the derivation of Eq. (5.22). One arrives at the same basic equation for the dependence of the fraction of occupied binding sites on substrate concentration. A ligand x that interacts with the regulatory site exerts its action in effect by changing Y_0. Thus, Y_0 in Eq. (5.22) is replaced by (Blangy *et al.*, 1968)

$$Y_0' = Y_0 \left(\frac{1 + K_{Rx}[x]}{1 + K_{Tx}[x]}\right)^n \tag{5.28}$$

The subscripts indicate binding constants for ligand x at the regulatory site in the T or R conformations; x will alter the dependence of enzyme activity on substrate concentration solely through a change in the effective equilibrium constant of the allosteric transition. The direction and magnitude of this effect ultimately depends on [x] and the relative magnitudes of K_{Rx} and K_{Tx}.

Equation (5.22) with Y_0 gives the concentration dependence of the velocity in the absence of an allosteric regulator (Fig. 5.10). The presence of enhancers and inhibitors dictates using Y_0' in Eq. (5.22) to generate velocity plots. The enzyme shows a steep cooperativity in the absence of a regulatory molecule. As an effector is added, Y_0' decreases or increases. The allosteric transition then becomes easier or harder for ligand binding to induce, and the concentration of substrate needed to induce the transition changes accordingly. The enzyme ultimately reaches the limiting behaviors of the R and T conformations. Pure T behavior reflects a protein in which Y_0' is so low that even when all the binding sites are occupied, the R state is still beyond reach. The absence of the allosteric transition uncouples the binding sites so that the concentration dependence has no

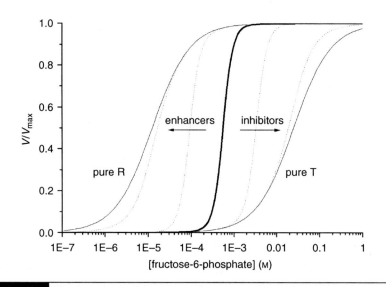

Fig. 5.10. The dependence of the velocity of enzyme catalysis on fructose-6-phosphate concentration, generated from Eq. (5.22) with different Y_0 values. The thick solid curve in the middle is for enzyme without an allosteric regulator: $Y_0 = 2.5 \times 10^{-7}$, $K_R = 12$ μM, and $K_T = 25$ mM. Shifts to the right reflect the action of the inhibitor phosphoenolpyruvate ($Y_0' = 2.5 \times 10^{-10}$ and 2.5×10^{-13}). Shifts to the left reflect the action of the enhancer ADP ($Y_0' = 2.5 \times 10^{-4}$ and 2.5×10^{-1}). The R and T curves reflect the limiting cases of those conformations in the absence of an allosteric transition (see Blangy et al., 1968).

sigmoidicity. Pure R behavior emerges when $Y_0' > 1$, so that the protein is always in the R state. Again there is no allosteric transition and no sigmoidicity. In each of these limits, the cooperativity is lost as the allosteric transition ceases to be a factor (Problem 6).

From the energetic perspective we can interpret these results by imagining upward and downward relative shifts of the two linear plots in Fig. 5.8. Making $Y_0' < Y_0$ makes the gap between R_0 and T_0 larger and shifts the crossover point to the right. If the gap becomes so great that the lines never cross, then we have the limiting pure T behavior. Making $Y_0' > Y_0$ does the opposite. It shifts the crossover point in Fig. 5.8 to the left, leading ultimately to pure R behavior.

5.12 | Ligand-gated channels

Ligand-gated channels are a class of membrane receptors that form ion channels in the plasma membrane. The gating of the channel is an allosteric transition, with the R and T states as open and closed, respectively. Figure 5.7 could become such a protein if a channel were added at the central juncture of the subunits. In the nicotinic acetylcholine receptor, allosteric transitions have been detected in both liganded and unliganded proteins using single-channel current recording techniques (Jackson, 1994). Channels

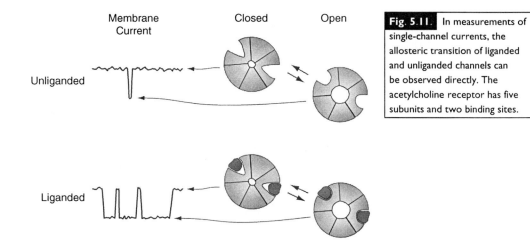

Fig. 5.11. In measurements of single-channel currents, the allosteric transition of liganded and unliganded channels can be observed directly. The acetylcholine receptor has five subunits and two binding sites.

open spontaneously with a very low frequency in the absence of ligand. In the presence of ligand, channels are mostly open, but briefly flicker closed without losing the ligand (Fig. 5.11). Furthermore, plots of the free energy of the open and closed states versus binding site occupancy are roughly linear, as expected for macroscopic additivity.

Shifts in K_{app} have been produced by mutations in many parts of ligand-gated channel proteins. The residues responsible for ligand binding have been clearly established in the nicotinic acetylcholine receptor by affinity labeling, and many of the mutations that shift K_{app} are separated from the binding sites by a considerable distance. In particular, K_{app} is very sensitive to changes in the residues that form the ion conduction pathway of the channel. The reason for this is that these are residues that make and break contacts during the channel gating transition. Hence they exert a strong effect on Y_0 (Jackson, 1997a). A large body of experimental data on the nicotinic acetylcholine receptor supports many of the predictions of the MWC theory (Jackson, 1998).

Most ligand-gated channels actually have at least three allosteric states. The closed and open states are obvious, but a third state is necessary to account for the finding that the continued presence of ligand causes the channels to close, or desensitize. Desensitization is an intrinsic braking mechanism that prevents the channel from being open for an excessive period of time. In some cases desensitization is very fast and limits a postsynaptic response to a few milliseconds. Desensitization requires a separate allosteric state of the receptor that binds ligand with an even higher affinity than the open state but which does not conduct ions. The high affinity of the desensitized state has been detected in binding measurements. A small amount of high affinity receptor is present at the beginning of a binding experiment, indicating that a small amount of receptor is in the desensitized state in the absence of ligand (Changeux, 1984).

5.13 | Subunit–subunit interactions: the Koshland–Nemethy–Filmer (KNF) model

It is important to know whether all of the subunits of a multisubunit protein undergo an allosteric transition in a concerted fashion, as was assumed in the MWC model. A model introduced by Koshland *et al.* (1966) is widely viewed as an alternative to the MWC because it does not postulate a concerted transition. The KNF model allows each subunit to undergo its transition individually.

Each subunit can assume different conformations with different binding affinities, but the free energy change during the transition in a given subunit depends on the state of its neighbors. This formulation leads to complicated mathematics. To simplify things, it was assumed that the conformational transition of each subunit is concomitant with ligand binding. In a trimeric protein we have the four states depicted in Fig. 5.12. The subunits fit together well if their neighbors are in the same state; otherwise the fit is poor and the interaction energy is positive.

In the KNF model we consider three energetic factors, the binding energy, the intrinsic energy difference between the two conformations of a subunit, and the subunit–subunit interaction energy. Since binding is assumed to be concomitant with a subunit's conformational transition, we can lump these two contributions together and take the energy per binding event as the free energy difference per subunit between the completely free and completely occupied states in Fig. 5.12. We denote this as $-RT \ln K_s$, where K_s is a binding constant that includes not only the energies directly relating to protein–ligand interactions, but also the energy of the obligatory conformational transition. For subunit-subunit interactions, we note that adjacent subunits in the same state fit together well (Fig. 5.12). When a liganded subunit abuts an unliganded subunit, the interaction energy is $-RT \ln \sigma$. So for a state with i occupied binding sites and j interfaces between subunits with different states of occupancy, the free energy relative to the completely unoccupied state is $\Delta G^{\circ} = -iRT \ln K_s - jRT \ln \sigma$.

A protein with one occupied binding site has $i = 1$ and $j = 2$ (Fig. 5.12). With a ligand concentration of [L], the equilibrium between a protein with zero and one occupied site can be expressed in terms of the two basic free energy parameters

Fig. 5.12. A trimeric protein in the KNF model. There are unfavorable contacts between liganded and unliganded subunits that push the system toward the fully bound or empty states.

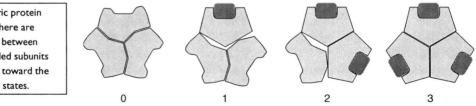

0   1   2   3

$$\frac{[P_1]}{[P_0][L]} = 3e^{-\Delta G^\circ/RT} = 3e^{\ln K_s + 2\ln\sigma} = 3K_s\sigma^2 \tag{5.29}$$

The factor of 3 accounts for the number of ways to achieve single occupancy. For the step from one to two occupied sites, there is no change in j (Fig. 5.12), and both P_1 and P_2 have the same multiplicity, so

$$\frac{[P_2]}{[P_1][L]} = K_s \tag{5.30}$$

Finally, for the third step there is a removal of two unfavorable contacts, so j is reduced by two. There are three sites that can lose ligand so the multiplicity factor is 1/3.

$$\frac{[P_3]}{[P_2][L]} = \frac{K_s}{3\sigma^2} \tag{5.31}$$

We now write the fraction of binding sites occupied

$$B = \frac{[\text{occupied sites}]}{[\text{total sites}]} = \frac{[P_1] + 2[P_2] + 3[P_3]}{3([P_0] + [P_1] + [P_2] + [P_3])} \tag{5.32}$$

Using Eqs. (5.29)–(5.31) to express $[P_1]$, $[P_2]$, and $[P_3]$ in terms of $[P_0]$, and making the substitutions in Eq. (5.32), gives

$$B = \frac{3[P_0][L]K_s\sigma^2 + 6[P_0]([L]K_s)^2\sigma^2 + 3[P_0]([L]K_s)^3}{3([P_0] + 3[P_0][L]K_s\sigma^2 + 3[P_0]([L]K_s)^2\sigma^2 + [P_0]([L]K_s)^3)}$$
$$= \frac{[L]K_s\sigma^2 + 2([L]K_s)^2\sigma^2 + ([L]K_s)^3}{1 + 3[L]K_s\sigma^2 + 3([L]K_s)^2\sigma^2 + ([L]K_s)^3} \tag{5.33}$$

Since σ represents subunit–subunit interactions, we can vary this parameter to see how these interactions influence the overall binding behavior. First of all, with zero interaction energy, $\sigma = 1$. This gives

$$B = \frac{[L]K_s + 2([L]K_s)^2 + ([L]K_s)^3}{1 + 3[L]K_s + 3([L]K_s)^2 + ([L]K_s)^3}$$
$$= \frac{[L]K_s(1 + [L]K_s)^2}{(1 + [L]K_s)^3}$$
$$= \frac{[L]K_s}{1 + [L]K_s} \tag{5.34}$$

This is what one would expect for completely independent subunits. On the other hand, if the subunit interaction energy is very large, σ goes to zero, leaving

$$B = \frac{([L]K_s)^3}{1 + ([L]K_s)^3} \tag{5.35}$$

This is the Hill equation for concerted binding (see Section 4.2.1 and Eq. (4.7)).

Plots of Eq. (5.33) for values of σ ranging from 0 to 2 show how increasing the energy of the interaction between occupied and empty subunits makes the binding curve more cooperative (Fig. 5.13). When $\sigma < 0.02$ (corresponding to a subunit–subunit interaction energy of

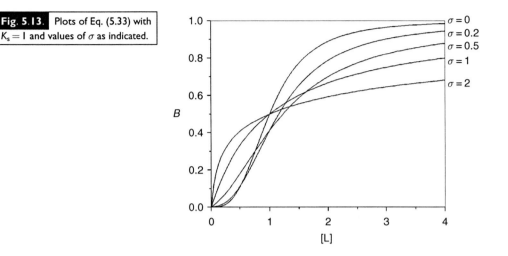

Fig. 5.13. Plots of Eq. (5.33) with $K_s = 1$ and values of σ as indicated.

2.4 kcal mole^{-1}) the plot becomes hard to distinguish from the limiting case of $\sigma = 0$ (Eq. (5.35)). On the other hand, $\sigma = 2$ corresponds to a negative (favorable) free energy for the contact between subunits with occupied and empty sites. This leads to negative cooperativity, a feature that is beyond the capability of the MWC model.

Analysis of the KNF model with four or more subunits is quite a bit more complicated. Because subunit–subunit interactions depend on the specific geometry of the protein, more assumptions must be made, and this makes it harder to see general trends. For example, the subunits of a tetramer could be arranged as a tetrahedron or a square, and the properties will vary depending on which of these alternatives is chosen. The rapid increase in complexity with increasing number of binding sites has made the KNF model less widely used compared to the MWC model. As a result, the assumptions regarding interactions between subunits often remain untested.

It has in fact proven difficult to resolve single subunit transitions within multisubunit proteins. One exception is voltage-gated channels. Although the functional switch in these proteins is induced by voltage rather than ligand binding, the principles embodied by the KNF model can still be recognized. There is abundant evidence that individual subunits within these tetrameric (and four-domain monomeric) proteins undergo conformational transitions on their own. The channel gating transition occurs only after all of the subunits have arrived in the state favored by positive voltage. Early modeling efforts treated the transitions in individual subunits as independent. However, the activation of a single K$^+$ channel was found to proceed through fractional increments that could be accounted for by transitions in individual voltage sensing domains (Chapman et al., 1997). It was later shown in tetrameric potassium channels that if only one of the subunits within a tetramer was mutated to shift the voltage dependence of its transition, the voltage dependence of the neighboring wild-type subunits was shifted as well (Tytgat and Hess, 1992). This reflects an interaction energy between adjacent subunits in different conformations.

---

*5.14 | The Szabo–Karplus (SK) model

X-ray crystallography has produced structures of hemoglobin crystallized in the presence and absence of oxygen (reviewed by Perutz *et al.*, 1998). This work indicates that the allosteric transition disrupts contacts, mostly salt bridges, between the subunits. Breaking these contacts allows the subunits to shift their positions and undergo changes in structure. This rearrangement leads to an increase in the O_2 affinity of the binding sites. This observation inspired an effort to extend allosteric theory to a more detailed level (Szabo and Karplus, 1972). The SK theory incorporates contact energies into a thermodynamic model in an approach conceptually related to the ideas of microscopic additivity discussed in Section 5.10. Here we will see how the simplest version of the SK model gives the same results as the MWC model, but with a different underlying physical picture. The SK model has the advantage of providing a robust framework within which new details are readily incorporated.

SK theory is built around a few key contacts between and within subunits. When O_2 binds to a site, a transition in the tertiary structure of that subunit is initiated which breaks the internal contacts and intersubunit contacts originating from that subunit. A global transition in quaternary structure is also assumed to occur between tense and relaxed states. This is essentially an allosteric transition, and it is closely related to the concerted transition between the T and R states in MWC theory. Here, the allosteric transition breaks all of the key intra- and intersubunit contacts. Thus, contacts can be broken either locally by a binding event or globally by the allosteric transition. Each contact is assigned the same energy value, E_S, for the sake of simplicity, and there are i contacts per subunit.

The O_2 binding energies always include the direct interaction between O_2 and its iron–heme binding site. This binding energy, $-E_B$, also includes the various statistical mechanical contributions discussed in Chapter 4. If inter- and intrasubunit contacts coupled to that binding site are intact, then they break upon binding and these energies must be counted. The energy for binding to a site with the protein in the R state is simply E_B, because there are no contacts to include. They were already broken during allosteric transition. The energy for binding to a site with the protein in the T state must include the contact breaking energies. It is thus $-(E_B - iE_S)$. This gives the following values for the binding constants

$$K_T = e^{(E_B - iE_S)/RT} \tag{5.36a}$$

$$K_R = e^{E_B/RT} \tag{5.36b}$$

The energy of the quaternary (allosteric) transition is divided into an intrinsic part, E_Q, and a part reflecting the contacts that can be broken. For the unliganded protein all $4i$ contacts have to be broken, so the energy is $E_Q + 4iE_S$. This gives a quantity equivalent to Y_0 of the MWC theory

$$Y_0 = e^{-(E_Q + 4iE_S)/RT} \tag{5.37}$$

As binding sites become occupied, the contacts coupled to that subunit break, and the allosteric transition becomes easier. When all the sites are occupied, there are no contacts left to break, so we are left with simply $e^{-E_Q/RT}$ corresponding to Y_0.

Based on the crystal structures, Szabo and Karplus counted the contacts as follows. Four contacts between the two α subunits break during the transition. One more contact breaks between each α subunit and its companion β subunit. With this assignment of contacts we have $i = 2$ for both the α and β subunits. Internal contacts within the β subunits break. No contacts were seen between the two β subunits. The states of occupancy can then be enumerated (nine in all), and the energies estimated by counting the number of contacts according to these rules (Fig. 5.14).

For each state represented in Fig. 5.14, the equilibrium constant relative to a reference state can be calculated by adding the number of occupied binding sites and the number of intact contacts. For each binding site occupied, the intrinsic energy contributes a statistical weight

$$k = e^{-E_B/RT} \tag{5.38}$$

Each broken contact contributes a statistical weight

$$s = e^{-E_S/RT} \tag{5.39}$$

The quaternary transition contributes a statistical weight

$$q = e^{-E_Q/RT} \tag{5.40}$$

For each state shown in Fig. 5.14 one can see which of these energy contributions is present, and then calculate an equilibrium constant as a product of the relevant statistical weights. An additional statistical factor has to be included for the number of different ways a particular state of occupancy can be achieved (its degeneracy). These multiplicities are 4 for singly liganded states, 6 for doubly liganded states, and 4 for triply liganded states. One can use all this to calculate the fraction of occupied binding sites as in Eq. (5.17). Writing out all the equilibrium constants and using them to express the concentration of each state in terms of $[O_2]$ gives

$$B = \frac{4ks^2[O_2] + 12k^2s^4[O_2]^2 + 12k^3s^6[O_2]^3 + 4k^4s^8[O_2]^4 + qs^8(4k[O_2] + 12k^2[O_2]^2 + 12k^3[O_2]^3 + 4k^4[O_2]^4)}{4(1 + 4ks^2[O_2] + 6k^2s^4[O_2]^2 + 4k^3s^6[O_2]^3 + k^4s^8[O_2]^4) + qs^8(1 + 4k[O_2] + 6k^2[O_2]^2 + 4k^3[O_2]^3 + k^4[O_2]^4)} \tag{5.41}$$

This can be factored with the aid of the binomial expansion (Eq. (A1.7)) to obtain an equation identical to Eq. (5.22), with $n = 4$.

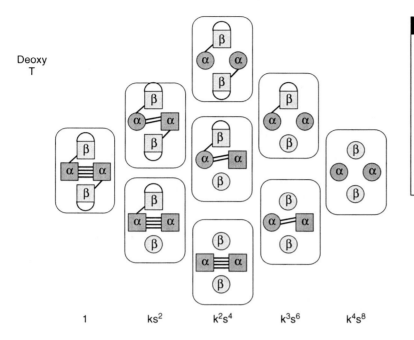

Deoxy
T

Fig. 5.14. Deoxy and oxy hemoglobin each have nine distinct occupancy states. Circles and squares denote subunits with and without O_2, respectively. Lines between subunits indicate the contacts, which can only form in the deoxy form. Half circles indicate the internal contact of the β subunit. As O_2 binds, contacts are broken. The statistical weights are given in the text.

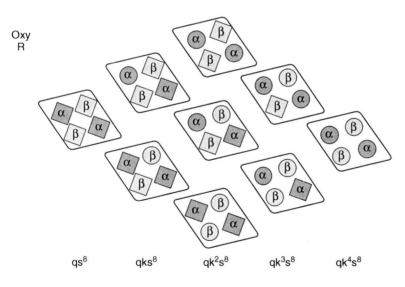

Oxy
R

$$B = \frac{[O_2]\,ks^2(1 + [O_2]\,ks^2)^3 + [O_2]\,k\,qs^8(1 + [O_2]k)^3}{(1 + [O_2]\,ks^2)^4 + qs^8(1 + [O_2]k)^4} \tag{5.42}$$

If more detail is added to specify different energies for some of the contacts, or different numbers of contacts breaking for ligand binding to different subunits, the result will resemble Eq. (5.41), but factorization and simplification to Eq. (5.42) is no longer possible.

Both the MWC and SK models use the quaternary transition to make the binding sites influence one another. Although both models give the same binding behavior, the SK model makes an important physical distinction. The MWC model does not assume that individual subunits can undergo isolated changes in tertiary structure. In the MWC model binding is strengthened during the allosteric transition by directly strengthening the interaction between the ligand and the protein. In the SK model, ligand binding causes a tertiary structural transition within one subunit of sufficient magnitude to alter interactions with other subunits. The enhanced binding to the R state is seen because contacts that oppose binding are gone. Structural studies of hemoglobin point to differences in the iron–heme binding sites between the T and R conformations that indicate a direct effect on the interaction with O_2, as assumed in the MWC model (Perutz et al., 1998). In the R state the iron atom lies in the plane of the heme, but in the T state the iron atom is pulled out of the plane so an oxygen molecule approaching from the other side cannot bind as tightly.

The contacts of the SK model are salt bridges formed by ionizable side chains. They can therefore change their state of protonation when they break, and this will depend on pH. Hemoglobin is very sensitive to pH. Low pH favors the T state and triggers the dissociation of O_2 (the Bohr effect, Section 5.8). The SK model provides a natural way to deal with this property. The original SK model could describe most of the results available at that time. Subsequent studies indicated that the α and β subunits had somewhat different affinities. Further, as noted above, the intrinsic component of the binding affinity, reflecting the direct interaction between O_2 and the iron atom, changes during the allosteric transition. The SK model has been extended to incorporate these and other features (Lee et al., 1988).

Problems for Chapter 5

1. Derive Eq. (5.12) using just the four equilibrium conditions in Eqs. (5.2a)–(5.2d).
2. Find the approximate form for Eq. (5.9) when $Y_1 \ll 1$. See how this justifies the use of K_T in Section 5.6 to describe the binding of the first of the two ligands.
3. Use energy balance to express Y_0/Y_{AB} in terms of the binding constants for the complete two-ligand model of Section 5.6.
4. In the MWC model, derive the fraction of protein in the R state as a function of ligand concentration. If biological activity is associated exclusively with the R state, then this expression is a dose–response function of the protein.
5. Show that the result of Problem 4 goes to Eq. (5.4) when $n = 1$ (hint: use Eq. (5.12)).
6. Show that Eq. (5.22) goes to noncooperative behavior of the R or T state in the limit of low and high Y_0.

7. Derive the equilibrium constant for the allosteric transition of a multisubunit protein such as phosphofructokinase in the presence of a regulatory ligand with binding constants K_{Rx} and K_{Tx}. The result is Eq. (5.28).

8. Write the concentration dependence of binding (fraction of sites occupied) for an MWC tetramer in which one subunit has the ligand tethered to its binding site so that it is permanently occupied.

9. Derive the KNF binding curve for a dimer. For large σ show that the two effective binding constants are $2\sigma K_s$ and $K_s/2\sigma$.

10. Derive the KNF binding curve for two kinds of tetramer, with the subunits arranged either as a tetrahedron. Write the expression for the cases $\sigma = 0$ and 1.

Chapter 6

Diffusion and Brownian motion

In the fluid world of biology, molecules are often free to move at random. We will now study the nature of this random motion. This will provide insight into the rates of change in many processes. In this chapter we take our first look at kinetics. The previous chapters all dealt with systems at equilibrium. We asked how conditions can alter an equilibrium, but we rarely worried about how long it would take to get there. There were no derivatives with respect to time. Now we will see many. We will develop the related theories of diffusion and Brownian motion. Because these theories are very general, they can be applied to many related situations. Subsequent chapters will explore other ways to study dynamic processes, and the ideas from this chapter will in many cases provide the starting point.

6.1 | Macroscopic diffusion: Fick's laws

The theory of diffusion can be developed from two simple and basic assumptions. The first of these is that a substance will move down its concentration gradient. The steeper the gradient the more movement of material. If the relation between gradient and flux is linear, then in one dimension we have what is known as Fick's first law

$$J = -D \frac{\partial C}{\partial x} \tag{6.1}$$

where x is the position, C is the concentration at that position, and D is the proportionality constant, called the diffusion constant. The variable J is the flux, and is defined as the amount of material passing across the point at x (or through a unit area perpendicular to the direction of flow) per unit time. The minus sign means that the flow is in the direction of decreasing concentration. The rationale for this equation is a common sense picture of how material should move from high concentrations to low.

The second assumption needed for the theory of diffusion is conservation of matter. If in a small element of length dx, the flux into the element from the left is different from the flux out of the element to the right, then $C(x)$ within this element will change. The difference between the two fluxes $J(x)$ and $J(x + \Delta x)$ determines how much material will accumulate within the region bounded by x and $x + dx$ in a time interval Δt

$$(J(x + \Delta x) - J(x))\Delta t = -\Delta C \Delta x \tag{6.2}$$

This equation ensures that all of the diffusing material is accounted for, and none is created or lost.

Equation (6.2) can be rearranged and converted to derivatives as follows

$$\frac{\partial C}{\partial t} = -\frac{\partial J}{\partial x} \tag{6.3}$$

The next task is to eliminate J by combining Eqs. (6.1) and (6.3). Differentiating Eq. (6.1) with respect to x gives an expression for the flux derivative that can then be equated with that in Eq. (6.3). The result is Fick's second law, which is the diffusion equation

$$\frac{\partial C}{\partial t} = D \frac{\partial^2 C}{\partial x^2} \tag{6.4}$$

From this derivation we see that any process will obey the diffusion equation if it obeys the two key assumptions. There must be a linear relation between the gradient and flux, and the diffusing substance must by conserved. Heat satisfies these same two conditions, as does passive spread of voltage in a cable (see Chapter 15). In both cases the diffusion equation (or something very similar) is obeyed.

Equation (6.4) is one dimensional. In three dimensions the spatial derivative in Eq. (6.1) is replaced by the gradient of the concentration. Fick's first law is now a vector equation

$$\mathbf{J} = -D\nabla C \tag{6.5}$$

The spatial derivative of the flux in one dimension (Eq. (6.3)) is replaced by the sum of the three spatial derivatives along the three different axes, or the divergence in vector notation

$$\frac{\partial C}{\partial t} = -\nabla \bullet \mathbf{J} \tag{6.6}$$

When $\mathbf{J}$ is eliminated in the same way by combining these two equations, the result is the diffusion equation in three dimensions

$$\frac{\partial C}{\partial t} = D\nabla^2 C \tag{6.7}$$

where ∇^2 is the Laplacian operator $(\partial^2/\partial x^2) + (\partial^2/\partial y^2) + (\partial^2/\partial z^2)$.

6.2 | Solving the diffusion equation

The diffusion equation is a partial differential equation that tells us how a spatial distribution of material changes over time. If we are given an initial distribution $C(x, y, z, t_0)$, then the solution to the diffusion equation, $C(x, y, z, t)$, will tell us the distribution at any later time. There are as many different solutions as there are initial distributions. We can imagine an initial distribution in which C is very high at $x = y = z = 0$, and zero everywhere else. Alternatively, C could be constant within a particular region, and zero elsewhere. Any arbitrary function is imaginable as an initial distribution for C, and it will evolve through time in a specific way dictated by the diffusion equation.

The initial distribution is not enough to specify the solution to the diffusion equation. The geometry in which diffusion is taking place can vary and that will also have an influence. There might be an infinite volume, or an infinitely long tube, or a closed container with a complicated shape. These geometric factors constitute boundary conditions, and are specified as mathematical constraints on C, or its spatial derivatives, or both, at the boundary of the container. To define a solution to the diffusion equation, we need both initial conditions and boundary conditions. If these are specified, then we can in principle find $C(x, y, z, t)$. If the equation cannot be solved analytically, then a computer can be used to solve the equation numerically.

There is a vast mathematical literature on the diffusion equation, and one can find solutions for many different initial and boundary conditions in the standard books such as Carslaw and Jaeger (1959) and Crank (1975). Here, we will consider a few important examples to help illuminate the basic principles.

6.2.1 One-dimensional diffusion from a point

Take as the initial condition that C is infinite at $x = 0$, and 0 everywhere else. In one-dimension this is expressed as the delta function $C = M\delta(x)$. The integral $\int_{-\infty}^{\infty} C(x)\mathrm{d}x = M$ at $t = 0$, where M specifies the total amount of diffusing substance. The infinitely high C in an infinitesimally small region may be artificial, but we can avoid the details of what this singularity looks like. The diffusion equation can be solved for this initial condition using the method of Fourier transforms.[1] The result is a Gaussian function

---

[1] Taking the Fourier transform of Eq. (6.4) in the spatial dimension gives (Eq. (A3.15))

$$\frac{\partial C_f(f, t)}{\partial t} = -Df^2 C_f(f, t)$$

where f is the frequency variable used in the Fourier transform and C_f is the Fourier transform of C. This has the solution

$$C_f(f, t) = C_f(f, 0)\mathrm{e}^{-Df^2 t}$$

where $C_f(f, 0)$ is the Fourier transform of the initial condition.

$$C_f(f, 0) = \frac{1}{\sqrt{2\pi}} \int_{-\infty}^{\infty} M\delta(x)\mathrm{e}^{\mathrm{i}fx}\mathrm{d}x = \frac{M}{\sqrt{2\pi}}$$

and C is computed from C_f by taking the inverse Fourier transform.

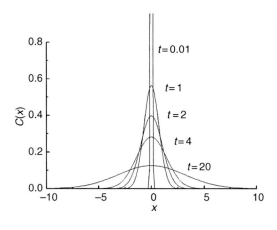

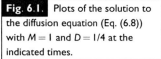
Fig. 6.1. Plots of the solution to the diffusion equation (Eq. (6.8)) with $M = 1$ and $D = 1/4$ at the indicated times.

$$C(x, t) = \frac{M}{\sqrt{4\pi Dt}} e^{-x^2/(4Dt)} \tag{6.8}$$

This can be substituted into Eq. (6.4) to verify that it is a solution. The integral over $C(x,t)$ will remain M for any value of t, as can be checked by using $\int_{-\infty}^{\infty} e^{-\alpha x^2} dx = \sqrt{(\pi/\alpha)}$ (Eq. (A4.4)). This verifies that mass is conserved. So as expected for a solution to the diffusion equation, Eq. (6.8) is consistent with one of the basic assumptions employed in the original derivation. This property is built into the diffusion equation, and it is wise to check a solution to be sure that this condition is satisfied.

Figure 6.1 shows how $C(x, t)$ evolves through time. As expected, the diffusing material spreads out as time passes. The curves remain centered at $x = 0$ for all time, with $C(0, t)$ going down as diffusion progresses. The curve at $t = 0.01$ illustrates how the function approaches a delta function as $t \to 0$.

An important hallmark of diffusion is that the rms displacement from the initial position increases in proportion to the square root of the time

$$\sqrt{\overline{x^2}} = \sqrt{2Dt} \tag{6.9}$$

$$C(x, t) = \frac{M}{2\pi} \int_{-\infty}^{\infty} e^{-Df^2 t} e^{ifx} df$$

and Eq. (6.8) is obtained by evaluating this integral as follows (see Eq. (A4.7))

$$C(x, t) = \frac{M}{2\pi} e^{-x^2/(4Dt)} \int_{-\infty}^{\infty} e^{-Df^2 t + ifx + x^2/(4Dt)} df$$

$$= \frac{M}{2\pi} e^{\frac{-x^2}{4Dt}} \int_{-\infty}^{\infty} e^{-\left(f\sqrt{Dt} - \frac{ix}{2\sqrt{Dt}}\right)^2} df$$

$$= \frac{M}{2\pi} e^{\frac{-x^2}{4Dt}} \sqrt{\frac{\pi}{Dt}}$$

$$= \frac{M}{\sqrt{4\pi Dt}} e^{-x^2/(4Dt)}$$

This result comes directly from Eq. (6.8) by evaluating the integral $\int_{-\infty}^{\infty} x^2 C(x)dx$. One can see by inspection of Eq. (6.8) that $4Dt$ is the width of this Gaussian function. One can also show that the value of x defined in Eq. (6.9) marks a position where the concentration is constant. At shorter distances the concentration is decreasing and at longer distances it is increasing (Problem 4).

Equation 6.9 is used very widely for order of magnitude estimates. It was derived for one special case here, but the initial condition of starting at one discrete location and spreading out is a useful general way to envision diffusion. For example, it is common to use this expression to estimate how long it will take a metabolite to diffuse through a cell when it is produced at one location. For a typical diffusion constant of 10^{-6} cm^2 s^{-1} we find that for $t = 5$ ms, $\sqrt{x^2} = 1\,\mu$m. For $t = 50$ ms, $\sqrt{x^2} = 3.2\mu$m. For $t = 500$ ms, $\sqrt{x^2} = 10\,\mu$m. Each factor of 10 in time gives a factor of $\sqrt{10} = 3.16$ in distance.

6.2.2 Three-dimensional diffusion from a point

A very similar result is obtained for diffusion in three dimensions. To see this we start with Eq. (6.7) instead of Eq. (6.4). We are interested only in the radial distance from the starting point so the Laplacian operator should be converted to spherical coordinates (Eq. (A6.5))

$$\frac{\partial C}{\partial t} = D \frac{1}{r^2} \frac{\partial}{\partial r}\left(r^2 \frac{\partial C}{\partial r}\right) \tag{6.10}$$

The solution is similar but not identical to the one-dimensional case

$$C(r,t) = \frac{M\,4\pi r^2}{(4\pi Dt)^{3/2}} e^{-r^2/(4Dt)} \tag{6.11}$$

In fact, if we write down Eq. (6.8) for x, y, and z, multiply them together, and replace $x^2 + y^2 + z^2 = r^2$, the result is Eq. (6.11). (Note that there is an implicit dx in Eq. (6.8), and so we must also replace the product $dxdydz$ with $4\pi r^2 dr$; Eq. (A6.4).) Thus, we still have a Gaussian distribution about the central starting point. With the aid of Eq. (6.9) we can obtain a spread $\sqrt{r^2} = \sqrt{6Dt}$. This is Eq. (6.9) times the square root of the number of dimensions. The Pythagorian manner by which the different components add together leads to this simple dependence of $\sqrt{r^2}$ on dimension number.

6.2.3 Diffusion across an interface

When solutions with different concentrations are in contact, molecules will diffuse across the interface. One can imagine a tube or pipe filled from one side with a solution and from the other side with just water. We will assume that the pipe is very long so that we do not need to worry about what is happening at the ends. If the tube is filled carefully so that there is no mixing, then immediately

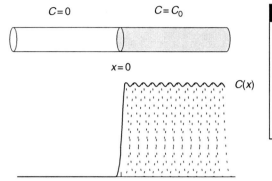

Fig. 6.2. A pipe (above) is filled from $x = 0$ to the right with a solution at concentration C_0 and to the left with concentration 0. The concentration profile (below) of the diffusing solute (solid curve) is represented as infinitely many closely spaced Gaussian functions (dotted curves).

after this delicate operation is complete we have $C = C_0$ for $x > 0$ and $C = 0$ for $x < 0$ (the interface is at $x = 0$).

We need to find a solution to the one-dimensional diffusion equation that satisfies this initial condition. This can be developed by using the Gaussian solution obtained above for the initial condition of a point source of diffusing particles. We consider the region to the right of the interface as an infinite row of infinitesimal point sources, each spreading according to Eq. (6.8). If these point sources are close enough together their sum will approach a constant function (Fig. 6.2).

The solution to this problem is then the superposition of all these point source solutions (Fig. 6.1). This depends on the assumption that solute starting in one of the point sources diffuses independently of the solute from all the other point sources. This turns out not to be a new physical assumption because if there were interference between solute molecules starting at different places then we would not have a simple linear relation between force and flow, and Eq. (6.1) would no longer hold. Because linear relations add, superimposability is automatic.

With this view of diffusion across an interface we can sum up the contributions from each point source. Using x' as the center of a point source, and dx' as the distance between adjacent point sources, we obtain $C(x)$ as an integral in the limit of small dx'

$$C(x) = \frac{C_0}{\sqrt{4\pi Dt}} \int_{x'=0}^{\infty} e^{-(x'-x)^2/(4Dt)} \, dx' \tag{6.12}$$

This integral cannot be solved analytically, but it can be converted to a common special function called the complementary error function, defined as

$$\text{erfc}(x) = \frac{2}{\sqrt{\pi}} \int_{x}^{\infty} e^{-y^2} \, dy \tag{6.13}$$

Now with a change in variable to $z = \frac{(x'-x)}{\sqrt{4Dt}}$, Eq. (6.12) becomes

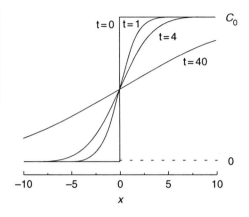

Fig. 6.3. Concentration profiles around an interface between a solution with C_0 and pure water. Equation (6.14) is plotted for a sequence of times, starting with a step function at $t = 0$.

$$C(x) = \frac{C_0}{\sqrt{\pi}} \int\limits_{-x/\sqrt{4Dt}}^{\infty} e^{-z^2}\,dz = \frac{C_0}{2} \text{erfc}(-x/\sqrt{4Dt}) \tag{6.14}$$

This result is plotted in Fig. 6.3. We see that with time the interface spreads. The sharp gradient at time $t = 0$ becomes less and less steep as time passes. This behavior has much in common with spread from a point source. If we note that the derivative of Eq. (6.14) with respect to x gives a Gaussian function, then we realize that with diffusion at an interface the concentration *gradient* looks just like the *concentration* for diffusion from a point source.

The method used to solve this problem can readily be generalized to an arbitrary initial concentration distribution, $C_0(x)$. Again, we treat each point as a spreading Gaussian, and integrate over the distribution, to give

$$C(x) = \frac{1}{\sqrt{4Dt}} \int\limits_{x'=0}^{\infty} C_0(x') e^{-(x'-x)^2/4Dt}\,dx' \tag{6.15}$$

Compare this result with Eq. (6.12).

6.2.4 Diffusion with boundary conditions

The problems treated so far were for infinitely large regions in space, so the boundary conditions did not require any special attention. But in a finite region of space the shape has to be taken into account by specifying what happens at the boundaries. To illustrate this, we look at a pipe that is open at both ends and filled with solution. The pipe is placed in water so that solute diffuses out the ends into a large surrounding reservoir (Fig. 6.4). There is an enormous excess of water surrounding the pipe. When a molecule reaches an open end at $x = 1$ or -1 it is immediately diluted and effectively disappears. The boundary is therefore called *absorbing*, and the mathematical representation of this condition is $C = 0$ at $x = 1$ and -1. This means that the solution to Eq. (6.4) must have a form which obeys this condition for all $t \geq 0$. It must also satisfy the initial distribution $C = C_0$ for $-1 < x < 1$ at $t = 0$.

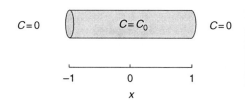

$C=0$ $C=C_0$ $C=0$

−1   0   1

x

Fig. 6.4. A pipe filled with a solution with concentration C_0 is placed into a large container full of pure water. Solute then diffuses out the of pipe into the water.

The Gaussian function obtained for point sources is difficult to use in this problem because it does not help us satisfy the boundary conditions. Instead we use the following general solution to the diffusion equation in one dimension, which can be obtained by the method of separation of variables.[2]

$$C(x,t) = (A\cos(\alpha x) + B\sin(\alpha x))e^{-\alpha^2 Dt} \tag{6.16}$$

This can be checked by substitution back into Eq. (6.4).

Equation (6.16) can readily be adapted to the boundary conditions by finding a set of As and Bs that make these trigonometric functions in Eq. (6.16) add up to C_0. The method of Fourier transforms does exactly that, providing a general method for solving diffusion problems with boundary conditions.

The condition $C = 0$ at $x = \pm 1$ means that as long as α has the form $n\pi/2$ (where n is an odd integer) any cosine term will satisfy the boundary conditions, and any sine term will not. We therefore let all the Bs in Eq. (6.16) be zero and build the initial distribution with cosine terms.

Before we put these cosine functions together to make the rectangular initial distribution, it is instructive to look at how one of these cosine functions behaves. Suppose we had an initial distribution $C_0 = M\cos(\pi x/2)$. Equation (6.16) then gives the solution to the diffusion equation satisfying this particular initial condition

$$C(x,t) = M\cos\left(\frac{\pi x}{2}\right)e^{-(\pi/2)^2 Dt} \tag{6.17}$$

---

[2] We express $C(x, t)$ as the product of one function that depends only on x and another function that depends only on t.

$$C(x,t) = X(x)T(t)$$

Substitution into Eq. (6.4) and rearranging gives

$$\frac{1}{DT}\frac{\partial T}{\partial t} = \frac{1}{X}\frac{\partial^2 X}{\partial x^2}$$

Varying t should not alter X or, for that matter, the entire right-hand side of this equation. The lack of dependence of T on x leads to a similar statement about the left-hand side. Thus, both sides of the equation should be constant. We can solve the two equations formed by setting each side equal to this constant, which will be denoted as $-\alpha^2$

$$T = e^{-\alpha^2 Dt}$$
$$X = A\cos(\alpha x) + B\sin(\alpha x)$$

where A and B are any constants. Taking the product of X and T gives Eq. (6.16).

This function decays uniformly within a pipe, maintaining the basic cosine shape as the decay progresses. At each point the concentration decays exponentially at the same rate. If we put together a sum of such functions, then the concentration at each point will decay as a sum of as many exponentials. At each point the proportion of the different exponentials will vary so that the time course seen at each position will look different.

Returning to our initial condition that $C = C_0$ everywhere within the pipe, the following infinite sum can be obtained by Fourier transformation of a rectangular function

$$C(x, 0) = C_0 \frac{4}{\pi} \left(\cos\left(\frac{\pi x}{2}\right) - \frac{1}{3}\cos\left(\frac{3\pi x}{2}\right) + \frac{1}{5}\cos\left(\frac{5\pi x}{2}\right) - \frac{1}{7}\cos\left(\frac{7\pi x}{2}\right) \cdots \right)$$

(6.18)

This sum has the rectangular shape desired. The solution to the problem is then obtained by using each element of this sum to create a term like Eq. (6.17) as follows

$$C(x, t) = C_0 \frac{4}{\pi} \left(\cos\left(\frac{\pi x}{2}\right) e^{-\left(\frac{\pi}{2}\right)^2 Dt} - \frac{1}{3}\cos\left(\frac{3\pi x}{2}\right) e^{-\left(\frac{3\pi}{2}\right)^2 Dt} \right.$$
$$\left. + \frac{1}{5}\cos\left(\frac{5\pi x}{2}\right) e^{-\left(\frac{5\pi}{2}\right)^2 Dt} - \frac{1}{7}\cos\left(\frac{7\pi x}{2}\right) e^{-\left(\frac{7\pi}{2}\right)^2 Dt} \cdots \right)$$

(6.19)

Since each term in this sum has the form of Eq. (6.16), it satisfies the diffusion equation. This sum reverts to Eq. (6.18) at $t = 0$, thus satisfying the initial condition. At very short times the edges look very much like the solution for diffusion at an interface (Fig. 6.3). At longer times the rapidly decaying components are gone so we are left with Eq. (6.17), a uniformly decaying cosine function. A point worth noting here is that the terms in Eq. (6.19) that represent higher spatial frequencies decay more rapidly. Thus, diffusion collapses sharp spatial gradients rapidly and weak spatial gradients slowly. This illustrates a general property of diffusion. Diffusion makes concentrations change more rapidly on shorter distance scales.

The above example illustrates absorbing boundary conditions, in which the solution to the diffusion equation must be zero at the boundary. If the ends are sealed we have what is called a *reflecting* boundary. Because the flux, J, is zero at such a boundary, Eqs. (6.3) or (6.5) dictate that the spatial concentration derivatives normal to the boundary are zero. In an enclosure like that of Fig. 6.4, but with sealed ends, this means that we use the sine terms of a Fourier series rather than the cosine terms (Problem 5).

6.3 | Diffusion at steady state

In the above examples the concentration changed with time. However, there is an important class of problems in which the concentration stays the same within a large region. The diffusing substance enters in one place and exits from another at the same rate. The substance thus diffuses from a *source* to a *sink*, with a continuous drop in concentration along the way. There will be a steady flux of material through the system, as the flow into each volume element perfectly balances the flow out. Such a system is said to be in a *steady state*, as defined by the condition that the time derivative of the concentration is zero throughout a region of interest

$$\frac{\partial C}{\partial t} = 0 \tag{6.20}$$

The diffusion equation, Eq. (6.4), now simplifies to the Laplace equation

$$\nabla^2 C = 0 \tag{6.21}$$

This equation can be solved to determine the steady-state concentration distribution for a particular geometrical arrangement of boundaries, sources, and sinks.

6.3.1 A long pipe
To illustrate the basic properties of a steady state, consider a pipe of length l connecting two vessels, one filled with a solution with concentration C_0, and the other filled with pure water. Taking Eq. (6.21) in one dimension, and integrating once gives

$$\frac{\partial C}{\partial x} = A \tag{6.22}$$

where A is a constant of integration. Integrating again gives

$$C = Ax + B \tag{6.23}$$

where B is another constant of integration. We take the end of the pipe at the water filled container to be $x = 0$, and since $C = 0$ at that point, $B = 0$. At the other end of the pipe $C = C_0$, so at $x = l$ we solve for $A = C_0/l$. With A and B thus determined, Eq. (6.23) gives the steady-state concentration in the pipe

$$C(x) = \frac{C_0 x}{l} \tag{6.24}$$

Thus, the concentration changes linearly from 0 to C_0 as the pipe is traversed.

The flux, J, at any point can be obtained from Eq. (6.24) by differentiating Eq. (6.1) to give $-C_0 D/l$. This is the flux through a unit area. With an area, a, then the total flux is $-C_0 Da/l$. This expression is just like that for the electrical current in a resistor. Recall that the electrical resistance of an object is proportional to length and inversely proportional to area, so dividing by the resistance gives a current proportional to a/l.

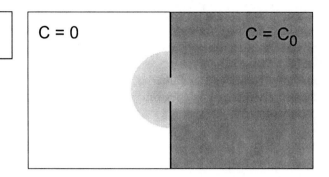

Fig. 6.5. Two solutions with $C=0$ and $C=C_0$ are separated by a divider with a hole.

6.3.2 A small hole

Now instead of a pipe connecting two containers, consider a small round hole connecting a solution of concentration C_0 with pure water (Fig. 6.5).

Assume that C depends only on the distance from the hole, i.e. it depends only on r. We envision hemispherical shells around the hole over which the concentration is uniform. We can then use the diffusion equation in spherical coordinates (Eq. (6.10)). At steady state this reduces to

$$\frac{1}{r^2}\frac{\partial}{\partial r}\left(r^2\frac{\partial C}{\partial r}\right) = 0 \tag{6.25}$$

Integrating once gives

$$r^2\frac{\partial C}{\partial r} = A \tag{6.26}$$

Integrating again gives

$$C = \frac{-A}{r} + B \tag{6.27}$$

where A and B once again are constants of integration.

As $r \to \infty$ on the left $C \to 0$, so for the solution on the left side of the divider, we have $B = 0$. At the center of the hole, C should be half way between 0 and C, i.e. $C_0/2$. We will make the reasonable approximation that C is $C_0/2$ very close to the pore, within its radius of r_a. Inserting this value in Eq. (6.27) allows us to determine A as $r_a C_0/2$. So on the left side of the hole the concentration is

$$C = \frac{r_a C_0}{2r} \tag{6.28}$$

Now we can apply Eq. (6.1) to obtain the flux at any value of r

$$J = -D\frac{\partial C}{\partial r} = D\frac{r_a C_0}{2r^2} \tag{6.29}$$

This is the flux per unit area. If we want to know the flux through any hemispherical shell we multiply by its area, $2\pi r^2$. Then the total rate of transport of solute through any hemispherical shell, T, is the same, as expected for a steady state

$$T = \pi r_a D C_0 \tag{6.30}$$

Because the surfaces of constant concentration are not exactly spherical, especially for small r, this result is approximate. The exact treatment involves solving the diffusion equation in cylindrical coordinates and gives Eq. (6.30), but with π replaced by 2 (Crank, 1975; Berg, 1983).

An important property of Eq. (6.30) is that the rate of transport is proportional to the radius of the hole rather than its area. This is not an intuitive result; recall that for the case just examined of a pipe the rate of transport is proportional to the area. For diffusion through holes transport is limited by the approach to the hole rather than passage through the hole. A general result for arbitrarily shaped holes is that the rate of transport scales with a linear dimension rather than with the area. A similar dependence arises in diffusion-limited association processes (Chapter 8). This scaling with size has important implications for transport through a porous membrane, as will be seen next.

6.3.3 A porous membrane

Picture a membrane covering the hole in Fig. 6.5. This membrane has many pores, each a miniature of the hole in the preceding example. For each pore the rate of transport will be given by Eq. (6.30), with the pore radius, r_p, in place of r_a. If there are N such pores, then we get a rate of transport

$$T = N\pi r_p D C_0 \tag{6.31}$$

However, there is something odd about this expression. When $N r_p = r_a$ the rate of transport through this porous membrane will be the same as in Eq. (6.30), which is what we got when there was no membrane at all, just a hole. Yet the fraction of the membrane area accounted for by these pores would be $N(r_p/r_a)^2$. If the pores are small compared to the entire hole, then because the ratio of the square of the radii is taken, the pores could cover a very small fraction of the total area of the membrane and still give the same flux as the open hole. For example, if the pores have diameters that are smaller than the hole by a factor of 100, then 100 of them would bring the rate of transport up to that of the unobstructed hole. But since each pore is a fraction $(0.01)^2 = 0.0001$ of the total area, 100 pores would account for only 1% of the membrane area. Indeed, we could get the impossible result that with more than 100 pores, transport would be greater than the unobstructed hole.

The problem arises from the overlap of the shells around each pore when they are close together. If we imagine two holes in the membrane, the spherical sections representing the gradients will overlap when the holes are close together. For a very small number of pores this would not matter and Eq. (6.31) would be reasonably accurate. But when more pores are added, the crowding of the shells makes Eq. (6.31) less accurate. A more complete treatment of this problem takes the overlap into account (Berg, 1983) and gives a rate of transport through a porous membrane as

$$\frac{T}{T_0} = \frac{1}{1 + (\pi r_a / N r_p)} \qquad (6.32)$$

where T_0 is the rate of transport for the unobstructed hole from Eq. (6.30).

The implication of this analysis is that a few pores can have a large impact on how quickly solute crosses a cell membrane. A cell membrane can harbor many different types of membrane transporters and allow the selective passage of many different kinds of molecules. Since the transporters for each molecule only need to occupy a small fraction of the membrane surface area, they will not crowd one another. A cell membrane can thus transport many different substances efficiently.

6.4 | Microscopic diffusion – random walks

In the above macroscopic treatment of diffusion, very little thought was given to what makes the molecules move (at least not at the molecular level). We know that molecules move randomly and collide with one another at a very high frequency. Now we will focus on this random motion to see what it can tell us about diffusion (Chandrasekhar, 1943).

The molecular approach to diffusion begins with the random-walk model. A molecule is said to move in discrete steps of fixed length, δ, and to take these steps at fixed intervals of time, τ. We will start with only one dimension, so steps are either plus or minus (i.e. right or left).

If a random walk has no directional bias, then the probabilities of steps in either direction will be equal. It is no more difficult to develop the theory with unequal probabilities of steps in different directions, so we will do so for the sake of generality. The probabilities of stepping right or left are taken as p and q, respectively, where $p + q = 1$. If N steps are taken, the probability of having m right steps (and simultaneously of $N - m$ left steps) is given by the binomial distribution

$$P(m) = \frac{N! p^m q^{N-m}}{(N-m)! m!} \qquad (6.33)$$

If right is positive and left negative, then m right steps and $N - m$ left steps give a net displacement of $x = (2m - N)\delta$. This means that $P(m)$ is the probability of a total displacement of this magnitude after N steps.

We can calculate the mean displacement from the origin as a function of time. First, we need the mean number of right steps, which is obtained by averaging over all possible values of m as

$$\overline{m} = \sum_{m=0}^{N} mP(m) = \sum_{m=0}^{N} \frac{m\,N!p^m q^{N-m}}{(N-m)!m!} \tag{6.34}$$

To evaluate this sum we start with the binomial expansion (Eq. (A1.7))

$$(p+q)^N = \sum_{m=0}^{N} \frac{N!p^m q^{N-m}}{(N-m)!m!} \tag{6.35}$$

Differentiate this expression with respect to p, and then multiply by p to give

$$Np(p+q)^{N-1} = \sum_{m=0}^{N} \frac{N!\,m\;p^m q^{N-m}}{(N-m)!m!} \tag{6.36}$$

Since $p+q=1$, the left-hand side is Np. The right-hand side is the sum in Eq. (6.34), so we have

$$\overline{m} = Np \tag{6.37}$$

Since $x = (2m - N)\delta$, the mean displacement is

$$\bar{x} = (2\overline{m} - N)\delta = (2Np - N)\delta = N(p - q)\delta \tag{6.38}$$

If $p = q = 1/2$, then $\bar{x} = 0$, which is expected because on average there will be the same number of steps in each direction.

The result that the mean displacement is zero for an unbiased random walk provides us with very little insight into the nature of the motion. The mean-square displacement provides a better way to quantify the progress of a random walk

$$\overline{x^2} = \delta^2 \,\overline{(2m - N)^2} \tag{6.39}$$

Squaring $2m - N$ and averaging each of the terms gives

$$\overline{x^2} = \delta^2 (4\overline{m^2} - 4\overline{m}N + N^2) \tag{6.40}$$

The mean square of m, $\overline{m^2}$, is calculated as

$$\overline{m^2} = \sum_{m=0}^{N} m^2 P(m) = \sum_{m=0}^{N} \frac{m^2 N!p^m q^{N-m}}{(N-m)!m!} \tag{6.41}$$

Differentiating Eq. (6.36) with respect to p and then multiplying by p gives an expression for this sum. Using the result in Eq. (6.41) gives

$$\overline{m^2} = (Np)^2 + Np(1 - p) = (Np)^2 + Npq \tag{6.42}$$

Substituting Eqs. (6.37) and (6.42) into Eq. (6.40) gives

$$\overline{x^2} = \delta^2 (4N^2 p^2 + 4Npq - 4pN^2 + N^2) \tag{6.43}$$

For an unbiased random walk, with $p = q = 1/2$, this expression becomes very simple

$$\overline{x^2} = \delta^2 N = \delta^2 \frac{t}{\tau} \tag{6.44}$$

where $t = N\tau$ was used for the total time to get the final answer.

Taking the square root of Eq. (6.44), we see that the rms displacement goes as the square root of the time, just as was found above from the solution of the diffusion equation (Eq. (6.9)). Furthermore, we can obtain an expression for D in terms of the basic parameters of the random-walk model by comparing the square root of Eq. (6.44) with Eq. (6.9)

$$D = \delta^2/2\tau \qquad (6.45)$$

With this expression for D we can estimate the effective step size and frequency for the motion of a diffusing particle during collisions with other molecules. Kinetic theory gives the rms velocity of a molecule as $\sqrt{(3kT/m)}$. For a 14 000 Da protein in solution (e.g. the enzyme lysozyme) this is $1.3 \times 10^3 \, \text{cm s}^{-1}$. At this velocity a molecule of protein would bounce around in a beaker dozens of times per second. It does not diffuse nearly that fast because it changes its direction so frequently.

Let's see how frequently. The diffusion constant of lysozyme is $\sim 10^{-6} \, \text{cm}^2 \, \text{s}^{-1}$. Taking the rms velocity as δ/τ and D as $\delta^2/2\tau$, we can solve these two equations for the two unknowns to obtain $\delta = 10^{-9} \, \text{cm}$ and $\tau = 10^{-12} \, \text{s}$. This means that the molecule changes direction every picosecond and moves just 0.1 Å between collisions. These time and distance scales are well worth remembering. They reflect the fundamental nature of molecular motions in aqueous solutions.

The above use of the binomial distribution was for a one-dimensional random walk, but it is easy to extend this to two or three dimensions. We can consider motion in each dimension to be an independent random walk. Then by the Pythagorean theorem we have $\overline{r^2} = \overline{x^2} + \overline{y^2} + \overline{z^2}$. With $\overline{x^2}$, $\overline{y^2}$, and $\overline{z^2}$ each equal to $\delta^2 t/\tau$ (Eq. (6.44)), we have $\sqrt{\overline{r^2}} = \delta \sqrt{(3t/\tau)}$.

The random walk has the property that the mean-square displacement increases linearly with the number of steps. This relation between a mean-square quantity and the number of items in a sum appears often in statistical problems. Recall from Chapter 3 that a polymer with monomers rotating in an uncorrelated fashion has an rms end-to-end distance that goes as the square root of the monomer number. These kinds of results are generalized in the central-limit theorem, an important basic result of mathematical probability.

*6.5 | Random walks and the Gaussian distribution

We saw that the random-walk model gives the same basic result as the diffusion equation for the mean-square displacement (Eq. (6.44) and square root of Eq. (6.9)). The parallel between the random walk and diffusion can be extended by showing that as N becomes large the probability distribution for the random walk goes to the Gaussian function obtained by solving the diffusion equation (Chandrasekhar, 1943). This is the appropriate solution of the

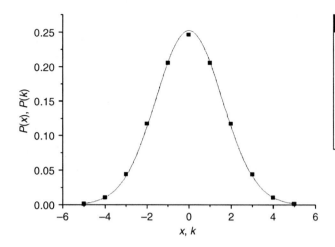

Fig. 6.6. The plotted points are probabilities for displacements according to a random walk of ten steps (Eq. (6.46) with $N = 10$). The Gaussian function obtained by solving the diffusion equation (Eq. (6.8)) is drawn as a solid curve for comparison. Note that $\delta = 0.5$ to give $x = 1$ at $k = 1$.

diffusion equation to compare with the random walk because the initial condition $C(x, 0) = \delta(x)$ corresponds to the random walk starting at $x = 0$.

First, we compare the two results graphically. If we take $p = 1/2$, then the most likely value of m is $N/2$. We can now introduce a new variable k to express the displacement in terms of the deviation from $N/2$, so that m is then $N/2 + k$. (The displacement x is $2k\delta$ in terms of k; it was $(2m - N)\delta$ in terms of m.) Equation (6.33) then becomes

$$P(k) = \frac{N!(1/2)^N}{(\frac{N}{2} - k)!(\frac{N}{2} + k)!} \tag{6.46}$$

A plot of this equation with $N = 10$ and $x = 2k\delta$ (Fig. 6.6) is almost exactly the same as the corresponding Gaussian.

This similarity between the Gaussian function and the binomial distribution can be demonstrated mathematically by taking the limit of Eq. (6.46) when N is large. We use Stirling's approximation for the factorial[3] $N! \sim \left(\frac{N}{e}\right)^N \sqrt{2\pi N}$.

---

[3] The logarithm of $N!$ is the sum

$$\ln N! = \sum_{m=1}^{N} \ln m$$

Approximating the sum as an integral gives

$$\ln N! \sim \int_{x=1}^{N} \ln x \, dx \sim N \ln N - N$$

where the term evaluated at $x = 1$ is discarded as it is negligibly small.

This expression is sufficient for most applications and would yield the desired Gaussian function here, but with a significant discrepancy. To achieve quantitative agreement with the solution to the diffusion equation we improve this formula by taking $(N + 1/2) \ln N - N$. This extra $1/2$ can be understood if the sum is viewed as the area of many rectangles and the integral only goes to the midpoint of the last rectangle. Then the sum differs from the integral by an extra half of the rectangle

$$P(x) \sim \frac{\left(\frac{N}{e}\right)^N \sqrt{2\pi N} \left(\frac{1}{2}\right)^N}{\left(\frac{\frac{N}{2}-k}{e}\right)^{\frac{N}{2}-k} \sqrt{2\pi \left(\frac{N}{2}-k\right)} \left(\frac{\frac{N}{2}+k}{e}\right)^{\frac{N}{2}+k} \sqrt{2\pi \left(\frac{N}{2}+k\right)}}$$

$$= \frac{\sqrt{2\pi N}}{2\pi \sqrt{\left(\left(\frac{N}{2}\right)^2 - k^2\right)}} \frac{\left(\frac{N}{e}\right)^N \left(\frac{1}{2}\right)^N}{\left(\frac{\frac{N}{2}-k}{e}\right)^{\frac{N}{2}-k} \left(\frac{\frac{N}{2}+k}{e}\right)^{\frac{N}{2}+k}} \tag{6.47}$$

Since the square root varies much less than the exponentials, the k^2 in the square root can be neglected. Furthermore, the factors of e^N cancel leaving

$$P(x) = \sqrt{\frac{2}{\pi N}} \frac{\left(\frac{N}{2}\right)^N}{\left(\frac{N}{2}-k\right)^{\frac{N}{2}-k} \left(\frac{N}{2}+k\right)^{\frac{N}{2}+k}}$$

$$= \sqrt{\frac{2}{\pi N}} \frac{\left(\frac{N}{2}\right)^N}{\left(\frac{N}{2}\right)^{\frac{N}{2}-k} \left(1-\frac{2k}{N}\right)^{\frac{N}{2}-k} \left(\frac{N}{2}\right)^{\frac{N}{2}+k} \left(1+\frac{2k}{N}\right)^{\frac{N}{2}+k}}$$

$$= \sqrt{\frac{2}{\pi N}} \left(1-\frac{2k}{N}\right)^{-N/2} \left(1-\frac{2k}{N}\right)^k \left(1+\frac{2k}{N}\right)^{-N/2} \left(1+\frac{2k}{N}\right)^{-k}$$

$$= \sqrt{\frac{2}{\pi N}} \left(1-\frac{4k^2}{N^2}\right)^{-N/2} \left(1-\frac{2k}{N}\right)^k \left(1+\frac{2k}{N}\right)^{-k} \tag{6.48}$$

Each factor containing a term of the form $1 \pm a$ can be approximated as $e^{\pm a}$ because $k << N$ (expansion to first term in Eq. (A1.4)).

$$P(x) = \sqrt{\frac{2}{\pi N}} e^{2k^2/N} e^{-2k^2/N} e^{-2k^2/N}$$

$$= \sqrt{\frac{2}{\pi N}} e^{-2k^2/N} \tag{6.49}$$

Thus, we have obtained the limiting Gaussian for the binomial distribution and can see how Eq. (6.8) is beginning to emerge. To complete the picture we must convert numbers of steps to units of distance and time. Using $k = x/2\delta$ and $N = t/\tau$ gives

$$P(x) = \sqrt{\frac{2\tau}{\pi t}} e^{-\tau x^2/2t\delta^2} \tag{6.50}$$

---

at the upper limit of N. An additional factor of $1/2 \ln(2\pi)$ does a good job of correcting for the accumulated error between the area of the continuous function and the rectangles. Incorporating these corrections to the above expression and taking the exponential gives

$$N! \sim \left(\frac{N}{e}\right)^N \sqrt{2\pi N}$$

Using Eq. (6.45) to replace τ and δ with the diffusion coefficient gives

$$
\begin{aligned}
P(x) &= \sqrt{\frac{\delta^2}{\pi Dt}}\, e^{-x^2/4Dt} \\
&= \sqrt{\frac{1}{4\pi Dt}}\, e^{-x^2/4Dt}\, 2\delta
\end{aligned}
\tag{6.51}
$$

We can interpret 2δ as dx because $x = 2k\delta$. That makes the result identical to Eq. (6.8).

As shown above in Fig. 6.6, N does not have to be very large for this limit to work extremely well. One region where the Gaussian limit clearly cannot be correct is for $x > N\delta$. The largest possible displacement occurs with all N steps in the same direction. The Gaussian function gives extremely low probabilities for larger displacements, but their probability is zero in the random walk. These errors rarely create problems and the Gaussian function provides a very useful representation for the random-walk model.

6.6 | The diffusion equation from microscopic theory

It was noted above that assuming each step is independent leads to the result that the rms distance is proportional to $\sqrt{t}$. The assumption of independent steps is at the heart of diffusion, and Einstein used it as the basis for a derivation of the diffusion equation. (See Einstein (1956) for a collection of Einstein's seminal papers on Brownian motion.)

The probability of arriving at position r at time $t + dt$ is denoted as $p(r, t + dt)$. It can be written as the integral over all products of two probabilities. The first is the probability of arriving at $r + \delta$ after time t, denoted as $p(r + \delta, t)$. The second is the probability of jumping from $r + \delta$ to r in the time dt, denoted as $\psi(\delta)$. We therefore write

$$
p(r, t + dt) = \int_{-\infty}^{\infty} p(r + \delta, t)\psi(\delta)\, d\delta
\tag{6.52}
$$

Note that dt and δ are both small, so the range for δ in the integral is irrelevant. Using Taylor expansions (see Appendix 1) for $p(r + \delta, t)$ and $p(r, t + dt)$ gives

$$
p(r, t + dt) = p(r, t) + dt\,\frac{\partial p(r, t)}{\partial t}
\tag{6.53}
$$

and

$$
p(r + \delta, t) = p(r, t) + \delta\,\frac{\partial p(r, t)}{\partial r} + \frac{1}{2}\delta^2\,\frac{\partial^2 p(r, t)}{\partial r^2}
\tag{6.54}
$$

We now replace the left-hand side of Eq. (6.52) with Eq. (6.53) and use Eq. (6.54) for $p(r + \delta, t)$ on the right-hand side.

$$p(r,t) + dt \frac{\partial p(r,t)}{\partial t} = \int_{-\infty}^{\infty} \left(p(r,t) + \delta \frac{\partial p(r,t)}{\partial r} + \frac{1}{2}\delta^2 \frac{\partial^2 p(r,t)}{\partial r^2} \right) \psi(\delta) \, d\delta$$

(6.55)

The first term of the integral on the right-hand side is simply $p(r, t)$, because the integral over all possible steps δ must be one. It then cancels with $p(r, t)$ on the left-hand side. The second term is zero because the probability of a plus step equals the probability of a minus step. The last term does not reduce to a simple expression, but Einstein assumed that this integral divided by dt has a definite value, and equated it with the diffusion coefficient

$$\frac{1}{dt} \int_{-\infty}^{\infty} \tfrac{1}{2}\delta^2 \psi(\delta) \, d\delta = D$$

(6.56)

Note the similarity with the expression from the random walk model $D = \delta^2/2\tau$ (from Eq. (6.45)). Equation (6.55) then becomes the diffusion equation

$$\frac{\partial p(r,t)}{\partial t} = D \frac{\partial^2 p(r,t)}{\partial r^2}$$

(6.57)

The last step of this derivation (Eq. (6.56)), in which it is assumed that the mean square displacement divided by dt remains constant, is equivalent to assuming that rms displacement increases with the square root of time. Thus, the logic is circular because this step actually assumes one of the pivotal results of diffusion theory.

Perhaps the most important assumption that goes into this derivation of the diffusion equation is embodied by Eq. (6.52), which expresses the independence of the two probabilities p and ψ. That is why they can be multiplied together to obtain a joint probability. No matter where in space the particle is, the probability of making a jump δ is given by $\psi(\delta)$, a function that does not depend on r. The jumping behavior is thus independent of past history. When such an assumption can be made we say that we are dealing with a *Markov* process. This is an important assumption that appears in many mathematical treatments of dynamic processes, and it will be seen again in Chapters 7 and 9.

6.7 | Friction

Friction and diffusion are closely related. The molecular collisions that generate Brownian motion will also retard sustained motion in any direction. We will now examine this relation and see how to express the diffusion constant in terms of the coefficient of friction.

Let's take a molecule in solution under the influence of an external force F_e. This could be a charged molecule in an electric field, or we could consider a gravitational or centrifugal field. As a molecule is accelerated by the field its movement is opposed by friction. This force, F_f, is taken as proportional to the molecule's velocity

$$F_f = -fv \tag{6.58}$$

where f denotes the coefficient of friction, and the minus sign indicates that friction opposes the motion. As the molecule speeds up, F_f increases until it equals F_e. At that point the net force, $F_f + F_e$ is zero, and the balance between the F_e and F_f leads to a motion with a constant velocity. So $F_e - fv = 0$, and we have $v = F_e/f$. With this average velocity, the flux of molecules will be $J = C(x)v = C(x)F_e/f$.

The external force creates a concentration gradient, which makes molecules diffuse in the opposite direction. If an equilibrium is reached, then $C(x)$ will be stable, reflecting a balance between these two opposing forces. This balance is expressed by adding the flux produced by the external force, $C(x)F_e/f$, to the flux produced by diffusion down the concentration gradient (Eq. (6.1)), and setting this sum equal to zero

$$\frac{C(x)F_e}{f} - D\frac{\partial C(x)}{\partial x} = 0 \tag{6.59}$$

The next step is to realize that in this equilibrium state where the external force is balanced by diffusion, $C(x)$ will obey the Boltzmann distribution. For a constant force independent of position, the potential energy is simply $F_e x$. The Boltzmann distribution for this potential energy function is then

$$C(x) = C(0)\, e^{-F_e x/kT} \tag{6.60}$$

where $C(0)$ is the concentration of some reference position. Differentiating with respect to x and comparing with the original expression leads to

$$\frac{C(x)\, F_e}{kT} - \frac{\partial C(x)}{\partial x} = 0 \tag{6.61}$$

Finally, comparing Eqs. (6.59) and (6.61) leads to the relation

$$D = \frac{kT}{f} \tag{6.62}$$

This equation was derived by Einstein, and bears his name. It tells us how diffusion and friction are related. Friction slows diffusion down, regardless of the shape of the molecule or the viscosity of the fluid. Recall that the random walk model gave us an expression for D in terms of the size and duration of steps underlying the random thermal motion of molecules. Equation (6.62) tells us that these fluctuations that give rise to diffusion also give rise to friction. Diffusion theory thus gives us a special physical picture of friction. It is a consequence of the randomness of the motion of many individual molecules colliding with an object and changing its direction of motion. This will be explored further in Section 12.14.

The summation of an external force with the force of friction gives us a general way to study diffusion in a force field. We take Eq. (6.59), but allow for an imbalance between F_f and F_e. The observed flux is then the sum of these two terms

$$J = \frac{C(x)F_e}{f} - D\frac{\partial C(x)}{\partial x} \tag{6.63}$$

And with the Einstein relation, we have

$$J = D\left(\frac{C(x)F_e}{kT} - \frac{\partial C(x)}{\partial x}\right) \tag{6.64}$$

This equation is an important starting point in the development of theories for a variety of rate processes in later chapters. When the external force is electrical, Eq. (6.64) is known as the Nernst–Planck equation.

6.8 | Stokes' law

The preceding section provided a strong incentive for understanding the friction on a small diffusing particle as it moves through a fluid. If the frictional coefficient can be measured, or a theory used to derive it, then Eq. (6.62) tells us that we will also know the diffusion constant. So we will now briefly comment on the frictional coefficient of a spherical particle.

The calculation of a frictional coefficient requires solving the equations for viscous fluid flow, and this can be quite complex. In the case of a spherical molecule, the fluid streams around as depicted in Fig. 6.7. A complete description of the fluid flow can be obtained, and from this the force exerted by the fluid on the particle can be calculated (Berg, 1983). From the expression for force, the coefficient of friction can be derived, and the result is known as Stokes' law

$$f = 6\pi\eta r \tag{6.65}$$

where r is the radius and η is the viscosity.

For irregularly shaped objects the calculation of f can be much more complicated, so Eq. (6.65) is used a great deal to make rough order of magnitude estimates of f in terms of molecular size. For example, it is common to assume that a protein is a sphere with a

Fig. 6.7. A sphere moving through a viscous fluid experiences friction as the fluid moves around it.

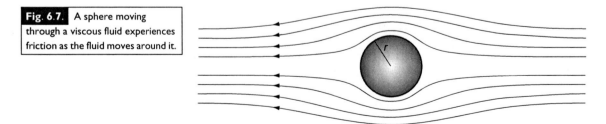

radius proportional to the cube root of the molecular weight, and estimate f in this way. For a number of different geometries, mathematical analysis has demonstrated that f is proportional to, or nearly proportional to, a linear dimension of the molecule. This means that we can expect f to scale as the cube root of the molecular weight for a series of molecules with the same overall shape. The following section explores this relationship.

6.9 | Diffusion constants of macromolecules

We can use Stokes' law and the Einstein relation to make useful estimates of the diffusion constants of proteins. Combining Eqs. (6.62) and (6.65) gives an expression for the diffusion constant in terms of the size of the molecule

$$D = \frac{kT}{6\pi\eta r} \tag{6.66}$$

This expression is referred to as the Stokes–Einstein relation. If we consider a globular protein as being approximately spherical, then the weight of a single molecule will be $(4/3)(\pi r^3 \rho)$, where ρ is the density. The density of globular proteins does not vary much, and is generally around $0.7\,\mathrm{g\,ml}^{-1}$. Equation (6.66) suggests that D will be inversely proportional to the cube root of the molecular weight.

Diffusion coefficients of many proteins have been measured, generally by setting up an interface like that in Figs. 6.2 and 6.3, recording the change in concentration with time, and fitting to Eq. (6.14) (Tanford, 1961). Some values are plotted versus molecular weight in Fig. 6.8, and the trend is very clear. The points are close to a line representing $M^{-1/3}$, indicating the appropriate proportionality. But in reality all the diffusion constants are slightly lower than that expected from the Stokes–Einstein relation. There are two reasons for this: (1) proteins are hydrated and this makes their size a bit bigger than that expected from their molecular weight; (2) globular proteins are usually not perfect spheres, and the irregular shape increases the frictional coefficient, thus reducing D.

Frictional coefficients have been calculated for other geometries such as ellipsoids. The results are fairly complex but a generalization emerges that for a given volume, a spherical shape has the lowest frictional coefficient. Deviations from spherical shape always increase f and decrease D. This effect is clearly illustrated by the triangles plotted in Fig. 6.8. These points are the diffusion coefficients of fibrous proteins. Stokes' law does not come close to describing their coefficients of friction, so they fall well below the values for globular proteins. Globular proteins with similar molecular weights have much higher diffusion constants.

Randomly coiled molecules also behave differently from globular proteins. In Chapter 3 we saw that many macromolecules behave like random coils, and have a mean-square end-to-end distance that

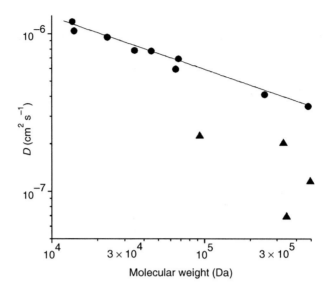

Fig. 6.8. The diffusion constants of a number of globular proteins are plotted versus molecular weight. In this log–log plot the slope is very close to −1/3, as illustrated by the line drawn through the filled circles ($\propto M^{-1/3}$). The globular proteins plotted here are (in order of increasing size) ribonuclease, lysozyme, chymotrypsinogen, β-lactoglobulin, ovalbumin, serum albumin, hemoglobin, catalase, and urease. The triangles are for fibrous proteins, tropomyosin, fibrinogen, collagen, and myosin. (Data from Table 21–1 of Tanford, 1961.)

scales as the number of segments. With $\sqrt{\overline{r^2}} \propto \sqrt{N}$ the frictional coefficient would then be proportional to $M^{1/2}$ rather than $M^{1/3}$. Combining Eqs. (6.62) and (6.65) then gives D inversely proportional to the square root of the molecular weight, and for random coils this has been confirmed experimentally (Tanford, 1961; Cantor and Schimmel, 1980). For charged molecules in an ionic solution the surrounding ions move in the opposite direction, pulling the water and slowing the motion of the charged molecule. This is studied in Section 11.8.

6.10 | Lateral diffusion in membranes

The emergence of the concept of membrane fluidity around 1970 sparked a great deal of interest in the two-dimensional diffusion of molecular components in biological membranes. Both lipid and protein molecules undergo random walks within the plane of a lipid bilayer membrane. This was demonstrated by using fluorescent labels to track movement over the surface of a cell. When a labeled cell fuses with an unlabeled cell, the fluorescence is initially confined to one side, but within a few minutes the fluorescence becomes uniform. When a laser photobleaches label in a small spot of cell surface, fluorescence recovery at that site reveals the diffusion of unbleached label into the bleached region. The technique of fluorescence recovery after photobleaching, commonly known as FRAP, was developed as a powerful quantitative tool in the study of lateral diffusion in membranes (Jacobson *et al.*, 1987). The technique of fluorescence correlation spectroscopy (Section 12.13) has also been widely used.

Two-dimensional diffusion coefficients of both lipids and proteins are generally at least two orders of magnitude lower than those of typical globular proteins discussed in the preceding section. Relating the diffusion coefficient to the size of the molecule and the viscosity of the lipid membrane is a harder problem in two dimensions than in three. The basic result in three dimensions, Stokes' law (Eq. (6.65)), depends on the existence of a steady state for the flow of the fluid around a sphere (Fig. 6.7). It turns out that in two dimensions there is no steady-state solution to the relevant equations for fluid flow, and that makes the mathematics much more challenging. An analysis of this problem by three more sophisticated alternative methods converged on the following expression for the lateral diffusion constant for a cylinder of radius a embedded in a membrane of thickness h (Saffman and Delbrück, 1975)

$$D_L = \frac{kT}{4\pi\eta_m h} \left(\ln \frac{\eta_m h}{\eta_w a} - 0.5772 \right)$$

(6.67)

The membrane viscosity is η_m and the viscosity of the surrounding aqueous medium is η_w. This result depends on the reasonable assumption that the former is the larger of the two.

Equation (6.67), known as the Saffman–Delbrück equation, differs in some important ways from Stokes' law. The logarithmic dependence on inverse radius is much weaker than the dependence on inverse radius in Stokes' law. So molecules of very different sizes have very similar lateral diffusion coefficients. In fact, proteins and lipids with diameters differing by a factor of ~10 have diffusion coefficients that differ by less than a factor of 2 (Peters and Cherry, 1982). Equation (6.67) indicates that D_L is also insensitive to the viscosity of the aqueous phase. Experiments in which η_w was varied over a 12-fold range changed D_L by only 50%. By contrast, increasing the membrane viscosity, for example by reducing the temperature below the gel-point at which the lipids become ordered, reduced D_L of both proteins and lipids by more than an order of magnitude.

The value of D_L for proteins in cell membranes is usually much lower than in artificial lipid bilayers (Jacobson *et al.*, 1987). One reason is that membrane proteins are often attached to cytoplasmic structures and cytoskeletal elements. Another important effect is molecular crowding. Cell membranes often have proteins occupying more than half of their surface area. Membrane proteins tend to get in each other's way, obstructing their random walks and reducing D_L.

It is often vital to the function of a receptor, channel, or transporter to be localized to a specific region of a cell surface. FRAP often reveals fractions of protein with different mobilities, indicating that some are anchored in some way and others are relatively free to diffuse. Recent studies in which single molecules were tracked as they moved in the cell membrane revealed surprising new forms of motion, indicating a high degree of organization in the way membranes direct protein traffic (Saxton and Jacobson, 1997). For membrane proteins, the freedom to wander makes it possible for them to form complexes, signal one another, regulate their function, and turn over through endocytosis.

Problems for Chapter 6

1. Verify that Eq. (6.11) satisfies conservation of mass.
2. Show by direct substitution into Eq. (6.4) that Eq. (6.14) is a solution to the diffusion equation.
3. Use Eq. (6.11) to derive the rms radial displacement from a point source in three dimensions, i.e. determine $\sqrt{\overline{r^2}}$ as a function of t.
4. For one-dimensional diffusion from a point source (Eq. (6.8)) determine where c is increasing, where it is decreasing, and where it is momentarily constant.
5. Solve the problem in Section 6.2.4, but with sealed ends (reflecting boundary conditions) and an initial condition of $C(x, 0) = \delta(x)$.
6. Solve the problem of diffusion in one dimension in a semi-infinite pipe with (a) a sealed end and (b) an open end at $x = -1$, and with an initial condition $C(x, 0) = \delta(x)$. (Hint: place another Gaussian at the image position, $x = -2$ to satisfy the boundary condition at $x = -1$.)
7. Write the Gaussian function plotted in Fig. 6.6.
8. During synaptic transmission, the transmitter released from a single vesicle (50 000 molecules) starts off at a highly concentrated point that can be approximated as a delta function. Diffusion then proceeds in two dimensions through the synaptic cleft (thickness 50 nm). Calculate the mean concentration as a function of time within a 2 μm disc centered at the release site (Khanin *et al.*, 1994).

Chapter 7

Fundamental rate processes

This chapter will continue the treatment of dynamic processes started in Chapter 6. In that chapter the dynamic process of diffusion was treated as transitions within a continuum of states. Here we will take the opposite extreme. We will look at interconversions between just two states, and model these interconversions as a one-step process. The same model was used for the global transitions of Chapter 1.

The kinetics of interconversion when a molecule simply flips back and forth between two states is mathematically straight-forward. The two-state model thus serves as an elementary building block for more complicated multi-state kinetic models (Chapter 9). Multi-state models represent a middle ground between two-state models and continuum models.

In Chapter 1 it was emphasized that two-state models are widely used in molecular biology. There, the point was to use thermo-dynamics to understand how experimental conditions influence the equilibrium between these two states. Now we want to understand how quickly such a system reaches that equilibrium. This chapter will start off with a phenomenological treatment of two-state transitions, and then explore some of the fundamental theories that give the transition rate physical meaning.

7.1 | Exponential relaxations

When we first analyzed the energetics of the global transition we simply assigned a free energy change to the process. Now we will assign rate constants to the forward and reverse transitions so that we can study the kinetics

$$A \underset{\beta}{\overset{\alpha}{\rightleftharpoons}} B \qquad (7A)$$

The rate constants in Scheme (7A), α and β, have units of s^{-1}. If the system is not at equilibrium, then the concentrations of A and B will change. Transitions from A to B will decrease the concentration of A, and the reverse transitions from B to A will increase the concentration of A. The rate of each process is equal to the rate constant times the concentration of the relevant species. This reasoning leads directly to a pair of differential equations, where the rates of change of A and B reflect the net effect of the forward and reverse transitions.

$$\frac{d[A]}{dt} = -\alpha[A] + \beta[B] \tag{7.1a}$$

$$\frac{d[B]}{dt} = \alpha[A] - \beta[B] \tag{7.1b}$$

The two coupled differential equations in the two variables [A] and [B] will be reduced to one differential equation in one variable in order to find a solution. Since the total protein, $T = [A] + [B]$, must be constant, we can eliminate [B] by replacing it with $T - [A]$. Equation (7.1a) then becomes a differential equation in one variable, [A]

$$\frac{d[A]}{dt} = \beta T - (\alpha + \beta)[A] \tag{7.2}$$

To simplify further we divide through by T and let x be $[A]/T$, the fraction of protein in the form of A. This gives

$$\frac{dx}{dt} = \beta - (\alpha + \beta)x \tag{7.3}$$

The general solution has an exponential with a decay constant equal to $\alpha + \beta$. The specific solution is determined for the initial condition that at $t = 0$, x equals its initial value, x_0

$$x = \left(x_0 - \frac{\beta}{\alpha + \beta}\right)e^{-(\alpha + \beta)t} + \frac{\beta}{\alpha + \beta} \tag{7.4}$$

Note that x goes to $\beta/(\alpha + \beta)$ after a long time. This is the fraction of protein present in the form of A when an equilibrium has been attained. Thus, any initial value of x different from the equilibrium value relaxes to the equilibrium value exponentially with a decay constant of $\alpha + \beta$. It is common to discuss the progress of this kind of process in terms of its time constant $\tau = 1/(\alpha + \beta)$. Note that τ is the time it takes for x to close the gap with equilibrium by a factor of e.

Many kinetic theories lead to mathematical expressions that are exponentials or sums of exponentials. If a protein undergoes a thermal or voltage-induced transition, then α and β depend on temperature or voltage, and a sudden change of the appropriate variable will change the rate constants to new values. This will make the protein relax with a characteristic rate to the new equilibrium. The behavior is illustrated in Fig. 7.1

In Fig. 7.1 we see how the system goes to equilibrium from the two starting points in the same amount of time. If we start out with $x < \beta/(\alpha + \beta)$ then x will increase, and if we start out with $x > \beta/(\alpha + \beta)$ then x will decrease. Either way the difference between

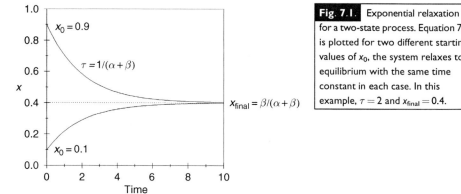

Fig. 7.1. Exponential relaxation for a two-state process. Equation 7.4 is plotted for two different starting values of x_0, the system relaxes to equilibrium with the same time constant in each case. In this example, $\tau = 2$ and $x_{final} = 0.4$.

x and $\beta/(\alpha + \beta)$ decreases e-fold as time increases by an increment of $\tau = 1/(\alpha + \beta)$. Note that the velocity, dx/dt, is not very informative. The initial slope of x can be either positive or negative, and vary in steepness, depending on whether x starts out near to or far from $\beta/(\alpha + \beta)$. Note that α and β are the fundamental quantities, and to learn something about them one needs the time constant.

7.2 | Activation energies

In the two-state model, there is a tacit assumption that the actual transitions are instantaneous. An insignificant amount of time is spent in transit, and the intermediate states have no bearing on the thermodynamic properties of the system. However, there must be a reaction pathway. Actually, there could be many but we will assume that only one carries the bulk of the traffic. As the molecule traverses this pathway its structure is neither A nor B. Because the molecule spends very little time on this pathway, the intermediate state must have a very high energy. This leads to the picture of an energy barrier of height $E^\dagger$ separating A from B. This *transition state* is the position along the reaction pathway with the highest energy. This is one of the most basic ideas in relating the rate constants α and β to the structure and energetics of a molecule as it undergoes a reaction. The rate of going over this barrier is then related to the probability of a molecule having that high energy, and this can be estimated from the Boltzmann distribution

$$\alpha = Ce^{-E^\dagger/RT} \tag{7.5}$$

This is the Arrhenius equation. It states that, everything else being equal, the rate will be slower if the energy of the barrier is higher. It also states that rates increase with temperature. The variable C is called the preexponential factor, and it can vary drastically depending on the nature of the reaction. More will be said about preexponential factors later. To estimate $E^\dagger$ from kinetic data one plots the log of the rate versus $1/T$ in what is commonly referred to as an Arrhenius plot. Arrhenius plots are often linear, and this supports the basic notion of barrier crossing as a good picture of a kinetic process. Barrier heights, $E^{\dagger}$, are generally referred to as activation energies. For protein conformational changes and enzymatic processes they are often in the range 10–20 kcal mole^{-1}.

There are a number of ways to think about C, the preexponential factor. If one imagines an activation free energy, $G^{\dagger}$, it would reflect both the potential energy of the barrier and the density of states in the vicinity of the barrier peak. With $G^{\dagger} = H^{\dagger} - TS^{\dagger}$ we then get

$$\alpha = C'e^{-G^{\dagger}/RT} = C'e^{-S^{\dagger}/R}e^{-H^{\dagger}/RT} \tag{7.6}$$

where the preexponential factor C of Eq. (7.5) becomes $C'e^{S^{\dagger}/R}$. This evokes a geometric picture of the transition state not just as a barrier but also as a saddle point, or as a low pass through a high mountain range. The molecule moves through the pass, and the activation entropy, $S^{\dagger}$, can be thought of as related to the width of the pass. Note that $e^{-S^{\dagger}/R}$ does not depend on temperature. This makes it hard to distinguish from the preexponential factor.

The temperature dependence of a rate is often characterized by the Q_{10}, which is the factor by which the rate changes for a 10 degree change in temperature. Starting with Eq. (7.5), the Q_{10} can be related to the activation energy by the following approximate expression

$$E^{\dagger} = \frac{RT^2\ln(Q_{10})}{10} \tag{7.7}$$

At typical physiological temperatures, a Q_{10} of 2 gives $E^{\dagger} = \sim 12$ kcal mole^{-1}.

The energy barrier is one of the most important concepts in kinetics. The structure of the transition state provides a visual picture of how a molecule gets from one state to another. As important as this transition state is for the study of rate processes, we often know very little about it. We can almost never isolate it or purify it to study it closely. In the vast majority of studies, the energy is all we know about the transition state. But the transition state energy is estimated from rate data, so it is easy to fall into a trap of circular reasoning.

7.3 | The reaction coordinate and detailed balance

The activation energy is taken as the height of a barrier in the plot of energy versus position along the reaction coordinate (Fig. 7.2). A diagram like this helps visualize an important relation between the relevant energies. For a molecule to cross from the left it needs an energy of $E^{\alpha\dagger}$, giving a rate constant

$$\alpha = Ce^{-E^{\alpha\dagger}/RT} \tag{7.8}$$

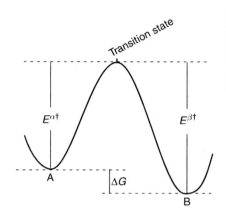

Fig. 7.2. The change in energy along the reaction coordinate. The wells at A and B are for the two reactants, and the peak occurs in the transition state.

And crossing from the right gives

$$\beta = Ce^{-E^{\beta\dagger}/RT} \tag{7.9}$$

At equilibrium the rate of conversion from A to B is balanced exactly by the reverse conversion from B to A. With zero time derivatives, either Eq. (7.1a) or (7.1b) then gives the following expression

$$\alpha[A] = \beta[B] \tag{7.10}$$

So we can express the equilibrium constant in terms of the rate constants

$$\frac{[B]}{[A]} = \frac{\alpha}{\beta} \tag{7.11}$$

This is an example of a very general relationship between kinetic and equilibrium theories. Any kinetic description of a system should also describe the equilibrium if you let the relevant time derivatives go to zero. From a practical perspective, this is a good way to check a kinetic theory. This kinetic constraint on the equilibrium leads to the condition of detailed balance (see Section 5.4).

With the condition of detailed balance the expressions for the rate constants, Eqs. (7.8) and (7.9), lead to the following

$$\frac{[A]}{[B]} = e^{(-E^{\beta\dagger}+E^{\alpha\dagger})/RT} = e^{-\Delta G/RT} \tag{7.12}$$

where the preexponential factor cancelled out of the ratio. Equation (7.12) is the equilibrium constant expressed in terms of the free energy difference between the two states. Thus, the energies are related,[1] as follows

$$E^{\beta\dagger} - E^{\alpha\dagger} = \Delta G \tag{7.13}$$

This relation can be readily visualized by examining Fig. 7.2. This demonstrates a basic relation between kinetics and energetics.

---

[1] The mixing of free energy and energy detracts from the rigor of this expression. Nevertheless a useful point is made about the relation between the various energy terms and the condition of detailed balance.

(a)

(b)

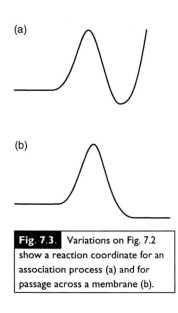

Fig. 7.3. Variations on Fig. 7.2 show a reaction coordinate for an association process (a) and for passage across a membrane (b).

When both forward and reverse rates can be studied well enough to get these energies, one can construct a rough picture of the reaction coordinate.

Figure 7.2 depicts the reaction coordinate for an isomerization. For an association, the reaction coordinate might look like Fig. 7.3a, reflecting the relative position of two molecules. When they collide with enough energy to cross the barrier, they subsequently find themselves in a restricted bound state represented by the potential energy well on the right. Another variation on the barrier-crossing idea is a molecule passing through a membrane. To get from one side to the other it must surmount an energy barrier. But it is free to move on either side of the barrier, and this is depicted by flat regions in the potential energy function (Fig. 7.3b).

If we tried to use Eqs. (7.8) and (7.9) for the situations in Fig. 7.3 we would run into trouble with the preexponential factor, C. This is because C has to take into account the frequency with which the barrier is encountered. Clearly, two associating molecules will encounter the barrier with a frequency proportional to the frequency of intermolecular collisions. Two bound molecules will spend their time in the energy well, and thus encounter the barrier much more often as the binary complex vibrates within these confines. In this case, a detailed balance relation like Eq. (7.13) can still be found relating the forward and reverse rates to the equilibrium constant. But since the equilibrium constant for an association includes translational and rotational entropies (Chapter 4) a relation between the energies of free and bound states must include these factors. On the other hand, for passage through a membrane the frequency of encounters is proportional to the concentrations on either side (Fig. 7.3b).

7.4 | Linear free energy relations

When molecular structure is altered, the typical result is that both the equilibrium constant and the rate constants change. Often these changes are well correlated, and this is the basis for linear free energy relations. The concept was developed as an approach to the study of reaction mechanisms in organic chemistry, and has found a number of applications in biophysics.

The reaction coordinate can be used to visualize how something influences the change in free energy of a process together with the rate constants. We might imagine that a structural change stabilizes state B relative to state A. Looking at Fig. 7.2, the right side of the reaction coordinate, corresponding to state B, will then be lowered. Now if the effect of this perturbation varies linearly with the reaction coordinate, then the energy profile along the barrier between the two states will be altered accordingly, and we can obtain the new reaction coordinate from the old one by subtracting a line.

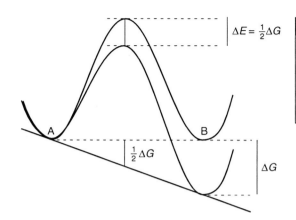

Fig. 7.4. An energy barrier is modified by the addition of a descending linear function. Since the top of the barrier is at the midpoint between the two sides, the change in $E^\dagger$ is half the change in ΔG.

Figure 7.4 shows a linear addition to the energy function. If we equate the energy difference between the two minima at A and B with the free energy change, ΔG, then the changes in ΔG and $E^\dagger$ will be proportional.

If the position of the peak does not change, then there will be a linear relation between the activation energy and the free energy change of the transition

$$E^\dagger = E^{\dagger 0} - \Phi \Delta G \tag{7.14}$$

Equation (7.14) takes into account the possibility that the top of the barrier is some arbitrary fraction of the way between the two sides of the barrier. This fraction is denoted by Φ, which varies between zero and one. For the drawing in Fig. 7.4 the top of the barrier is right in the middle, so $\Phi = 1/2$.

Different barrier positions are shown in Fig. 7.5. If the top of the barrier is close to A, then Φ will be small and the barrier for crossing from A to B will be only weakly sensitive to the free energy change. On the other hand, if the top of the barrier is closer to B, Φ will be large and the barrier for crossing will be more sensitive to ΔG.

Linear free energy relations can be applied to studies of a series of mutations, where one can measure both the equilibrium constant and rate constant for the same process. In order to make use of these kinds of measurements it is convenient to convert Eq. (7.14) into an expression in terms of these quantities. Combining Eq. (7.14)

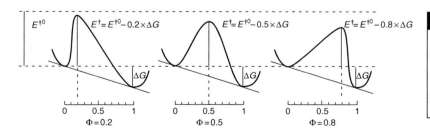

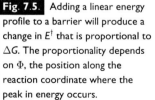

Fig. 7.5. Adding a linear energy profile to a barrier will produce a change in $E^\dagger$ that is proportional to ΔG. The proportionality depends on Φ, the position along the reaction coordinate where the peak in energy occurs.

with Eq. (7.5), replacing ΔG by $RT \ln K$, and taking the natural logarithm leads to

$$\ln \alpha = \ln C - E^{\ddagger 0}/RT + \Phi \ln K \tag{7.15}$$

Equation (7.15) expresses the linear dependence of the logarithm of the rate constant on the logarithm of the equilibrium constant. This relation has been applied to a wide range of kinetic phenomena. Equation (7.15) is especially well known in the context of chemical catalysis by acids and bases, where it is known as the Brønsted equation. In enzyme catalysis the Brønsted equation has been used to investigate how catalytic side chains in enzymes react with substrates (Sections 10.12 and 10.13).

Hemoglobin provides an excellent example of the analysis of a linear free energy relation (Eaton et al., 1991). This allosteric protein undergoes a transition between the T state, with a low affinity for oxygen, and the R state, with a high affinity for oxygen (Section 5.8). This transition can be studied under a variety of conditions including variations in number of binding sites occupied, variations in choice of ligand (O_2, CO, N_3^-), and pH. For each condition both the equilibrium constant between the two allosteric states and the rate of the transition can be measured. A plot of $\ln \alpha$ versus $\ln K$ shows that they follow a linear free energy relation fairly closely (Fig. 7.6). From the fit to Eq. (7.15) we obtain $\Phi = 0.17$. This is interpreted as meaning that the peak of the energy barrier is closer to the R state. It therefore resembles the left-most plot in Fig. 7.5. Taking the reasoning one step further, we might suspect that the structure of hemoglobin near the top of the barrier is closer to that of the R state than the T state.

Linear free energy plots are particularly attractive given the ease of using mutagenesis to produce large numbers of molecular variants for kinetic analysis. However, a word of caution is in order. Very little is known about actual reaction coordinates and the shape of the potential energy function. Indeed, one should question whether a single reaction coordinate exists in a large molecule with a very large number

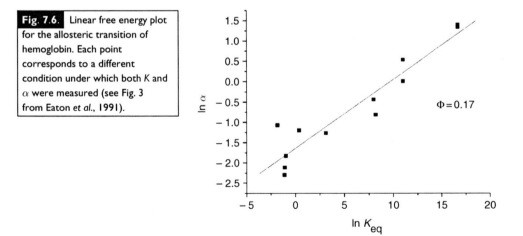

Fig. 7.6. Linear free energy plot for the allosteric transition of hemoglobin. Each point corresponds to a different condition under which both K and α were measured (see Fig. 3 from Eaton et al., 1991).

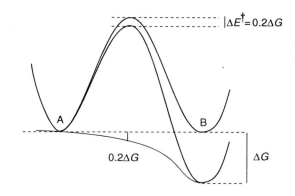

of internal degrees of freedom (see Section 7.10). Furthermore, the notion that a structural change produces a linear perturbation of a reaction coordinate is quite naïve. Figure 7.7 is a variation on Fig. 7.4 in which the perturbations are not linear. This could result from a modification that relieves some steric repulsion in B. Since steric repulsions are short range (Section 2.11), this perturbation will not reach very far. The energy at the top of the barrier will not change very much. The value of Φ will therefore be small (0.2 in Fig. 7.7). We would then conclude erroneously that the top of the barrier is closer to A. Our present state of understanding of the structure and energetics of molecules along transition pathways is inadequate for dealing with these sorts of questions.

7.5 | Voltage-dependent rate constants

The linear free energy concept can be applied to the kinetics of a voltage-induced transition in a membrane protein (Section 1.8). Recall that the quintessential process is the transition of an ion channel between the open and closed states, as shown in Scheme (7B)

$$C \underset{\beta}{\overset{\alpha}{\rightleftharpoons}} O \qquad (7B)$$

If the voltage drop across the membrane is linear, and if it is added to the energy barrier separating the two states, we can draw a diagram that looks like Fig. 7.8.

The fraction of voltage drop between the two minima is δ, so the voltage dependent part of the energy difference is δV. If the barrier is exactly half way between the two states (i.e. $\Phi = 1/2$), then the barriers change by adding or subtracting half this energy difference. This specifies the two rate constants

$$\alpha = C e^{-(E^\dagger - \frac{1}{2}\delta V)/RT} \qquad (7.16a)$$

$$\beta = C e^{-(E^\dagger + \frac{1}{2}\delta V)/RT} \qquad (7.16b)$$

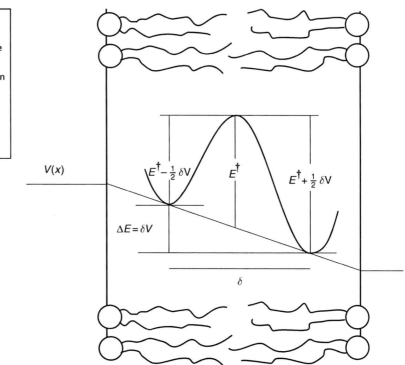

Fig. 7.8. Voltage influences the energy difference between two states of a membrane protein. The barriers for the transition in each direction are changed in proportion with the change in the energy difference (see Fig 1.4), and δ represents the fraction of the voltage drop between the two minima.

Earlier in this chapter it was shown that a two-state model shows exponential kinetics with a time constant $\tau = 1/(\alpha + \beta)$. Substituting the above expressions for α and β gives the time constant for the voltage-induced transition

$$\tau = \frac{e^{E^{\dagger}}}{C(e^{\delta V/2RT} + e^{-\delta V/2RT})}$$

$$= \frac{e^{E^{\dagger}}}{2C} \operatorname{sech}\left(\frac{\delta V}{2RT}\right) \tag{7.17}$$

where $\operatorname{sech}(x) = 1/\cosh(x)$ (see Appendix 5).

This function peaks at $V = 0$ (Fig. 7.9). In the thermodynamic analysis of this model (Section 1.8), the voltage midpoint of the transition reflected the intrinsic energy difference between the two states. The steepness reflected the amount of gating charge. When these parameters are incorporated into the kinetic analysis there are parallel effects on τ. The maximum in τ occurs at the voltage of the midpoint of the transition, and the peak is sharper if the transition is steeper.

When the time constants for voltage-induced transitions are measured experimentally, they often vary in this way. This behavior has been found in classical studies of the ion channels of neurons (Fig. 16.6b) as well as channels in artificial membranes (Ehrenstein *et al.*, 1974). Similar behavior has been seen for the flipping motions of hydrophobic ions within lipid membranes (Oberhauser and Fernandez, 1995).

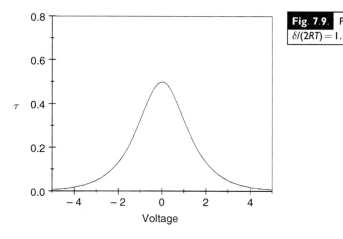

Fig. 7.9. Plot of Eq. (7.17) with $\delta/(2RT) = 1$.

7.6 | The Marcus free energy relation

How a reaction rate varies with its driving force depends on the detailed shape of the reaction coordinate. The models above produce linear relationships, but there is an important theory that yields a quadratic relationship. This theory was initially developed to investigate reactions involving electron transfer (Marcus, 1964), and subsequently in a modified form for proton transfer (Marcus, 1968). Marcus' theory has been successfully applied to a wide range of processes involving charge transfer, and will be applied later in this book to understand the activity of the enzyme carbonic anhydrase (see Section 10.15).

The potential energy function is represented in Fig. 7.10. There are two parabolic potential energy minima, one at $x = -1$ for the reactant and one at $x = 1$ for the product. The cusp at the intersection of the two parabolas is the transition state, and as the reaction coordinate passes through this point the potential energy function abruptly switches from one parabola to the other. It is reasonable to use a parabolic function in the vicinity of a potential energy minimum (Section 2.12). However, the original use of the quadratic function in Marcus' theory was actually based on Coulombic interactions, which vary as the square of the charge. Self-energy terms for charging a sphere in a dielectric medium were also included (see Section 2.2 and Eq. (2.4)). We then obtain a quadratic expression for the energy by taking the reaction coordinate as the amount of charge transferred, or the amount of solvent polarization that has to occur in order to equalize the electrostatic energy of charge jumping from one center to another.

The quadratic dependence of potential energy in each of the two wells is written for the displacement from each minimum

$$G_{-1} = G_0^{\dagger}(x + 1)^2 \tag{7.18a}$$

$$G_1 = G_0^{\dagger}(x - 1)^2 + \Delta G^{\circ} \tag{7.18b}$$

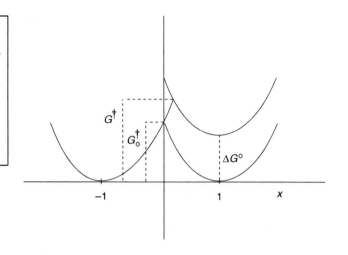

Fig. 7.10. Parabolas at $x = \pm 1$ define the potential energy as a function of reaction coordinate for the reactant and product. The intersection defines the transition state with an energy $G^\dagger$. Raising the energy on the product side changes the driving force and alters the position and energy of the transition state.

The letter G is used to denote the energy, and it is taken as a free energy of the same form as the driving force of the reaction, $\Delta G^{\rm o}$. The force constants in Eqs. (7.18a) and (7.18b) are taken as $G_{\rm o}^\dagger$ to produce the very convenient result that when the two potential energy minima are the same ($\Delta G^{\rm o} = 0$), the energy at the intersection is the activation energy $G_{\rm o}^\dagger$. When $\Delta G^{\rm o}$ is not zero, the position of the intersection can be found by setting the two functions (Eqs. (7.18a) and (7.18b)) equal to one another.

$$G_{\rm o}^\dagger(x+1)^2 = G_{\rm o}^\dagger(x-1)^2 + \Delta G^{\rm o} \tag{7.19}$$

Solving for x gives the position of the transition state.

$$x^\dagger = \frac{\Delta G^{\rm o}}{4G_{\rm o}^\dagger} \tag{7.20}$$

The energy of the transition state is then determined by evaluating $G_{-1}(x^\dagger)$ or $G_1(x^\dagger)$

$$G^\dagger = G_{\rm o}^\dagger \left(1 + \frac{\Delta G^{\rm o}}{4G_{\rm o}^\dagger}\right)^2 \tag{7.21}$$

Thus, we have a quadratic dependence of energy barrier height on driving force.

The situation depicted in Fig. 7.10 lacks an important feature needed for a quantitative description of charge transfer experiments. An additional term must be added to account for the work done to bring the two reacting groups into the correct position and orientation. These work terms, denoted as $w^{\rm r}$ for the reactants and $w^{\rm p}$ for the products, are independent of the driving force. The complete free energy relation is thus obtained by adding $w^{\rm r}$ to Eq. (7.21).

7.7 | Eyring theory

Now we will look at the barrier crossing problem in more depth. Statistical mechanics lets us develop this idea into a quantitative expression for the transition rate. The result is the absolute reaction rate theory of Eyring. It is typically derived for the case of two reacting molecules that collide, but the derivation given here will be for the two-state model. We start by looking at a reaction coordinate diagram such as Fig. 7.2 and asking what is the probability of a molecule being at the top of the barrier, relative to the probability of being somewhere in the well on the left side. Denoting the reaction coordinate with the variable ξ, then the probability of being within an element $d\xi$ is given by the Boltzmann distribution

$$p(\xi)d\xi = e^{-U(\xi)/kT}d\xi/\Lambda \tag{7.22}$$

where $U(\xi)$ is the potential energy plotted in Fig. 7.2, and $\Lambda = \sqrt{h^2/(2\pi mkT)}$ is the thermal De Broglie wavelength introduced in Section 4.6 to deal with the statistical mechanics of translational freedom. Recall that it is the length of one edge of a cubicle of space defined by quantum mechanics. Thus, the form of the Boltzmann distribution in Eq. (7.22) normalizes the length element $d\xi$ to this basic quantum of length. It is necessary to do it this way to obtain an absolute probability relative to the probability of being in the potential energy well on the left side.

Now consider a position just to the left of the top of the barrier, so close that the potential energy is almost flat. We define an element of length, $d\xi$, such that molecules in this element moving to the right with velocity v will cross the actual midpoint of the barrier in a time interval dt. These molecules will then undergo the reaction. For the velocity we take the rms velocity $\sqrt{\overline{v^2}} = \sqrt{(kT/m)}$. The length element containing these molecules destined to cross the barrier is then the distance traveled in dt, or vdt

$$d\xi = dt\sqrt{\frac{kT}{m}} \tag{7.23}$$

Substitution into Eq. (7.22) then gives half the number of molecules that undergo the transition in a time dt (the other half are going the wrong way). We call this number dN

$$dN = \frac{dt}{2\Lambda}\sqrt{\frac{kT}{m}}\,e^{-E^\dagger/kT} \tag{7.24}$$

Substitution of the explicit form of Λ given above simplifies the result to give

$$dN = dt\frac{\sqrt{\pi/2}\,kT}{h}\,e^{-E^\dagger/kT} \tag{7.25}$$

Since the rate constant is the number of molecules per unit time that cross the barrier, we obtain the reaction rate

$$\alpha = \frac{\kappa\, kT}{h}\; e^{-E^{\dagger}/kT} \tag{7.26}$$

The factor κ replaces $\sqrt{\pi/2}$. It is a general fudge factor called the transmission coefficient that takes into account a number of poorly understood aspects of the process. A more rigorous derivation, which uses the distribution of velocities rather than the rms used here can be performed, but it still does not give an exact value for κ (see Hill, 1960, Chapter 11).

Equation (7.26) is an important result in the theory of reaction kinetics. It has been extensively tested in reactions in the gas phase, giving excellent agreement with experiment (see Moore, 1972, Chapter 9). However, it is much more difficult to apply this theory to rate processes in liquids. The key is whether the motion along the ξ coordinate changes direction rapidly near the top of the barrier. For gas phase reactions, motion in ξ does not reverse very often. In liquids collisions and thermal agitation reverse the direction frequently. The velocity, here taken as $\sqrt{\overline{v^2}} = \sqrt{(kT)/m}$, will there-fore not be useful in predicting how many molecules cross the barrier. Recall from Section 6.4 that an analysis of the random walk indicated that on average a molecule will change directions after moving only 0.1 Å. This randomness in the motion along the reaction coordinate must be taken into account. One way to do this is to develop a theory in which the reaction coordinate is coupled to an infinite number of other coordinates representing the positions of the surrounding solvent molecules. This approach converges with the result of a simpler theory based on diffusion (Hänggi et al., 1990).

The diffusion theory of chemical reactions will be developed in the next section. But before leaving the subject of Eyring theory, it is important to emphasize that the preexponential factor of Eq. (7.26), $(\kappa kT)/h$, is of no value whatsoever in dealing with reactions in the liquid milieu commonly encountered in biology. This must be sta-ted emphatically because the literature contains a large number of inappropriate uses of Eyring theory.

7.8 | Diffusion over a barrier – Kramers' theory

The mathematics of diffusion provides a very powerful approach to kinetic processes involving a jump over a barrier (Kramers, 1940; Hänggi et al., 1990). Recall from Chapter 6 how an external force F produces a flux of $FC(\xi)/f$, where f is the coefficient of friction. Here we will take F as the force from the steepness of the function drawn in Fig. 7.2. This can be replaced by minus the derivative of the potential energy, $-dU(\xi)/d\xi$. A concentration gradient also produces a flux, and it is equal to $-D(dC(\xi)/d\xi)$, where D is the diffusion constant. The concentration gradient in this case is actually the gradient in probability of a molecule being at a particular position along the reaction coordinate. For this reason we should switch variables from $C(\xi)$ to $P(\xi)$. We can now use Eq. (6.64) to express the flux of probability

$$J = D\left(-\frac{P}{kT}\frac{dU}{d\xi} - \frac{dP}{d\xi}\right)$$ (7.27)

This equation describes how $P(\xi)$ flows under the influence of the potential energy function as well as the random collisions with solvent molecules and other parts of the reacting molecule. The two derivatives can be combined into a single derivative of a product

$$J = -De^{-U/kT}\frac{d}{d\xi}(Pe^{U/kT})$$ (7.28)

This can be verified by using the product rule to differentiate $Pe^{U/kT}$. We now assume that there is a steady flux across the barrier (a steady state). In this steady state J is constant along the entire energy barrier, and is thus independent of ξ; and P will not change with time because flux into any point is balanced by flux out.

We will now derive an expression for J subject to the assumption of a steady state. First, we multiply Eq. (7.28) through by $e^{U/kT}$ and integrate from one side of the barrier to the other. Since J is constant it can be factored out of the integral on the left

$$J\int_a^b e^{U/kT}d\xi = D(P(a)e^{U_a/kT} - P(b)e^{U_b/kT})$$ (7.29)

With reference to Fig. 7.2, $\xi = a$ is the midpoint of the well centered at A and $\xi = b$ is the midpoint of the well centered at B. At $t = 0$ none of the molecules has made the transition, so all of the molecules are on the left somewhere in the well at A. This means that $P(b) = 0$. If we set our zero energy point at $\xi = a$, Eq. (7.29) simplifies to the following expression for the flux across the barrier

$$J = \frac{DP(a)}{\int_a^b e^{U/kT}d\xi}$$ (7.30)

If we had a mathematical function for the barrier we could integrate it to get a complete solution. We will use a reasonable mathematical representation of the reaction coordinate at the top of the barrier, $U(\xi) = E^\dagger - \phi^\dagger(\xi - \xi^\dagger)^2$, where $E^\dagger$ is the height of the barrier and $-\phi^\dagger(\xi - \xi^\dagger)^2$ makes the function fall off quadratically on either side. Factoring out $e^{-E^\dagger/kT}$, leaves the integral of a Gaussian function, which can be integrated to give $\int_{-\infty}^\infty e^{-\phi^\dagger(\xi-\xi^\dagger)^2/kT}d\xi = \sqrt{\pi kT/\phi^\dagger}$

(Eq. (A4.4), the limits of $\pm\infty$ do not matter because the function is very small away from the peak). The steady-state flux is now

$$J = \frac{DP(a)}{\sqrt{\pi kT/\phi^\dagger}} e^{-E^\dagger/kT} \tag{7.31}$$

The final task is to evaluate $P(a)$. This is the probability of being at the very center of the potential energy well on the left side of the barrier. At the start of the reaction the probability equals one that the molecule is somewhere in the well on the left side of the barrier. We assume that the molecules are distributed within the well according to the Boltzmann distribution. This is not exactly correct because there is a steady flux to the right, but transitions across the barrier are infrequent, so the Boltzmann distribution is a reasonable approximation. We take the potential energy as an upward facing quadratic function centered at $\xi = a$, $U(\xi) = \phi_a(\xi - a)^2$. So in this region $P(\xi) = e^{-\phi_a(\xi-a)^2/kT}$; $P(a)$ is then $1/\int_{-\infty}^{\infty} e^{-\phi_a(\xi-a)^2/kT} d\xi = 1/\sqrt{\pi kT/\phi_a}$. This can be substituted into Eq. (7.31) to give the rate constant predicted for diffusion over a barrier

$$\alpha = \frac{D\sqrt{\phi_a\phi^\dagger}}{\pi kT} e^{-E^\dagger/kT} \tag{7.32}$$

The comparison with the expression from Eyring theory (Eq. (7.26)) makes some important points. We still have the exponential dependence on barrier height, so the basic idea of the Boltzmann probability of being at the top of an energy barrier is preserved. However, the preexponential term is very different. The dependence of α on D in Eq. (7.32) provides the intuitively reasonable result that the ease of diffusion along the reaction coordinate will make a difference. In fact, if we replaced D with $kT/(6\pi\eta r)$ (Eq. (6.66)), we find that α is inversely proportional to the viscosity. Thus, the friction experienced by the molecule for motion along the reaction coordinate has a direct effect on the reaction rate. The dependence on $\phi^\dagger$ and ϕ_a are also of interest. These parameters reflect the inverse width of the barrier and of the left potential energy well, respectively. So making the barrier narrower increases the rate, as does making the potential energy well narrower.

Equation (7.32) gives the rate of barrier crossing activated by thermal motions. This is the definitive theory for elementary rate processes in liquids where random molecular motions cause changes in the direction of motion on a rapid time scale. From a more general physical perspective, this situation is referred to as the high-viscosity or over-damped limit. In the low-viscosity limit the situation is very different. A molecule oscillates back and forth within a potential energy well many times without crossing the barrier. Collisions with other molecules increase or decrease the energy of the oscillation, and by this random process a molecule rises to an energy level where barrier crossing is possible. Thus, in this case more collisions make the reaction faster. In the high-viscosity case collisions increase the coefficient of friction and reduce the coefficient of diffusion. Thus, collisions slow the reaction down. A good understanding of the relation between the rate of a reaction and thermal fluctuations depends on a grasp of these concepts, which are accessible in the original paper of Kramers (1940), and thoroughly expounded on by Hänggi *et al.* (1990).

7.9 | Single-channel kinetics

When the current through a small patch of membrane is measured, the opening and closing of single channels is readily visible as upward and downward steps (Fig. 7.11). These steps represent conformational transitions of the channel protein between its open and closed states. Such transitions occur haphazardly in time because of the inherent randomness of molecular motion. To apply kinetics to these single-molecule transitions requires a stochastic version of kinetic theory.

This analysis resembles the analysis of the two-state model presented above, but with some interesting twists. The different form of single-channel data forces us to ask fundamentally different questions. In a single-channel experiment one must think in terms of probabilities, rather than concentrations. What is the probability that a channel will remain open or closed for a certain time? or remain open or closed longer than a certain time? We define the probability density function, $p(t)$, which gives the probability as $p(t)dt$ that an event has a duration in the interval $[t, t+dt]$. The distribution function, $P(t)$, gives the probability that the event will have a duration longer than t.[2] The two are related as follows

$$\int_t^\infty p(s)ds = P(t) \qquad (7.33)$$

Once again we start with the basic two-state model

$$C \underset{\beta}{\overset{\alpha}{\rightleftharpoons}} O \qquad (7C)$$

1 pA
1 s

Fig. 7.11. Single-channel currents of a GABA-activated Cl⁻ channel in an outside-out patch held at a voltage of –80 mV (from Zhang and Jackson, 1995).

---

[2] The use of this form for the distribution is especially convenient for single-channel kinetics. In other applications this quantity is often referred to as a "survivor" function, and the term distribution is used to denote one minus the quantity in Eq. (7.33), or the integral of $p(s)$ from 0 to t.

In Scheme (7C), C represents the closed state and O represents the open state. We will now derive an equation for $P_o(t)$, the open time distribution, by asking what is the relationship between $P_o(t)$ and $P_o(t+dt)$. Obviously, if a channel stays open until $t+dt$, it must also have stayed open until t. A closure can occur in the time between t and $t+dt$, so $P_o(t+dt)$ should be smaller than $P_o(t)$. The probability of closing in this interval is βdt. The only other possibility is not closing, so the probability is $1-\beta dt$ that the channel will not close in the interval $[t, t+dt]$. The probability of an opening lasting until $t+dt$ is then taken as the product of the probability of it lasting until t, times the probability of it not closing in the ensuing short interval

$$P_o(t+dt) = (1-\beta dt)P_o(t) \tag{7.34}$$

Before continuing a few key assumptions must be noted. First of all, the interval dt must be small enough so that the probability of more transitions during dt is negligibly small. Second, this derivation depends on the assumption that this is a Markov process (Section 6.6). It does not matter how long the channel has been open, the probability that a closing transition will take place depends only on the intrinsic rate, β, and the duration of the small interval, dt.

A simple rearrangement of Eq. (7.34) leads to a first order differential equation

$$\frac{dP_o}{dt} = -\beta P_o \tag{7.35}$$

The solution is an exponential plus a constant, but the constant is found to be zero by imposing the starting condition $P(0)=1$. So we have

$$P_o(t) = e^{-\beta t} \tag{7.36}$$

where $P_o(t)$ starts off at one when $t=0$, and decays to zero after a sufficiently long time. This reflects the fact that any open channel must stay open at least an infinitesimally short time, and all open channels must close eventually. This differential equation (Eq. (7.35)) is simpler than the pair used to describe the kinetics of a macroscopic two-state system (Eqs. (7.1a) and 7.1b)). This is because once a channel closes, it is out of the picture. This is analogous to absorbing boundary conditions as discussed in the context of diffusion (Section 6.2.4).

The macroscopic analog of the single-channel probability can be visualized by imagining a large number of single-channel open-state intervals lined up at a common starting time (Fig. 7.12). If N such channels open at once, the initial current will be Ni, where i is the single-channel current. The total current will decay exponentially as $iNP(t) = iNe^{-\beta t}$ (Eq (7.36)). In fact, this situation is well approximated at many synapses. Neurotransmitter is released into the synaptic cleft and causes the simultaneous opening of many channels. Unbound

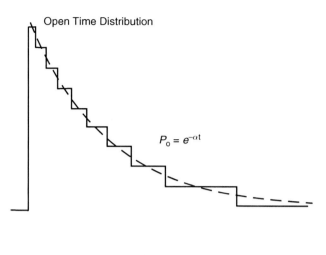

Open Time Distribution

$$P_o = e^{-\alpha t}$$

Fig. 7.12. The open-time distribution can be pictured as the sum of an assortment of single-channel currents aligned at the same start time. At a synapse the channels are opened almost simultaneously and close at random to replicate the single-channel open-time distribution.

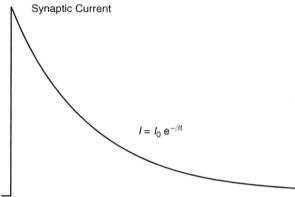

Synaptic Current

$$I = I_0\, e^{-\beta t}$$

neurotransmitter is removed very rapidly so that there is no chance of rebinding a receptor and making the channel open a second time. The synaptic current thus follows the single-channel open probability, decaying exponentially with a decay constant of β.

Equation (7.36) should be compared with Eq. (7.4), which is the analogous expression for a macroscopic measurement. The initial and final conditions are simpler for single channels, so the exponential function does not have the additive constant. Another interesting difference between Eq. (7.36) and Eq. (7.4) is that the decay constant is β rather than the sum $\alpha + \beta$.

By the same logic used to derive the open time distribution, the closed-time distribution is

$$P_c(t) = e^{-\alpha t} \qquad (7.37)$$

Thus, one can determine α and β separately from the closed- and open-time distributions, respectively. Note that the probability density functions are obtained by differentiating and changing the sign (see Eq. (7.33)).

$$p_o(t) = \beta e^{-\beta t} \qquad\qquad (7.38a)$$

$$p_c(t) = \alpha e^{-\alpha t} \qquad\qquad (7.38b)$$

This simple model for the time course of a synaptic current determined by the rate of channel closing was developed in early analyses of the channel properties at the neuromuscular junction (Anderson and Stevens, 1973). This view was modified by later investigators to take into account the fact that the channel can close and reopen repeatedly while the binding sites are occupied. Thus, the decay of a synaptic current is determined by the lifetime of a burst of openings. This type of model requires more kinetic states and can be treated with the methods developed in Sections 9.7–9.9.

It is important to appreciate that single-channel analysis provides a way of measuring the rate constants, α and β, without applying some stimulus to initiate the kinetic process. There is no $t = 0$ marking the start. Instead, one simply follows the time course of a rather long record of current as it fluctuates in a stepwise fashion around an equilibrium state. The open and closed times are then tallied and distributions constructed. When the steps are too small to see, the collective fluctuations of many channels may still give rise to channel noise. This noise can also be studied to learn something about the properties of channels (Section 12.8 and 12.11).

7.10 | The reaction coordinate for a global transition

The rate theories considered so far focused on a single reaction coordinate. But for a transition in a protein this is a gross over-simplification. There is a vast number of internal coordinates as thousands of atoms change their interatomic distances. We have to wonder whether the theories based on a single reaction coordinate have any applicability to proteins.

For a global transition we have a problem involving motion in a space with a very large number of dimensions. Within this multi-dimensional framework a global state was viewed as a collection of microstates residing in the neighborhood of a minimum in a multi-dimensional potential energy surface (Section 1.1). There might be many pathways through this space between the two energy minima of different global states. The energy must rise to escape from one energy minimum and then come back down to land in another. Along any particular pathway, there must be an energy maximum, and the pathway most often traveled should be that pathway with the lowest maximum. As noted in Section 7.2, this pathway with the lowest maximum can be viewed as a pass through a mountain range. A surface shaped like this is said to have a saddle point (Fig. 7.13).

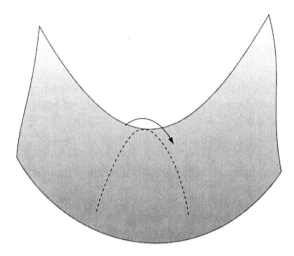

Fig. 7.13. In a multidimensional potential energy function, the pathway for a global transition traverses a saddle point. The potential energy function has a minimum with respect to all of the internal coordinates except for one, the reaction coordinate. The progress of the reaction is indicated by the arrow. The reaction coordinate goes through a maximum at the saddle point.

The theory for passage over a barrier in one dimension can be extended to many dimensions by adding in some structural information about the energy surface in the vicinity of the saddle point, along with information about the energy well escaped from (Hänggi et al., 1990). We now take $E^{\dagger}$ as the energy at the center of the saddle point. This makes a plausible case for using barrier-crossing theories to understand global transitions in proteins, and puts the structure of the transition state in a multidimensional perspective.

The structure of the transition state will now be examined for the global transition between the native state of a protein and its unfolded denatured state. This transition can be induced either with elevated temperature (Section 1.3) or with denaturants (Section 1.7). A denaturant, D, alters the free energy difference between the folded and unfolded states according to Eq. (1.22)

$$\Delta G_{u} = \Delta G_{u}^{o} - m_{u}[D] \tag{7.39}$$

where ΔG_{u}^{o} is the free energy of unfolding in the absence of denaturant; and ΔG_{u} is the free energy for a denaturant concentration of [D]. The parameter m_{u} gauges the denaturing action of D and can be interpreted as the interaction energy between D and the protein interior. Viewing unfolding as a two-state global transition, we can write down the equilibrium constant for the folded and unfolded proteins, P_{f} and P_{u}, in terms of ΔG_{u}. The manipulations used in Chapter 1 lead to an expression for the extent of unfolding

$$\frac{[P_{u}]}{[P_{f}] + [P_{u}]} = \frac{1}{1 + e^{\Delta G_{u}^{o} - m_{u}[D]}} \tag{7.40}$$

This has the same basic form as Eqs. (1.13) and (1.27). Measurements of $[P_{u}]$ versus [D] can be fitted to this equation to determine ΔG_{u}^{o} and m_{u}.

The rate of unfolding is taken as an exponential function of the free energy of the transition state

$$k_u = Ce^{-\Delta G_u^{\ddagger}/RT} \qquad (7.41)$$

If we assume that the denaturant acts only on $\Delta G_u^{\ddagger}$, with no effect on the preexponential factor C, then we assess the effect of denaturant on the rate of unfolding using the relation

$$\Delta G_u^{\ddagger} = \Delta G_u^{\ddagger o} - m_u^{\ddagger}[D] \qquad (7.42)$$

Taking the logarithm of Eq. (7.41) and using Eq. (7.42) for $\Delta G_u^{\ddagger}$ leads to

$$\ln k_u = \ln k_u^o + m_u^{\ddagger}[D] \qquad (7.43)$$

where k_u^o is the rate at $[D] = 0$. With the aid of this equation, measurements of the unfolding rate versus [D] provide an estimate of $m_u^{\ddagger}$. Since m_u and $m_u^{\ddagger}$ both reflect the amount of protein that is buried in the interior, the ratio $m_u^{\ddagger}/m_u$, known as the Tanford β_T value. This quantity is very much like Φ of Eq. (7.15), and tells us what fraction of the interior is buried at the unfolding transition state (Fersht, 1998). The value of β_T ranges from 0.28 to 0.88, indicating that proteins vary quite a bit in terms of the amount of their interior that remains inaccessible to solvent in the transition state (Tanford, 1970).

Using solvent denaturation in combination with protein engineering provides a structural map of the unfolding transition state. A mutation that incorporates a side chain into the protein interior that is attracted to D will increase m_u, allowing lower concentrations of D to unfold the protein. Thus, we measure the specific effect of a mutation on the energy balance between the native and unfolded states.

For k_u measurements from wild type and mutant proteins, we can divide the rates in Eq. (7.41) and take the logarithm to obtain the effect of the mutation on the transition state free energy

$$\Delta\Delta G_u^{\ddagger} = -RT \ln (k_u/k_u') \qquad (7.44)$$

Likewise, the effect of the mutation on ΔG_u^o in Eq. (7.40) can be measured to give $\Delta\Delta G_u^o$. Looking at Fig. 7.5, we can expect the perturbation of the transition state free energy, $\Delta\Delta G_u^{\ddagger}$, to be a fraction of the perturbation of the unfolding free energy $\Delta\Delta G_u^o$ as follows

$$\Delta\Delta G_u^{\ddagger} = \Phi\Delta\Delta G_u^o \qquad (7.45)$$

So we can determine Φ from measurements of $\Delta\Delta G_u^o$ and $\Delta\Delta G_u^{\ddagger}$.

The quantity Φ takes on a different meaning in the context of a multidimensional kinetic process. Most importantly, it has different values for different mutants. The transition state in different parts of the protein may be unfolded to different degrees. One part of the protein may be completely folded in the transition state. In that case a mutation at that site will give $\Phi = 0$. This mutation will change the equilibrium constant without changing the rate of unfolding. Another part of the protein may be completely unfolded in the transition state. That would give $\Phi = 1$, so the rate of unfolding will be changed by the same factor as the equilibrium constant. The value of Φ for a mutation thus tells us something about the structure of that part of the protein in the transition state.

The values of Φ are presented for the enzyme barnase, a bacterial ribonuclease (Fig. (7.14b)). There are some clear trends. Values of Φ are near one in the second and third α-helical elements and the first and second strands of β-sheet. This indicates that these parts of the protein are folded in the transition state. Thus, when an unfolded protein folds, these elements fold first. This part of the protein acts as a nucleus for the folding of the rest of the protein. Looking at the structure (Fig. 7.14a), the segments $\alpha2$, $\alpha3$, $\beta1$, and $\beta2$ are in contact. These contacts bury a number of hydrophobic residues so that

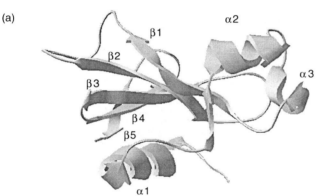

(a)

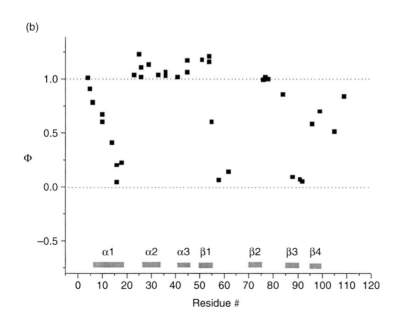

(b)

Fig. 7.14. (a) A ribbon diagram of barnase shows the principal elements of secondary structure. (b) Values of Φ show which part of barnase is folded in the transition state (data from Serrano *et al.*, 1992).

secondary structure formation is cooperative with the formation of this hydrophobic complex. The beginning of α1 is also part of this cluster and its Φ values are also near one. This analysis shows that the reaction coordinate for the folding of barnase proceeds through a state in which a few key elements fold first, and the rest of the protein then folds up around this core. Molecular dynamics simulations of the unfolding of barnase reveal a transition state that agrees well with this experimental analysis of Φ values (Daggett and Fersht, 2003).

Not all the Φ values are zero or one. Figure 7.14b shows that a number of residues have fractional Φ values. Another protein, chymotrypsin inhibitor 2, has almost no Φ values near zero or one. Φ is scattered over a wide range of fractional values, averaging a bit below 0.5 over the entire length of the protein (Fig. 7.15b). There is a hint of higher values in the helical stretch, and in fact, one residue, alanine 16 has $\Phi = 1$.

In contrast to Φ values of zero and one, which reflect fully folded or fully unfolded regions of a protein in the transition state, fractional values of Φ are ambiguous. They could mean that the relevant part of a protein is in a twilight zone of partial folding in the transition state. Alternatively, a fractional Φ value could reflect an average over multiple transition states seen along different folding pathways. This can be sorted out by examining a series of mutants concentrated in one site (Fersht et al., 1994). Consider two distinct folding pathways in which different domains of the protein, A and B, fold up in alternating sequences. The pathway that goes A-then-B has a rate k_a and the pathway that goes B-then-A has a rate k_b. The observed rate will be the sum over both pathways, $k_u = k_a + k_b$. If mutations are concentrated in a region that influences A, then only the rate k_a will change. For this group of mutants we would have

$$k_u = k_a e^{-\Delta\Delta G_u/RT} + k_b \tag{7.46}$$

If k_b starts off being faster than k_a in the wild type protein, then mutants with small $\Delta\Delta G_u$ effects will not produce a noticeable change. But as $\Delta\Delta G_u$ gets larger the k_a pathway will start to take over and dominate. Thus, a plot of k_u versus $\Delta\Delta G_u$ will start flat and then turn into a line with a slope of one. On the other hand, a single pathway of unfolding predicts a linear plot of k_u versus $\Delta\Delta G_u$ over an entire set of mutants in this region.

This analysis was performed for chymotrypsin inhibitor 2 and barnase, and the results exhibited two different patterns (Fig. 7.16b). Chymotrypsin inhibitor 2 was mutated repeatedly in a cluster of three residues that form the minicore (space filling residues in Fig. 7.15a). Progressive destabilization of the folded state by reducing the hydrophobic contacts among these residues produces a linear increase in the rate of unfolding (Fig. 7.16b). This argues for a single unfolding pathway for chymotrypsin inhibitor 2.

The situation is quite different for barnase (Fig. 7.16b). Here the plot was gently sloped for small perturbations but became

(a)

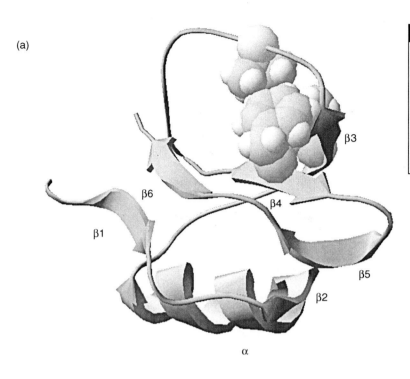

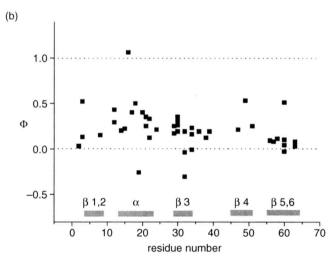

Fig. 7.15. (a) A ribbon diagram of chymotrypsin inhibitor 2 shows the principle elements of secondary structure. Three residues that form a cluster referred to as the "minicore" are shown as space-filling. (b) The Φ values from neutral mutations of chymotrypsin inhibitor 2 are fractional through most of the protein (data from Itzhaki *et al.*, 1995).

steeper for larger perturbations. The result is more like Eq. (7.46), so multiple pathways are indicated. A quantitative analysis of this plot did not fit a simple two-pathway model, but it does indicate that not every mutant unfolds along the same pathway. It is reasonable that the larger size of barnase (110 amino acids) provides more options for unfolding than a small protein like chymotrypsin inhibitor 2 (64 amino acids). In fact, barnase can be envisioned as having modules that unfold with some degree of autonomy. Each of these modules unfolds through a transition state like that of chymotrypsin

Fig. 7.16. Mutations that destabilize the folded state increase the rate of unfolding. (a) In chymotrypsin inhibitor 2 a linear plot of ln k_u versus $\Delta\Delta G_u$ indicates that unfolding proceeds through a single pathway. Mutations were in the minicore indicated in Fig. 7.15a. (b) A plot for mutations in the major α-helix of barnase ($\alpha 1$ in Fig. 7.14a) shows a departure from linearity, suggesting some redistribution between pathways (Fersht et al., 1994).

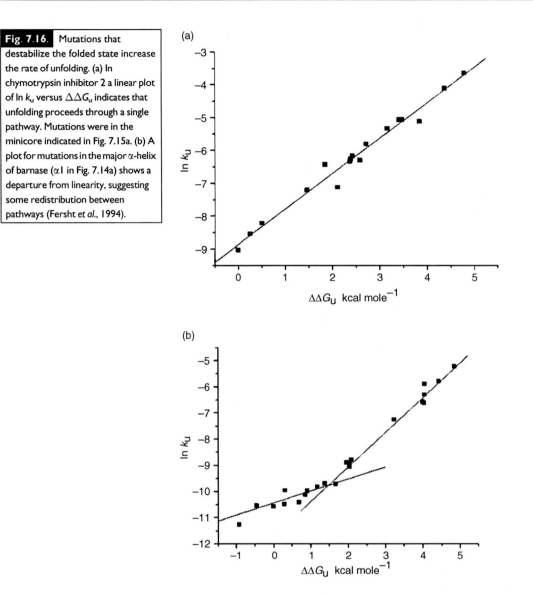

inhibitor 2, in which there is partially formed secondary structure distributed over a wide region.

Protein engineering and linear-free-energy analysis have also been combined to map the reaction pathway in the gating transition of ion channels. The equilibrium and rate constants for opening and closing of the acetylcholine receptor channel were studied for a series of mutants using single-channel current recording. For transitions initiated by acetylcholine binding, residues near the binding site gave Φ values near one and residues near the channel gave Φ values near zero. This suggests a transition pathway in which the binding site first moves to its high affinity configuration while the channel has barely opened. The channel opening is thus completed after the binding site has accommodated the binding of acetylcholine (Grosman *et al.*, 2000). By contrast, when the channel openings occurred spontaneously, the Φ values showed a very different pattern (Grosman, 2003). Values of Φ throughout the protein are then all near one. This indicates that the reaction pathway differs substantially for unliganded channel openings. Here, the transition state is very close to the open conformation. The shift in the reaction pathway may be a key element in the $\sim 10^7$-fold acceleration of channel opening resulting from acetylcholine binding.

Problems for Chapter 7

1. Solve for α and β using the values of τ and x_{final} in Fig. 7.1.

2. Derive a Marcus-like free-energy relation by assuming that the top of the energy barrier has the form $E^\dagger - \phi(x-x_0)^2$, and that a linear potential connects the left and right sides as in Fig. 7.4.

3. Write voltage-dependent rates (Eqs. (7.16a), (7.16b) and (7.17)) for the case that the transition midpoint is not zero.

4. Write down the equilibrium constant for a two-state process in terms of the forward and reverse rates from Kramers' theory. Interpret the result with the aid of the partition function for a harmonic potential (Section 2.12).

5. In a membrane with a single channel, consider the case where a recording starts at an arbitrary time when the channel is closed. Derive the waiting time distribution until the channel opens. Compare this with the closed time distribution (Jackson, 1985).

6. Derive the open time distribution for a channel for which openings can end either by closing, with a rate β, or by being chemically destroyed, with a rate η.

7. For the channel in Problem 6, derive the open time probability density function for openings that end by closing and compare with the open time probability density for openings that end in destruction.

8. The mean lifetime of a channel state can be obtained from the probability density function as $\bar{t} = \int_0^\infty t p(t) dt$. Show that the distribution gives the mean lifetime as $\bar{t} = \int_0^\infty P(t) dt$.

Chapter 8

Association kinetics

Chapter 4 introduced the subject of molecular associations, pointing out their role in the initiation of a host of important biological signaling processes. That chapter focused on thermodynamic aspects of associations, and the factors that influence their strength. This chapter will focus on their kinetics, and the factors that influence their speed. As in Chapter 7, we start by assuming the rate constant has a particular value and develop phenomenological equations. We then turn to the question of how fundamental physical processes determine what the value of an association rate constant will be. Since associations depend on two reactants finding each other through random motion in solution, the kinetics of this process is a problem of random walks and diffusion.

8.1 | Bimolecular association

Consider two molecules, A and B, colliding and sticking together in a bimolecular reaction that produces a complex C (see Scheme (8A)). For starters we will ignore the reverse process to keep it simple. That leaves only the second order rate constant for association, α (with units of $M^{-1}s^{-1}$)

$$A + B \xrightarrow{\alpha} C \tag{8A}$$

Define the extent of the reaction, x, as the amount of A that has combined with B. Then x will increase with a velocity equal to $\alpha[A][B]$. If the starting concentrations are $[A]_0$ and $[B]_0$, then $[A] = [A]_0 - x$. As $[B]$ will decrease stoichiometrically, $[B] = [B]_0 - x$. This leads to a differential equation for the rate of change of x

$$\frac{dx}{dt} = \alpha[A][B] = \alpha([A]_0 - x)([B]_0 - x) \tag{8.1}$$

which can be rearranged to give

$$\frac{dx}{([A]_0 - x)([B]_0 - x)} = \alpha dt \tag{8.2}$$

A particularly simple result is obtained if we specify that $[A]_0 = [B]_0$

$$\frac{dx}{([A]_0 - x)^2} = \alpha dt \tag{8.3}$$

This is easily integrated as follows

$$\frac{1}{[A]_0 - x} = \alpha t + C \tag{8.4}$$

where C is the constant of integration. At $t = 0$, $x = 0$, giving $C = 1/[A]_0$. After making this substitution, we solve for x

$$x = \frac{\alpha [A]_0^2 t}{1 + \alpha [A]_0 t} \tag{8.5}$$

This expression provides an intuitively reasonable picture of the time course of association. The value of x starts out at zero and approaches $[A]_0$ after a sufficiently long time. When $t = 1/(\alpha [A]_0)$ we see that $x = [A]_0/2$, so $1/(\alpha [A]_0)$ is the half-time of the process. This number can often be compared with the time constant of exponential kinetics found for an isomerization (Section 7.1). However, Eq. (8.5) shows that the time course of an association process has a fundamentally different mathematical form. It is hyperbolic rather than exponential.

Not setting $[A]_0 = [B]_0$ makes Eq. (8.2) more difficult to integrate, but it can still be carried out using the method of partial fractions. After determining a constant of integration in the same way as above, and solving for x, we obtain

$$x = [A]_0 [B]_0 \frac{e^{-\alpha([A]_0 - [B]_0)t} - 1}{[B]_0 e^{-\alpha([A]_0 - [B]_0)t} - [A]_0} \tag{8.6}$$

In this expression we see exponentials, as in the solution for an isomerization, but the form is clearly more complicated. Regardless of whether we use Eq. (8.5) or Eq. (8.6), the time course for an association process is qualitatively different from the exponential seen for isomerization. In general, for higher order kinetic processes the time course is not exponential. However, when the reactants are near equilibrium, the kinetics become approximately exponential, as will be seen next.

8.2 | Small perturbations

Often a kinetic process is studied near its equilibrium. A small perturbation is applied to move the system away from equilibrium. An equilibrium can be perturbed with a small jump in temperature, pressure, concentration, voltage, etc. A small quick step in an experimental variable will shift the equilibrium constant.

The concentrations will change in response, and the time course of this adjustment will depend on the rate constants.

Here we must consider both the forward and reverse processes, because it is stipulated that the reaction is near equilibrium

$$A + B \underset{\beta}{\overset{\alpha}{\rightleftharpoons}} C \qquad (8B)$$

In Scheme (8B), [A], [B], and [C] are assumed to be near their equilibrium concentrations, $[A]_{eq}$, $[B]_{eq}$, and $[C]_{eq}$. The deviation from equilibrium is denoted with the variable x. We then have $[A] = [A]_{eq} + x$, $[B] = [B]_{eq} + x$, and $[C] = [C]_{eq} - x$. We now write an expression for the rate of change of [A]

$$\frac{d[A]}{dt} = -\alpha[A][B] + \beta[C] \qquad (8.7)$$

Expressing all the concentrations in terms of x gives

$$\frac{d([A]_{eq} + x)}{dt} = \frac{dx}{dt} = -\alpha([A]_{eq} + x)([B]_{eq} + x) + \beta([C]_{eq} - x)$$
$$= -\alpha[A]_{eq}[B]_{eq} + \beta[C]_{eq} - (\alpha([A]_{eq} + [B]_{eq}) + \beta)x - \alpha x^2 \qquad (8.8)$$

Since x is small compared to everything else, we can ignore x^2. The differential equation then assumes the same form as Eq. (7.3). This means that the solution will be an exponential with a decay constant equal to $\alpha([A]_{eq} + [B]_{eq}) + \beta$ (the factor multiplying x). The constant of integration can be determined by noting that x will go to zero for long times because that is its equilibrium value. If the initial value of x is x_0, then the solution to Eq. (8.8) is

$$x = x_0 e^{-(\alpha([A]_{eq} + [B]_{eq}) + \beta)t} \qquad (8.9)$$

So we see that a higher order kinetic process will have an exponential time course near equilibrium.

Perhaps the most common application of Eq. (8.9) is in temperature-jump experiments. Equilibrium constants are usually temperature dependent, and a jump in temperature of a degree or so will cause a small change, which is made to order for this kind of analysis. It is common to study the kinetics for different values of the concentration of one of the reactants. The decay constant in Eq. (8.9) will show a linear dependence on concentration if Scheme (8B) is correct. This is illustrated in Fig. 8.1 for the binding of Ca^{2+} to the Ca^{2+}-sensitive fluorescent dye fura-2. The slope gives α and the zero $[Ca^{2+}]$ intercept gives $\beta + \alpha$ [fura-2]. So an analysis of such a plot yields both the forward and reverse rate constants.

The decision to use Eq. (8.9), Eq. (8.5), or Eq. (8.6) to study the kinetics of an association process will depend on the conditions at the start of the reaction, and on how far the process is from equilibrium.

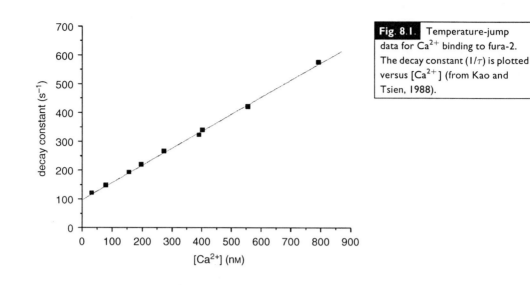

Fig. 8.1. Temperature-jump data for Ca^{2+} binding to fura-2. The decay constant ($1/\tau$) is plotted versus $[Ca^{2+}]$ (from Kao and Tsien, 1988).

8.3 | Diffusion-limited association

The rate of association of two molecules depends on how frequently they collide as they move about randomly in solution. In fact, we can envision two molecules with especially strong affinities for one another, so that they associate into a complex whenever they collide. This would define an upper limit to the rate of association. A reaction could be slower when not every collision succeeds, but a reaction cannot be faster. This concept of diffusion-limited association is very useful in evaluating the kinetics of bimolecular reactions.

Diffusion theory can be used to calculate the frequency of collisions, and thus provide an estimate of the diffusion-limited rate of association (Hammes, 1978). Consider two spherical molecules A and B (Fig. 8.2). Focus on A, and fix its center at the origin of a spherical coordinate system. As molecules of B diffuse, they will undergo collisions with A. We would like to know how frequently diffusion brings molecules of B into contact with molecules of A.

Immediately after the reactants are mixed, the molecule of A is surrounded by a solution of the other molecule, B, with a uniform concentration, C_b. As the reaction gets underway, the molecules of B that are near the molecule of A will be more likely to have reacted, so the concentration of B will no longer be uniform, but will depend on the distance from the center of A. We thus have C_b varying as a function of this distance, r. A description of the frequency of collisions then depends on knowing the function, $C_b(r)$.

Although $C_b(r)$ changes rapidly at the start of the reaction, a steady state quickly forms in which $C_b(r)$ no longer depends on time. That means we can set the time derivative equal to zero in the diffusion equation (Section 6.3). In spherical coordinates (Appendix 6), this equation is (Eq. (6.25))

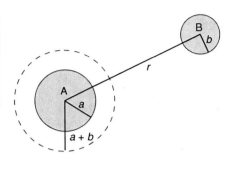

Fig. 8.2. Diffusion produces collisions between two spherical molecules, A and B, with radii a and b, respectively. The dashed circle around A marks the distance of closest approach, $r = a + b$.

$$\frac{1}{r^2}\frac{\partial}{\partial r}\left(r^2 \frac{\partial C_b(r)}{\partial r}\right) = 0 \tag{8.10}$$

In fact, the analysis is almost identical to that for the problem of diffusion through a small hole (Section 6.3.2). So we can jump to the solution we obtained prior to the imposition of boundary conditions (Eq. (6.27))

$$C_b(r) = \frac{-A}{r} + B \tag{8.11}$$

The boundary conditions are different from those used in Section 6.3.2. In the present situation, once a molecule of B has collided with the molecule of A it has associated and disappeared from the system. This makes the concentration of B zero at the surface, so $C_b(a + b) = 0$. (Recall the absorbing boundary conditions of Section 6.2.4.) Far from A, with $r \to \infty$, we have the bulk value $C_b(\infty)$. These two boundary conditions determine the values for the constants of integration in Eq. (8.11) as $A = (a + b)C_b(\infty)$ and $B = C_b(\infty)$. So we have the solution

$$C_b(r) = \frac{-(a + b)C_b(\infty)}{r} + C_b(\infty) \tag{8.12}$$

The frequency of collisions is proportional to the flux of molecules of B at $r = a + b$. This is determined by taking the derivative with respect to r (Eq. (6.1)).

$$J = -D_b \frac{\partial C_b}{\partial r} = -D_b \frac{(a + b)C_b(\infty)}{r^2} = -\frac{D_b C_b(\infty)}{a + b} \tag{8.13}$$

This flux is the rate at which molecules pass through a unit area at $r = a + b$. To obtain the total flux over the entire surface, we must multiply Eq. (8.13) by the surface area $4\pi(a + b)^2$. This total flux is the collision frequency, given by

$$\text{collision frequency} = 4\pi(a + b)D_b C_b(\infty) \tag{8.14}$$

So the frequency is proportional to the bulk concentration, $C_b(\infty)$, as expected, and the association rate constant is the factor

$$\alpha = 4\pi(a + b)D_b \tag{8.15}$$

This was calculated for a single molecule of A, so with D_b in units of $cm^2 s^{-1}$, and a and b in units of cm, the units of α are cm^3 molecule^{-1} s^{-1}. The conversion to the more familiar units of $M^{-1} s^{-1}$ will be performed shortly.

This derivation overlooked the fact that molecules of A diffuse as well as molecules of B. This will increase the frequency of collisions. This problem is resolved by replacing the diffusion coefficient of B with the sum of the two diffusion coefficients

$$\alpha = 4\pi(a+b)(D_a + D_b) \tag{8.16}$$

Equations (8.15) and (8.16) are important results in the theory of bimolecular reaction kinetics. The ideas originated with Smoluchowski in 1917 and provide an accurate estimate of rate constants for chemical reactions that are limited by the frequency of collisions. The theory has many ramifications for problems in chemical kinetics (Keizer, 1987).

These results are also important for understanding association problems in biophysics, but the collisions are almost always restricted in some way. For example, the dependence of an association rate on the size of the molecules is artificial because binding was assumed to occur anywhere on the surface. In reality, binding in most chemical and biochemical interactions occurs at specific sites. There are additional constraints on binding imposed by the orientation of the two molecules relative to one another. These issues will be taken up below, but for now it can be seen that all of these neglected complications can only reduce the rate to a value below that specified in Eq. (8.16), so it is a good upper limit.

To get an idea of how fast association rates can be, take a typical diffusion constant for a small molecule, $D_a = D_b = 5 \times 10^{-6}$ cm^2 s^{-1}, and let $a = b = 2.5 \times 10^{-8}$ cm (2.5 Å). Equation (8.16) then gives $\alpha = 6 \times 10^{-12}$ cm^3 molecule^{-1} s^{-1} or 6×10^{-15} liter molecule^{-1} s^{-1}. Multiplying by Avogadro's number converts this to 4×10^9 $M^{-1} s^{-1}$.

A more general diffusion-limited association rate can be derived by eliminating the diffusion constant and molecular size in Eq. (8.15) with the Stokes–Einstein relation $D = RT/6\pi r\eta$ (Eq. (6.66)) and letting $r = a = b$

$$\alpha = 8RT/3\eta \tag{8.17}$$

Evaluating this expression for the viscosity of water and room temperature gives about 10^{10} $M^{-1} s^{-1}$. The lack of dependence on molecular parameters (size and diffusion coefficient) reflects a fortuitous cancellation. The fastest measured association rates are generally in the range of these theoretical limits, i.e. 10^9–10^{10} $M^{-1} s^{-1}$ (Section 8.6; but see Section 8.7 on protonation). This range defines the fastest possible rates of association.

The derivations above assumed that the molecules are hard spheres with no interactive forces other than the repulsion at their surfaces. Attractions can increase the rate of association, and this can be incorporated into the rate equation by using a flux equation that includes the effect of a force (Eq. (6.64), with $F = \partial U(r)/\partial r$; see also Eq. (7.27)). The collision frequency is the flux (J from Eq. (6.64)) times the area of the sphere, $4\pi r^2$. So we evaluate the total flux through an entire spherical shell as $4\pi r^2 J$

$$\text{total flux} = -4\pi r^2 D_b \left(\frac{C_b(r)}{kT} \frac{\partial U(r)}{\partial r} + \frac{\partial C_b(r)}{\partial r} \right) \tag{8.18}$$

The potential energy function, $U(r)$, can assume any form, and need not be a barrier as it was in Section 7.8. It could be attractive, repulsive, or flat.

Assume that a steady state has been reached, so the total flux is constant. Following the same logic that led from Eq. (7.27) to Eq. (7.30), we obtain

$$\text{total flux} = \frac{4\pi D_b (C_b(\infty) e^{U(\infty)/kT} - C_b(a+b) e^{U(a+b)/kT})}{\int\limits_{a+b}^{\infty} \frac{1}{r^2} e^{U(r)/kT} dr} \tag{8.19}$$

The limits of integration are from the distance of closest approach ($r = a + b$) to infinitely far away ($r = \infty$).

As before in determining the constants of integration in Eq. (8.11), collisions remove molecules of B so $C_b(a+b) = 0$. At $r = \infty$ we use the bulk concentration, $C_b(\infty)$ and take $U(\infty) = 0$. Equation (8.19) then gives the collision frequency as

$$\text{collision frequency} = \frac{4\pi D_b C_b(\infty)}{\int\limits_{a+b}^{\infty} \frac{1}{r^2} e^{U(r)/kT} dr} \tag{8.20}$$

We can return to the hard-sphere case by setting $U(r) = 0$. Then the integral in the numerator of Eq. (8.20) is easily evaluated as $1/(a+b)$. Equation (8.20) then reduces to Eq. (8.15).

8.4 | Diffusion-limited dissociation

The same theory that was used above to obtain a diffusion-controlled association rate can be used to derive a dissociation rate. Start with Eq. (8.11), which was a general steady-state result for spherical geometry. Recall that for association we set $C_b(a+b) = 0$, and $C_b(\infty) =$ the bulk concentration. For dissociation we turn the problem around, and take $C_b(\infty) = 0$ and $C(a+b) = $(one molecule)/(volume around A). The volume around A is $(4/3)\pi(a+b)^3$ to give $C_b(a+b) = 1/((4/3)\pi(a+b)^3)$. With these boundary conditions the constants of integration in Eq. (8.11) are $B = 0$ and $A = -3/(4\pi(a+b)^2)$, so we have the steady-state distribution

$$C_b(r) = \frac{3}{4\pi(a+b)^2 r} \tag{8.21}$$

Now the flux is evaluated at $r = a + b$ as follows

$$J = -D_b \frac{\partial C_b}{\partial r} = D_b \frac{3}{4\pi(a+b)^2 r^2} = \frac{3D_b}{4\pi(a+b)^4} \tag{8.22}$$

The sign is opposite to that in Eq. (8.13) because with dissociation the flux is in the opposite direction. The frequency is obtained by multiplying by the surface area

$$\text{frequency} = \frac{3D_b}{(a+b)^2} \tag{8.23}$$

This is the first order rate constant for dissociation in units of molecules per unit time. As in Eq. (8.15), we can replace D_b with the sum $D_a + D_b$ to take into account the diffusion of both molecules.

This is the rate constant for dissociation in the complete absence of binding forces that stabilize the complex. To include these we perform the analogous operation on the equation for diffusive flux in a force field (Eq. (8.19)). The steady-state solution with the boundary conditions just specified is

$$\text{frequency} = \frac{4\pi D_b e^{U(a+b)/kT}}{\frac{4}{3}\pi(a+b)^3 \int\limits_{a+b}^{\infty} e^{U(r)/kT} dr/r^2}$$

$$= \frac{3D_b e^{U(a+b)/kT}}{(a+b)^3 \int\limits_{a+b}^{\infty} e^{U(r)/kT} dr/r^2} \tag{8.24}$$

If the potential energy function $U(r)$ is zero, then the integral is easily performed to recover Eq. (8.23).

Equation (8.23) is not widely used in biophysics because the diffusion limit applies only for dissociation when the attractive forces are very weak, and in that case no complex will form in the first place. However, the derivation here is useful for the sake of logical completeness. It shows that the dissociation rate also depends on the diffusion constants. This is conceptually important because equilibrium constants generally do not depend on transport coefficients such as the diffusion constant. The equilibrium constant equals the ratio of the forward and reverse rate constants. Since both are proportional to the diffusion constant, it cancels when the ratio is taken (Problem 2).

8.5 | Site binding

The derivation of the diffusion-limited association rate was based on the assumption that a collision between any two parts of the reacting molecules will produce a complex. That might be true for the combination of some small molecules (e.g. iodine with CCl_4; see Keizer, 1987), or for two droplets of liquid, but for most molecules of biological interest it matters which parts of the molecules meet during a collision.

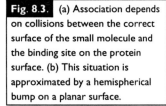

Fig. 8.3. (a) Association depends on collisions between the correct surface of the small molecule and the binding site on the protein surface. (b) This situation is approximated by a hemispherical bump on a planar surface.

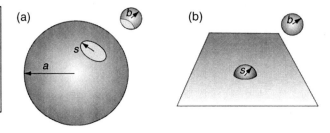

To improve on the model pictured in Fig. 8.2 we can stipulate that the collision must be between reactive sites on the two molecules. Clearly, this will reduce our estimate of the rate of association.

To address this problem consider a small molecule with a reactive surface colliding with a reactive site on the surface of a protein (Fig. 8.3a). We would like to know the rate of collisions between these two sites. An intuitive approach would be to start with Eq. (8.16), make the calculation using the radii of the protein and small molecule, and scale this result by the fraction of the surface area covered by each binding site. If s is the radius of a site on the protein (as shown in Fig. 8.3a), then we take the ratio of the areas, s^2/a^2, as our scaling factor, and use a similar factor for the ligand. However, intuition is misleading in this case; the rate is actually proportional to the linear dimension of the binding site and not its area, just as the rate of association in Eq. (8.16) is proportional to linear dimensions of the reacting molecules.

We can make a reasonable estimate of the frequency with which ligands collide with a site by using the model depicted in Fig. 8.3b. Forget about the reactive site on the smaller molecule, b, because the small size of the protein's binding site has a much greater impact. Then replace the circular site on the surface of a sphere (Fig. 8.3a) with a hemispherical bump on the surface of a plane (Fig. 8.3b). This problem can be solved by a slight modification of the derivation used in Section 8.3. Equation (8.10) was based on the steady-state flux at the surface of a spherical shell. The steady-state flux through a hemispherical shell is simply half of this.[1] The analysis is then the same as that used to derive Eq. (8.15). Ignoring the relatively small diffusion coefficient of the protein, the rate constant for encounters in the problem pictured in Fig. 8.3b is

$$\alpha = 2\pi(s + b)D_b \tag{8.25}$$

---

[1] The presence of the unreactive surface around the bump turns out not to be a problem, although, more generally, adding a boundary can complicate solving the diffusion equation. The surrounding flat surface has zero flux, and that imposes the boundary condition that the concentration gradient normal to the surface is zero (Section 6.2.4). In the steady-state solution to the spherical problem the gradients are purely radial, and the concentration is constant within hemispherical shells centered around the bump. The gradients normal to the flat surface in Fig. 8.3b are therefore zero, so there is no flux in that direction. That means that Eq. (8.10) scaled down by a factor of 2 gives the correct solution to this problem.

This problem can also be solved for a circular site on a plane rather than the hemispherical bump used here. The mathematics for solving this problem is a bit above the level used in this book. The result is $4sD_b$ (Shoup et al., 1981), changing Eq. (8.25) by a factor of $\pi/2 = 1.57$. A similar correction was noted for the problem of diffusion through a small hole (Section 6.3.2).

A common modification of this result is to assume that the ligand must fit entirely into the binding site. This reduces the effective radius of the binding site by the radius of the ligand. The result is then (Shoup et al., 1981)

$$\alpha = 4(s - b)D_b \qquad (8.26)$$

At this point it should be recognized that assumptions about the size match between the ligand and binding site are rather arbitrary. Thus, it is just as reasonable to take the binding site and ligand as similar in size in Eq. (8.25). This reduces to $8bD_b$ (with the replacement of 2π by 4 based on the result for a circular site just mentioned). Now using the Stokes–Einstein relation, just as we did to obtain Eq. (8.17), gives

$$\alpha = 4RT/3\pi\eta \qquad (8.27)$$

which is a factor of 2π less than Eq. (8.17), the expression for collisions between equal-sized spheres. For the viscosity of water and room temperature this gives a useful general diffusion limit of $1.6 \times 10^9 \, \text{M}^{-1} \, \text{s}^{-1}$ for ligand binding to a protein.

Restriction of the reactive surface of the ligand (Fig. 8.3a) will further reduce the frequency of successful collisions. The above analysis suggests that we use the size of the ligand's binding site in place of the ligand's size. However, if the ligand rotates rapidly, an unfavorable collision can turn into a favorable collision. Rotational diffusion thus increases the reaction rate, and if rotational diffusion is much faster than translational diffusion, it can eliminate the effect of a ligand's small interaction site. This brings us back to using the size of the whole ligand in the calculation of the diffusion-limited association rate. However, rotation is generally not that fast, leaving an intermediate situation where the limited size of the ligand's binding site and ligand's rotation combine to produce a rate that falls between the two extremes (Shoup et al., 1981).

8.6 | Protein–ligand association rates

The theories for diffusion-limited association are approximate, and there is no exact value for how fast a diffusion limited association really can be. The numbers calculated above can provide a rough

Table 8.1. | *Rapid protein–ligand association rates*

Protein	Ligand	Association rate constant ($M^{-1} s^{-1}$)
acetylcholinesterase[a]	N-methyacridinium	1.1×10^{10}
horseradish peroxidase[b]	p-nitrobenzoic acid	1.3×10^{8}
superoxide dismutase[c]	superoxide ion	2.4×10^{9}
triose phosphate isomerase[d]	glyceraldhyde phosphate	2.4×10^{8}
β-lactamase[e]	benzylpenicillin	7.6×10^{7}
chymotrypsin[f]	p-nitrophenyl ester	9×10^{7}
calmodulin[g], N-domain	Ca^{2+}	5×10^{8}
C-domain		1×10^{8}
acetylcholine receptor[h],	acetylcholine	
1st site		6×10^{7}
2nd site		1×10^{8}

[a] Nolte *et al.* (1980).
[b] Nakatani and Dunford (1979).
[c] Fielder *et al.* (1974).
[d] Putnam *et al.* (1972).
[e] Hardy and Kirsch (1984).
[f] Brouwer and Kirsch (1982).
[g] Falke *et al.* (1994).
[h] Sine *et al.* (1990).

estimate. Many proteins bind their preferred ligands very rapidly. Examples are given in Table 8.1. The association rates often approach the diffusion limits estimated above in Section 8.3. In fact, the measurement of a rate in excess of $\sim 10^8$ M^{-1} s^{-1} often prompts a speculation that the association process goes as fast as diffusion will allow. However, additional criteria can be used to test this hypothesis more rigorously.

Because of the uncertainties in the theoretical upper bounds to association rates, experimental tests of the diffusion limit are very important. The viscosity appears in Eqs. (8.17) and (8.27), and diffusion constants generally are inversely proportional to solution viscosity (Eq. (6.66)). However, using the viscosity dependence of a reaction rate to determine whether it is diffusion limited has a pitfall. As we saw in Section 7.8, the diffusion constant along the reaction coordinate enters into the rate equation for a barrier crossing mechanism. If the motions within the protein are coupled to the molecular motions of the solvent, then the rate of barrier crossing will show a viscosity dependence (see Berg and von Hippel (1985) for a discussion of different forms of viscosity dependence; see Section 10.11 for a discussion of this issue in enzyme catalysis). To address this problem, one can compare the viscosity dependence of two related ligands with different association rates. If a more slowly binding ligand has a weak viscosity dependence, then the barrier to this association is weakly coupled to the solution motions. The viscosity dependence of the association of the more rapidly binding ligand thus cannot be attributed to crossing the same barrier. So for this ligand association is diffusion limited.

The temperature dependence is also informative. If the temperature dependence is weak, or follows that of the solvent viscosity, then the association does not depend on crossing an energy barrier. A weak temperature dependence thus can also indicate that the association is diffusion limited.

8.6.1 Evolution of speed

Evolutionary pressure will optimize the speed of many ligand binding processes. For almost any enzyme, speed is a clear advantage. An organism will economize energy in protein synthesis if a smaller number of faster enzymes can get the same job done. If an enzyme were able to catalyze its reaction instantaneously, then its velocity would be limited by diffusion. So the rate of diffusional encounters represents the absolute maximum rate of catalysis for any enzyme. In fact, nearly all enzymes investigated exhibit catalytic efficiencies that fall within two orders of magnitude of the diffusion limit (Miller and Wolfenden, 2002). This probably reflects a nearly universal advantage of speed in enzymes.

For an enzyme that works near its diffusion limit, there is little incentive for further improvement of its catalytic action on a bound substrate. This idea has been developed for the enzyme triose phosphate isomerase, which binds its substrate very rapidly (Table 8.1). Substrate binding is slowed by increasing the solution viscosity, thus satisfying a basic condition of the diffusion limit (Blacklow *et al.*, 1988). The velocities of the subsequent catalytic steps are in the same range as that of the substrate binding step seen with physiological concentrations of substrate. Making these steps faster offers little improvement in overall speed. Triose phosphate isomerase has therefore been described as a "perfect" enzyme from an evolutionary perspective (Albery and Knowles, 1976).

8.6.2 Acetylcholinesterase

Acetylcholinesterase hydrolyzes the neurotransmitter acetylcholine to terminate the synaptic response at the neuromuscular junction. This enzyme is incredibly fast, and the viscosity dependence indicates that the rate of substrate binding is diffusion limited. In fact, the rate of 1.1×10^{10} M^{-1} s^{-1} (Table 8.1), is quite a bit faster than the relevant diffusion limit of 1.6×10^9 M^{-1} s^{-1} derived from Eq. (8.27).

The crystal structure of acetylcholinesterase provided a reasonable explanation for this extraordinary binding rate (Fig. 8.4). The active catalytic site is located deep in a long narrow gorge. About 40% of the surface of this gorge is lined with the aromatic side chains of phenylalanine, tyrosine, and tryptophan. These groups strongly attract cations (Section 2.6), and acetylcholine has a positively charged quaternary amine. This effectively extends the reach

Fig. 8.4. Structure of acetylcholinesterase with aromatic residues in dark gray and acidic residues in light gray. The white circle encloses the catalytic site (Sussman *et al.*, 1991).

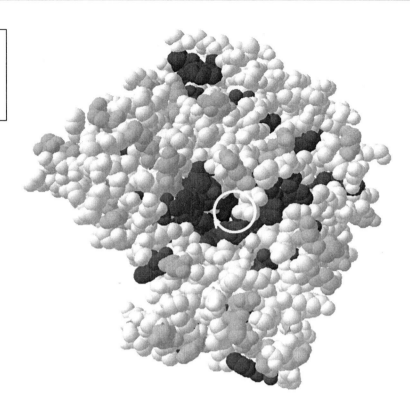

of the binding site. The gorge acts like an electrostatic vacuum cleaner. So acetylcholine molecules colliding with the extended length of this gorge stick to this site and are drawn in toward the catalytic residues in the active site. This attraction is aided by some negatively charged glutamate and aspartate side chains strategically located around the gorge.

If we allow that the gorge of acetylcholinesterase gives us a linear extent for the binding site that is ∼5 times as large as the ligand, then Eq. (8.25) gives about a 3-fold increase over Eq. (8.27). Even with this enhancement, the rate is still a factor of ∼2 less than what can be accounted for by diffusion. It is likely that the long-range Coulombic attraction of the acidic residues reaches out and pulls the substrate into the gorge. This is supported by a high sensitivity of the binding rate to salt (Nolte *et al.*, 1980), which neutralizes and weakens electrostatic interactions (Chapter 11).

Superoxide dismutase also does somewhat better than the diffusion limit (Table 8.1). In this enzyme an array of positively charged side chains extend the length of the binding site for the negatively charged substrate, O_2^- (Tainer *et al.* 1983).

8.6.3 Horseradish peroxidase
Horseradish peroxidase is another very fast enzyme, as indicated by the binding rate in Table 8.1. A critical comparison was made between the binding rate of $1.3 \times 10^8 \, M^{-1} s^{-1}$ and various

theoretical estimates of the diffusion-limited collision rate. An equation closely related to Eq. (8.16) gave 1.7×10^{10} M^{-1} s^{-1} for the rate of encounters with the entire protein molecule. Using the size of a potential substrate that fails to undergo catalysis because it is too large, a size of the binding site was estimated as 4 Å. Scaling the association rate linearly with this length (according to Eq. (8.25)) gave a rate that was in very reasonable agreement with experiment (Nakatani and Dunford, 1979).

8.7 | Proton transfer

Biological molecules exchange protons with their aqueous environment, and these proton transfers can be extraordinarily fast (Eigen and Hammes, 1963; Gutman and Nachliel, 1997). The rates for proton association with various acceptors often exceed the diffusion limit discussed above (Table 8.2). The primary reason is the exceptionally high mobility of the proton in liquid water. Conductivity measurements indicate that the mobility of a proton is >5 times greater than that of other monovalent cations (Moore, 1972).

It is actually not the proton itself that diffuses so rapidly, but rather its charge. Proton movement entails hopping between charged species of the form H_3O^+, $H_5O_2^+$, and $H_9O_4^+$. When a water molecule accepts a proton, it can pass off one of its other H atoms as a proton to a neighboring water molecule on the other side. The charge thus gets a free ride. The charge is translocated in one step that is about as long as the size of a water molecule. This mechanism shuttles protons around and delivers them to their acceptors with unusual speed.

The high intrinsic speed of proton transfer can be thought of as a sort of compensation for the generally low abundance of protons in water. At a physiological pH of 7, the concentration is 10^{-7} M. Thus, the lifetime of a free base proton acceptor will typically be slightly less than 1 ms for most of the rate constants in Table 8.2.

Table 8.2. | *Rates of protonation*

Proton acceptor	Association rate constant (M^{-1} s^{-1})
OH^- [a]	1.3×10^{11}
$CH_3CO_2^-$ [a]	4.5×10^{10}
imidazole [a]	1.5×10^{10}
bacteriorhodopsin [b]	
cytoplasmic site	5.8×10^{10}
extracellular site	1×10^9
calcium channel [c]	4.1×10^{11}

[a] Eigen and Hammes, 1963.
[b] Checover *et al.* (1997).
[c] Prod'hom *et al.* (1987).

The protonation of sites on proteins, especially membrane proteins, can be enhanced by clustering of negative charges to increase the electrostatic attraction. In the light-driven proton pump bacteriorhodopsin, a cluster of three negatively charged carboxyl groups helps speed up proton uptake at the cytoplasmic side of the protein (Gutman and Nachliel, 1997). This is the first step in the active pumping of protons out of a cell following the absorption of a photon.

The rate of protonation of a Ca^{2+} channel was measured using single-channel kinetics (Table 8.2). This is the highest known association rate constant. The speed of this transfer process has yet to be explained in terms of the structure of the protein.

8.8 | Binding to membrane receptors

The binding of a ligand to a site on a protein was examined in terms of the frequency of collisions with a single reactive spot (Section 8.5). Receptors for hormones often reside on the cell surface, and the speed with which these binding sites fill up would appear to be closely related to the site-binding problem. The first idea that comes to mind is to take Eqs. (8.25)–(8.27) and scale them by the number of receptors on the cell. But this ignores an important effect. If a molecule collides with any part of the cell surface, the cell restricts subsequent diffusion so that the molecule has a high probability of bumping into the cell surface again. With many binding sites scattered over the surface, subsequent bumps can easily produce a hit with a receptor, even if the first does not. This tends to make the whole cell the effective target, even though the true target is the receptors scattered over the surface.

We can evaluate the effect of these repeated encounters by using a model of Berg and Purcell (1977). The first step is to look at a molecule found at a distance y from the center of a cell, depicted as a spherical target of radius a. We consider two outcomes: (1) the molecule diffuses off to infinity without ever touching the surface of the target; or (2) the molecule encounters the target. Since the encounter with the target is the defining event for outcome (2), the surface of the target is treated as an absorbing boundary.

The random walk version of this problem is difficult, but fortunately there is an equivalent problem in steady-state diffusion that is straightforward. If the molecules are produced at a constant rate in a spherical shell at a distance y from the center of the target, then there will be fluxes in toward the center and out to infinity (Fig. 8.5). The ratio of these two steady-state fluxes is equal to the ratio of the probabilities of the two outcomes specified above. The reason for this is that the steady-state diffusion problem reflects the average result of all possible random walks starting at y.

We now write out the total flux through a shell at a distance r. Denoting these quantities as G_i or G_o for the regions indicated in Fig. 8.5, they are equal to the surface area times the flux per

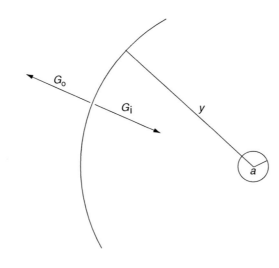

Fig. 8.5. Molecules are produced at a constant rate in a spherical shell at a distance y from the center of a spherical target of radius a. The total inward flux is G_i and the total outward flux is G_o.

unit area (denoted as J in the above analysis). Here J equals the diffusion constant, D, times the concentration gradient, $\partial C_i / \partial r$. This gives

$$G_i = 4\pi r^2 J_i = -4\pi r^2 D \frac{\partial C_i}{\partial r} \qquad (8.28a)$$

$$G_o = 4\pi r^2 J_o = -4\pi r^2 D \frac{\partial C_o}{\partial r} \qquad (8.28b)$$

where C_i and C_o indicate the respective concentrations in the regions inside and outside the shell. Note that G_i and G_o are constant within these regions, because molecules are not being produced there. That makes it easy to integrate the equations and obtain the general solution $C = G/4\pi D + A$. The boundary conditions are $C_i = 0$ at $r = a$, and $C_o = 0$ at $r = \infty$. With A determined from these conditions we have

$$C_i = \frac{G_i}{4\pi r D} - \frac{G_i}{4\pi a D} \qquad (8.29a)$$

$$C_o = \frac{G_o}{4\pi r D} \qquad (8.29b)$$

Although G_i and G_o have not been determined, we know they are constant. We also know that they are opposite in sign, with G_i negative to make C_i positive. The two concentrations must be equal at $r = y$, so we equate Eqs. (8.29a) and (8.29b) to give

$$\frac{G_i}{4\pi y D} - \frac{G_i}{4\pi a D} = \frac{G_o}{4\pi y D} \qquad (8.30)$$

Rearranging gives

$$\frac{G_i}{G_i - G_o} = \frac{a}{y} \qquad (8.31)$$

Since G_i is negative, the quantity on the left is $|G_i|/(|G_i| + |G_o|)$. This is the fraction of molecules produced at y that move in toward the target.

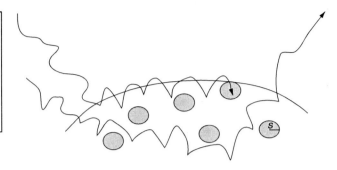

Fig. 8.6. Two molecules undergoing random walks approach a cell from the left. After repeated encounters with the cell surface, one lands on a binding site and is bound. The other bumps along the surface and escapes without colliding with a binding site.

This simple result is thus the probability that a molecule starting at distance y will hit the target rather than diffuse away to infinity.

We will now use Eq. (8.31) to estimate the probability that a molecule bumping along the cell surface will hit a receptor before diffusing away (Fig. 8.6). Consider a random walk bouncing along the cell surface. For an excursion from the cell surface to be far enough away for a return to sample a new area of the cell, the molecule must diffuse a distance away from the surface comparable to the size of the binding site, s. If a molecule has wandered to a distance s, then according to Eq. (8.31), the probability of the molecule having another collision with the cell surface is

$$P_s = \frac{a}{a+s} \tag{8.32}$$

The probability of not having another collision is $s/(a+s)$ so that the ratio of the two is s/a (Eq. (8.31)).

With each collision we can take the probability of not hitting a receptor as one minus the fraction of the cell surface covered by receptors, or $\beta = 1 - (Ns^2/4a^2)$, where N is the number of receptors on the cell. The probability of an encounter occurring without binding to a receptor is then the product βP_s. The probability of n such encounters followed by an escape is then $P_s^n \beta^n (1 - P_s)$. The probability of a molecule escaping (P_{esc}) without ever binding a receptor is then the sum

$$
\begin{aligned}
P_{esc} &= \sum_{n=0}^{\infty} P_s^n \beta^n (1 - P_s) \\
&= \frac{s}{a+s} \sum_{n=0}^{\infty} \left(\frac{a\beta}{a+s} \right)^n \\
&= \frac{s}{a+s} \frac{1}{(1 - (a\beta/(a+s)))} \\
&= \frac{s}{a+s-a\beta} \\
&= \frac{4a}{4a+Ns}
\end{aligned}
\tag{8.33}
$$

where the sum was evaluated as a geometric series with Eq. (A1.11).

Equation (8.33) expresses the fraction of molecules that encounter the cell and escape without binding a receptor. The frequency of collisions with the cell can be taken from Eq. (8.15) as $4\pi aDC$ (we neglect b because $a \gg b$; the cell is much larger than the ligand). The probability of not escaping is $1 - P_{esc}$. Calculating this from Eq. (8.33), and multiplying it by the frequency of collisions with the cell gives the frequency with which ligands bind receptors as

$$\text{frequency} = 4\pi aDC \frac{Ns}{4a + Ns} \qquad (8.34)$$

This equation has important implications for the rate with which receptor binding sites are filled by ligands (Abbott and Nelsestuen, 1988; Nelsestuen and Martinez, 1997). There are two important limiting cases: low receptor density and high receptor density. When the density is low, $Ns \ll 4a$, so Eq. (8.34) goes to the limit

$$\text{frequency} \sim \pi NsDC \qquad (8.35)$$

In this limit the cell size is irrelevant; the receptor is the target. It is as though the receptors are free in solution. When the density is high, $Ns \gg 4a$, so we have

$$\text{frequency} \sim 4\pi aDC \qquad (8.36)$$

Now we have the diffusion limit for collisions with the cell. The cell is the target.

This analysis shows how the behavior varies between two extremes for cells with high versus low receptor densities. This has been studied in detail for the enzyme alkaline phosphatase, which is confined to the periplasmic space between the inner and outer membranes of E. coli (Martinez et al., 1996). At higher substrate concentrations there are very few available binding sites so molecules that wander into the periplasmic space encounter fewer free enzymes. The activity of the enzyme is then determined by Eq. (8.35). At lower substrate concentrations there are many vacant enzyme binding sites, so it is likely that any molecule colliding with the cell and entering the periplasmic space will be captured. Equation (8.36) then applies.

It is important to appreciate how this kind of association process can alter the kinetics from that described in Section 8.1. The rate of decrease of available cell surface receptors will be given by Eq. (8.34) as

$$\frac{dN}{dt} = -4\pi aDC \frac{Ns}{4a + Ns} \qquad (8.37)$$

This integrates to

$$4a \ln (N/N_0) + N - N_0 = -4\pi aDCt \qquad (8.38)$$

where N_0 is the initial number of receptors. Although this transcendental equation cannot be solved for N, it is clear that the time course of receptor occupation will deviate from the forms derived in Sections 8.1 and 8.2. The non-exponential and non-hyperbolic behavior in this situation does not reflect a complex association mechanism, but reflects the transition between these two regimens of high and low receptor density.

8.9 | Reduction in dimensionality

We have seen how placing receptors on a surface enhances a ligand's search for its binding site. One can view this enhancement as a reduction in the number of dimensions of the space in which the search is occurring. Three-dimensional diffusion is made easier by increasing the size of a preliminary target and breaking the process up into two steps. The first step is a collision with a very large preliminary target. The second is a search in two dimensions along the surface of the cell for the final small target. The reduction in the dimensions in which diffusion occurs is an important general kinetic principle for diffusion-limited recognition events in biology (Berg and Purcell, 1977; Adam and Delbrück, 1968).

To explore this effect of dimensionality, we would like to repeat the analysis of Section 8.3 in one and two dimensions. This would tell us how fast diffusing in a plane produces encounters with a circular target, and how fast diffusing along a line produces encounters with a point target. However, the kinetic equations in one and two dimensions do not have a steady-state solution. Another strategy is needed, so we turn to the method of mean capture times. We try to calculate the mean time a molecule takes to find its target by random motion. We will solve the problem in one dimension and state the results for two and three (Berg, 1983).

The random walk in one dimension is as laid out in Chapter 6. Hops of length δ to the right or left occur at regular intervals of time, τ. We define the quantity $W(x)$ as the mean capture time, the average time it takes for the molecule starting at position x to reach some specified location for the first time. Every trajectory from x to the target must proceed through either $x + \delta$ or at $x - \delta$ (with equal probability) at the very next time step. This means that $W(x)$ can be expressed in terms of $W(x + \delta)$ and $W(x - \delta)$ as

$$W(x) = \tau + (W(x + \delta) + W(x - \delta))/2 \tag{8.39}$$

This states that starting from x is like starting from $x + \delta$ or $x - \delta$ at a time τ later. Multiplying by $2/\delta$, Eq. (8.39) can be rearranged as

$$(W(x + \delta) - W(x))/\delta - (W(x) - W(x - \delta))/\delta + 2\tau/\delta = 0 \tag{8.40}$$

Now we recognize the two derivatives with respect to x

$$\left.\frac{dW}{dx}\right|_{x+\delta/2} - \left.\frac{dW}{dx}\right|_{x-\delta/2} + \frac{2\tau}{\delta} = 0 \tag{8.41}$$

Dividing through by δ allows us to combine the derivatives into a second derivative. Furthermore, we can make use of Eq. (6.45) to replace $2\tau/\delta^2$ by $1/D$ as follows

$$\frac{d^2W}{dx^2} + \frac{1}{D} = 0 \tag{8.42}$$

Solving Eq. (8.42) then gives us W, the mean capture time. Boundary conditions have to be used, and they are very similar to those used in solving the diffusion equation (Section 6.2.4). If the target is at $x = a$, then we have an absorbing boundary, with $W(a) = 0$. On the other hand, if there is a barrier at $x = a$, then we have a reflecting boundary, with $dW/dx = 0$.

We can now derive the mean capture time for a one-dimensional diffusion-controlled association. Place a molecule in a region of length b, with the target at $x = 0$. The general solution of Eq. (8.42) is

$$W(x) = -\frac{x^2}{2D} + Ax + B \tag{8.43}$$

where A and B are constants of integration. Since $W(0) = 0$ (absorbing boundary), $B = 0$. With $dW/dx = 0$ (reflecting boundary) at b, $A = b/D$. The specific solution is then

$$W(x) = \frac{1}{D}(xb - x^2/2) \tag{8.44}$$

Now averaging over all starting positions of the molecule between 0 and b gives the desired result

$$\bar{t} = \frac{1}{bD} \int_0^b (xb - x^2/2) \, dx$$
$$= \frac{1}{bD}(b^3/2 - b^3/6)$$
$$= \frac{b^2}{3D} \tag{8.45}$$

Extending this analysis to two and three dimensions is straightforward, although the size of the target, a, now matters (Adam and Delbrück, 1968; Berg and Purcell, 1977; Berg and von Hippel, 1985). In two dimensions we have

$$\bar{t} = \frac{b^2}{2D} \ln \frac{b}{a} \tag{8.46}$$

and in three dimensions we have

$$\bar{t} = \frac{b^2}{3D} \frac{b}{a} \tag{8.47}$$

Note that in each of these equations for $\bar{t}$, D is for diffusion in the relevant number of dimensions.

These three equations illustrate how the number of dimensions qualitatively alters the timescale for a process involving diffusion and collisions. The factors b/a (three dimensions) and $\ln (b/a)$ (two dimensions) can greatly lengthen the time for an association to occur. The fact that binding of ligands to receptors on a cell surface can be described by the rate of collisions with the cell is thus a manifestation of the reduction in the dimensionality of the search from three to two after the cell has been encountered. A related example of this principle can be found in the binding of a protein to a specific site on a long strand of DNA.

8.10 | Binding to DNA

Some DNA-binding proteins can find a specific site on a long DNA molecule far more rapidly than can be accounted for by diffusion-limited encounters with the small region of DNA harboring the actual site (Berg and von Hippel, 1985). It turns out that the protein finds its binding site in a two-stage process. First, the protein binds nonspecifically to any part of the DNA. Then the protein diffuses in one dimension along the DNA molecule in search of the site containing the specific recognition sequence. If the protein were just searching randomly in three dimensions, the maximum, diffusion-limited rate constant would be given by Eq. (8.15). We can ignore the diffusion coefficient of the DNA molecule because it is much smaller than that of the protein.

According to the two-stage model, if the protein binds nonspecifically to a random part of the DNA, it will then diffuse in one dimension along the DNA chain. The protein will either find the recognition sequence or fall off the DNA. The mean time until falling off is the inverse of the dissociation rate, $1/k_{diss}$. During this time the protein will diffuse a distance, d, along the chain specified by a one-dimensional random walk (Eq. (6.9))

$$d = \sqrt{2D_1/k_{diss}} \tag{8.48}$$

where D_1 is the one-dimensional diffusion coefficient. Nonspecific binding to a site that is less than a distance d from the recognition site thus has a good chance of leading to a stable association. This effectively increases the target size to d, from the size of the actual DNA recognition sequence. If D_1 is large and k_{diss} is small, d can be quite a bit larger than the binding site, b. This increases the effective target size so the association will be correspondingly large. Thus, the weak nonspecific binding reduces the number of dimensions from three to one, and this assists the association, much as reducing the dimensionality makes it easier for a ligand to find the receptors on the surface of a cell.

An important requirement of this mechanism is that the protein can move easily along the DNA while bound. In fact, the high density of negative charge of the phosphates on DNA makes cations bind in a nonspecific manner referred to as counterion condensation (Section 11.10). This form of binding is quite strong, so k_{diss} will be low. But since all sequences of DNA are equivalent in terms of charge density, a trapped protein will be free to move in either direction. This two-stage model has been very helpful in explaining the speed with which proteins such as the lac repressor find the operator region of DNA (Berg and von Hippel, 1985).

Problems for Chapter 8

1. Use Fig. 8.1 to estimate the association and dissociation rate constants for Ca^{2+} and fura-2 ([fura-2] = 5.33 μM in these experiments).

2. Use Eqs. (8.15) and (8.23) to derive the binding equilibrium constant.

3. Incorporate a simple electrostatic attraction or repulsion into the diffusion-limited association rate by taking $U(r)$ in Eq. (8.19) as the Coulomb potential (Hammes, 1978).

4. For a dissociation equilibrium constant of 10^{-6} M, and diffusion-limited association with a rate constant of 10^{10} M^{-1} s^{-1} (calculated from Eq. (8.17)), what is the rate constant of dissociation?

5. Derive Eq. (8.46). Assume a small circular target of radius a in the center of a circular area of radius b, and approximate based on $b \gg a$.

6. Use Eq. (8.20) to derive the association rate constant when $U(r) = -U_0$ for $r < c$ and $U(r) = 0$ for $r > c$ (where $c > a + b$). What happens when $U_0 \rightarrow \infty$? Interpret this result.

Chapter 9

Multi-state kinetics

The two-state model of Chapter 7 can be used as a basic building block for more complicated processes. When a system has more than two states, conversions between different pairs can occur, and then the kinetics reflects the aggregate behavior of those various transitions. The time course is no longer a simple exponential function, and in this chapter we will see how multi-state models lead to multi-exponential kinetics.

Much of Chapter 7 probed the fundamental physical processes that govern the speed of a transition. Now we will accept the basic phenomenon of a transition with a given rate, put some of these transitions together, and work out the dynamic behavior of these more complicated systems. The mathematical method for handling multi-state kinetics is very robust and powerful. With them one can develop quantitative descriptions for a general class of models. Aside from kinetic problems in biophysics, the mathematics introduced in this chapter has been applied to an extraordinarily wide range of fields from stochastic processes in physics to population dynamics in ecology.

Multi-state kinetics has considerable practical value, because most models for molecular mechanisms involve interconversions between a few distinct states. Kinetic behavior often provides the most direct experimental tool for testing the predictions of such a model.

9.1 | The three-state model

Many important features of multi-state kinetics can be illustrated with this example. We take the two-state model and add one more state (Scheme (9A))

$$A \underset{\beta}{\overset{\alpha}{\rightleftharpoons}} B \underset{\delta}{\overset{\gamma}{\rightleftharpoons}} C \tag{9A}$$

As in the analysis of the two-state model, we write down differential equations for the rates of change of concentrations of the different species, A, B, and C. Each rate constant represents a process that increases one species at the expense of another. Thus, [A] decreases as $-\alpha[A]$ and increases as $\beta[B]$. Likewise, [B] increases as $\alpha[A] + \gamma[C]$ and decreases as $-(\beta + \gamma)[B]$. This reasoning leads to three differential equations

$$\frac{d[A]}{dt} = -\alpha[A] + \beta[B] \tag{9.1a}$$

$$\frac{d[B]}{dt} = \alpha[A] - (\beta + \gamma)[B] + \delta[C] \tag{9.1b}$$

$$\frac{d[C]}{dt} = \gamma[B] - \delta[C] \tag{9.1c}$$

These differential equations can be solved to obtain [A], [B], and [C] as functions of time. We will see that the general solution is a sum of exponentials. Exponentials decaying at different rates are combined in different proportions depending on the initial conditions. The mathematical approaches to these kinds of kinetics problems proceed through two broad steps. First, the decay constants of the exponentials are determined. Then initial conditions are used to determine the coefficients multiplying the exponentials.

The general method for solving these systems of equations will be presented shortly, but working through this example by a more direct approach will provide a useful orientation for how the general method works. The strategy is to eliminate two of the variables by substitution and thereby obtain a differential equation in one variable. First, note that because the total amount of interconverting material is conserved, we have $[A] + [B] + [C] = T$, where T is constant. This can be used to eliminate [C] from Eq. (9.1b) as follows

$$\frac{d[B]}{dt} = \alpha[A] - \beta[B] - \gamma[B] + \delta(T - [A] - [B]) \tag{9.2}$$

Now we rearrange Eq. (9.1a) to obtain $[B] = (1/\beta)\,((d[A]/dt) + \alpha\,[A])$. This can be differentiated to obtain $d[B]/dt$. Substituting these two expressions eliminates [B] from Eq. (9.2) to yield a differential equation with [A] as the only remaining dependent variable

$$\frac{d^2[A]}{dt^2} + (\alpha + \beta + \gamma + \delta)\frac{d[A]}{dt} + (\alpha\gamma + \alpha\delta + \beta\delta)[A] - \beta\delta T = 0 \tag{9.3}$$

The last term, $\beta\delta T$, can be removed by replacing [A] with a new variable $[A]' = [A] - \beta\delta T/(\alpha\gamma + \alpha\delta + \beta\delta)$

$$\frac{d^2[A]'}{dt^2} + (\alpha + \beta + \gamma + \delta)\frac{d[A]'}{dt} + (\alpha\gamma + \alpha\delta + \beta\delta)[A]' = 0 \tag{9.4}$$

An exponential function satisfies this differential equation, so we insert $e^{\lambda t}$ for [A]'. All the exponential factors then cancel to give a quadratic equation in λ

$$\lambda^2 + (\alpha + \beta + \gamma + \delta)\lambda + \alpha\gamma + \alpha\delta + \beta\delta = 0 \tag{9.5}$$

If λ satisfies this equation then $e^{\lambda t}$ satisfies Eq. (9.4). Equation (9.5) has two solutions

$$\lambda_1 = \frac{-(\alpha + \beta + \gamma + \delta) + \sqrt{(\alpha + \beta + \gamma + \delta)^2 - 4(\alpha\gamma + \alpha\delta + \beta\delta)}}{2}$$

(9.6a)

$$\lambda_2 = \frac{-(\alpha + \beta + \gamma + \delta) - \sqrt{(\alpha + \beta + \gamma + \delta)^2 - 4(\alpha\gamma + \alpha\delta + \beta\delta)}}{2}$$

(9.6b)

The expression for $[A]'$ which satisfies Eq. (9.4) thus has the following general form

$$[A]' = a_1 e^{\lambda_1 t} + a_2 e^{\lambda_2 t}$$

(9.7)

where a_1 and a_2 are constants that will be determined in the following section. For now we observe that $[A]'$ changes with time as a sum of two exponentials. The two exponential decay constants are λ_1 and λ_2 given by Eqs. (9.6a) and (9.6b); λ_1 and λ_2 are always negative and real for positive values of the rate constants, so $[A]'$ will always decay to zero.

Recall that $[A]' = [A] - \beta\delta T/(\alpha\gamma + \alpha\delta + \beta\delta)$. Thus, when $[A]'$ has decayed to zero, $[A] = \beta\delta T/(\alpha\gamma + \alpha\delta + \beta\delta)$. In fact, this is the equilibrium value of $[A]$ that can be obtained from Scheme (9A) by expressing the equilibrium constants in terms of the ratios of the rate constants and solving for $[A]$. Defining this as $[A]_{eq}$ gives the solution as

$$[A] = a_1 e^{\lambda_1 t} + a_2 e^{\lambda_2 t} + [A]_{eq}$$

(9.8)

Thus, whatever $[A]$ starts out as, after a sufficiently long time it ends up being $[A]_{eq}$.

We could have chosen to eliminate $[A]$ and $[C]$ to obtain a differential equation for $[B]$, or eliminate $[A]$ and $[B]$ to obtain a differential equation for $[C]$ (Problem 1). Either way, we obtain a double exponential with the same two decay constants specified in Eqs. (9.6a) and (9.6b). The time-dependent expressions for each species will be distinguished from one another by the factors a_1 and a_2.

This analysis makes the point that adding a third state to the kinetic scheme adds a second exponential term to the solution. What if the scheme had four states? The system of differential equations would increase in number to four. The same elimination process could be performed, but when we finally arrived at a differential equation in one variable, it would be third order, with a third derivative. Using $e^{\lambda t}$ as a test function would give a cubic polynomial instead of the quadratic polynomial seen in Eq. (9.5). A cubic polynomial has three roots, so the general solution to the set of differential equations is a sum of three exponentials. By extension, adding another state would give us a fourth exponential, and so on. Thus, each time we add a state to a kinetic model, we add another exponential to the time course. This is a very general and important property of multi-state kinetic systems, which will be examined further below.

9.2 | Initial conditions

So far, we have seen how double exponential kinetics arises, but we still have to complete the solution by determining the coefficients a_1 and a_2 in Eq. (9.8). Equation (9.8) is a general solution, and it becomes a specific solution once this is done. The specific solution depends on the initial state of the system, or the initial conditions. We have to know the values for [A], [B], and [C] at $t = 0$. Here we will take the case of $[A]_0 = T$ and $[B]_0 = [C]_0 = 0$. Thus, at $t = 0$, Eq. (9.8) reduces to

$$T = a_1 + a_2 + [A]_{eq} \qquad (9.9)$$

We need one more equation with a_1 and a_2 to solve for these two unknowns. Using Eq. (9.1a), we can write [B] in terms of [A] as

$$\beta[B] = \alpha[A] + \frac{d[A]}{dt} = \alpha(a_1 e^{\lambda_1 t} + a_2 e^{\lambda_2 t} + [A]_{eq}) + (a_1 \lambda_1 e^{\lambda_1 t} + a_2 \lambda_2 e^{\lambda_2 t}) \qquad (9.10)$$

Setting $t = 0$ and $[B] = 0$ for the initial condition gives another expression with a_1 and a_2

$$0 = \alpha a_1 + \alpha a_2 + \alpha[A]_{eq} + a_1 \lambda_1 + a_2 \lambda_2 \qquad (9.11)$$

Equations (9.9) and (9.11) are then solved for the two unknowns

$$a_1 = \frac{-\alpha T - \lambda_2 (T - [A]_{eq})}{\lambda_1 - \lambda_2} \qquad (9.12a)$$

$$a_2 = \frac{\alpha T + \lambda_1 (T - [A]_{eq})}{\lambda_1 - \lambda_2} \qquad (9.12b)$$

Using these expressions in Eq. (9.9) gives the specific solution for this initial condition

$$[A] = \frac{-\alpha T - \lambda_2 (T - [A]_{eq})}{\lambda_1 - \lambda_2} e^{\lambda_1 t} + \frac{\alpha T + \lambda_1 (T - [A]_{eq})}{\lambda_1 - \lambda_2} e^{\lambda_2 t} + [A]_{eq} \qquad (9.13)$$

This provides an essentially complete description of the system. With the aid of Eq. (9.1a), an expression for [B] is easily obtained from Eq. (9.13). An expression for [C] follows by using either Eq. (9.1b), or (9.1c), or the conservation relation, $[A] + [B] + [C] = T$.

The behavior of the three-state model can be illustrated by giving the rate constants specific values. With $\alpha = \gamma = 4$ and $\beta = \delta = T = 1$, we obtain the following expressions

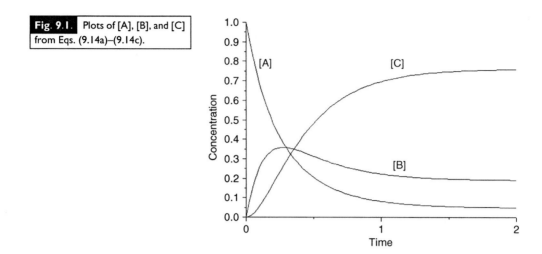

Fig. 9.1. Plots of [A], [B], and [C] from Eqs. (9.14a)–(9.14c).

$$[A] = 0.667e^{-3t} + 0.286e^{-7t} + 0.048 \qquad (9.14a)$$

$$[B] = 0.667e^{-3t} - 0.857e^{-7t} + 0.19 \qquad (9.14b)$$

$$[C] = -1.333e^{-3t} + 0.571e^{-7t} + 0.762 \qquad (9.14c)$$

These are plotted to illustrate how all three species vary with time as sums of the same two exponentials (Fig. 9.1).

Equations (9.14a–c) and Fig. 9.1 show the expected behavior. With [A] initially at 1 there is a steady double-exponential decay to the equilibrium level of 0.048. We see that [B] starts out at zero, goes through a maximum, and then decays to its equilibrium value. The maximum is possible because of the negative coefficient of the more rapidly decaying exponential in Eq. (9.14b). Similarly [C] starts off at zero and initially shows no increase. The lag is caused by the need for [B] to rise first. Thus, initially the derivative of [C] is zero (Problem 2).

The equations derived here allow one to predict the time course of a kinetic process when the rate constants are known. However, experimenters usually want to do the opposite, i.e. determine the rate constants from a kinetic experiment. In fact, a single kinetic experiment that gives a time course of the form in Eq. (9.13) provides sufficient information to solve for α, β, γ, and δ (Problem 4).

9.3 | Separation of timescales

It is easy to envision how a three-state kinetic model can give rise to double exponential kinetics. For example, if transitions between A and B are fast and transitions between B and C are slow, a system that starts out in A will show an initial rapid conversion of A to B. However, the reverse process will lead to a quasi-stable situation. Now if B interconverts slowly with C, then as this process progresses there will be a further slow decay in [A]. Intuitively, we might think that the first decay constant will reflect the transitions between A and B, so we would expect one decay constant to be $\sim -(\alpha + \beta)$. Likewise the second exponential should reflect the rates between B and C, so we might guess that the other decay constant will be $\sim -(\gamma + \delta)$.

We can check our intuition by looking at the exact expression for the decay constants (Eqs. (9.6a) and (9.6b)). In general, if $4(\alpha\delta + \beta\delta + \beta\gamma)$ is relatively small, then $(\alpha + \beta + \gamma + \delta)^2$ will dominate the square root term. Equation (9.6a) then gives $\lambda_1 \sim 0$ and Eq. (9.6b) gives $\lambda_2 \sim -(\alpha + \beta + \gamma + \delta)$. Thus, the timescales separate. If α and $\beta >> \gamma$ and δ, this gives one of expected decay constants.

$$\lambda_2 \sim -(\alpha + \beta) \tag{9.15}$$

To obtain an approximate expression for λ_1 in this limit, we factor $\alpha + \beta + \gamma + \delta$ out of the radical in Eq. (9.6a) as follows

$$\lambda_1 = -\frac{\alpha + \beta + \gamma + \delta}{2}\left(1 - \sqrt{1 - \frac{4(\alpha\gamma + \alpha\delta + \beta\delta)}{(\alpha + \beta + \gamma + \delta)^2}}\right) \tag{9.16}$$

Now the Taylor expansion $\sqrt{1 + x} \sim 1 + x/2$ (Eq. (A1.6)) reduces λ_1 to

$$\lambda_1 \sim -\frac{\alpha\gamma + \alpha\delta + \beta\delta}{\alpha + \beta + \gamma + \delta} \tag{9.17}$$

If β is much larger than all the other rate constants, then we get $\lambda_1 \sim -\delta$. If α is larger, we have $\lambda_1 \sim -(\gamma + \delta)$. So depending on the specifics, the approximate form for the slower decay constant can vary. Thus, the simplistic approximation of a three-state system as two separate two-state processes does not work out exactly as expected. Nevertheless, separation of timescales is a common feature of complex multi-state kinetic models. Simple approximate expressions for some of the decay constants often exist, but they must be employed with caution, and verified by careful analysis.

It is also worth noting the conditions where the timescales do not separate. When the square root term in Eqs. (9.6a) and (9.6b) is small $((\alpha + \beta + \gamma + \delta)^2 \sim 4\alpha\gamma + 4\alpha\delta + 4\beta\delta)$, then λ_1 and λ_2 will be similar, and the decay constants will not be easily approximated in terms of one transition or the other. Clearly, when the term in the radical of Eqs. (9.6a) and (9.6b) is zero, then $\lambda_1 = \lambda_2$. When this happens the three-state system will exhibit single-exponential kinetics. In this case, the decay constants are said to be degenerate. We must therefore qualify the general rule about state number. The number of exponentials is actually the minimum number of states.

9.4 | General solution to multi-state systems

The method used to solve the three-state model could in principle be extended to models with more states, but the elimination method used to solve the problem in Section 9.1 becomes too complicated to be practical. Multi-state systems become very manageable with the aid of matrices and vectors (Cox and Miller, 1965). This approach actually systematizes the elimination method used above to solve the three-state model. The analysis of this section may seem abstract, but the subsequent section provides a bridge to practical applications by repeating the analysis of the three-state model in these terms.

There are n species, and we define the concentration of the ith species as $[A_i]$. A column vector composed of the concentrations of all these species can then represent the state of a system.

$$\mathbf{A} = \begin{pmatrix} [A_1] \\ [A_2] \\ \vdots \\ [A_n] \end{pmatrix} \tag{9.18}$$

The derivative of this vector is another column vector $d\mathbf{A}/dt$ with elements $d[A_i]/dt$.

Interconversion rates will still be given by a rate constant multiplied by the concentration of the species undergoing the transition, e.g. $\alpha[A]$ or $\gamma[B]$. A differential equation can be written down for the time derivative of each species as a sum of such terms. Eqs. (9.1a)–(9.1c) illustrate this, and we can extend this set of equations to accommodate additional species and transitions. Each transition makes a contribution to the time derivatives of the two participating states. A transition from A_i to A_j produces a loss of A_i and a gain of A_j; the reverse transition does the opposite. If we take all the transitions to and from state A_i into account, the concentration derivative of $[A_i]$ will be a sum of the form $\alpha_{i1}[A_1] + \alpha_{i2}[A_2] + \cdots + \alpha_{in}[A_n]$, where the αs are obtained from the rate constants. This allows us to define a matrix $\mathbf{Q}$, and write the entire system of n differential equations as a single matrix-vector equation

$$\frac{d\mathbf{A}}{dt} = \mathbf{Q}\mathbf{A} \tag{9.19}$$

The elements of $\mathbf{Q}$ are determined by the following general rule. The diagonal elements Q_{ii} are minus the sum of the exit transitions from A_i to all the other states. Each off-diagonal element Q_{ji} is the rate of the transition from state A_j to A_i.

This compact expression for a system of kinetic differential equations is referred to as the master equation in the theory of stochastic processes (Van Kampen, 1981). It represents the total effect of all the transitions relevant to a given system, summing up all the gains and losses for each participating species.

To solve Eq. (9.19) we do what we did in solving Eq. (9.4): we try out an exponential function. Take $\mathbf{A} = \mathbf{U}e^{\lambda t}$, where $\mathbf{U}$ is an unknown vector of the same dimension as $\mathbf{A}$, and substitute it into Eq. (9.19)

$$\lambda \mathbf{U}e^{\lambda t} = \mathbf{Q}\mathbf{U}e^{\lambda t} \tag{9.20}$$

Thus, $\lambda \mathbf{U} = \mathbf{Q}\mathbf{U}$ (9.21)

This is rearranged to obtain

$$(\mathbf{Q} - \lambda \mathbf{I})\mathbf{U} = 0$$ (9.22)

where $\mathbf{I}$ is the identity matrix, with ones along the diagonal and zeros everywhere else. Setting all the elements of $\mathbf{U}$ equal to zero is one way to solve the equation, but this trivial solution is worthless because our state vector $\mathbf{A}$ must have nonzero elements. To find vectors for which $\mathbf{U} \neq 0$, note that Eq. (9.22) places an important restriction on the matrix $\mathbf{Q} - \lambda \mathbf{I}$. A nonzero vector $\mathbf{U}$ that solves the equation gives a way to combine the columns of $\mathbf{Q} - \lambda \mathbf{I}$ to obtain an all-zero vector. The columns are then not linearly independent, so the determinant is zero (Appendix 2)

$$|\mathbf{Q} - \lambda \mathbf{I}| = 0$$ (9.23)

This is the *characteristic* equation (Eq. (A2.17)) of $\mathbf{Q}$. Writing out the determinant of $\mathbf{Q} - \lambda \mathbf{I}$ produces a polynomial in λ of order n (the number of states). The roots of this equation are the *characteristic values* or *eigenvalues* of $\mathbf{Q}$. Since the number of roots of a polynomial is equal to its order, this tells us that there are as many values for λ as there are states in the kinetic system. With n states, Eq. (9.23) defines a set of n exponential decay constants, $\lambda_1, \lambda_2 \dots \lambda_n$. There will then be n exponential components. This constitutes a formal proof of the assertion made above. Each time a state is added to a system, there will be another exponential component added to the solution of the kinetics problem.

One might think that there is a discrepancy because the *three*-state system in Section 9.1 gave only *two* exponentials. With three states, Eq. (9.23) should give a cubic polynomial, and that means three exponentials. The issue is resolved by realizing that one of the values is $\lambda = 0$. This corresponds to the constant term, $[A]_{eq}$, in Eq. (9.13).

With this analysis we see that the kinetics of any species in an n-state kinetic system will be described by a general version of Eq. (9.8)

$$[A_i] = \sum_{k=1}^{n} a_{ik} e^{\lambda_k t}$$ (9.24)

where k is used to index the eigenvalues and it is distinct from i, the index used for the species. Each $[A_i]$ has its own set, a_{ik}, of coefficients that are used to combine the same set of exponentials in different proportions. As with the three-state model, these coefficients are determined by the initial conditions. In an n-state model, the initial condition takes the form $\mathbf{A}(0) = \mathbf{A}_0$, where $\mathbf{A}_0$ specifies the initial concentrations of all the species.

To calculate the a_{ik} in Eq. (9.24) from $\mathbf{A}_0$ we return to Eq. (9.22). For each eigenvalue, Eq. (9.22) is solved for the vector, $\mathbf{U}$, using one of the λ_ks. These vectors are the characteristic vectors or *eigenvectors* of $\mathbf{Q}$. Each of the n eigenvalues has its own eigenvector, so there are also n eigenvectors. The kth eigenvector is paired with the kth eigenvalue to give $\mathbf{QU}_k = \lambda_k \mathbf{U}_k$ (Eq. (9.21)).

Solving for $\mathbf{U}$ in Eq. (9.22) involves solving a system of n linear equations for n unknowns. Since Eq. (9.22) was obtained by testing the trial solution $\mathbf{A} = \mathbf{U}e^{\lambda t}$ in Eq. (9.19), $\mathbf{U}_k e^{\lambda_k t}$ is a solution. If our initial condition happened to be one of the eigenvectors ($\mathbf{A}_0 = \mathbf{U}_k$), we would have a special case where every species (every component of $\mathbf{A}$) changes as a single exponential. This is a special property of eigenvectors. For the particular set of concentrations in an eigenvector, interconversions all balance out to give a single exponential. In practice this almost never happens. It is extremely unlikely that all concentrations will coincidentally have the values of one of the eigenvectors, and eigenvectors often have negative values making such a state physically impossible. Real solutions are sums of all n fundamental solutions, $\sum_{k=1}^{n} \mathbf{U}_k e^{\lambda_k t}$. Thus, each species has n exponential terms.

To obtain the a_{ik} in Eq. (9.24), the eigenvectors must be combined in the right proportion to get the desired initial state vector, $\mathbf{A}_0$. This can be done because linear algebra tells us the eigenvectors form an orthogonal set that can be combined to represent any n-dimensional vector. There must therefore be a set of coefficients, b_k, that combine the eigenvectors to give

$$\sum_{k=1}^{n} b_k \mathbf{U}_k = \mathbf{A}_0 \tag{9.25}$$

The left-hand side can be written as the product of a matrix and vector

$$\mathbf{UB} = \mathbf{A}_0 \tag{9.26}$$

The eigenvectors have been used to create the matrix $\mathbf{U}$, in which the kth column is $\mathbf{U}_k$, and where $\mathbf{B}$ is a column vector with elements, b_k. We can solve Eq. (9.26) to obtain the set of coefficients in $\mathbf{B}$ as follows

$$\mathbf{B} = \mathbf{U}^{-1}\mathbf{A}_0 \tag{9.27}$$

Now we combine the n functions, $\mathbf{U}_k e^{\lambda_k t}$, to obtain a specific solution to Eq. (9.19)

$$\mathbf{A} = \sum_{k=1}^{n} b_k \mathbf{U}_k e^{\lambda_k t} \tag{9.28}$$

where b_k is a component of the vector $\mathbf{B}$. So each species is $[A_i] = \sum_{k=1}^{n} b_k U_{ik} e^{\lambda_k t}$ and each coefficient in Eq. (9.24) is $a_{ik} = b_k U_{ik} = (\mathbf{U}^{-1}\mathbf{A}_0)_k U_{ik}$. Note that at $t = 0$, we recover the initial condition Eq. (9.25).

Computers make this approach easy to use. For small matrices, the eigenvalues and eigenvectors are computed rapidly. The matrix of eigenvectors can be inverted to implement Eq. (9.27) and obtain the coefficients. It should be noted that $\mathbf{U}$ can be used to convert $\mathbf{Q}$ to a diagonal form

$$\mathbf{U}^{-1}\mathbf{Q}\mathbf{U} = \mathbf{\Lambda} \tag{9.29}$$

where $\mathbf{\Lambda}$ is a diagonal matrix with $\Lambda_{kk} = \lambda_k$ (Eq. (A2.20)).

9.5 | The three-state model in matrix notation

To see how the matrix formalism works we will now re-solve the three-state model introduced at the start of the chapter. The system of three equations for the rates of change of the three states (Eqs. ((9.1a)–(9.1c)) is written in matrix format as

$$\begin{pmatrix} \frac{d[A]}{dt} \\ \frac{d[B]}{dt} \\ \frac{d[A]}{dt} \end{pmatrix} = \begin{pmatrix} -\alpha & \beta & 0 \\ \alpha & -\beta-\gamma & \delta \\ 0 & \gamma & -\delta \end{pmatrix} \begin{pmatrix} A \\ B \\ C \end{pmatrix} \tag{9.30}$$

The characteristic equation then follows from Eq. (9.23)

$$\begin{vmatrix} -\alpha-\lambda & \beta & 0 \\ \alpha & -\beta-\gamma-\lambda & \delta \\ 0 & \gamma & -\delta-\lambda \end{vmatrix} = 0 \tag{9.31}$$

Expanding this determinant leads to a cubic polynomial

$$\lambda^3 + (\alpha+\beta+\gamma+\delta)\lambda^2 + (\alpha\gamma+\alpha\delta+\beta\delta)\lambda = 0 \tag{9.32}$$

Note that $\lambda = 0$ is a solution, and factoring it out yields Eq. (9.5). So Eqs. (9.6a) and (9.6b) give the other roots.

To obtain expressions for the eigenvectors we write out Eq. (9.21)

$$\lambda\mathbf{U} = \begin{pmatrix} -\alpha & \beta & 0 \\ \alpha & -\beta-\gamma & \delta \\ 0 & \gamma & -\delta \end{pmatrix} \mathbf{U} \tag{9.33}$$

With the test solution $\begin{pmatrix} U_{k1} \\ U_{k2} \\ U_{k3} \end{pmatrix} e^{\lambda_k t}$ we obtain the following three equations.

$$-\alpha U_{k1} + \beta U_{k2} = \lambda_k U_{k1} \tag{9.34a}$$

$$\alpha U_{k1} - (\beta+\gamma)U_{k2} + \delta U_{k3} = \lambda_k U_{k2} \tag{9.34b}$$

$$\gamma U_{k2} - \delta U_{k3} = \lambda_k U_{k3} \tag{9.34c}$$

At first glance one might try to use these three equations to solve for the three unknowns, U_{k1}, U_{k2}, and U_{k3}, and thus determine the elements of the eigenvector. This does not work because the rows of the matrix are not linearly independent (Eq. (9.34b) is just minus the sum of (9.34a) and (9.34c)). Determination of the elements of $\mathbf{U}$ requires an additional piece of information. We specify that the eigenvectors have a unit length

$$U_{k1}^2 + U_{k2}^2 + U_{k3}^2 = 1 \tag{9.35}$$

Combining this with two of the three equations (Eq. (9.34)) allows us to solve for the elements of $\mathbf{U}_k$

$$U_{1k} = \frac{\dfrac{\beta}{\lambda_k + \alpha}}{\sqrt{1 + \left(\dfrac{\beta}{\lambda_k + \alpha}\right)^2 + \left(\dfrac{\gamma}{\lambda_k + \delta}\right)^2}} \qquad (9.36a)$$

$$U_{2k} = \frac{1}{\sqrt{1 + \left(\dfrac{\beta}{\lambda_k + \alpha}\right)^2 + \left(\dfrac{\gamma}{\lambda_k + \delta}\right)^2}} \qquad (9.36b)$$

$$U_{3k} = \frac{\dfrac{\gamma}{\lambda_k + \delta}}{\sqrt{1 + \left(\dfrac{\beta}{\lambda_k + \alpha}\right)^2 + \left(\dfrac{\gamma}{\lambda_k + \delta}\right)^2}} \qquad (9.36c)$$

This is actually three equations, one for each eigenvalue ($k = 1, 2$, and 3). They can be checked as solutions of Eqs. (9.34a)–(9.34c) and (9.35).

---

*9.6 | Stationarity, conservation, and detailed balance

Kinetic systems often exhibit some form of stability, and can settle into a stable state. Stability means that the concentrations are constant and their time derivatives are zero. This requires that the rate constants obey certain relationships. A system might have a stationary state if the number of moles is conserved, or if it is at thermodynamic equilibrium. We saw in Section 7.3 that a thermodynamic equilibrium leads to the condition of detailed balance. This can be generalized to multi-state kinetic systems where the relations between rate constants determine important properties of the Q-matrix.

First we look at stationarity. This means that a system has a state that is stable and unchanging. It could be a thermodynamic equilibrium, but not all stationary states are equilibrium states, and the distinction will be explained shortly. In a stationary state all of the time derivatives are zero. We can use the vector $\mathbf{A}_{st}$ to denote a particular set of concentrations where this condition is met. Equation (9.19) then gives

$$\mathbf{QA}_{st} = 0 \qquad (9.37)$$

Note that this equation can be solved to obtain $\mathbf{A}_{st}$ in terms of the rate constants.[1]

---

[1] Inverting $\mathbf{Q}$ does not work because we will soon see that $|\mathbf{Q}| = 0$, making it non-invertible. To solve for $\mathbf{A}_{st}$ requires another condition such as conservation of number of moles.

Now, consider Eq. (9.21). Taking $\mathbf{A}_{st}$ as an eigenvector gives

$$\mathbf{Q}\mathbf{A}_{st} = \lambda\mathbf{A}_{st} \tag{9.38}$$

From Eq. (9.37) it is clear that $\lambda = 0$. Thus, $\mathbf{Q}$ must have one eigen-value equal to zero, and $\mathbf{A}_{st}$ is the corresponding eigenvector. This is an important property of a kinetic system with a stationary state. The converse of this statement is also true. If a system does not have a stationary state, none of the eigenvalues can be zero.

This situation was noted above for the three-state model. It has an equilibrium state, and one of the solutions to Eq. (9.32) is zero. Look at the solution of the three-state model (Eq. (9.13)), and imagine that the final term $[A]_{eq}$ is really $[A]_{eq}e^{\lambda_3 t}$. This does not change anything because with $\lambda_3 = 0$, $e^{\lambda_3 t} = 1$. The zero eigenvalue guarantees that when all of the other kinetic processes are over and done with, the system can come to rest at its stationary state with $\mathbf{A} = \mathbf{A}_{st}$.

Now we consider conservation of the total number of moles of all interconverting species. For a kinetic system to have this property, there can be no destructive or creative processes, and no fluxes into or out of the system. Associations and dissociations are not allowed, as they change the number of moles. Aside from conservation issues, association processes give rise to nonlinear terms of the form $[A][B]$, and the mathematical methods used here to solve these problems no longer work.

If the total number of moles never changes, the time derivatives must sum to zero

$$\sum_{i=1}^{n} \frac{d[A_i]}{dt} = 0 \tag{9.39}$$

Looking at the right-hand side of Eq. (9.19), the product $\mathbf{Q}\mathbf{A}$ can be multiplied out and the elements of the resulting vector added together. Equation (9.39) tells us this must be zero

$$\sum_{j=1}^{n} \sum_{i=1}^{n} Q_{ij}[A_i] = 0 \tag{9.40}$$

Now, reverse the summation order and collect factors of each A_i

$$\sum_{j=1}^{n} Q_{1j}[A_1] + \sum_{j=1}^{n} Q_{2j}[A_2] + \cdots + \sum_{j=1}^{n} Q_{nj}[A_n] = 0 \tag{9.41}$$

Each term of Eq. (9.41) contains a sum over all the elements of one column of $\mathbf{Q}$. This equation holds for any $[A_1]$, $[A_2]$, $\ldots$ $[A_n]$, and that means that the sums multiplying each $[A_i]$ must also be zero. Hence, we have the result that each column of $\mathbf{Q}$ must sum to zero. It is easy to check that $\mathbf{Q}$ in Eq. (9.30) obeys this rule. This is a general property of multi-state kinetic systems that conserve moles. We will see in the analysis of single-channel kinetics that the matrices do not have this property because probability is not conserved (Section 9.11).

Both stationarity and conservation lead to the result that $\mathbf{Q}$ has a determinant equal to zero. For stationarity the matrix has an eigenvalue equal to zero, so $|\mathbf{Q}| = 0$ by Eq. (A2.23). Conservation also gives us $|\mathbf{Q}| = 0$; this is a consequence of columns adding up to zero. Because any row of $\mathbf{Q}$ can be expressed as minus the sum of all the other rows, the rows are not linearly independent, and the determinant is zero. This means that a kinetic system that obeys conservation must have at least one zero eigenvalue, and so it must also have a stationary state.

Finally, we look at equilibrium and the condition of detailed balance. As noted in Section 7.3, when the two-state system reaches equilibrium the forward and reverse rates balance out perfectly. If a multi-state system is at equilibrium, then this same condition will hold for each interconversion

$$\alpha_{ij}[A_i] = \alpha_{ji}[A_j] \tag{9.42}$$

where α_{ij} denotes the rate constant by which A_i converts to A_j, and α_{ji} denotes the rate constant for the reverse process. In words, this means that the rate of direct conversion from A_i to A_j equals the rate of direct conversion in the opposite direction. This is exactly what was stated for the two-state system in Section 7.3.

The ith diagonal element of $\mathbf{Q}$ is the sum of the exit rates from A_i, i.e. $\sum_{j=1}^{n} a_{ij}$. Each off-diagonal element of column i will be the rate constant of one of the reverse steps, i.e. one of the α_{ji}. So with the aid of Eq. (9.42), we see that the columns will sum to zero, as already deduced from the constraint of conservation. Further, since a system that obeys detailed balance will have a stationary state, the matrix will have one eigenvalue equal to zero. Hence, the condition of detailed balance (as expressed in Eq. (9.42)) imposes the same constraints on the rate matrix derived from stationarity and conservation.

Conserved systems and systems with a stationary state need not obey detailed balance. The columns of $\mathbf{Q}$ sum to zero, but this can be done without satisfying Eq. (9.42). This means that the system will cycle through a sequence of states repetitively. A molecule may go from state A to B to C to A more often than from state A to C to B to A. In general, for this to happen, a source of energy is required. For example, going from A to B may be accompanied by ATP hydrolysis, so that cycling in the order A, B, C, A consumes energy. The reverse cycle would require ATP synthesis, and would be energetically unfavorable. The system would keep cycling in the energetically favored direction, consuming ATP. Finally, when the ATP–ADP reaction reaches equilibrium, detailed balance will then be obeyed. These issues will be taken up below in relation to single-channel kinetics (Section 9.11).

9.7 | Single-channel kinetics: the three-state model

In Section 7.9 we saw that single-channel kinetics relates rate constants to the stochastic behavior of ion channels. Now we will return to the single-molecule realm to see how the multi-state kinetic methods are adapted to this class of problems. This means that we must change from a macroscopic level, where concentrations are the variables of interest, to the microscopic level, where the variables are probabilities. In fact, this change is trivial. If the sum of the concentrations of all the different interconverting species is T, then the concentration of species A will be $P_A T$, where P_A is the probability of a molecule being in state A. As a result the kinetic equations for changes in probability have the same basic form as the equations for changes in concentration. The mathematical strategies for finding solutions are also quite similar.

The ideas in this section are applicable to any single-molecule measurement, but we will use ion channel terminology here and refer to states as open and closed. We will start with a three-state model (see Scheme (9B)), like that used at the beginning of the chapter, but here we will specify that the two right-most states are open channels

$$\text{C} \underset{\beta}{\overset{\alpha}{\rightleftarrows}} \text{O}_1 \underset{\delta}{\overset{\gamma}{\rightleftarrows}} \text{O}_2 \tag{9B}$$

Both open states have the same conductance, so that one cannot tell by measuring current which open state the channel is in.

It should be clear from the two-state model of Section 7.9 that the closed-time distribution of Scheme (9B) will be a single exponential, $e^{-\alpha t}$. This is because there is only a single closed state. However, it is clear that a single exponential cannot describe the open-time distribution of Scheme (9B). Openings begin in state O_1, but from there they can either close directly or flip to O_2. Thus, we would expect to see a population of brief events that go straight to C, and a population of long events that spend some time in O_2.

The derivation of the open-time distribution for Scheme (9B) parallels the treatment of the macroscopic kinetics of the three-state model. Differential equations are again written down for the different species, but now they express gains and losses in probability instead of concentration. We only need differential equations for the two open states. The probability of being closed is irrelevant because once a channel closes, subsequent activity has no bearing on the problem. Taking the probability of being in state O_1 as P_{o1}, and the probability of being in state O_2 as P_{o2}, we have

$$\frac{dP_{o1}}{dt} = -(\beta + \gamma)P_{o1} + \delta P_{o2} \tag{9.43a}$$

$$\frac{dP_{o2}}{dt} = \gamma P_{o1} - \delta P_{o2} \qquad (9.43b)$$

Note that the opening rate, α, does not appear in these equations. This reaffirms the irrelevance of the closed state.

At this point we might think that we should eliminate one of the variables with a conservation constraint, as was done for the macroscopic behavior of both the two-state and three-state models. However, as time evolves the probability of the channel being open decreases. Conservation does not apply here, and this will be examined further below. By a process of substitution and elimination between Eqs. (9.43a) and (9.43b), similar to that used to obtain Eq. (9.3) from Eqs. (9.1a)–(9.1c), we can obtain a second order differential equation in one of the variables. Using the trial solution $e^{\lambda t}$ leads to a quadratic equation similar in form to Eq. (9.5)

$$\lambda^2 + (\beta + \gamma + \delta)\lambda + \beta\delta = 0 \qquad (9.44)$$

There are two solutions

$$\lambda_1 = \frac{-(\beta + \gamma + \delta) + \sqrt{(\beta + \gamma + \delta)^2 - 4\beta\delta}}{2} \qquad (9.45a)$$

$$\lambda_2 = \frac{-(\beta + \gamma + \delta) - \sqrt{(\beta + \gamma + \delta)^2 - 4\beta\delta}}{2} \qquad (9.45b)$$

So the general solution to Eqs. (9.43a) and (9.43b) is a sum of two exponentials

$$P_{o1} = a_1 e^{\lambda_1 t} + a_2 e^{\lambda_2 t} \qquad (9.46)$$

As for the three-state model for macroscopic concentrations, going from the general solution to a specific solution requires an initial condition. However, now the initial condition is not arbitrary. At the start of a channel opening, the channel must be in O_1, because that is the only state accessible from the closed state. Thus, at $t = 0$, $P_{o1} = 1$ and $P_{o2} = 0$. Accordingly, Eq. (9.46) gives

$$a_1 + a_2 = 1 \qquad (9.47)$$

Substitution of Eq. (9.46) into Eq. (9.43a) gives

$$\delta P_{o2} = a_1 \lambda_1 e^{\lambda_1 t} + a_2 \lambda_2 e^{\lambda_2 t} + (\beta + \gamma)(a_1 e^{\lambda_1 t} + a_2 e^{\lambda_2 t}) \qquad (9.48)$$

Since $P_{o2} = 0$ at $t = 0$, we have

$$a_1(\lambda_1 + \beta + \gamma) + a_2(\lambda_2 + \beta + \gamma) = 0 \qquad (9.49)$$

Equations (9.47) and (9.49) are then solved for a_1 and a_2

$$a_1 = \frac{-\lambda_2 - \beta - \gamma}{\lambda_1 - \lambda_2} \qquad (9.50a)$$

$$a_2 = \frac{\lambda_1 + \beta + \gamma}{\lambda_1 - \lambda_2} \qquad (9.50b)$$

We can now write out Eq. (9.46) explicitly as

$$P_{o1} = \frac{-\lambda_2 - \beta - \gamma}{\lambda_1 - \lambda_2} e^{\lambda_1 t} + \frac{\lambda_1 + \beta + \gamma}{\lambda_1 - \lambda_2} e^{\lambda_2 t} \qquad (9.51)$$

And Eq. (9.48) can then be used to obtain P_{o2} from Eq. (9.43a)

$$P_{o2} = \frac{\gamma}{\lambda_1 - \lambda_2} e^{\lambda_1 t} - \frac{\gamma}{\lambda_1 - \lambda_2} e^{\lambda_2 t} \qquad (9.52)$$

(Simplification of Eq. (9.52) was achieved with the aid of the relations $\lambda_1 \lambda_2 = \beta \delta$, and $\lambda_1 + \lambda_2 = -(\beta + \gamma + \delta)$, which are easily derived from Eqs. (9.45a) and (9.45b).)

Now there are two ways to proceed to an expression for the open-time probability. We can realize that the probability of a closing transition occurring at time t is equal to the closing rate constant, β, times the probability of being in state O_1. This is βP_{o1}, and the open-time probability distribution is obtained by integrating this expression from 0 to t. (Recall that the probability distribution is the integral of the probability density – Eq. (7.33).)

$$P_o = \frac{-\lambda_2 - \beta}{\lambda_1 - \lambda_2} e^{\lambda_1 t} + \frac{\lambda_1 + \beta}{\lambda_1 - \lambda_2} e^{\lambda_2 t} \qquad (9.53)$$

Alternatively, we could take the open-time distribution as $P_{o1} + P_{o2}$. This means that the probability of a channel still being open at time t is the sum of the probabilities of the channel being in either of the two open states. Both strategies lead to the same expression for the open-time distribution (Problem 6).

A plot of P_{o1}, P_{o2}, and their sum helps visualize how these different probability functions change with time (Fig. 9.2). We see that P_{o1} starts out at one and decays rapidly. This reflects the

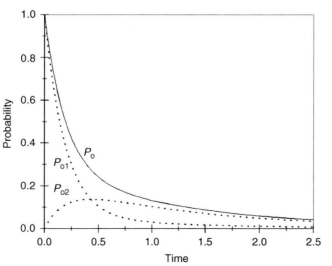

Fig. 9.2. Plots of Eqs. (9.51)–(9.53) with $\beta = 4$, and $\gamma = \delta = 1$ (giving $\lambda_1 = -0.76$ and $\lambda_2 = -5.24$).

predominant fast component. Because all openings begin in O_1, P_{o2} is initially zero and rises to a maximum before decaying. We see that P_o reflects both of these quantities.

All of these functions are sums of two exponentials with the same decay constants, but in different proportions. The maximum in P_{o2} reflects the fact that one of the coefficients in Eq. (9.52) is negative, as was seen in the case of [B] plotted in Fig. 9.1.

9.8 | Separation of timescales in single channels: burst analysis

It is very common in single-channel experiments to observe channel openings in clusters or bursts. A long period with no openings will be followed by many openings in rapid succession, and then another long period with no openings (Fig. 9.3).

A three-state model with two closed states can account for this behavior, provided that the transitions between the open and closed states are fast and the transitions between the two closed states are slow. The closed-time distribution will have two components. The brief closed times within the bursts will make up a fast component, and the long closed times between the bursts will make up the slow component.

The two-closed state model has the following form (Scheme (9C))

$$C_1 \underset{\beta}{\overset{\alpha}{\rightleftarrows}} C_2 \underset{\delta}{\overset{\gamma}{\rightleftarrows}} O \tag{9C}$$

We just solved essentially the same problem in the preceding section. We just need to interchange β with γ, and α with δ. From Eqs. (9.45a) and (9.45b), the closed-time distribution then has the following exponential decay constants

$$\lambda_1 = \frac{-(\alpha + \beta + \gamma) + \sqrt{(\alpha + \beta + \gamma)^2 - 4\alpha\gamma}}{2} \tag{9.54a}$$

$$\lambda_2 = \frac{-(\alpha + \beta + \gamma) - \sqrt{(\alpha + \beta + \gamma)^2 - 4\alpha\gamma}}{2} \tag{9.54b}$$

Fig. 9.3. Three bursts of single-channel openings. The middle burst is expanded.

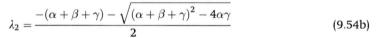

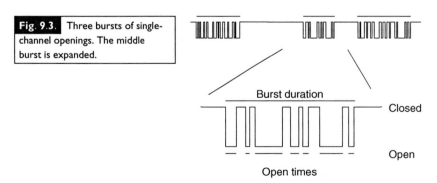

Burst duration

Closed

Open

Open times

The closed-time distribution is obtained by using these expressions for λ_1 and λ_2 and making the corresponding replacements in the coefficients of Eq. (9.53).

As in Section 9.3, the condition that leads to a separation of timescales is that the second term in the square root must be much smaller than the first: $(\alpha + \beta + \gamma^2) >> 4\alpha\gamma$. The situation that gives rise to bursting is when α is slower than the other rates. A transition to C_1 will then begin a long-lasting inter-burst interval. The brief closed times result from the channel entering C_2 from O, and returning directly to O. With the stipulation that β and $\gamma >> \alpha$, the same reasoning that led to Eqs. (9.15) and (9.17) gives the following approximate expressions for λ_1 and λ_2

$$\lambda_1 \sim \frac{-\alpha\gamma}{\beta + \gamma} \tag{9.55a}$$

$$\lambda_2 \sim -(\beta + \gamma) \tag{9.55b}$$

First consider λ_2. The lifetime distribution of state C_2, is $e^{-(\beta+\gamma)t}$. The two pathways out of that state have a perfectly additive effect in decreasing its lifetime. We can now determine the fraction of departures from C_2 into the two destinations, O and C_1. They should be in the proportion γ/β. The fraction of exits from C_2 to O is $\gamma/(\gamma + \beta)$, and the fraction of exits from C_2 to C_1 is $\beta/(\gamma + \beta)$. These fractions can be used in the closed-time distribution to give the proportion of brief and long closures, respectively

$$P_C \sim \frac{\gamma}{\gamma + \beta} e^{-(\beta+\gamma)t} + \frac{\beta}{\gamma + \beta} e^{-\alpha\gamma t/(\beta+\gamma)} \tag{9.56}$$

Viewing the process in this way allows us to derive a few additional interesting features of bursts. We can ask how many openings are likely to appear in a burst. First we determine the probability of a burst having only one opening. The channel will be in C_2, and for the burst to end, the channel has to flip to C_1. According to the reasoning of the preceding paragraph, the probability of doing so is $\beta/(\gamma + \beta)$. For a two-opening burst the channel must return from C_2 back to O; the probability of this happening is $\gamma/(\gamma + \beta)$. After the second opening is over the channel is once again in C_2, and this time there must be a transition to C_1, with a probability of $\beta/(\gamma + \beta)$. The product of these two separate events gives the probability of a two-opening burst as $\beta\gamma/(\gamma + \beta)^2$. Repeating these steps leads to a general expression for the probability of an n-opening burst

$$P(n) = \frac{\beta\gamma^{n-1}}{(\beta + \gamma)^n} \tag{9.57}$$

The mean value of n can be written out and the sums evaluated with Eqs. (A1.11) and (A1.12)

$$\bar{n} = \frac{\sum\limits_{n=1}^{\infty} nP(n)}{\sum\limits_{n=1}^{\infty} P(n)} = \frac{\beta + \gamma}{\beta} \tag{9.58}$$

It is easy to count the number of openings per burst in a single-channel record and determine the average. The mean lifetime of a closure within a burst is $1/(\beta + \gamma)$, and together with Eq. (9.58) one can solve for the rate constants β and γ (Colquhoun and Hawkes, 1982).

One notable application of burst analysis was an investigation of channel blockade by local anesthetics (Neher and Steinbach, 1978). These drugs enter open channels and act like plugs to prevent ion flow. In the presence of a local anesthetic, recordings of acetyl-choline receptor single-channel currents look very much like Fig. 9.3. But the closures within bursts are drug binding events rather than channel gating transitions.

Neher and Steinbach used the following model (Scheme (9D)) to interpret their data

$$ C \underset{\beta}{\overset{\alpha}{\rightleftharpoons}} O \underset{k_d}{\overset{k_b[D]}{\rightleftharpoons}} B \tag{9D} $$

The state denoted by B is a blocked state. The local anesthetic binds to the open state at a rate equal to the binding rate constant, k_b, times the drug concentration [D], and dissociates with a rate k_d. The constant β was determined from the open-time distribution in the absence of the local anesthetic (but in the presence of acetyl-choline). When the local anesthetic was added the open-time distribution showed a more rapid exponential decay. The rate of decay increased linearly with concentration, and the slope gave k_b. In the presence of the local anesthetic brief closures were seen, reflecting blocking events. These brief events had an exponential distribution of lifetimes, and this gave an estimate of k_d.

These results were consistent with Scheme (9D), but a more rigorous test of this model was sought. The bursts of openings became longer as the local anesthetic concentration was increased. Neher and Steinbach (1978) made a separation of timescales argument to obtain an approximation for the mean burst duration. They assumed that O and B equilibrated rapidly so that within a burst the fraction of time spent in O was

$$ x_o = \frac{k_d}{k_d + k_b[D]} \tag{9.59} $$

a burst must end by transitions from O to C, so the rate of burst termination is βx_o. The mean burst duration is then obtained as the inverse of this rate

$$ \overline{t_b} = \frac{1}{\beta}\left(1 + \frac{k_b[D]}{k_d}\right) \tag{9.60} $$

This equation predicted the measured mean burst duration very well over a range of local anesthetic concentrations, providing strong support for the open-channel block model represented by Scheme (9D).

9.9 | General treatment of single-channel kinetics: state counting

A channel gating scheme is represented by a set of differential equations, as exemplified by Eqs. (9.43a) and (9.43b) for the two-open-state model, and these equations can be written in the matrix-vector form of Section 9.4. Our state vector is now $\mathbf{P}_o$, which represents a set of open probabilities for an arbitrary number of open states.

$$\mathbf{P}_o = \begin{pmatrix} P_{o1} \\ P_{o2} \\ \vdots \\ \vdots \\ P_{on} \end{pmatrix} \tag{9.61}$$

The matrix is formulated from the rates for transitions between and out of the various open states. Return transitions from the closed to the open states are irrelevant. The two-open-state model has the following matrix

$$\mathbf{Q}_o = \begin{pmatrix} -\beta - \gamma & \delta \\ \gamma & -\delta \end{pmatrix} \tag{9.62}$$

The subscript o means that this matrix is for open-state probabilities. For closed states we would use the subscript c. In terms of $\mathbf{P}_o$ and $\mathbf{Q}_o$, the evolution through time of the open-state probability vector obeys a differential equation of the same form as Eq. (9.19)

$$\frac{d\mathbf{P}_o}{dt} = \mathbf{Q}_o \mathbf{P}_o \tag{9.63}$$

This equation can be solved in the same way as Eq. (9.19). The exponential decay constants are the eigenvalues of $\mathbf{Q}_o$. The number of eigenvalues gives us the number of exponentials in the open-time distribution. This number in turn will be set by the dimension of $\mathbf{P}_o$, which is equal to the number of open states. Of course, the same method can be used for closed times where a matrix $\mathbf{Q}_c$ replaces $\mathbf{Q}_o$.

This leads to an important rule for single-channel kinetics. The number of exponentials that a gating scheme predicts in the lifetime distribution for one particular conductance level is equal to the number of states with exactly that conductance. So, for example, when one observes that an open-time distribution is well described by a sum of three exponentials, one can conclude that the open state of that channel possesses three kinetically distinguishable conformations. Figure 9.4 shows an example of lifetime distributions for both open times and closed times from a Cl^- channel activated by the neurotransmitter GABA. These distributions were created from a long record of channel transitions such as that shown in Fig. 7.11. Each distribution was well fitted by a sum of two exponentials, implying that this channel has two open states and two closed states.

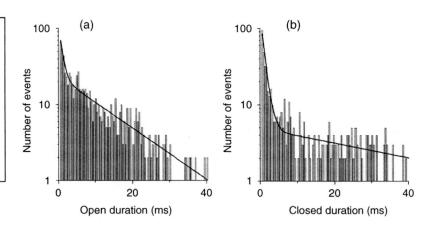

Fig. 9.4. Lifetime distributions for open times and closed times, from single-channel recordings of the GABA receptor channel (Fig. 7.11). A semilogarithmic plot shows exponentials as straight lines. Fits gave $97e^{-t/0.81} + 23e^{-t/12.9}$ for the open-time distribution (a) and $145e^{-t/0.99} + 4.9e^{-t/45.3}$ for the closed-time distribution (b) (Zhang and Jackson, 1995).

In practice, one must recognize the possibility of processes too fast or too slow to detect, as well as similar or degenerate eigenvalues. So the number of exponentials is generally taken as the lower bound to the number of states with a given conductance.

The coefficients of the exponentials are determined by the same mathematical method used for macroscopic kinetics. One determines the eigenvectors of the matrix, and combines them to satisfy the initial condition. In the case of a single-channel lifetime distribution the initial condition is again dictated by the distribution among open states at the start of an opening. For a full treatment of single-channel kinetics with matrix methods see Colquhoun and Hawkes (1995).

*9.10 | Relation between single-channel and macroscopic kinetics

The equations for multi-state kinetics are clearly very similar for macroscopic and single-molecule systems. The two realms can be connected explicitly. Consider the kinetic scheme for a channel with multiple open and closed states. We start with the full matrix, $\mathbf{Q}$, for interconversions between all these states (Eq. (9.19)), and then partition it into submatrices based on the channel conductance states. Within $\mathbf{Q}$ we will find the submatrix $\mathbf{Q_o}$ that accounts for transitions between the various open states as well as transitions out of the collection of open states. We will also find the submatrix $\mathbf{Q_c}$ that accounts for transitions between the various closed states as well as transitions out of the collection of closed states. Finally, we can also recognize the submatrices $\mathbf{Q_{c-o}}$ and $\mathbf{Q_{o-c}}$ that account for transitions between open and closed states. In these terms, Eq. (9.19) is written as

$$\begin{pmatrix} \mathbf{Q_c} & \mathbf{Q_{c-o}} \\ \mathbf{Q_{o-c}} & \mathbf{Q_o} \end{pmatrix} \begin{pmatrix} \mathbf{P_c} \\ \mathbf{P_o} \end{pmatrix} = \begin{pmatrix} \frac{d\mathbf{P_c}}{dt} \\ \frac{d\mathbf{P_o}}{dt} \end{pmatrix} \qquad (9.64)$$

Thus, we have Eq. (9.19), but with $\mathbf{Q}$ partitioned as just described, and with $\mathbf{A}$ partitioned into $\mathbf{P}_c$ and $\mathbf{P}_o$. Multiplying out the left-hand side gives two matrix equations

$$\mathbf{Q}_c\mathbf{P}_c + \mathbf{Q}_{c-o}\mathbf{P}_o = \frac{d\mathbf{P}_c}{dt} \tag{9.65}$$

$$\mathbf{Q}_{o-c}\mathbf{P}_c + \mathbf{Q}_o\mathbf{P}_o = \frac{d\mathbf{P}_o}{dt} \tag{9.66}$$

The matrix equation for calculating the closed-time distribution is obtained by discarding the term $\mathbf{Q}_{c-o}\mathbf{P}_o$ from Eq. (9.65), and the matrix equation for calculating the open-time distribution is obtained by discarding the term $\mathbf{Q}_{o-c}\mathbf{P}_c$ from Eq. (9.66). These two discarded terms represent returns to the state in question, and in single-channel kinetics these processes are irrelevant. This outlines a general method for going back and forth between the full kinetic scheme and the sub-schemes for the lifetime distributions at the single-molecule level.

*9.11 | Loss of stationarity, conservation, and detailed balance

The full kinetic scheme for the conformational transitions of a protein will have an equilibrium stationary state, provided that all the transitions satisfy detailed balance (Section 9.6). However, because single-channel analysis focuses on a limited part of the full scheme, there are fundamental differences. In single-channel kinetics, the probability of one channel staying in any particular state must decay to zero with time; no state endures forever. This means that there is no stationary state. The probability decays with time so it cannot be a conserved quantity. Within the confines of single-channel kinetics, detailed balance no longer holds because transitions out of the state of interest (e.g. from closed to open) are not balanced by return transitions.

All this has a bearing on the properties of $\mathbf{Q}_o$ and $\mathbf{Q}_c$. They generally have nonzero determinants, and all nonzero eigenvalues. If they had an eigenvalue equal to zero, then the lifetime probability would have a constant, non-decaying term (c.f. the last term of Eq. (9.13)), and this would give the impossible result of a nonzero limit as $t \to \infty$. The zero eigenvalue of the transition matrix is a requirement of stationarity, conservation, and detailed balance, and it is important to realize that the partitioning process of the foregoing section, by which one extracts the relevant single-channel submatrices from the Q-matrix, leads to matrices with fundamental differences.

This is generally true for single channels, regardless of whether the scheme obeys detailed balance or not. When detailed balance holds for the full kinetic scheme, more can be said about the shape of the lifetime distribution. For a single-channel kinetic scheme with reversible transitions, the lifetime distribution is restricted to monotonic decay from one to zero (see P_o in Fig. 9.2). The lifetime distribution is a sum of decaying exponentials in which each term has a positive coefficient. If detailed balance is violated, this is no longer true. Negative coefficients are possible and lifetime distributions can have inflections as probabilities rise and fall.

To see how this comes about, let's look at the following cyclic scheme (Scheme (9E))

$$\begin{array}{c}
\alpha \nearrow \quad O_1 \\
C \overset{\beta}{\underset{\phi}{\swarrow}} \delta \updownarrow \quad \gamma \\
\varepsilon \searrow \quad O_2
\end{array}$$

(9E)

This scheme has two open states, like Scheme (9B), except that both open states connect to the closed state. This creates an interesting situation where violation of detailed balance leads to a preference for clockwise or counterclockwise movement through the three states.

First, we note that the rate constants are no longer independent. We can see this by calculating the ratios of concentrations (much like the cyclic schemes of Chapter 5). The ratio $[O_2]/[C]$ determined by direct interconversion is ϕ/ε, and this same ratio can also be found to be $\alpha\gamma/\beta\delta$ by considering transitions from O_2 to C passing through O_1. At equilibrium the two expressions must be equal

$$\alpha\gamma\varepsilon = \beta\delta\phi \tag{9.67}$$

The idea is the same as that for Eq. (5.13), but there we used equilibrium constants. If the rates do not obey Eq. (9.67), then detailed balance is violated, and the system will show a preference for cycling in one direction. This means that some source of energy drives at least one of the steps, so that it is not at equilibrium.

To derive the open-time distribution of Scheme (9E) we determine the matrix for open times

$$\mathbf{Q}_o = \begin{pmatrix} -\beta - \gamma & \delta \\ \gamma & -\delta - \varepsilon \end{pmatrix} \tag{9.68}$$

where $\mathbf{Q}_o$ has the following eigenvalues

$$\lambda_1 = \frac{-(\beta + \gamma + \delta + \varepsilon) + \sqrt{(\beta + \gamma + \delta + \varepsilon)^2 - 4(\beta\delta + \beta\varepsilon + \gamma\varepsilon)}}{2} \tag{9.69a}$$

$$\lambda_2 = \frac{-(\beta + \gamma + \delta + \varepsilon) - \sqrt{(\beta + \gamma + \delta + \varepsilon)^2 - 4(\beta\delta + \beta\varepsilon + \gamma\varepsilon)}}{2} \tag{9.69b}$$

The initial condition is determined by the relative frequency of openings into O_1 versus O_2, and they are clearly in proportion with the rates α and ϕ. Thus, $P_{o1}(0) = \alpha/(\alpha + \phi)$ and $P_{o2}(0) = \phi/(\alpha + \phi)$. For the coefficients of the exponentials in P_{o1}, this leads to

$$a_1 + a_2 = \frac{\alpha}{\alpha + \phi} \tag{9.70}$$

We can then calculate P_{o2} from P_{o1} and obtain the second condition

$$\frac{\delta\phi}{\alpha + \phi} = (\beta + \gamma - \lambda_1)\, a_1 + (\beta + \gamma - \lambda_2)\, a_2 \tag{9.71}$$

Now Eqs. (9.70) and (9.71) can be solved for a_1 and a_2. This leads to expressions for P_{o1} and P_{o2} that can be added together to obtain the open-time distribution

$$P_o = \frac{1}{(\alpha + \phi)(\lambda_1 - \lambda_2)}\left(-((\alpha + \phi)\lambda_2 + \alpha\beta + \phi\varepsilon)e^{\lambda_1 t} + ((\alpha + \phi)\lambda_1 + \alpha\beta + \phi\varepsilon)\, e^{\lambda_2 t}\right) \tag{9.72}$$

Now we can examine the consequences of violating detailed balance by taking an extreme case and letting β and ϕ go to zero. Looking back at Scheme (9E) we see that this makes all openings start in O_1. Further, they must return to C from O_2. This forces the channel to cycle through all three states for every opening. Eq. (9.72) then reduces to

$$P_o = \frac{-\lambda_2}{\lambda_1 - \lambda_2}\, e^{\lambda_1 t} + \frac{\lambda_1}{\lambda_1 - \lambda_2}\, e^{\lambda_2 t} \tag{9.73}$$

There is an important difference between Eq. (9.72) and Eq. (9.73). It may not be obvious, but the coefficients of Eq. (9.72) are always positive if Eq. (9.67) is obeyed. By contrast, it is quite easy to see that the coefficients of Eq. (9.73) are opposite in sign. Thus, the violation of detailed balance leads to a qualitative difference. This is illustrated by plotting the two expressions (Fig. 9.5), first with $\beta = 1$, $\phi = 10$ (values chosen to satisfy Eq. (9.67)), and then with $\beta = \phi = 0$ (see Colquhoun and Hawkes, 1995).

The slope at $t = 0$ is zero for $\beta = \phi = 0$. This means that the probability of a channel closing immediately after opening is zero. The probability density function (the derivative of the distribution) would thus start at zero (where the $\beta = \phi = 0$ curve in Fig. 9.5 is flat), go through a maximum, and then decay to zero. This kind of behavior is occasionally observed in channel studies. One good example is the cystic fibrosis transmembrane conductance regulator. The binding of ATP drives this Cl^- channel into an open state. Closure of the channel follows hydrolysis of the bound ATP. Thus, there is a delay before an open channel can close (Zeltwanger et al., 1999).

The result illustrated above with the three-state model has been generalized to show that for any single-channel scheme in which the individual steps obey detailed balance, the coefficients are necessarily positive (Kijima and Kijima, 1987). Thus, the distribution decays

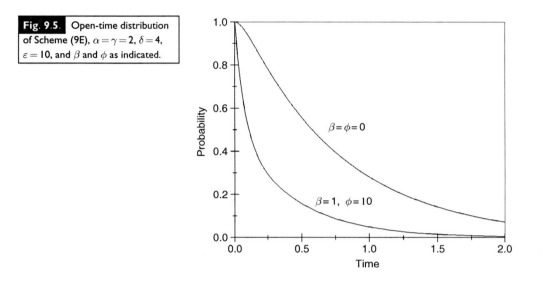

Fig. 9.5. Open-time distribution of Scheme (9E), $\alpha = \gamma = 2$, $\delta = 4$, $\varepsilon = 10$, and β and ϕ as indicated.

monotonically with no peaks or inflections. The observation of such behavior in single-channel data is thus evidence that the channel gating scheme has at least one energy-consuming step such as ATP hydrolysis. However, the converse is not true: when the observed lifetime distribution has all positive coefficients, an energy-consuming step cannot be ruled out. It is then more difficult to detect nonequilibrium behavior and more sophisticated analytical methods must be used (Steinberg, 1987; Rothberg and Magleby, 2001).

9.12 | Single-channel correlations: pathway counting

So far we have focused on lifetimes of individual states of a channel. However, for some kinetic schemes the lifetimes of successive intervals are interdependent, i.e. they are correlated. This means that our expectation for the duration of one state will depend on the duration of the preceding state. Indeed, correlations can extend over many intervals. We can express this interdependence in terms of a two-dimensional probability density $p_{c,o}(t_c, t_o)$. The mathematical methods used above for single-channel lifetime probabilities can be extended to this more complex problem, although the derivation is rather involved (Fredkin *et al.*, 1985). Here we will summarize some of the interesting results without delving into the mathematical details.

The very existence of correlations in channel lifetimes often establishes an important property of the underlying kinetic scheme. Imagine that a recording of single-channel current shows clear correlations (Fig. 9.6). We see long closures next to short openings, and vice versa. Imagine further that one has already analyzed the open-time and closed-time distributions and found that both are fitted by double exponentials (Fig. 9.4). Thus, we know that there are

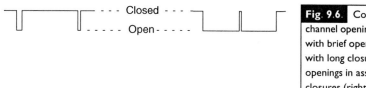

two open states and two closed states. The existence of correlations tells us that two separate gating pathways must exist between the open and closed states. A model with only one gating pathway cannot account for correlations.

To see how this works, consider the following sequential gating scheme (Scheme (9F))

$$C_1 \rightleftharpoons C_2 \rightleftharpoons O_1 \rightleftharpoons O_2 \tag{9F}$$

A long-duration opening may result if O_1 converts to O_2. But when the opening ends with a transition from O_1 to C_2, the information about the duration of the opening is irrelevant to whether the closure will end quickly with a return to O_1, or endure for a long time with a transition to C_1. In fact, if one derives the two-dimensional probability for an open time and subsequent closed time, one obtains a product of the open-time and closed-time distributions. The probabilities are independent.

Now look at a model (Scheme (9G)) with different behavior

$$O_1 \rightleftharpoons C_1 \rightleftharpoons C_2 \rightleftharpoons O_2 \tag{9G}$$

An opening in O_1 will end in C_1 so closed times will be defined by this initial condition. An opening in O_2 will end in C_2 and so the closed-time probability will be different. The duration of a closed time will now be correlated with the duration of the preceding opening.

In fact, many channels show clear correlations between successive state lifetimes. This is true of the acetylcholine receptor, and an analysis along these lines was used to eliminate kinetic schemes with only one gating pathway between the open and closed states (Jackson *et al.*, 1983).

Mathematical analysis of correlations in single-channel lifetimes produces a correlation coefficient that is a function of the number of intervening intervals, n. The result is a sum of exponentials decaying in n, and the number of exponential terms is one less than the number of gating pathways (Fredkin *et al.*, 1985). Thus, an analysis of single-channel data in this way reduces to a process of performing a count of the number of gating pathways by which open and closed states are connected.

More information can be obtained about a channel's gating mechanism by using matrix methods to calculate the probability, or likelihood, of a long sequence of lifetimes constituting many minutes of channel activity and thousands of single-channel transitions. This calculation can be repeated for several different candidate kinetic schemes, and the scheme that gives the highest calculated likelihood is the one that best represents the channel's gating activity (Horn and Lange, 1983). This method of likelihood maximization has provided a great deal of insight into the mechanisms of channel gating, particularly of the voltage-gated Na^+ channel (Horn and Vandenberg, 1984).

9.13 | Multisubunit kinetics

Multisubunit proteins can exhibit cooperative behavior. In Chapters 1 and 5 thermodynamic analysis showed how the equilibrium state of a multisubunit protein varied in a characteristic way. Now we will look at the kinetics of a multisubunit protein. Take the case where identical subunits undergo conformational transitions independently. The simplest is a two-subunit protein described by Scheme (9H)

$$A_2 \underset{\beta}{\overset{2\alpha}{\rightleftharpoons}} AB \underset{2\beta}{\overset{\alpha}{\rightleftharpoons}} B_2 \tag{9H}$$

The state vector for this system is $([A_2], [AB], [B_2])$. Each subunit in the A conformation undergoes transitions to B with a rate constant α; the reverse rate constant is β. From A_2, either subunit can switch, so the rate constant for going from A_2 to AB is 2α. The rate from AB to B_2 is α, because in AB there is only one subunit that can make the transition. Likewise, the rate constant for going from B_2 to AB is 2β, and for AB to A_2 is β. Using these rates for the various transitions leads to the following matrix

$$\mathbf{Q} = \begin{pmatrix} -2\alpha & \beta & 0 \\ 2\alpha & -\alpha - \beta & 2\beta \\ 0 & \alpha & -2\beta \end{pmatrix} \tag{9.74}$$

The characteristic equation (Eq. 9.23) is

$$\lambda^3 - 3(\alpha + \beta)\lambda^2 - 2(\alpha^2 + 2\alpha\beta + \beta^2)\lambda = 0 \tag{9.75}$$

This can be solved by factoring to give the following three eigenvalues

$$\lambda_1 = 0 \tag{9.76a}$$

$$\lambda_2 = -(\alpha + \beta) \tag{9.76b}$$

$$\lambda_3 = -2(\alpha + \beta) \tag{9.76c}$$

Thus, the two nonzero decay constants will be multiples of the fundamental single-subunit decay constant, $-(\alpha + \beta)$.

For n subunits, the model takes the form of Scheme (9I) below

$$A_n \underset{\beta}{\overset{n\alpha}{\rightleftharpoons}} A_{n-1}B \underset{2\beta}{\overset{(n-1)\alpha}{\rightleftharpoons}} A_{n-2}B_2 \underset{3\beta}{\overset{(n-2)\alpha}{\rightleftharpoons}} \cdots AB_{n-1} \underset{n\beta}{\overset{\alpha}{\rightleftharpoons}} B_n$$

(9I)

The MWC model has two rows like this (Chapter 5, Scheme (9F)). We take the initial condition as all subunits in the A conformation, so A_n is the only species present. If a sudden change initiates transitions, then B will start to form. Individual subunits will relax to a new equilibrium according to Eq. (7.4)

$$x = \frac{\alpha}{\alpha + \beta}\,e^{-(\alpha+\beta)t} + \frac{\beta}{\alpha + \beta}$$

(9.77)

where x is the fraction of subunits as A. Note that $x_0 = 1$.

Because the subunits are independent, the probability of finding a particular combination A_iB_{n-i} is then given by the binomial distribution, with x as the elementary probability taken from Eq. (9.77)

$$P_i(t) = \frac{n!}{(n-i)!i!}x^i(1-x)^{n-i}$$

(9.78)

In this expression powers of x produce powers of the basic exponential function $e^{-(\alpha+\beta)t}$, i.e. $e^{-(\alpha+\beta)t}$, $e^{-2(\alpha+\beta)t}$, $e^{-3(\alpha+\beta)t}$, up to $e^{-n(\alpha+\beta)t}$. Thus, without actually writing out and solving the characteristic equation, we can still see that the eigenvalues take the form $-j(\alpha+\beta)$, for j ranging from 0 to n (Chen and Hill, 1973).

If we were looking at P_{B_n}, we could use Eq. (9.78) with $i=0$ to give

$$P_{B_n} = \left(\frac{\alpha}{\alpha + \beta}(1 - e^{-(\alpha+\beta)t})\right)^n$$

(9.79)

This is plotted for different values of n to show that the time course has a characteristic sigmoidal shape (Fig. 9.7).

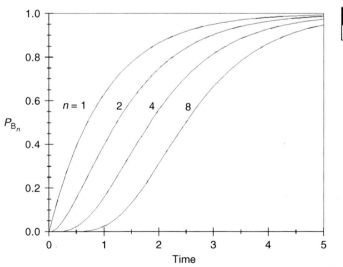

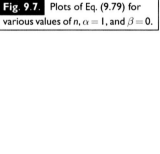

Fig. 9.7. Plots of Eq. (9.79) for various values of n, $\alpha = 1$, and $\beta = 0$.

This form of kinetic behavior was found in the gating of voltage-activated channels (Chapter 16), and structural work has verified that these channels are formed from the expected number of independent or nearly independent structural domains.

This model has an elegant mathematical structure with eigenvalues as integral multiples of $\alpha + \beta$. However, a very simple modification produces a model for conformational transitions in proteins with qualitatively different behavior (Jackson, 1993c). If we imagine that each of the transitions in Scheme (9I) is the isomerization of a bond within a protein, we can stipulate that each bond sterically obstructs a global transition when it is in the A position. The global transition is represented by a step from B_n to a new state, i.e. $B_n \rightarrow C$, and can occur only when all n bonds have rotated out of the way.

If the transitions to B are slightly disfavored, or if there are many of these bonds, then a protein will rarely be in B_n. The rarity of this situation means that the global transition will occur infrequently; it will be slow compared to the individual bond isomerization rates. With values of α and β of 10^9 s^{-1}, typical of bond isomerization rates in macromolecules, one obtains n eigenvalues that are very close to integral multiples of $\alpha + \beta = 2 \times 10^9$ s^{-1}. These eigenvalues reflect the isomerization kinetics of these bonds. However, the $B_n \rightarrow C$ transition prevents the model from having a stationary state, so the eigenvalue of zero seen in Schemes (9H) and (9I) (Eq. (9.76a)) is lost. Instead we obtain an eigenvalue that is very close to zero. For a variety of reasonable values of α, β, and n, the slowest eigenvalue can be as low as 1 s^{-1}. This is consistent with the generally slower rates of global transitions. Furthermore, the separation into very different timescales means that the global transition will be dominated by the slow eigenvalue. The overall process appears to follow simple single-exponential kinetics, even though the underlying model has many states.

9.14 | Random walks and "stretched kinetics"

A kinetic model with a sequence of identical steps often arises (Scheme (9J))

$$A_0 \underset{\beta}{\overset{\alpha}{\rightleftharpoons}} A_1 \underset{\beta}{\overset{\alpha}{\rightleftharpoons}} A_2 \underset{\beta}{\overset{\alpha}{\rightleftharpoons}} \cdots A_{n-1} \underset{\beta}{\overset{\alpha}{\rightleftharpoons}} A_n \tag{9J}$$

In fact, this is the random-walk model of Chapter 6, except that the steps occur continuously in time rather than at discrete intervals of τ. This continuous-time random walk can be solved, but the result is mathematically complex (Van Kampen, 1981). Here, we will revert to the discrete time model developed in Chapter 6.

An interesting property of this model emerges when we take the initial state as A_0, and ask how the probability of being in that state decays with time. If we let $\alpha = \beta$, then $p = q = 1/2$ and the random walk is symmetric with regard to hopping probability. However, there are no states to the left of A_0. In that sense the model is asymmetric.

This asymmetry forces the first step to be from A_0 to A_1. We can determine the probability of being at state A_1 after $N-2$ steps. This probability times $1/2$ gives the desired result, the probability of being in A_0 after N steps. The probability of going from A_1 back to A_1 in $N-2$ steps can be solved with the method of image points (Chandrasekhar, 1943). The state A_0 is a reflecting barrier and cannot be crossed. It can be shown that each pathway from A_1 to A_1 that crosses the barrier corresponds with one pathway from A_1 to A_{-1}; A_{-1} is the image point of A_1. So the probability of arriving at A_1 in the presence of the barrier is the sum of the probabilities of arriving at A_1 and A_{-1} when the barrier is absent. Adding these two together gives

$$
\begin{aligned}
P_{A_1}(N-1) &= \frac{(N-2)!(1/2)^{N-2}}{\left(\frac{N}{2}-1\right)!^2} + \frac{(N-2)!(1/2)^{N-2}}{\left(\frac{N}{2}\right)!\left(\frac{N}{2}-2\right)!} \\
&= \frac{(N-2)!(1/2)^{N-2}}{\left(\frac{N}{2}\right)!\left(\frac{N}{2}-1\right)!}\left(\frac{N}{2}+\frac{N}{2}-1\right) \\
&= \frac{(N-1)!(1/2)^{N-2}}{\left(\frac{N}{2}\right)!\left(\frac{N}{2}-1\right)!}
\end{aligned}
\tag{9.80}
$$

Multiplying the numerator and denominator by $N/2$ gives

$$
P_{A_1}(N-1) = \frac{N!(1/2)^{N-1}}{\left(\frac{N}{2}\right)!^2}
\tag{9.81}
$$

The probability of arriving at A_0 after N steps is half this result, raising $1/2$ to the power N. This is Eq. (6.46) with $k = 0$. Now the Gaussian approximation (Eq. (6.49)) can be used

$$
P_{A_0}(N) = \sqrt{\frac{2}{\pi N}}
\tag{9.82}
$$

Since steps are taken at regular intervals in time, N is proportional to time. Equation (9.82) therefore predicts a $t^{-1/2}$ time dependence for the decay of A_0 in Scheme (9J). This is a unique form of dynamic behavior that is qualitatively different from exponential or multiexponential decays predicted for models with just a few states.

This result illustrates a general feature of the dynamics of random walk models. They predict a survival probability that decreases as an inverse power of time. Because the factor of $t^{-1/2}$ is raised to the power of the number of dimensions in the random walk (Section 6.2.2), it is common to write the survival probability as $\propto t^{-d/2}$, where d is the number of dimensions. Note that d is determined as a parameter from a fit to the time course, and is often referred to as the dimension of the kinetic process.

The interesting feature of $t^{-d/2}$ kinetics is that it allows a process to keep going over an incredibly broad range of times. An exponential process decays very rapidly as t increases beyond its characteristic

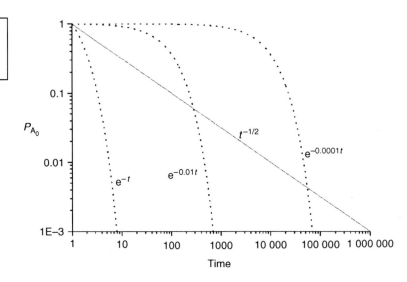

Fig. 9.8. The curves plotted illustrate the slower change in $t^{-1/2}$ compared to exponentials with various decay constants.

time constant. With $t > 10\tau = 10/(\alpha + \beta)$ the exponential function is down to $e^{-10} \sim 5 \times 10^{-5}$. On the other hand, $t^{-1/2}$ kinetics means a process will decrease by a factor of only $\sqrt{10} = 3.1$ for every 10-fold increase in t (Fig. 9.8).

Kinetic processes stretched out over such a broad timescale are not seen very often, but one well-characterized example is the rebinding of the ligand CO to myoglobin after photodissociation. This typically is modeled not as $t^{-d/2}$ but as $\exp(t^{-d/2})$, which has its origin in a mechanism related to this one. The favored interpretation for the stretched kinetics of this recombination process is that there are closely spaced energy levels (Hagen and Eaton, 1996). Hopping between energy levels is a kind of random walk, with behavior similar to that predicted by Scheme (9J).

Problems for Chapter 9

1. For the three-state model, obtain the second order differential equation in [C] by the same method used to derive Eq. (9.4) for [A]. Then derive the characteristic equation for the eigenvalues.

2. For the three-state model with the initial condition treated in Section 9.2, derive the general time-dependent expression for [C] in terms of the rate constants.

3. For the three-state model, obtain the solution to the time dependence of [A] with initial conditions [B] = 1 and [A] = [C] = 0.

4. Consider a_1, a_2, λ_1, λ_2, and $[A]_{eq}$ to have been determined from a fit of Eq. (9.13) to an experimental time course. Normalize all concentrations to T, and solve for α, β, γ, and δ (hint: calculate $\lambda_1\lambda_2$ and $\lambda_1 + \lambda_2$ and use these results).

5. Determine the $\lambda = 0$ eigenvector for the three-state model (Scheme (9A)) using Eqs. (9.36a)–(9.36c). Determine the equilibrium fractions of [A], [B], and [C] directly and compare these two results.

6. For Scheme (9B) show that $P_{o1} + P_{o2}$ gives P_o in Eq. (9.53).

7. For Scheme (9B) show that the time derivative of P_o at $t = 0$ is $-\beta$. In a model with three sequential open states, show that this relationship still holds (Jackson, 1997b).

8. Derive the open and closed time distributions for a channel governed by the following kinetic model

$$O_1 \underset{\beta}{\overset{\alpha}{\rightleftharpoons}} C \underset{\delta}{\overset{\gamma}{\rightleftharpoons}} O_2$$

9. Derive the open-time and closed-time distribution for a channel governed by the following kinetic model

$$O_1 \underset{\beta}{\overset{\alpha}{\rightleftharpoons}} C_1 \underset{\delta}{\overset{\gamma}{\rightleftharpoons}} C_2 \underset{\phi}{\overset{\varepsilon}{\rightleftharpoons}} O_2$$

10. Use Eq. (9.63) to show that the mean lifetime of state r of a single-molecule can be expressed as the sum of the elements of the vector $\mathbf{Q}_r^{-2} \mathbf{A}_0$.

11. In Fig. 9.4 what is the total number of open-channel events and closed-channel events?

Chapter 10

Enzyme catalysis

How enzymes accelerate biochemical reactions is one of the oldest and most challenging problems in biophysics. An enzyme binds with high specificity to a substrate molecule, chemically modifies it, releases the product, and then repeats the cycle. Without the enzyme, the same chemical reaction can still take place, but at a vastly slower rate. The most impressive enzymatic accelerations approach 10^{20} (Miller and Wolfenden, 2002). A value in the region of 10^8-10^{12} is more typical, but that still represents a remarkable enhancement. Enzymes are responsible for virtually all of the metabolic chemistry in the biological world. However, there is another point worth mentioning. Nearly all enzymes are proteins (the exception is ribozymes - catalytic RNA), so a study of enzyme catalysis provides a window into the basic mechanics of proteins carrying out their functions. Enzyme catalysis provides excellent examples of how the structure and dynamics of proteins relate to their activity.

We know a great deal about the chemical mechanisms employed by enzymes. X-ray crystallography has given us atomic-level pictures of enzyme–substrate complexes in which many important contacts are evident. Some of these contacts are strictly for binding, and enable the recognition of specific substrates. The binding of substrate is the first step of enzyme catalysis, and this provides an important application of the physics of molecular associations (Chapter 4). The next step is the chemical reaction, and this brings the rate processes of Chapter 7 into the picture. We will see how enzyme catalysis can be formulated in terms of how an enzyme alters the free energy profile along a reaction coordinate.

10.1 | Basic mechanisms – serine proteases

It will be very helpful to start with an example that illustrates the basic sequence of events. For this we will look at a class of enzymes known as serine proteases. These enzymes include trypsin, chymotrypsin, and elastase, all of which digest proteins into small peptides and amino acids. They are called serine proteases because of the essential role played by a key serine residue in the enzyme's active site. The general chemical reaction catalyzed by serine proteases is the hydrolysis of a peptide bond, as shown in Scheme (10A) below

$$X - \overset{\overset{\displaystyle O}{\|}}{C} - \overset{\displaystyle H}{N} - Y + H_2O \longrightarrow X - \overset{\overset{\displaystyle O}{\|}}{C} - OH \ H_2N - Y \qquad (10A)$$

This happens spontaneously in water, but very slowly. In water an OH^- anion attacks the carbonyl and displaces the amino group in what is generally referred to as an SN2 displacement. Serine proteases perform the same task by a general mechanism illustrated in Fig. 10.1. The enzyme first binds the substrate through a highly specific interaction largely determined by a special binding pocket that recognizes an amino acid side chain on the substrate (R in Fig. 10.1). The hydroxyl group of the key serine residue (serine 195 of chymotrypsin) becomes chemically activated when a proton is drawn away by nearby histidine (histidine 57 of chymotrypsin), leaving a reactive $-O^-$ behind. This $-O^-$ then attacks the carbonyl carbon of the substrate in the same way that a hydroxyl anion might in water. An intermediate then forms in which the

Fig. 10.1. The hydrolysis of a peptide bond by a serine protease proceeds through (a) substrate binding, (b) serine attack of the substrate carbonyl to form the first tetrahedral intermediate, (c) release of amine product to form an acyl enzyme, (d) attack of the acyl enzyme by water to form the second tetrahedral intermediate, and (e) release of carboxylic acid product. Residue numbers are for chymotrypsin but the mechanism is general.

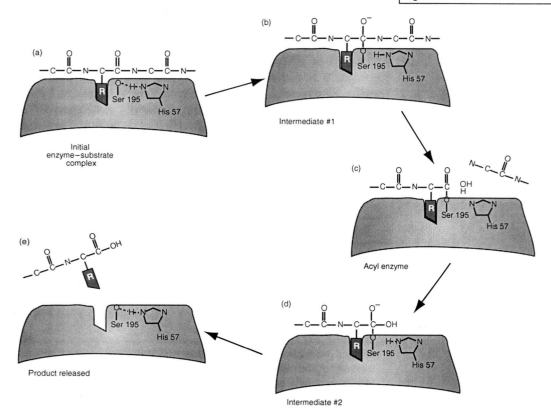

targeted C atom assumes a tetrahedral bonding geometry. The peptide bond is broken as the amino group leaves and the carbonyl C=O double bond forms again. The —O⁻ of the serine thus displaces the amino group to form an acyl enzyme.

At this stage half the job is done. The amino end of the peptide has been released, but the carboxyl end of the peptide is still covalently attached to the enzyme by an ester linkage to the serine oxygen atom (acyl enzyme in Fig. 10.1). This ester bond is then hydrolyzed by water, with assistance from the same histidine that activated the serine hydroxyl group. It now pulls a proton away from a water to create an activated hydroxyl, which attacks the carbonyl group coupled to the serine side chain (an example of general base catalysis – see Section 10.13). A second tetrahedral intermediate forms and decomposes as serine 195 releases the carboxyl group of the substrate.

Serine proteases illustrate some of the key principles of enzyme catalysis. Firstly, the function of substrate recognition can be distinguished from the function of catalytic activity. Serine proteases are very choosy about their substrates. The most important determinant on the substrate is the side chain of the residue at the carboxyl side of the peptide bond destined for cleavage. Trypsin preferentially cleaves peptides with positively charged arginine or lysine in this position. Chymotrypsin preferentially cleaves peptides with an aromatic residue in this position. Another serine protease, elastase, preferentially cleaves peptides with small aliphatic side chains in this position.

Examples of this specificity are illustrated with some rates of hydrolysis of different substrates in Table 10.1. A peptide with tryptophan is cleaved 10^6 times faster by chymotrypsin than by trypsin. Conversely, when the tryptophan is replaced by lysine, the tables are turned, and now trypsin works more than 10^4 times better. An explanation for this specificity can be found in the structure of the protein. Serine proteases have a pocket that pulls in the side chain, as roughly depicted in Fig. 10.1. Trypsin has a negatively charged aspartate at the base of this pocket that interacts with the

Table 10.1	*Rate of hydrolysis ($M^{-1} s^{-1}$)*	
	Substrate with X = tryptophan	Substrate with X = lysine
Chymotrypsin	1.3×10^6	69
Trypsin	1.5	1.9×10^5
Trypsin (D189S)	4.8	7.3

Rates of hydrolysis of peptide substrates by chymotrypsin, trypsin, and a mutant trypsin with aspartate 189 replaced by serine. The substrate is succinate–alanine–alanine–proline–X–aminomethylcoumarin, with X either tryptophan or lysine as indicated (from Hedstrom et al., 1994).

positively charged lysine or arginine in the substrate. Mutation of this aspartate to serine makes trypsin even slower than chymotrypsin in cleaving the lysine substrate.

This is far from the whole story for substrate recognition by serine proteases. Several residues of the enzyme help shape the binding pocket, and adjacent residues on the substrate interact with other sites (Hedstrom *et al.*, 1994). But these residues do not participate directly in catalysis. Catalysis is performed by a separate group of residues that includes the serine after which serine proteases are named in addition to the histidine shown in Fig. 10.1 and an aspartate (see Fig. 10.12). These residues define the catalyzed chemical reaction, and are conserved among serine proteases with very different substrate specificities. We will see in Section 10.14 how the side chains of these amino acids can work together to form and break covalent bonds.

Another important principle of enzyme catalysis illustrated by the serine proteases is that the substrate is bound in such a way that the targeted part of the substrate, the carbon atom of the peptide bond, is held in position adjacent to the attacking catalytic group of the enzyme. Holding two chemically reactive groups in close proximity allows the reaction to proceed far more rapidly than it would if the same two groups were free in solution. This proximity effect plays a major role in catalysis, and its impact will be estimated in Section 10.8.

Another point to be taken up in Section 10.14 is that the environment of the enzyme helps to enhance the reactivity of the group carrying out the catalysis. The hydroxyl side chain of serine is ordinarily rather benign, but through interactions with other groups in the protein it is converted into a highly reactive oxyanion ($-O^-$). Without the special environment of an enzyme it takes extreme conditions such as high or low pH to create a species with comparable reactivity.

It is one thing to write out the chemical mechanism used by an enzyme, and quite another to explain how the reaction is actually accelerated. The chemical mechanism is an essential starting point in posing questions about the physical factors that lead to rate enhancement. We will now turn to the formulation of the kinetic equations, and use this as a framework for analyzing the physical factors that make enzymes such effective catalysts.

10.2 | Michaelis–Menten kinetics

Various enzymes employ different chemical mechanisms involving different numbers of binding, conversion, and release steps. In spite of this, a simple model captures the qualitative behavior quite generally. If we recognize that it all starts with substrate binding, and lump the subsequent steps together, we can envision a basic two-step process (Scheme (10B))

$$E + S \overset{K_S}{\rightleftharpoons} ES \overset{k_{cat}}{\longrightarrow} E + P \tag{10B}$$

This is a minimal scheme. Substrate binding is described as a chemical equilibrium and assigned the dissociation equilibrium constant K_S. The conversion of substrate S into product P, and the ensuing dissociation step are combined into one step to which the rate constant k_{cat} is assigned. We assume that the first step of enzyme–substrate binding is rapid compared to the second step so that K_S can be used to relate the concentrations, [E], [S], and [ES]

$$\frac{[ES]}{[E][S]} = \frac{1}{K_S} \tag{10.1}$$

With the total amount of enzyme fixed at $[E_t] = [E] + [ES]$, we can eliminate [E] and solve for [ES]

$$[ES] = \frac{[E_t][S]}{[S] + K_S} \tag{10.2}$$

The production of P from ES is a first order reaction with a velocity $v = k_{cat}[ES]$ (Scheme (10B)). Replacing [ES] with the expression in Eq. (10.2) allows us to write the reaction velocity as a function of [S]

$$v = \frac{k_{cat}[E_t][S]}{[S] + K_S} \tag{10.3}$$

This is the Michaelis–Menten equation. It makes the important prediction that the velocity of the reaction first increases with [S] when [S] is low, and then saturates when [S] goes above K_S (the Michaelis constant, generally referred to as K_m). Thus, enzyme catalysis is second order at low [S], with a rate of $\sim (k_{cat}[E_t]/K_S)[S]$, and first order at high [S], with a rate of $\sim k_{cat}[E_t]$. This can be seen in a plot of Eq. (10.3) (see Fig. 10.2). Saturation behavior is almost universally observed for enzyme kinetics, and an equation of the form of Eq. (10.3) describes the dependence of the rate of product formation as a function of substrate for most enzymes. Enzymatic processes can be more complex, particularly where there are multiple substrates, but saturation is still observed. In rare cases where saturation is not observed, it is usually because the solubility of a substrate prevents experiments with high enough concentrations.

In performing a kinetic analysis of an enzyme, Eq. (10.3) is usually fitted to a plot of velocity versus [S]. In this way the basic parameters K_S and k_{cat} are determined. These parameters are of fundamental importance in quantifying enzyme activity, and provide the starting point for more detailed questions about catalytic mechanism.

At saturating substrate concentration there is essentially no free enzyme; 100% of the enzyme has substrate bound. Product will then be formed at a rate of $k_{cat}[E]$. For this reason k_{cat} is often referred to as the turnover number of the enzyme; k_{cat} can be

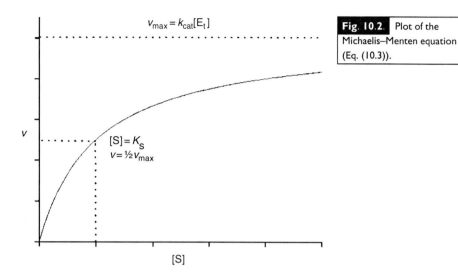

$v_{max} = k_{cat}[E_t]$

$[S] = K_S$
$v = \frac{1}{2}v_{max}$

v

$[S]$

Fig. 10.2. Plot of the Michaelis–Menten equation (Eq. (10.3)).

interpreted in terms of catalytic efficacy, i.e. as the rate constant for the actual catalytic event in which bound substrate is converted to product. However, the assumptions that go into the derivation of the Michaelis–Menten equation can be violated in ways that leave the basic form of the equation intact but undermine this interpretation of k_{cat}. For example, if the dissociation of product is rate limiting, then k_{cat} will have more to do with the release step then the actual catalytic conversion. The catalytic step could then be much faster, but it would appear to be slower because of the time required for dissociation. For this reason, k_{cat} is taken as a lower bound to the conversion rate of the enzyme–substrate complex.

When k_{cat} is close to the true turnover rate of the enzyme–substrate complex, it can be compared with the rate of the uncatalyzed reaction to provide a measure of the catalytic power of an enzyme. The rate of the reaction without the enzyme is defined in terms of a rate constant k_0 as $k_0[S]$. If we take k_{cat} as a corresponding rate of the S → P reaction within the enzyme, then the rate enhancement by the enzyme is the ratio k_{cat}/k_0.

The rate constant k_{cat} can be divided by the Michaelis constant to give k_{cat}/K_S, and this ratio has an important meaning as the apparent second order rate constant at low, non-saturating substrate concentrations. The ratio k_{cat}/K_S is often referred to as the catalytic efficiency, and for many different enzymes falls in the range 10^5–$10^9 \text{M}^{-1} \text{ s}^{-1}$ (Miller and Wolfenden, 2002). This may seem like a wide range, but the uncatalyzed reaction rates range from 10^{-17} to 10^{-1}s^{-1}. The range of observed k_{cat}/K_S values may indicate that there is a physical limit to the catalytic power that can be achieved by an enzyme. Alternatively, this may reflect the result of evolutionary pressure to make enzymatic reactions go at biologically useful rates.

Another interesting generalization has been made about K_S (Fersht, 1998). For most enzymes K_S is 1–100 times greater than the typical value of [S] seen in a cell. This means that most enzymes are not saturated under biological conditions. This allows the enzyme to regulate the concentration of the substrate more effectively. If the concentration goes up the rate of the reaction increases in a compensatory manner.

Enzymes do not appear to maximize the strength of substrate binding. The value of K_S is usually high compared to the dissociation constant of the same molecule to a protein with no catalytic function (Fersht, 1998). For example, K_S is about 0.1–10 mM for ATP for many enzymes. However, other proteins bind ATP with dissociation constants $\sim 10^{10}$ lower (for myosin it is 10^{-13} M). This suggests that the evolutionary pressure on a substrate binding site does not favor binding with maximal strength. Indeed, very tight binding might counteract catalysis, either by freezing the structure of the substrate to prevent the transition state from forming, or by retarding product release. We will see that one reason that binding does not reach its maximum strength is because some of the potential binding energy is utilized for catalytic functions (Sections 10.6–10.10)

10.3 | Steady-state approximations

Most enzymes employ mechanisms that are more complex than that assumed above for the Michaelis–Menten equation. Nevertheless, they can usually be treated with a steady-state approximation, in which the concentrations of intermediates do not change with time. To see how this works in a simple case we will modify the Michaelis–Menten mechanism (Scheme (10B)) to include the forward and reverse substrate binding steps

$$E + S \underset{k_{-1}}{\overset{k_1}{\rightleftharpoons}} ES \xrightarrow{k_{cat}} E + P \tag{10C}$$

Scheme (10C) is the Briggs–Haldane mechanism. The rate of change of [ES] depends on the rates of association, dissociation, and catalysis

$$\frac{d[ES]}{dt} = k_1[E][S] - k_{-1}[ES] - k_{cat}[ES] \tag{10.4}$$

If binding is rapid, then we can assume that the concentration of ES will quickly reach a steady state, and remain constant as the reaction proceeds. Setting the above derivative equal to zero, solving for [ES], and using the condition $[E_t] = [E] = [ES]$, leads to an expression similar in form to Eq. (10.3)

$$v = \frac{k_{cat}[E_t][S]}{[S] + (k_{cat} + k_{-1})/k_1} \tag{10.5}$$

Note that if $k_{cat} << k_{-1}$, then Eq. (10.3) is recovered (with $K_S = k_{-1}/k_1$).

For a more complex example we can take the serine protease mechanism (Scheme (10D)) as shown below

$$E + S \underset{k_{-1}}{\overset{k_1}{\rightleftharpoons}} ES_1 \xrightarrow{k_2} ES_2 + P_1 \xrightarrow{k_3} E + P_2 \tag{10D}$$

In this mechanism ES_1 corresponds with the initial complex and ES_2 corresponds with the acyl enzyme (Fig. 10.1). The two products released in successive steps, P_1 and P_2, are the amino and carboxy peptides.
We have the following kinetic equations for ES_1 and ES_2

$$\frac{d[ES_1]}{dt} = k_1[E][S] - (k_{-1} + k_2)[ES_1] \tag{10.6}$$

$$\frac{d[ES_2]}{dt} = k_2[ES_1] - k_3[ES_2] \tag{10.7}$$

At steady state these derivatives are zero, so we have

$$\frac{[ES_1]}{[E][S]} = \frac{k_1}{k_{-1} + k_2} \tag{10.8}$$

$$\frac{[ES_2]}{[ES_1]} = \frac{k_2}{k_3} \tag{10.9}$$

The rate of product formation is k_3 times the concentration of ES_2

$$v = k_3[E_t] \frac{[ES_2]}{[E] + [ES_1] + [ES_2]} = \frac{k_3[E_t]}{\left(\frac{[E]}{[ES_2]} + \frac{[ES_1]}{[ES_2]} + 1\right)} \tag{10.10}$$

Using Eqs. (10.8) and (10.9) and rearranging yields the velocity as a function of [S]

$$v = \frac{k_1 k_2 k_3 [E_t][S]}{k_1[S](k_2 + k_3) + k_3(k_{-1} + k_2)} \tag{10.11}$$

The important point to note here is that the velocity has the same qualitative dependence on [S] as the Michaelis–Menten and Briggs–Haldane mechanisms. But the values of the parameters are now very different. This is an important point because the concentration for half saturation is no longer a simple equilibrium constant and the maximum velocity depends on the rates of two steps.

Steady-state approximations are very useful in the study of enzyme kinetics (Cleland, 1970; Plowman, 1972). Even with a long string of intermediates and with multiple substrates, the steady-state approximation can still provide a useful expression for the velocity. The concentration dependence usually has the form of a rectangular hyperbola (Fig. 10.2), but with parameters depending on combinations of rate constants. The steady-state approximation is especially important in enzyme kinetics because the exact method of solving multi-state kinetics problems of Chapter 9 runs into problems in treating binding steps. The second order terms in the rate equations prevent the use of matrix methods and make the mathematics much more difficult.

10.4 | Pre-steady-state kinetics

It takes time for an enzymatic reaction to reach a steady state. The approach to steady state is often detectible, and can provide important information about a reaction mechanism. From Scheme (10C) we can write a differential equation for [ES]

$$\frac{d[ES]}{dt} = k_1[E][S] - (k_{-1} + k_{cat})[ES] \tag{10.12}$$

We use $[E] = [E_t] - [ES]$ to eliminate [E] as follows

$$\frac{d[ES]}{dt} = k_1[E_t][S] - (k_1[S] + k_{-1} + k_{cat})[ES] \tag{10.13}$$

This equation is still difficult to solve, but if we neglect changes in [S] during the approach to steady state it becomes easy. This approximation is reasonable because there is much more substrate than enzyme (after all, the enzyme is the catalyst). So a small amount of substrate can completely tie up all of the enzyme. With [S] constant, Eq. (10.13) becomes a simple first order differential equation, for which the solution is an exponential function with a decay constant $k_1[S] + k_{-1} + k_{cat}$. The specific solution depends on initial conditions. At $t = 0$, $[ES] = 0$, and at longer times a steady state is reached where [ES] is constant. Thus, [ES] has the following dependence on time

$$[ES] = \frac{[E_t][S]}{[S] + (k_{cat} + k_{-1})/k_1} \left(1 - e^{-(k_1[S] + k_{cat} + k_{-1})t} \right) \tag{10.14}$$

This expression starts off with $[ES] = 0$, and then rises to a steady state where $[ES] = [E_t]([S]/([S] + (k_{cat} + k_{-1})/k_1))$. The rate of product formation is $k_{cat}[ES]$, so there is a lag before product formation gets up to speed. This lag has the same decay constant as that for [ES] to reach steady state.

Note that when [S] is large compared to $(k_{cat} + k_{-1})/k_1$, the steady-state level of [ES] equals the total enzyme concentration. This reflects mass action, where high substrate pushes the equilibrium to the right, so that nearly all the enzyme is complexed with substrate. This can be very useful because sometimes the enzyme concentration is not known. There are practical problems such as an unknown fraction of denatured protein, and variations in the fraction of active enzyme from batch to batch. This creates some uncertainty about the exact concentration of active enzyme, and if [ES] can be accurately measured in the pre-steady-state domain, then a useful estimate of $[E_t]$ can be made.

An important example of pre-steady-state kinetics is the hydrolysis of p-nitrophenyl esters by chymotrypsin. The product of the first step is p-nitrophenol, which is liberated during the acylation of the enzyme.

The kinetics of this step is especially easy to follow because *p*-nitrophenol has a strong absorbance of visible light, and the ester does not. The analysis leading to Eq. (10.14) can be adapted to this particular case to estimate the time course of formation of the acyl enzyme (Problem 3). There is an initial burst of *p*-nitrophenol production because it is the product of the first step of the reaction. But as the acyl enzyme intermediate builds up the rate of *p*-nitrophenol production slows down to a steady-state level. The time constant for reaching this steady state is the same as that in Eq. (10.14). The burst of *p*-nitrophenol production at the beginning of the measurement is then equal to the amount of enzyme that can be acylated, and so the concentration of active enzyme can be determined (Fersht, 1998).

An analysis of pre-steady-state kinetics often provides crucial data that allow a complete determination of a set of rate constants. Analysis of the steady-state rate (Eq. (10.5)), provides determinations of the ratio $(k_{cat} + k_{-1})/k_1$ and of $[E_t]k_{cat}$. With $[E]_t$ determined from the magnitude of the pre-steady-state burst, we then have k_{cat}. To solve for the two remaining unknowns, k_1 and k_{-1}, we can use the time constant for approach to steady state from Eq. (10.14), and the concentration of half-maximal rate.

10.5 | Allosteric enzymes

In the analysis of the steady-state approximation it was noted that more complex mechanisms still lead to the same simple qualitative dependence on substrate concentration. This is because many enzymes have a single substrate binding site. Enzymes with multiple substrate binding sites can exhibit cooperative behavior, and a sigmoidal dependence of velocity on [S]. For a concerted binding model such as in Section 4.2.1 we might expect something like Eq. (4.7)

$$v = k_{cat}[E_t]\frac{[S]^n}{[S]^n + K_S} \qquad (10.15)$$

We saw how the activity of the enzyme phosphofructokinase shows cooperative activation, and can be regulated by various effectors (Section 5.11). The Monod–Wyman–Changeux model accounts very well for this behavior, with activity as a function of substrate concentration going as the fraction of occupied binding sites (Eq. (5.22)).

For mechanisms of catalysis Monod *et al.* identified two distinct classes of allosteric enzymes, which they called "K systems" and "V systems". In K systems, it is the affinity, K, which is regulated. Substrates and allosteric effectors bind with different affinities to the T and R conformations. In these enzymes the catalytic efficacy of the T and R conformations is the same. The function of the protein is regulated solely through binding, and the sigmoidal dependence of rate on concentration perfectly mirrors the sigmoidal binding behavior. In V systems the velocity is regulated. Substrate has the same affinity for the T and R conformations, so there will be no sigmoidicity as described in Eq. (10.15). The value of k_{cat} for the R state is higher than that of the T state, and an allosteric activator preferentially binds the R conformation. By shifting the conformational equilibrium toward the catalytically active R state, the rate of enzyme catalysis is accelerated, but K_S does not change. Allosteric inhibitors bind preferentially to the catalytically inactive T state and shift the equilibrium the other way.

Monod *et al.* found a number of examples of both K and V systems, and the classification is clearly useful. But it is not entirely clear why there are no mixed K and V systems, with allosteric states differing in both affinities and catalytic efficacy. Nevertheless, the classification takes on added significance in providing another example of the fundamental importance of the distinction between the binding function and catalytic function in enzymes.

10.6 | Utilization of binding energy

The presentation so far has been phenomenological. Enzymes were assumed to enable a reaction to go faster and the consequences were examined. But the real question about enzymes is how do they enhance these rates. We will now turn to this question and explore the fundamental physics of catalysis.

Most ideas about enzyme catalysis invoke some form of utilization of binding energy. Stronger binding to the transition state is one example. Binding energy is envisioned as reducing the height of the energy barrier $E^{\dagger}$. However, we will soon see that there are entropy effects as well.

When we think about the utilization of binding energy, the binding process can no longer be viewed as completely distinct from the catalyzed reaction. Stronger contacts between the enzyme and substrate can be harnessed for catalysis. However, it is also possible for stronger contacts between the substrate and enzyme to oppose motion in the reaction coordinate toward the top of the energy barrier. In this case stronger binding would slow the reaction down.

Measurements on a series of substrates support the idea that binding energy and catalytic effectiveness are related. These measurements show a general trend that K_S and k_{cat} are inversely correlated (Fersht, 1998; Hedstrom *et al.*, 1994). Thus, tighter binding reduces K_S and increases k_{cat}. This is not a hard and fast rule because there are so many ways in which structural variations can influence binding and catalysis. But the broad trend makes a case for utilization of binding energy.

In Fig. 10.3, k_2, the rate of the first acylation step of chymotrypsin (Fig. 10.1 and Scheme (10D)) is plotted logarithmically versus the substrate binding energy ($- RT \ln K_S$). This plot shows a wide scatter as some variations in substrate structure strengthen binding with or without increasing k_{cat}. However, this scatter is roughly delimited to

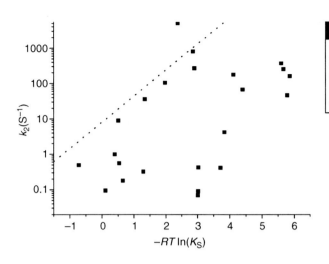

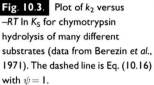

Fig. 10.3. Plot of k_2 versus $-RT \ln K_S$ for chymotrypsin hydrolysis of many different substrates (data from Berezin et al., 1971). The dashed line is Eq. (10.16) with $\psi = 1$.

the left by an exponential increase in rate with binding energy (dashed line). This represents an envelope corresponding to structural modifications in which binding energy is utilized with greatest effect to accelerate the reaction.

A literal interpretation of the idea of utilization of binding energy would suggest that increments in binding energy produce an exponential increase in rate

$$k_{cat} \propto e^{-\psi RT \ln K_S} \tag{10.16}$$

where ψ is a parameter introduced to denote the fraction of binding energy that is utilized. The dashed line in Fig. 10.3 illustrates the case for $\psi \sim 1$, representing maximally efficient utilization of binding energy in catalysis. This may indicate that some contacts involved in substrate binding are very strongly coupled to the reaction. The utilization of binding energy represents a partial breakdown of the distinction between the two basic functions of an enzyme, binding and catalysis. Binding energy aids catalysis, and we will now explore some of the ways this can happen.

10.7 | Kramers' rate theory and catalysis

Understanding the utilization of binding energy requires a theoretical framework, so we turn to Kramers' theory for diffusion over an energy barrier (Section 7.8). This theory provides the ideal starting point for examining most of the key hypotheses for enzymatic catalysis.

We start with Eq. (7.31)

$$J = \frac{DP(a)}{\sqrt{\pi kT/\phi^\dagger}} \, e^{-E^\dagger/kT} \tag{10.17}$$

This expression was obtained by taking the steady-state solution of the equation for diffusion in a reaction coordinate. The reaction coordinate was denoted by ξ, and D was the diffusion constant for movement along this coordinate. The potential energy as a function of ξ has a barrier with a height $E^\dagger$ and a width $1/\phi^\dagger$. The flux over the barrier is given by J; this flux can be divided by moles or molecules to obtain a rate constant.

We will use Eq. (10.17) to examine each of the key factors that control the rate of a reaction. Of course, the height of the energy barrier, $E^\dagger$, is an obvious focus. We can also look at the distribution of molecules at the foot of the barrier. In fact, some of the most important mechanisms of enzymatic rate enhancement can be understood in terms of altering this distribution, as will be seen in the next two sections. We can also focus on D and try to obtain some insight into how an enzyme might alter the ease of movement along the reaction coordinate.

10.8 | Proximity and translational entropy

Binding a substrate restricts its motion. The substrate's chemically targeted bonds are held near an enzyme's catalytic side chains. The catalytic group, e.g. the serine oxygen atom of a serine protease, could react with a substrate if it were free in solution, but this reaction would depend on collisions between the two molecules. The enzyme assists this process by drawing the catalytic group and the substrate together. This in effect converts a second order reaction, dependent on the concentration of substrate and chemical catalyst, into a first order reaction dependent on the concentration of the enzyme–substrate complex.

This effect can be visualized in terms of the reaction coordinate at the foot of the energy barrier. Figure 10.4 shows what the reaction coordinate looks like for intramolecular and intermolecular processes. For the intramolecular case, ξ is confined to a potential energy well on the reactant side. For the intermolecular case the reactant side of the barrier is a flat line; diffusion will proceed along this line, with the molecule reaching the foot of the barrier and crossing it.

The shape of these curves at the foot of the barrier determines the value of $P(a)$ in Eq. (10.17). This equation was actually the penultimate step in the derivation of the Kramers expression for the rate constant in Chapter 7. The final step of the derivation entailed evaluating $P(a)$ for an equilibrium distribution in ξ within a parabolic energy well on the left side of the barrier. Having the reactant bound in this way corresponds to the intramolecular case. The energy in the vicinity of the minimum at $\xi = a$ is $U(\xi) = \phi_a(\xi - a)^2$, and the probability of ξ having a particular value is given by the Boltzmann distribution, $P(\xi) = e^{-U(\xi)/kT}$. Evaluating the Boltzmann integral for this potential energy function gave $P(a) = \sqrt{\phi_a/\pi kT}$. By substituting this expression into Eq. (10.17), we retrace the derivation of Eq. (7.32)

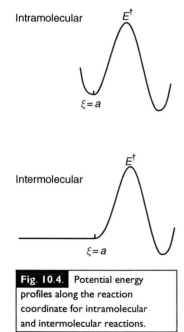

Fig. 10.4. Potential energy profiles along the reaction coordinate for intramolecular and intermolecular reactions.

$$k_{\text{intra}} = \frac{D_{\text{intra}}\sqrt{\phi_a\phi^\dagger}}{\pi kT}\, e^{-E_{\text{intra}}^\dagger/kT} \tag{10.18}$$

where D has been given the subscript intra to take into account the possibility that the diffusion constant takes on different values in different environments. This will be considered further in Section 10.11.

For the intermolecular case $P(a)$ is simply the bulk concentration, C, which should be the same everywhere to the left of $\xi = a$ in the lower sketch of Fig. 10.4. Equation 10.17 then yields the corresponding intermolecular rate constant

$$k_{\text{inter}} = D_{\text{inter}}\sqrt{\frac{\phi^\dagger}{\pi kT}}\, e^{-E_{\text{inter}}^\dagger/kT} \tag{10.19}$$

To compare the speed of the intra and intermolecular reactions, we must multiply k_{inter} by the concentration, C (which for now is one dimensional), and then take the ratio of Eqs. (10.18) and (10.19)

$$\frac{k_{\text{intra}}}{Ck_{\text{inter}}} = \frac{D_{\text{intra}}}{CD_{\text{inter}}}\sqrt{\frac{\phi_a}{\pi kT}}\, e^{\left(E_{\text{inter}}^\dagger - E_{\text{intra}}^\dagger\right)/kT} \tag{10.20}$$

The width of the barrier $1/\phi^\dagger$ was assumed to be the same for both and so it cancelled (if one had a basis for believing that an enzyme altered this parameter, then it could be retained in Eq. (10.20)). This equation can now be used to compare the rates of the uncatalyzed and enzyme catalyzed reactions. We will use it not just for evaluating proximity but for other factors as well.

To focus on proximity, we take $D_{\text{intra}} = D_{\text{inter}}$ and $E^\dagger_{\text{intra}} = E^\dagger_{\text{inter}}$. The ratio in Eq. (10.20) then reduces to $(1/C)\sqrt{(\phi_a/\pi kT)}$. This key quantity is the rate enhancement afforded by forcing the substrate and catalytic side chain into close proximity. It is actually a ratio of two reciprocal lengths. The variable C is molecules per unit length to the left of the barrier, and ϕ_a is a force constant, with units of energy per square length. So dividing by kT (which has units of energy), and taking the square root leaves units of inverse length. This ratio of $k_{\text{intra}}/(Ck_{\text{inter}})$ is in effect the length per molecule when free in solution divided by the length over which the molecule can move in the bound state.

We now estimate these two lengths. For C we use the standard concentration of one molar. The volume per molecule at one molar is 0.166×10^{-23} liter. Since C in the present discussion is one dimensional, we take the edge along a cube with this volume, 1.18×10^{-7} cm.

Now we turn to ϕ_a, the force constant that holds the reactant in its potential energy well around $\xi = a$. A few values were mentioned in Section 2.12. Stretching a C–C bond has a force constant of 660 kcal Å^{-2} mole^{-1}, and for a typical hydrogen bond the force constant is 30 kcal Å^{-2} mole^{-1}. It is unlikely that an enzyme would hold a substrate in position as tightly as a C–C bond so we can use this force constant to get an upper bound, and $\sqrt{(\phi_a/\pi kT)}$ then becomes $19\,\text{Å}^{-1}$. So the corresponding length available to ξ within the energy well is $5.3 \times 10^{-10}\,\text{cm}$. The ratio of this quantity to the value of $1.18 \times 10^{-7}\,\text{cm}$ based on the standard concentration for the intermolecular rate gives an enhancement factor of 223. The hydrogen bond force constant can be used to make a more plausible estimate, because this is typical of the noncovalent interactions that stabilize protein–ligand complexes. This gives a somewhat smaller factor of 47.5.

These factors are for the one-dimensional reaction coordinate drawn in Fig. 10.4. To go to the three-dimensional case we can think of a three-dimensional potential energy well centered at a, and the intermolecular reaction as a three-dimensional volume with a at a corner. So we can simply cube the one-dimensional factor to obtain acceleration factors of 1.1×10^7 for the C–C bond and 10^5 for the hydrogen bond force constants.

It is helpful to view these rate enhancement factors in terms of effective concentrations. The intermolecular rate is for a standard state of $1\,\text{M}$, so we can think of the intramolecular reaction proceeding as though it were intermolecular, but with a concentration that gives the intramolecular rate. Thus, a rate enhancement factor of 10^5 of the intramolecular rate over the intermolecular rate means that one of the reactants has an effective concentration of $10^5\,\text{M}$. It was once thought that the concentration of water represents a physical limit to how high the effective concentration can be. This gave a maximum rate enhancement factor of 55, which seriously underestimates the accelerating power of proximity. The reason is that the reaction coordinate can be confined to a much smaller volume than the average volume per molecule of water.

This same idea emerged earlier when we calculated the translational entropy change of an association process (Section 4.6). In fact, the evaluation of $P(a)$ in the Kramers rate expression (leading to Eq. (10.18)) entails the same calculation made in the evaluation of the translational entropy of a bound molecule. To calculate $P(a)$ for a potential energy function we evaluated an integral that is essentially a partition function; it is the integral of a Boltzmann factor (Eq. (1.4)). This establishes a formal connection between the proximity effect in catalysis and the entropy loss upon formation of the activated complex. The reacting complex has a lower entropy than the two separate species. By pulling the substrate into position, the enzyme overcomes this entropy obstacle and accelerates the reaction.

An experimental approach to the evaluation of the proximity effect is to synthesize an organic molecule that undergoes an intramolecular reaction equivalent to a reaction that occurs between two separate molecules. An early example was the nucleophilic attack on an ester by a carboxyl oxygen atom (Fig. 10.5) (Bruice, 1970). In succinyl ester, the free carboxyl group attacks the esterified carboxyl group; the reaction is intramolecular. The corresponding intermolecular reaction between acetic acid and acetyl ester is slower by a factor of 10^5 at concentrations of $1\,\text{M}$ of each reactant.

Intramolecular

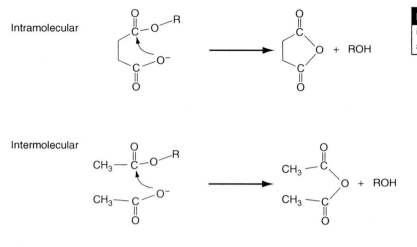

Fig. 10.5. Intramolecular versus intermolecular mechanisms of anhydride formation.

Intermolecular

In the intramolecular case, the carboxyl group is within easy reach of its target. For the comparable intermolecular reaction the acetyl group must encounter the ester by diffusion, and so the reaction is slower. The observed rate in the intramolecular case suggests that the carboxyl group has an effective concentration of 10^5 M. There are many other examples of comparisons between intramolecular and inter-molecular reactions, and for the most part, the rate enhancements approach but rarely exceed 10^5–10^6 (Bruice, 1970). However, the rate enhancement can be even higher when rotation is restricted, and this can also be taken into account.

10.9 | Rotational entropy

The formation of an enzyme–substrate complex restricts rotation as well as translation. Two reactants usually must be correctly oriented for the reaction to occur. The enzyme–substrate complex can bring the reacting groups into the correct orientation as well as the correct position. So one can view rotational restriction as a proximity effect in rotational coordinates. The rotational factor will not be derived here as directly as the proximity factor of the preceding section. Instead, the parallel will be exploited between translational and rotational entropy. This parallel allows us to use the analysis of rotational con-tributions to binding energy (Section 4.7) to assess how restricting rotation contributes to enzyme catalysis.

Consider the various contributions to binding free energy treated in Chapter 4. The formation of a complex involves the loss of three translational degrees of freedom, the loss of three rotational degrees of freedom, and the gain of six vibrational degrees of freedom. The translational and rotational entropy lost during a molecular associa-tion (Eqs. (4.25) and (4.30)) are opposed by a gain in vibrational entropy (Eq. (4.33)). The translational factor of the preceding section can be seen as a part of that balance. The rotational contribution to the binding entropy was similar in magnitude (Eq. (4.30)) so the translational and rotational contributions to enzyme catalysis should also be similar. Restricted rotation should therefore give about the same enhancement as proximity, a factor of 10^5-10^7.

Combining the effects of proximity and restricted rotation together gives a total enhancement of 10^{10}-10^{14}. Thus, proximity combined with restricted rotation can contribute an enormous amount to the rate enhancement by enzymes, and these entropy effects are generally regarded as among the most important factors in rate enhancement by enzymes (Jencks, 1975).

10.10 | Reducing $E^{\dagger}$: transition state complementarity

In spite of the importance of the entropy effects just discussed, one naturally thinks about catalysis in terms of a reduction in the energy of the transition state. This old idea in enzyme catalysis introduced by Pauling in 1946 has guided a great deal of experimental and theoretical work. In general, the rates of reactions catalyzed by enzymes have Q_{10} values of 1.5–2, and the Q_{10}s of the corresponding uncatalyzed reactions are generally quite a bit higher (Miller and Wolfenden, 2002). This demonstrates that enzymes reduce $E^{\dagger}$.

Crystal structures of enzyme–substrate and enzyme–inhibitor complexes are full of suggestions about how an enzyme reduces $E^{\dagger}$. The first enzyme whose structure was determined, lysozyme, showed an aspartate in a position where, if ionized, it could stabilize a positively charged oxycarbenium ($C-O^+=C$) intermediate that forms during the cleavage of the polysaccharide substrate. The bond angles of the cyclic sugar that contains these atoms must also change to accommodate this intermediate, and the enzyme could strain the substrate to bend the bonds.

Proving the ideas for transition state stabilization based on crystal structures is quite difficult. If one wants to calculate the reduction in transition state energy resulting from binding to an enzyme, the force fields discussed in Section 2.14 are inadequate. They do not include terms for changes in covalent bonding. Quantum mechanics must be used for this critical energetic contribution, and quantum mechanical calculations are so demanding on computer power that they must be limited to a small part of the enzyme–substrate complex. Thus, hybrid force fields have been developed in which a small part of the system is selected for more quantitative analysis. The first such effort made a strong case for the role of electrostatic interactions in the reduction of $E^{\dagger}$ by lysozyme, with bond strain playing a relatively minor role (Warshel and Levitt, 1976). More work with such hybrid models shows great promise in dissecting the contributions of various enzyme–substrate interactions in the reduction of $E^{\dagger}$ (Wang et al., 2001).

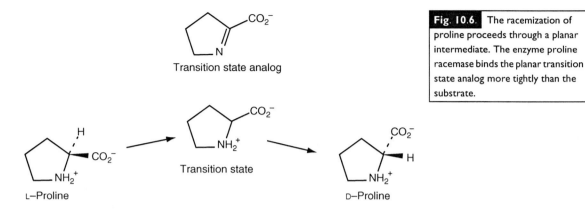

Fig. 10.6. The racemization of proline proceeds through a planar intermediate. The enzyme proline racemase binds the planar transition state analog more tightly than the substrate.

A number of experiments have suggested that enzymes bind more tightly to the transition state. Transition state analogues have been synthesized and found to bind the active site of an enzyme more tightly than substrate molecules. The enzyme proline racemase inverts the symmetry of the α carbon of proline. The reaction involves the removal of the hydrogen atom from the α carbon to form a planar intermediate (Fig. 10.6). The transition state analog shown in the top part of Fig. 10.6 has the same planar geometry as the transition state. This molecule is a very potent inhibitor of the enzyme (Cardinale and Abeles, 1968). Indeed, proline racemase binds this compound 160 times more tightly than proline. It is tempting to say that this means that the enzyme reduces $E^\dagger$ by the corresponding amount ($RT \ln 160 = 3\,\text{kcal mole}^{-1}$), but the double bond in the transition state analog changes its electronic structure and alters its interactions with the binding site. We do not know how to tease apart the effect of geometry from the effect of electronic structure, but it is nevertheless clear that enzymes show preferential binding to structures that mimic the transition state of the reaction they catalyze.

Transition state analogues such as that shown in Fig. 10.6 are among the most potent enzyme inhibitors known. X-ray crystallography of complexes with these compounds has confirmed the complementarity between the active sites of enzymes and the transition states. In general, these compounds bind about 10^2–10^4 times more tightly than substrates (Fersht, 1998). Without knowing how much of this is due to differences in electronic structure and how much to differences in geometry, we can guess that enzyme-substrate complementarity can reduce $E^\dagger$ by $\sim$2.8–5.6 kcal mole^{-1}. This should be viewed as a lower bound, because the analogues only approximate the transition state. A real transition state could bind more tightly to reduce $E^\dagger$ by even more.

The alternative to varying the structure of the ligand is varying the structure of the enzyme by genetic engineering. This approach has been used extensively by Fersht and colleagues in the study of the enzyme tyrosyl-tRNA synthase. This enzyme catalyzes the formation of an ester between the amino acid carboxyl and an OH of the

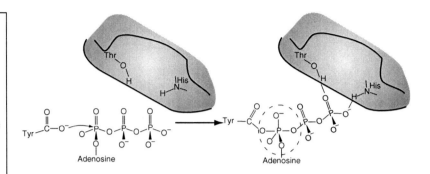

Fig. 10.7. Tyrosyl-tRNA synthase catalyzes nucleophilic attack of the tyrosine carboxyl group on the α phosphate of ATP. Threonine and histidine residues form H-bonds with the γ phosphate, and this distorts the angles of bonds to the α phosphate to assist in the formation of a penta-coordinated P atom in the transition state (dashed ellipse).

terminal ribose of the tRNA. These studies revealed two residues which when mutated reduce k_{cat} but do not significantly alter the binding of the substrates tyrosine and ATP (Wells and Fersht, 1986). These residues, threonine 40 and histidine 45, therefore play a role in the enzyme's tighter binding to the transition state. A model based on the crystal structure has shown that these residues are situated such that they can form H-bonds with the γ phosphate of ATP, provided that the α phosphate–oxygen bond angles are distorted to allow this phosphorous atom to assume a penta-coordinated state. This is what should happen during nucleophilic attack by the tyrosine carboxyl oxygen (Fig. 10.7). Based on the factor by which the rate is changed, it can be estimated that each hydrogen bond between the protein and γ phosphate contributes 3–5 kcal mole^{-1} of stabilization to the transition state, a range consistent with typical H-bond energies (Section 2.10)

Antibodies against transition state analogs provided a particularly dramatic demonstration of transition state complementarity (Schultz and Lerner, 1995). Synthetic analogs of the transition state were used to immunize an animal, and the antibodies that were produced acted like enzymes. For example, the tetrahedral intermediate in an ester hydrolysis reaction is mimicked by a phosphonate ester (Fig. 10.8). Antibodies against this analog hydrolyze esters. In general, the rate enhancement by these catalysts is rather modest, ranging from 10^2 to 10^4. The amount of acceleration is difficult to interpret quantitatively in terms of a specific mechanism. Nevertheless, reducing $E^{\dagger}$ must be the major factor. The catalytic activity of these antibodies provides an independent line of evidence for the importance of transition state complementarity as a factor in catalysis.

Finally, it should be emphasized that binding to the transition state is not the only way to reduce $E^{\dagger}$. The enzyme can also interact with the substrate in its ground state to raise its energy and close the gap with the transition state. An example of such a mechanism was suggested by a combined structural and theoretical study of the enzyme orotidine monophosphate decarboxylase (ODCase) (Wu et al., 2000). ODCase removes a carboxyl group from the nucleotide orotidine monophosphate to produce uridine monophosphate. The

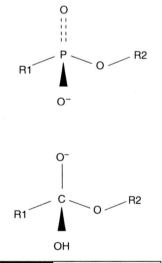

Fig. 10.8. The phosphonate ester (above) resembles the transition state of a carboxylate ester during hydrolysis (below). Antibodies against the phosphonate ester hydrolyze the carboxylate ester by stabilizing the transition state.

carboxyl group targeted for removal has a negative charge. When the enzyme binds, it pushes a negatively charged aspartate, located in its active site, close to the carboxyl group of the substrate. The resulting electrostatic repulsion helps expel the carboxyl group. Of course, this strong electrostatic repulsion strongly opposes binding, but it is overwhelmed by a large number of attractive interactions between the enzyme and other parts of the substrate. Such an unfavorable interaction in an initial enzyme–substrate complex has been referred to as "electrostatic stress" (Fersht, 1998), and this idea has been invoked in many studies of enzymes.

10.11 | Friction in an enzyme–substrate complex

Kramers' theory tells us that if the environment of the enzyme alters the constant for diffusion along the reaction coordinate, then the rate will change. Rapid random motions generate friction (Section 6.7) and such motions within a protein generate friction along the reaction pathway. The solution can contribute to this friction through the random motions of solvent and solute molecules. Now we will consider whether the random motions within a protein are different from molecular motions in solution, and what consequences these differences have for D_{intra} versus D_{inter} (see Eqs. (10.18)–(10.20)).

Internal motions in proteins can be studied using molecular dynamics simulations. We will look at the results of one such study of the small protein pancreatic trypsin inhibitor (McCammon et al., 1979). This work focused on the torsional motions of tyrosine side chains. The flat aromatic ring of tyrosine rotates around an axis formed by the C–C bond connecting the phenol to the peptide backbone. These random motions are strongly damped by the environment, and from the time course of this damping a coefficient of friction was estimated. From there the Einstein formula $D = kT/f$ (Eq. (6.62)) gave a diffusion constant $D = 2.3 \times 10^{11}$ s^{-1}. This value is 3 or 4 times larger than rotational diffusion constants for benzene, suggesting that the environment of a protein can isolate its internal motion from the solution. Thus, there is a modest shielding of solvent fluctuations and a reduction of friction for tyrosine side-chain rotation within the protein. An effect like this on a reaction coordinate would accelerate the reaction in a similar way, but compared to the many orders of magnitude enhancement of rates provided by changing $E^{\dagger}$, and translational and rotational entropy, this effect is not impressive.

Experimentally, the influence of friction on motions within a protein can be studied by varying the solvent viscosity. Because some of the random motions within a protein involve moving parts of the protein through the solvent, the solvent viscosity slows these motions down. The rates of CO and O_2 binding to hemoglobin and myoglobin decrease with increasing viscosity, in accord with Kramers' theory (Beece et al., 1980). But the dependence was weaker than expected, indicating once again that the protein partially shields internal motions from solvent fluctuations. The value of D for these motions will therefore be larger.

In a similar study, the rate of catalysis by the enzyme carboxypeptidase was found to depend on solvent viscosity. These results were interpreted in detail in terms of Kramers' theory (Gavish and Werben, 1979). The variable k_{cat} for hydrolysis of the peptide substrate was reduced as the viscosity increased to a degree that suggests that in this case the protein does not shield the reaction coordinate from solvent movements. If the enzyme were able to screen out solvent motions very well then one would expect a weaker dependence of k_{cat} on viscosity. Thus, although the studies of this question are rather limited, it appears that enhancing rates by shielding out solvent fluctuations and thus reducing the effective friction along the reaction coordinate is not a major factor in enzyme catalysis. Nevertheless, these studies establish the role of solvent friction in the dynamic processes within proteins.

Clear thinking about the relation between fluctuations and friction can help explain what is wrong with some fanciful ideas about protein fluctuations accelerating enzymatic reactions. Enzymes cannot harness fluctuations in the internal motions of proteins because their random nature causes rapid reversal of the direction of any motion within distances of ~ 0.1 Å and times of 1 ps (Section 6.4). The intrinsic randomness of such motions generates friction, and increasing these fluctuations increases friction to make any kinetic process slower. Under some conditions Kramers' theory can generate a rate constant that increases with increasing friction. But this is for a low-density, low-viscosity limit that applies to gases. In the high-density, high-viscosity limit applicable to liquids, fluctuations cannot make anything speed up. Furthermore, an enzyme cannot release stored potential energy to kick a molecule over its transition state because kinetic energy in any one mode is dissipated very rapidly into a vast number of other modes.

10.12 | General-acid–base catalysis and Brønsted slopes

Returning to the question of how an enzyme utilizes binding energy, we examine the relationship between driving force and transition state energy. For this we turn to the linear free energy relations for kinetic processes (Section 7.4). If binding energy can change the driving force for a reaction, then we can expect a proportional change in the transition state energy. Acids and bases serve as catalysts in a wide range of chemical and biochemical reactions, and there is an extensive body of literature on how to relate kinetics to energetics for such reactions.

Amino acid side chains are often reluctant catalysts. For example, the serine oxyanion of a serine protease does not readily form in solution. The situation is illustrated best by considering ester hydrolysis by water. Either OH^- or H_2O can act as a nucleophile and attack a carbonyl to form a tetrahedral transition state. OH^- readily reacts with esters, but at neutral pH the OH^- concentration is very low. H_2O attacks an ester too, but extremely slowly compared to OH^-. So the situation in solution is that the reactive species is not abundant and the abundant species is not reactive. Likewise, in a serine protease the serine hydroxyl predominates and it is a weak nucleophile. The serine oxyanion, which is a strong nucleophile, should be abundant only at very unphysiological pH.

A weakly reactive H_2O molecule will behave more like a strongly reactive OH^- ion in the presence of a general-base catalyst. A general-base catalyst is a buffer that pulls a proton away from a water molecule. Ester hydrolysis is accelerated by increasing the buffer concentration, even as the pH is kept fixed. The B^- form of the buffer activates a water molecule to increase the nucleophilicity of the aqueous oxygen atom (Fig. 10.9). Buffers can also act as general-acid catalysts by activating water to donate H^+. These two forms of rate enhancement are called "general" to distinguish them from the "specific" pH dependent variety involving direct action of H^+ or OH^-.

To measure the effectiveness of general-acid or general-base catalysis, one varies the concentration of the buffer and measures the second order rate constant of the catalyzed reaction. It turns out that different buffers show considerable variation in their effectiveness as general-acid or general-base catalysts. Strong acids or bases work better than weak acids or bases. For example, the strong base pyridine works much better than the weak base acetate as a catalyst of ester hydrolysis. When the logarithm of the second order rate constant (k_2) for each catalyst is plotted versus pK_a for several buffers, the relation is close to linear (Fig. 10.10).

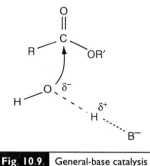

Fig. 10.9. General-base catalysis of nucleophilic attack by water.

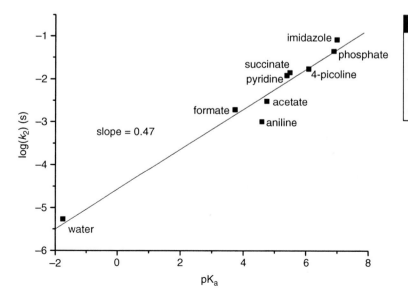

Fig. 10.10. The log of the rate of ester hydrolysis is plotted versus pK_a of the conjugate acid of the buffer used as general-base catalyst. The slope of 0.47 is the parameter β in Eq. (10.21). Data from Jencks and Carriuolo (1961).

This is an example of a linear free energy relation and the equation that describes this relation for acid–base catalysis is known as the Brønsted equation. For general-base catalysis

$$\log k_2 = A + \beta \, \mathrm{pK_a} \tag{10.21}$$

For general-acid catalysis

$$\log k_2 = A - \alpha \, \mathrm{pK_a} \tag{10.22}$$

Different reactions have their own characteristic value of α or β. These parameters tell us how sensitive or susceptible a particular reaction is to one of these forms of catalysis.

The $\mathrm{pK_a}$ is the logarithm of an ionization equilibrium constant, so it is the free energy of ionization divided by RT. The Brønsted equations tell us that some of the ionization energy of the buffer can be used to lower the energy of the transition state. Accordingly, α and β reflect the fractions of ionization energy that can be used in this way.

The fact that buffers with more extreme values of pK can make better catalysts offers a strategy by which enzymes can use side chains with higher or lower pKs in their active sites to enhance a reaction rate. But an additional problem has to be overcome. The strongest acids, which make the best catalysts, are mostly ionized at neutral pH, and in the ionized state they are useless in catalysis. Likewise, the strongest bases are mostly protonated and cannot pull a proton away from water. These factors make it difficult to observe general acid–base catalysis under many conditions. Enzymes overcome this problem by providing an environment in which a very strong acid or base can be prevented from exchanging a proton with the surrounding water. This will be illustrated first with β-galactosidase and then with chymotrypsin.

10.13 | Acid–base catalysis in β-galactosidase

β-Galactosidase splits disaccharides containing galactose. The enzyme attacks the O-linked carbon atom of galactose using a glutamate side chain carboxyl as a nucleophile. As the bond between the enzyme and galactose forms, the bond with the other sugar is broken. The resulting covalent enzyme–galactose complex is unstable and is hydrolyzed by water, releasing free galactose into solution. An interesting side note on this mechanism is that in these two steps, the symmetry of the O-linked carbon atom of the galactose is inverted twice. The net result is no change in the symmetry of the sugar. Other enzymes perform the same chemical task with a one-step mechanism, and these enzymes invert the symmetry of the carbon atom.

In the first step, the nucleophilic attack of the sugar by a glutamate side chain, a second nearby glutamate residue provides assistance, serving as a general-acid catalyst. As the nucleophilic glutamate attacks the galactose carbon, the second, catalytic glutamate provides

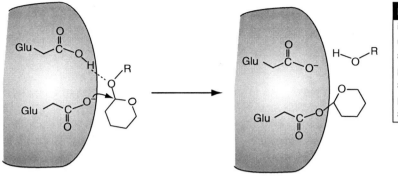

Fig. 10.11. In β-galactosidase, an ionized glutamate carboxyl acts as a nucleophile, attacking the carbon atom on the galactose ring. A protonated glutamic acid (above) accelerates the process by providing a proton for the oxygen atom of the departing sugar.

a proton to the departing oxygen atom (Fig. 10.11). By receiving a proton from the catalytic glutamic acid, the sugar oxygen atom is much more easily displaced by the nucleophilic glutamate carboxyl group (Richard, 1998).

Normally, a glutamate side chain carboxyl, with a pK of around 4, is ionized at neutral pH. The glutamate that attacks the galactose ring is ionized, but the other catalytic glutamate close by is not. Ionization of the second glutamate would place two negative charges right next to each other, and this would have a very high energetic cost. The pK for ionization of the second glutamate is thus raised to a value above 7. As a result, this second glutamate retains its proton at neutral pH, giving us an ideal acid catalyst, a strong acid that is protonated.

The environment of the second glutamate changes as the galactosyl–enzyme intermediate forms. Once the first glutamate has formed a bond with the galactose, it is no longer charged. This removes the obstacle to ionization of the second glutamate, so it loses its proton. This deprotonation during the first step adds a substantial energy to the driving force of the second step. We can use the Brønsted relation (Eq. (10.22)) to estimate the impact of the increased driving force (Richard, 1998). For reactions in appropriate model compounds, α for this reaction was estimated as $\sim$0.9. An analysis of mutant enzymes lacking the catalytic glutamate residue indicated that glutamate protonation contributed 5.3 kcal mole^{-1} to the ionization energy. Thus, according to Eq. (10.22) log k_2 will increase by $0.9 \times 5.3 = 4.8$, and the rate will be enhanced by a factor of 63 000.

For the second step of the reaction, the hydrolysis of the enzyme–galactose complex, the catalytic glutamate's role is reversed. Having ionized, it is now perfectly poised to function as a general-base catalyst. It draws a proton away from a water molecule to enhance its reactivity toward the carbon atom that links the galactose to the enzyme. This accelerates the second step of the process. Thus, we have the same residue serving first as an acid catalyst when it is protonated, and then as a general-base catalyst when it is ionized.

10.14 | Catalysis in serine proteases and strong H-bonds

The serine protease example used at the beginning of this chapter would seem like a perfect illustration of base catalysis (Fig. 10.1). In chymotrypsin, histidine 57 should serve as a base to withdraw the proton from serine 195, thus activating it for nucleophilic attack on the peptide carbonyl. However, the pKs don't work out. Histidine generally has a pK of ~7, which makes it far too weak to activate the serine hydroxyl (its pK is ~14). To resolve this issue it has been proposed that the environment of the histidine side chain (histidine 57 in Fig. 10.1) changes when a substrate binds, raising its pK and improving the ability of histidine to catalyze the reaction.

A key element in the proposed mechanism for the pK shift of histidine 57 is the special character of an H-bond with aspartate 102 (Cassidy *et al.*, 1997). The proton that forms this H-bond has an unusually large upfield chemical shift in NMR. Furthermore, in complexes between the enzyme and certain transition-state analog inhibitors, the distance between the $N^{\delta 1}$ atom of histidine 57 and the carboxyl O atom of aspartate 102 is unusually short (2.52 Å compared to ~2.8–2.9 Å in typical H-bonds; Section 2.10). These properties suggest that this H-bond is extremely strong; the energy was estimated as 11 kcal mole^{-1}. Such strong H-bonds are formed when the H atom is shared more equally by the donor and acceptor. The H atom then assumes a position roughly equidistant between the donor and acceptor, and there is little or no barrier to proton transfer between the two. This is then called a low-barrier H-bond.

Formation of a low-barrier H-bond depends on matching the pKs of the donor and acceptor. A low dielectric constant is also necessary, and the interior of a protein provides this. The pKs of aspartate and histidine become more closely matched provided that both of the ring N atoms of histidine are protonated. If the histidine $N^{\varepsilon 1}$ is ionized, then $N^{\delta 1}$ will have a much higher affinity for its proton than the aspartate carboxyl group (Fig. 10.12). The mismatch of pKs gives

Fig. 10.12. Substrate binding to chymotrypsin brings histidine 57 and aspartate 102 close together, allowing them to form a very strong low-barrier H-bond. This increases the pK of the $N^{\varepsilon 1}$ atom of histidine 57, allowing it to act as a general-base catalyst and pull the hydroxyl proton off serine 195.

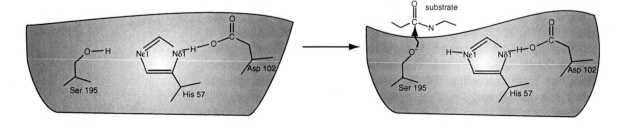

rise to an H-bond with a typical energy of about 3–4 kcal mole^{-1}. Substrate binding has been proposed to induce a conformational change in the protein, shortening the distance between the N and O atoms, and allowing them to form a strong H-bond. This increase in the energy of the H-bond must then be added to the energy of protonation of $N^{\varepsilon 1}$. The increase in bond energy was estimated as ~ 7 kcal mole^{-1}, and the pK of $N^{\varepsilon 1}$ was shown to go up to ~ 12.

This is a substantial increase in the pK of the side chain of histidine 57, and as a result it becomes a more effective catalyst (Fig. 10.12). In fact, the increase in pK varies with different side chains on the substrate, and as the pK increases, acylation gets faster. This was demonstrated by investigating a series of chymotrypsin inhibitors that induce the conformational transition and increase the histidine pK to varying degrees. The log of the rate enhancement for substrates with the same side chains as the inhibitors was plotted versus the pK, giving a slope of 0.68 (Lin et al., 1998). Thus, we have a Brønsted slope, β, which applies to the catalysis of the nucleophilic attack of a peptide bond by a serine hydroxyl within the unique environment of chymotrypsin's active site. Low-barrier H-bonds have been proposed to play a role in catalysis by a number of other enzymes as well as chymotrypsin (Cleland et al., 1998).

10.15 | Marcus' theory and proton transfer in carbonic anhydrase

Carbonic anhydrase catalyzes the reaction shown in Scheme (10E)

$$HCO_3^- \rightleftharpoons CO_2 + OH^- \tag{10E}$$

Unlike most enzymes, both the forward and reverse directions of the reaction catalyzed by carbonic anhydrase are physiologically important. This enzyme assists in the solvation of CO_2 produced during respiration, so in tissues the reaction runs to the left, but in the lungs dissolved HCO_3^- must be converted to CO_2 and released into the air so the reaction runs to the right.

Carbonic anhydrase has a Zn atom to which a water molecule is tightly bound. The bound water rapidly donates a proton to bicarbonate (Scheme (10F))

$$E - Zn - H_2O + HCO_3^- \rightleftharpoons E - Zn - OH^- + CO_2 + H_2O \tag{10F}$$

The rate-limiting step in the overall process is not this one but the change in protonation state of the Zn-bound water through proton exchange with a histidine residue in the active site of the enzyme (Scheme (10G))

$$E - Zn - OH^- + HisH \rightleftharpoons E - Zn - H_2O + His^- \tag{10G}$$

Mutating this histidine to alanine reduces k_{cat} 20–50 fold, and the reason that the reduction is not greater is that buffers in solution can supply the proton formerly provided by the histidine.

Structure–activity studies with mutant forms of carbonic anhydrase have shed light on the reaction coordinate for this proton transfer step (Silverman *et al.*, 1993). Mutants were constructed in which residues near the proton donor histidine were altered. The change in charge and polarity at these locations changed the histidine's pK. Additional data were provided by using proton donors in solution. For each case the rate of proton transfer and the pK of the donor and acceptor were measured. The pK difference between the donor and acceptor provides a measure of the driving force of the reaction. These data thus can be used to explore the dependence of rate on driving force.

Marcus' theory provides a relation between the rate and driving force that describes a wide range of charge-transfer processes (Section 7.6), including proton transfer. Taking Eq. (7.21) and adding a parameter, w, gives an expression that can be applied to Scheme (10G) as follows

$$G^{\dagger} = w + G_0{}^{\dagger}\left(1 + \frac{\Delta G^{\circ}}{4G_0{}^{\dagger}}\right)^2 \tag{10.23}$$

where $G^{\dagger}$ is the activation free energy for a particular driving force, ΔG°, and $G_0{}^{\dagger}$ is the value of the activation energy at zero driving force. We refer to w as the "work term," and it represents contributions to the energy barrier that remain the same as the driving force changes.

Plotting the logarithm of the rate of proton transfer versus the driving force revealed a curvature predicted by the Marcus free energy relation (Fig. 10.13). This plot gave an excellent fit to Eq. (10.23), indicating that Marcus' theory can account for the kinetics of proton transfer in carbonic anhydrase.

The fit of Eq. (10.23) indicates that once the donor and acceptor are in position for proton transfer, the intrinsic barrier for the reaction is quite low ($\Delta G_0{}^{\dagger} = 1.4$ kcal mole^{-1}). This is similar to the intrinsic barrier for non-enzymatic proton transfers between similar groups. However, although carbonic anhydrase is considered a fast enzyme (k_{cat} is over 10^5 s^{-1}), proton transfer reactions in

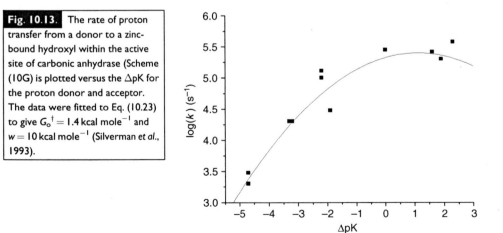

Fig. 10.13. The rate of proton transfer from a donor to a zinc-bound hydroxyl within the active site of carbonic anhydrase (Scheme (10G) is plotted versus the ΔpK for the proton donor and acceptor. The data were fitted to Eq. (10.23) to give $G_0{}^{\dagger} = 1.4$ kcal mole^{-1} and $w = 10$ kcal mole^{-1} (Silverman *et al.*, 1993).

solution can be considerably faster (10^{11} s^{-1}; Section 8.7). The difference is the large value of w (10 kcal mole^{-1}), which represents a contribution to the barrier unrelated to the driving force. This has been interpreted in terms of the need for organization of the water molecules within the catalytic pocket of the enzyme. The histidine proton donor and the zinc-bound hydroxyl acceptor are separated by 9–12 Å. Water molecules in between form a shuttle for the proton. It has been suggested that these waters must align themselves correctly for proton transfer to occur. The probability of attaining the correct alignment would then relate to w, and this probability does not appear to depend on the pK values of the donor and acceptor (Silverman, 2000).

Problems for Chapter 10

1. Work out the time course for a substrate consumed in an enzymatic reaction that follows Michaelis–Menten kinetics (the result will be in the form of t as a function of [s]). Plot and compare the results for an initial concentration of $5 \times K_S$ and $0.2 \times K_S$; take $V_{max} = 1$ for both.

2. Start with Scheme (10D) and incorporate the binding of a second substrate into the second step. Derive the steady-state expression for the rate of product formation as a function of the concentrations of the two substrates.

3. Derive an expression for the pre-steady-state kinetics of p-nitrophenol production by chymotrypsin. Assume that p-nitrophenol is produced in the first step of Scheme (10C) and take this step as irreversible.

4. Consider the consequences of multiplying ϕ_a in Eq. (10.18) by a factor. Compare the rate enhancement with the change in free energy of a harmonic oscillator. (Use the result of Problem 11 from Chapter 1.)

5. One expects that lowering $E^{\dagger}$ of a reaction by $\Delta E^{\dagger}$ should increase the reaction rate by $e^{-\Delta E^{\dagger}/kT}$. If the entire expression for the activation energy ($U(\xi) = E^{\dagger} - \phi^{\dagger}(\xi - \xi^{\dagger})^2$; Section 7.8) were divided by a factor, what does Kramers' theory tell us about the acceleration of the rate?

6. Combine all of the factors for rate enhancement discussed in this chapter into an estimate of the maximal catalytic power of the ultimate enzyme.

Chapter 11

Ions and counterions

The biological milieu is a salty aqueous solution. Ions dissolved in water are pushed and pulled by the electrical forces from other ions and charged macromolecules. Charged molecules tend to be surrounded by ions of the opposite sign. These surrounding ions neutralize electrostatic interactions and screen them out. Dissolved ions dramatically alter the energetic landscape in a solution, and this chapter will discuss these effects and develop the theories that explain them.

To get a feel for the energies involved, consider the Coulomb potential (Eq. (2.1))

$$U = \frac{q_1 q_2}{\varepsilon r} \tag{11.1}$$

For Na^+ and Cl^-, the distance of closest approach (center-to-center) is about 3 Å. Using the dielectric constant of water ($\varepsilon \sim 80$), we obtain a potential energy of -9.3×10^{-14} erg when these two ions are in direct contact. The actual interaction is somewhat stronger than predicted by Eq. (11.1) because the field is so strong around an ion that the solvent becomes ordered, effectively lowering ε. At 298 K, $kT = 4.1 \times 10^{-14}$ erg. So the attraction between Na^+ and Cl^- has a $\sim$ two-fold advantage over thermal fluctuations. Recalling the amount of energy it takes to overcome translational entropy (Chapter 4), we see that this attraction is far too weak to cause a stable association.

The interaction between Na^+ and Cl^- is weak enough to allow them to dissociate and dissolve in water. Although electrostatic interactions in water are often not very strong, they are long ranged. Compared with all other intermolecular potential energy functions, the Coulomb potential decreases slowly with increasing r. The combined properties of weakness and long range create a situation in which the spatial distribution of ions becomes especially important. Though they do not associate, on average Na^+ and Cl^- spend more time in each other's vicinity. And ions with the same charge tend to avoid each other. The distribution reflects a balance between electrostatic forces and thermal agitation. To explore this balance we will develop the Poisson–Boltzmann equation, and solve it in various forms and contexts to understand a wide range of interesting and important ionic effects in biology.

11.1 | The Poisson–Boltzmann equation and the Debye length

The Poisson–Boltzmann equation combines the Poisson equation of electrostatics with the Boltzmann distribution. The Poisson equation expresses the general relation between a charge distribution, $\rho(\mathbf{r})$, and the electrical potential $\varphi(\mathbf{r})$

$$\nabla^2 \varphi(\mathbf{r}) = -\frac{4\pi}{\varepsilon} \rho(\mathbf{r}) \tag{11.2}$$

This equation generalizes the Coulomb potential to an arbitrary distribution of charge, and reduces to the Coulomb potential when $\rho(\mathbf{r})$ is a single point of charge in a uniform dielectric medium. Because $\rho(\mathbf{r})$ is closely related to the distribution of ions that we would like to derive, the Poisson equation is ideal for the present purpose. We take $\mathbf{r}$ as the position with respect to some charged object of interest, which could be an ion, a charged protein, a membrane, or biopolymer. For each of these cases, the Poisson equation will be used, but the geometry will determine the coordinate system, and accordingly, the form for $\nabla^2 \varphi(\mathbf{r})$.

Focusing on the charge density, we can express it in terms of the concentrations of the various ions. For a simple binary salt with concentrations $c_+(\mathbf{r})$ for cations and $c_-(\mathbf{r})$ for anions, $\rho(\mathbf{r}) = e\tilde{A}10^{-3}(c_+(\mathbf{r}) - c_-(\mathbf{r}))$. Since ρ is in units of esu cm^{-3}, and c is in units of mole liter^{-1}, we must multiply by e, the unitary electronic charge, $\tilde{A}$, Avogadro's number, and 10^{-3} liter cm^{-3}. For an arbitrary collection of ions with valences z_i and concentrations c_i, this generalizes to

$$\rho(\mathbf{r}) = e\tilde{A}10^{-3} \sum_i z_i c_i(\mathbf{r}) \tag{11.3}$$

The concentration at $\mathbf{r}$ can be related to the potential energy with the aid of the Boltzmann distribution

$$c_i(\mathbf{r}) = c_i(\infty) e^{-ez_i\varphi(\mathbf{r})/kT} \tag{11.4}$$

Here $c_i(\infty)$ is the bulk concentration at a far off reference position where φ is taken as 0. Substituting this into Eq. (11.3) leads to

$$\rho(\mathbf{r}) = e\tilde{A}10^{-3} \sum_i z_i c_i(\infty) e^{-ez_i\varphi(\mathbf{r})/kT} \tag{11.5}$$

Using this expression for $\rho(\mathbf{r})$ in Eq. (11.2) leads to the Poisson–Boltzmann equation

$$\nabla^2 \varphi(\mathbf{r}) = -\frac{4\pi e \tilde{A} 10^{-3}}{\varepsilon} \sum_i z_i c_i(\infty) e^{-ez_i\varphi(\mathbf{r})/kT} \tag{11.6}$$

Thus, we have eliminated $\rho(\mathbf{r})$ to obtain a differential equation for $\varphi(\mathbf{r})$. The solution to Eq. (11.6) provides an expression for the electrical potential that takes into account the screening by ions. Once known, $\varphi(\mathbf{r})$ can then be used to calculate the charge density with Eq. (11.5). Depending on the geometry, solving this equation requires different strategies and approximations, as we will soon see.

Although solving the Poisson–Boltzmann equation is a major goal of this chapter, we can obtain one particularly useful insight without going that far. Since the interaction is often weak the quantity in the exponential, $ez_i\varphi(\mathbf{r})/kT$, is often $\ll 1$. Thus, we can approximate the exponential in Eq. (11.6) as $1 - ez_i\varphi(\mathbf{r})/kT$, with the result

$$\nabla^2 \varphi(\mathbf{r}) = -\frac{4\pi e \tilde{A} 10^{-3}}{\varepsilon} \sum_i z_i c_i(\infty) \left(1 - \frac{ez_i \varphi(\mathbf{r})}{kT}\right) \tag{11.7}$$

At the reference point where concentrations assume their bulk values, the solution is electrically neutral, so $\Sigma_i ez_ic_i(\infty) = 0$. This simplifies Eq. (11.7) to

$$\nabla^2 \varphi(\mathbf{r}) = \frac{4\pi e^2 \tilde{A} 10^{-3}}{\varepsilon kT} \sum_i z_i^2 c_i(\infty) \varphi(\mathbf{r}) \tag{11.8}$$

This is referred to as the linearized Poisson–Boltzmann equation. It will soon be solved, but first we look at the factor multiplying $\varphi(\mathbf{r})$ on the right-hand side. It defines a key parameter as follows

$$\frac{1}{\lambda_D^2} = \frac{8\pi e^2 \tilde{A} 10^{-3} I}{\varepsilon kT} \tag{11.9}$$

The symbol I denotes the ionic strength, $I = \frac{1}{2}\Sigma_i z_i^2 c_i(\infty)$. Now Eq. (11.8) is simply

$$\nabla^2 \varphi(\mathbf{r}) = \frac{1}{\lambda_D^2} \varphi(\mathbf{r}) \tag{11.10}$$

The variable λ_D has the dimension of length and is known as the Debye length. If Eq. (11.10) were transformed to a variable with units of λ_D, then there would be no parameters in the equation. That means that λ_D is the fundamental unit of length for an ionic solution.

For 1 M NaCl ($I = 1$) and $T = 298$ K, $\lambda_D = 3.05$ Å. Note that $\lambda_D \propto I^{-1/2}$, so reducing I makes λ_D longer. For example, with 10 mM NaCl, $\lambda_D = 30.5$ Å. At distances shorter than λ_D electrostatic interactions will be strong, and at longer distances they will be effectively screened out by ions. Therefore λ_D is essentially the ionic screening distance, and so it makes sense that it decreases with increasing ionic strength. Note that I depends on the valence as z^2. Thus, I increases more rapidly when multivalent ions are added to a solution. Multivalent ions are more effective than monovalent ions at screening electrostatic interactions.

11.2 | Activity coefficient of an ion

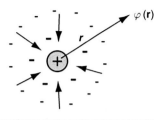

Ionic solutions are not ideal. For example, a 100 mM NaCl solution has an osmolarity of 187 mOsm rather than 200 mOsm. A major factor in this nonideal behavior is the attraction between the oppositely charged ions. The Poisson–Boltzmann equation, as employed in the classical work of Debye and Hückel, provides a means for understanding this nonideality. We will illustrate this by deriving an expression for the ionic activity coefficient. We first solve the linearized Poisson–Boltzmann equation to obtain the potential energy function, $\varphi(\mathbf{r})$, and then use this result to calculate the energy of placing an ion in a solution with other ions.

Fig. 11.1. A positive ion is surrounded by a negative "atmosphere," with charge density, ρ. This atmosphere screens the field of the central positive charge and reduces $\varphi(\mathbf{r})$. Note that the minus signs depict a distribution rather than individual ions.

The relevant geometry is illustrated in Fig. 11.1. A cation will tend to attract anions so that their average density close in will be higher than the density further out. This surrounding charge is viewed as the ionic "atmosphere" of the central ion.

The distance to the center of the ion, r, is the only relevant coordinate. That means that ∇^2 should be taken in spherical coordinates (Appendix 6). The Poisson–Boltzmann equation cannot be solved analytically in spherical coordinates, so we use the linearized form (Eq. (11.10))

$$\frac{1}{r}\frac{d^2(r\varphi(r))}{dr^2} = \frac{1}{\lambda_D{}^2}\varphi(r) \tag{11.11}$$

Multiplying through by r gives a differential equation in the function $r\varphi(r)$, which is easily integrated

$$r\varphi(r) = Ae^{-r/\lambda_D} + Be^{r/\lambda_D} \tag{11.12}$$

In Eq. (11.12), A and B are constants of integration that must be determined by boundary conditions. The coefficient B is clearly zero to allow $\varphi \rightarrow 0$ as $r \rightarrow \infty$. The constant A is determined by considering the electric field at the ion's surface, which must be a continuous function of r. At the surface there is no ionic screening, so the field can be computed from the Coulomb potential for a charge q. With an ionic radius of a, the field at the surface is

$$E = \frac{q}{\varepsilon a^2} \tag{11.13}$$

We can use this boundary condition to determine the remaining coefficient A in Eq. (11.12) by expressing the potential

$$\varphi(r) = \frac{Ae^{-r/\lambda_D}}{r} \tag{11.14}$$

and differentiating it with respect to r to obtain the electric field

$$E(r) = \frac{Ae^{-r/\lambda_D}}{r}\left(\frac{1}{\lambda_D} + \frac{1}{r}\right) \tag{11.15}$$

We now set $r = a$, equate Eq. (11.13) with Eq. (11.15), and solve for A as follows

$$A = \frac{q}{\varepsilon} \left(\frac{e^{a/\lambda_D}}{1 + a/\lambda_D} \right) \tag{11.16}$$

Substituting this expression for A into Eq. (11.14) yields the solution

$$\varphi(r) = \frac{q}{\varepsilon r} \left(\frac{e^{-(r-a)/\lambda_D}}{1 + a/\lambda_D} \right) \tag{11.17}$$

This is the potential energy around an ion. It can be broken up into two parts, a Coulombic part and an ionic screening part. The Coulombic part is $q/\varepsilon r$, so we rewrite Eq. (11.17) as the following sum

$$\varphi(r) = \frac{q}{\varepsilon r} + \frac{q}{\varepsilon r} \left(\frac{e^{-(r-a)/\lambda_D}}{1 + a/\lambda_D} - 1 \right) \tag{11.18}$$

The second term is the ionic screening contribution.

The stated goal of this section was to evaluate how ionic screening contributes to the nonideality of an ionic solution. The electrical potential in Eq. (11.18) enables us to do this, since it can be used to calculate the work, and hence free energy, of placing an ion in the solution. The work of placing the charge, q, at $r = 0$ is envisioned as an incremental charging process. The idea is the same as that used to calculate the self-energy of an ion in a dielectric medium (Section 2.2), and in fact, the charging of the first term in Eq. (11.18), the unscreened Coulomb potential of the ion, gives the familiar self-energy (Eq. (2.4)). This quantity is independent of other ions so it is irrelevant. The second term does depend on ion concentration (because λ_D does), so the integration of this term gives us the work done on the other ions in the solution by charging the central ion at $r = 0$

$$W = \frac{1}{\varepsilon a} \left(\frac{1}{1 + a/\lambda_D} - 1 \right) \int_0^q q' \, dq' = -\frac{q^2}{2\varepsilon} \left(\frac{1}{\lambda_D + a} \right) \tag{11.19}$$

This term is added to the expression for the free energy of an ideal solution to correct for ionic screening

$$G = G^\circ + RT \ln c + W \tag{11.20}$$

The activity coefficient, γ, for a single ion is defined as $kT \ln \gamma = W$. Thus, we have

$$\ln \gamma = -\left(\frac{q^2}{2\varepsilon kT} \right) \frac{1}{\lambda_D + a} \tag{11.21}$$

The value of λ_D is considerably larger than that of the radius of the ion at low salt concentrations, so we can drop a to obtain what is

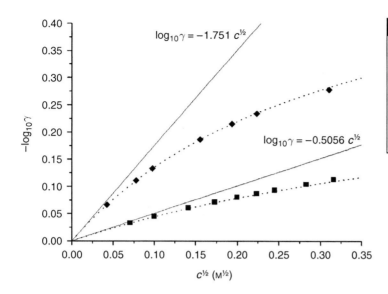

Fig. 11.2. The log of the activity coefficient versus $\sqrt{c}$ for NaCl (squares) and CaCl$_2$ (diamonds). The lines are the limiting expressions (Eq. (11.22)), and the dotted curves are Eq. (11.21) with $a = 4.45$ for NaCl and 5.24 for CaCl$_2$ (data plotted from MacInnes, 1961).

commonly referred to as the Debye–Hückel limiting law for the activity coefficient of an ionic solution.

$$\ln \gamma = -\frac{q^2}{2\varepsilon kT\lambda_D} \tag{11.22}$$

At 298 K this works out to $\log_{10}\gamma = -0.5056\sqrt{c}$ for a salt of monovalents (such as NaCl), and to $-1.751\sqrt{c}$ for a salt of a divalent and a monovalent (such as CaCl$_2$). Note that the activity coefficients go to zero as c approaches zero, as expected since all solutions are ideal at infinite dilution.

Figure 11.2 compares the theory with experiment, showing that the limiting law works as long as the concentration is low (up to $\sim$10 mM). Note that the limiting law gives an activity coefficient that depends only on an ion's charge, and not on any other specific properties. Thus, all ions of the same charge are the same in the limiting case. This is reasonable, since the Coulombic potential depends only on charge.

Equation (11.21) improves on the limiting law, providing agreement with experiments up to $\sim$100 mM. In contrast to the limiting law, this expression has an ion specific quantity, a, the ionic radius. The influence of this part is not electrostatic but rather reflects the inability of two ions to occupy the same space. This excluded volume effect would be present even for neutral molecules. The best value of a obtained by fitting Eq. (11.21) (legend of Fig. 11.2) tends to be larger than the sum of the crystal radii of the anion and cation. This reflects the hydration shell that prevents the ions from getting too close.

The Debye–Hückel theory provides a good approximate picture of the ionic atmosphere that screens the Coulomb potential of an ion. Using Eq. (11.17) to substitute for $\varphi(r)$ in Eq. (11.2) (with ∇^2 as in Eq. (11.11)) gives

$$\rho(r) = \frac{q}{4\pi r\lambda_D{}^2}\left(\frac{e^{-(r-a)/\lambda_D}}{1 + a/\lambda_D}\right) \tag{11.23}$$

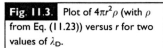

Fig. 11.3. Plot of $4\pi r^2 \rho$ (with ρ from Eq. (11.23)) versus r for two values of λ_D.

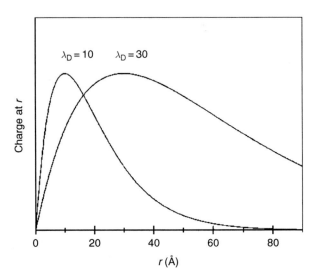

The amount of charge in a shell of thickness dr at r is $4\pi r^2 \rho(r)dr$. A graph of $4\pi r^2 \rho(r)$ is plotted in Fig. 11.3.

We see that the maximum counterion charge density occurs at $r = \lambda_D$ (Problem 3). For longer λ_D (corresponding to lower ionic strength) the ionic atmosphere is more spread out. This illustrates the importance of λ_D as a basic unit of length in the evaluation of ionic interactions in solution.

The success of Debye–Hückel theory indicates that it captures the essence of ionic interactions in dilute solutions. However, the theory fails at higher concentrations owing to the breakdown of two major approximations. The first of these is the linearization of the Poisson–Boltzmann equation, used to obtain Eq. (11.7). This approximation depends on the exponent of Eq. (11.6), $ez\varphi/kT$, being small. Recall at the beginning of this chapter that the Coulomb interaction energy was estimated as 9.3×10^{-14} erg $\sim 2kT$ for $r = 3$ Å. However, for $r = \lambda_D$ (30 Å for $I = 10$ mM), $e\varphi/kT \sim 1/5$. Thus, the region with the highest ion density (Fig. 11.3) gives a potential that is low enough for the linearization approximation to work. The concentration at which the theory starts to deviate from experiment in Fig. 11.2 corresponds with the point at which $e\varphi(\lambda_D)/kT$ becomes significant compared to one.

The other important approximation used in Debye–Hückel theory is a bit more subtle. This involves using a potential based on the average charge density, $\rho(r)$. The true mean potential at a particular distance from a selected ion requires taking an average over the positions of all the other ions. This includes configurations in which there are many ions in the same region. The mean charge density does not give the correct weight to these multi-ion configurations. Multi-ion effects are more important at high concentrations, and that is when calculating the potential from the mean charge density becomes a poor approximation. The linearization approximation discussed in the preceding paragraph can be overcome by using a computer to obtain a numerical solution. However, the problems arising from the use of the mean charge density are more difficult. They require more sophisticated theoretical techniques or Monte Carlo computations to treat the complex behavior of a system with many molecules.

Rigorous analysis shows that the Debye–Hückel theory becomes exact in the limit of infinite dilution. Excellent intermediate level discussions of ionic solutions, including the approximations of Debye–Hückel theory and some of the methods for overcoming these problems, can be found in Chapter 18 of Hill (1960) and Chapter 15 of McQuarrie (1976).

An additional problem with the Poisson–Boltzmann equation is especially relevant to certain biophysical systems. If one considers the distribution of counterions in water near a low dielectric region, the ions will be repelled by the image force (Section 2.3). It is difficult to incorporate this effect into the Poisson–Boltzmann equation. This situation was studied by performing computer simulations of the Brownian motion of ions in small spherical and cylindrical water-filled cavities within a low dielectric medium (Moy *et al.*, 2000). As the cavity shrinks down to the Debye length of the solution, there is less and less ionic screening. In fact, under typical physiological conditions where $\lambda_D \sim 10$ Å, ionic screening is essentially eliminated inside a cylinder the size of large ion channels, ~ 10 Å. Ions feel the full image force of the low dielectric medium and ionic screening is irrelevant. Only when the cylinder was more than 2 Debye lengths wide did the effect of ionic screening for an ion in the center of the cylinder approach that of a bulk solution.

11.3 | Ionization of proteins

Ionizable amino acids such as glutamate, lysine, and arginine speckle the surface of a protein. When one of their side chains ionizes, the total charge of the protein changes, and electrostatic interactions alter the ionization energies of the other amino acid side chains. Ionizing one side chain makes it harder to ionize others, and this spreads out a protein's titration curve over a wider pH range than if the same groups are independent in solution. This effect is reduced by the addition of salt, because salt screens the electrostatic interaction between the charges on the protein. Debye–Hückel theory provides a satisfactory explanation for this effect.

The basic model, due to Linderstrøm-Lang (1924), assumes that the charge on a protein is smeared out uniformly over the surface of a sphere. The ionization energy is then taken as the electrostatic work done in charging the sphere. If there were no salt then we could use the self-energy (Eq. (2.4)). In an ionic solution, we use the integral computed in Eq. (11.19), which gives the energy for charging a sphere in the presence of salt. Following Scatchard (1949), we write the free energy change for ionizing one particular group on a protein, with a change in valence from z to $z + 1$, as follows

$$\Delta G^{\circ} = -RT \ln K_0 + w(z+1)^2 - wz^2 \tag{11.24}$$

where K_0 is the intrinsic equilibrium constant for the group in the absence of electrostatic interactions and $w(z+1)^2 - wz^2$ is the change in electrostatic energy. It depends on the square of the charge, and this is a hallmark of electrostatic self-energies. Here z^2 corresponds to q^2 in Eq. (11.19), so w in Eq. (11.24) is

$$w = \frac{1}{2\varepsilon} \frac{1}{\lambda_D + a} \tag{11.25}$$

To relate these equations to a protein titration experiment we consider an ionization equilibrium in which the charge increases by one. We divide the charges into two groups, p permanent charges (or at least charges completely ionized in the pH range of interest) and n ionizable charges. According to Eq. (11.24), when the number of ionizable charges increases from i to $i+1$ the free energy change is

$$\begin{aligned} \Delta G^{\circ} &= -\frac{RT(n-i)}{i} \ln K_0 + w(p+i+1)^2 - w(p+i)^2 \\ &= -\frac{RT(n-i)}{i} \ln K_0 + w(2p+2i+1) \\ &= -\frac{RT(n-i)}{i} \ln \left(K_0 e^{-\frac{w(2p+2i+1)}{RT} \frac{i}{n-i}} \right) \end{aligned} \tag{11.26}$$

Here $(n-i)/i$ is a statistical factor that reflects the number of sites at which the forward and reverse ionization processes can occur (the same factor appears in the analysis of the MWC model of Section 5.7). It was absent from Eq. (11.24) because this equation is for just one particular ionizing group on the protein.

The equilibrium constant of ionization is then the term in parentheses in Eq. (11.26)

$$\frac{[H^+][A_i]}{[A_{i+1}]} = K_0 e^{-\frac{w(2p+2i+1)}{RT} \frac{i}{n-i}} \tag{11.27}$$

where A_i denotes the protein with ionization state i. Taking the logarithm (base 10) and using $\log(e) = 0.434$ gives

$$\log[H^+] + \log\frac{[A_{i+1}]}{[A_i]} = \log K_0 - \frac{0.434 wi}{RT(n-i)}(2p+2i+1) \tag{11.28}$$

$$\text{or} \quad pH - \log\frac{[A_{i+1}]}{[A_i]} = pK + \frac{0.434 wi}{RT(n-i)}(2p+2i+1) \tag{11.29}$$

To know the ratio $[A_{i+1}]/[A_i]$ we would have to make very difficult measurements of the number of proteins with i and $i+1$ titrated groups. To avoid this problem we focus on the center of the titration curve, where about half the groups are ionized and $i \sim n - i$.

The statistical factor, $i/(n - i)$, is then about one, so the same equilibrium constant applies to any group. The ionization of the species, A_i going to A_{i+1} is then representative of $A \leftrightarrow A^-$ interconversions

$$\frac{[A^-][H^+]}{[AH]} \sim \frac{[A_i][H^+]}{[A_{i+1}]} \tag{11.30}$$

We now define the extent of the ionization process as $\alpha = [A^-]/([AH] + [A^-])$. This allows us to express i as αn and $[A^-]/[AH] = [A_{i+1}]/[A_i]$ as $\alpha/(1 - \alpha)$. Equation (11.29) thus becomes

$$\text{pH} - \log\frac{\alpha}{1 - \alpha} = \text{pK} + 0.434(2p + 2\alpha n + 1)w/RT \tag{11.31}$$

This equation has been widely used in the analysis of protein titration experiments. One varies pH and measures α. A plot of $\text{pH} - \log(\alpha/(1 - \alpha))$ versus α will be linear, and the slope can be used to calculate w. Values of w measured in this way agree reasonably well with estimates based on Eq. (11.25) (Tanford, 1955; Breslow and Gurd, 1962; Stigter and Dill, 1990). Note that in the absence of electrostatic interactions w is zero, and Eq. (11.31) simplifies to the Henderson–Hasselbach equation, as expected for independent and equivalent ionization events.

11.4 | Gouy–Chapman theory and membrane surface charge

Biological membranes usually have some charge at their surface. This charge arises from the polar head groups of phospholipids, charged amino acids on membrane proteins, and charged sugars on membrane glycoproteins. Membrane surface charge will influence the distribution of dissolved ions, and this is described by the Poisson–Boltzmann equation in a theory developed by Gouy and Chapman. This theory was actually developed around 10 years earlier than the Debye–Hückel theory, and the two theories show a striking parallel (for references to original articles see Overbeek, 1952, and Latorre *et al.*, 1992).

In applying the Poisson–Boltzmann equation to membranes, we note that the only important spatial dimension is the distance from the surface, denoted here as x. In one dimension the Laplacian operator in Eq. (11.2) is simply the second derivative, so the Poisson equation becomes

$$\frac{d^2\varphi(x)}{dx^2} = -\frac{4\pi}{\varepsilon}\rho(x) \tag{11.32}$$

and the Poisson–Boltzmann equation is

$$\frac{d^2\varphi(x)}{dx^2} = -\frac{4\pi\tilde{A}e10^{-3}}{\varepsilon}\sum_i z_i c_i(\infty)e^{-ez_i\varphi(x)/kT} \tag{11.33}$$

At this point in the above development of Debye–Hückel theory the linearization approximation was used. This was necessary because the mathematics is too difficult when the Laplacian is taken in spherical coordinates. However, in one dimension the Poisson–Boltzmann equation can be solved exactly, without resorting to linearization.

First, Eq. (11.33) is multiplied by the integration factor, $2\mathrm{d}\varphi(x)/\mathrm{d}x$ to give

$$2\frac{\mathrm{d}\varphi(x)}{\mathrm{d}x}\frac{\mathrm{d}^2\varphi(x)}{\mathrm{d}x^2} = -2\frac{\mathrm{d}\varphi(x)}{\mathrm{d}x}\frac{4\pi e\tilde{A}10^{-3}}{\varepsilon}\sum_i z_i c_i(\infty)e^{-ez_i\varphi(x)/kT} \qquad (11.34)$$

Both sides can now be integrated from x to ∞ to give

$$\left(\frac{\mathrm{d}\varphi(x)}{\mathrm{d}x}\right)^2 - \left(\frac{\mathrm{d}\varphi(\infty)}{\mathrm{d}x}\right)^2 = \frac{8\pi kT\tilde{A}10^{-3}}{\varepsilon}\sum_i c_i(\infty)\left(e^{-ez_i\varphi(x)/kT} - e^{-ez_i\varphi(\infty)/kT}\right)$$

$$(11.35)$$

Since φ and its derivative are zero at $x=\infty$,

$$\left(\frac{\mathrm{d}\varphi(x)}{\mathrm{d}x}\right)^2 = \frac{8\pi kT\tilde{A}10^{-3}}{\varepsilon}\sum_i c_i(\infty)\left(e^{-ez_i\varphi(x)/kT} - 1\right) \qquad (11.36)$$

For a simple binary salt such as NaCl, $z=\pm1$, and we can write out the sum

$$\left(\frac{\mathrm{d}\varphi(x)}{\mathrm{d}x}\right)^2 = \frac{8\pi kT\tilde{A}10^{-3}\,c}{\varepsilon}(e^{-e\varphi(x)/kT} + e^{e\varphi(x)/kT} - 2) \qquad (11.37)$$

Taking the square root of both sides gives

$$\frac{\mathrm{d}\varphi(x)}{\mathrm{d}x} = \sqrt{\frac{8\pi kT\tilde{A}10^{-3}\,c}{\varepsilon}}(e^{-e\varphi(x)/2kT} - e^{e\varphi(x)/2kT})$$

$$= 2\sqrt{\frac{8\pi kT\tilde{A}10^{-3}\,c}{\varepsilon}}\sinh\left(\frac{-e\varphi(x)}{2kT}\right) \qquad (11.38)$$

where we used $(e^{-a/2} - e^{a/2})^2 = e^{-a} + e^{a} - 2$ in the first step.

The obvious next step is to integrate Eq. (11.38) to obtain $\varphi(x)$. However, it is worth pausing here and using this result for another important insight. We define a surface charge density, σ, and note that it must be balanced by an equivalent amount of ionic charge of opposite sign in the solution. This charge is the integral over $\rho(x)$, so the balance is expressed as

$$\sigma = -\int_0^\infty \rho(x)\mathrm{d}x \qquad (11.39)$$

Using Eq. (11.32) to express $\rho(x)$ in terms of φ gives

$$\sigma = \frac{\varepsilon}{4\pi}\int_0^\infty \frac{\mathrm{d}^2\varphi}{\mathrm{d}x^2}\mathrm{d}x = -\frac{\varepsilon}{4\pi}\frac{\mathrm{d}\varphi(0)}{\mathrm{d}x} \qquad (11.40)$$

where once again we have exploited the fact that the derivative of φ goes to zero at $x = \infty$.

Equation (11.40) relates the surface charge to the electric field (the gradient of the potential) at the membrane surface. Now, we use Eq. (11.38) to evaluate $d\varphi(0)/dx$ in Eq. (11.40) and obtain an important result known as the Gouy–Chapman equation

$$\sigma = -\sqrt{\frac{\varepsilon k T \tilde{A} 10^{-3} c}{2\pi}} \left(e^{-e\varphi_0/2kT} - e^{e\varphi_0/2kT}\right)$$

$$= \sqrt{\frac{2\varepsilon k T \tilde{A} 10^{-3} c}{\pi}} \sinh\left(\frac{e\varphi_0}{2kT}\right) \tag{11.41}$$

where φ_0, which replaces $\varphi(0)$, is referred to as the *surface potential*. The constant in front of the exponential factors works out to $\sqrt{c}/136$, giving σ in units of electronic charges per Å^2 (Latorre et al., 1992).

If the exponential in Eq. (11.41) is linearized, and the definition of the Debye length (Eq. (11.9)) is used, the equation reduces to a very simple form

$$\sigma = \frac{\varepsilon}{4\pi\lambda_\mathrm{D}} \varphi_0 \tag{11.42}$$

Thus, surface potential and surface charge are directly proportional to one another. This proportionality is analogous to the relation between voltage and charge on an electrical capacitor. Recall that capacitance, C is q/V. Likewise, Eq. (11.42) gives $\varepsilon/4\pi\lambda_\mathrm{D} = \sigma/\varphi_0$, so $\varepsilon/4\pi\lambda_\mathrm{D}$ is an effective capacitance per unit area. Its dependence on the dielectric constant, ε, is the same as that of a capacitor, and λ_D can be taken as the distance between the capacitor's two plates. The picture of two opposing layers of charge has lead to the term *ionic double layer* for this kind of charge distribution at a surface.

Now, we return to Eq. (11.38). Integration followed by some rearrangement leads to

$$\frac{x}{\lambda_\mathrm{D}} = \ln\left(\frac{(e^{e\varphi(x)/2kT} + 1)(e^{e\varphi_0/2kT} - 1)}{(e^{e\varphi(x)/2kT} - 1)(e^{e\varphi_0/2kT} + 1)}\right) \tag{11.43}$$

When $e\varphi/2kT \ll 1$, linearization of the exponentials again simplifies things quite a bit.

$$\frac{x}{\lambda_\mathrm{D}} = \ln\left(\frac{\varphi_0}{\varphi}\right) \tag{11.44}$$

or $\quad \varphi = \varphi_0 e^{-x/\lambda_\mathrm{D}} \tag{11.45}$

This can be compared to Eq. (11.17). Both potential functions decrease exponentially in units of λ_D. The factor of $1/r$ in Eq. (11.17) reflects the spherical geometry of the ion's environment.

The case of low ionic strength is of interest. As the salt is diluted, some ions must remain near the membrane to maintain electroneutrality. As a result, the concentration right at the surface goes to a nonzero limit as the bulk concentration is reduced. We can see how this comes about by noting that when φ_0 is large, only one of the exponential terms in Eq. (11.41) needs to be considered. For a negatively charged surface

$$\sigma = -\sqrt{\frac{\varepsilon k T \tilde{A} 10^{-3} c}{2\pi}} (e^{-e\varphi_0/2kT}) \qquad (11.46)$$

This is then squared to give

$$\sigma^2 = \frac{\varepsilon k T \tilde{A} 10^{-3}}{2\pi} c e^{-e\varphi_0/kT} \qquad (11.47)$$

Now, we note that the concentration of ions is given by the Boltzmann distribution

$$c(x) = c(\infty)e^{-e\varphi(x)/kT} \qquad (11.48)$$

So $c e^{-e\varphi_0/kT}$ in Eq. (11.47) is actually c_0, the concentration at the membrane surface. Making this substitution gives

$$c_0 = \frac{2\pi\sigma^2}{\varepsilon k T \tilde{A} 10^{-3}} \qquad (11.49)$$

Thus, c_0 is independent of $c(\infty)$, the bulk concentration, in this low c limit. This result has interesting consequences for permeation in ion channels, as we will soon see.

The Gouy–Chapman theory makes some very important predictions that are readily tested. This will be illustrated after a discussion of the limitations of Gouy–Chapman theory, and how these limitations are overcome by the modifications of Stern.

11.5 | Stern's improvements of Gouy–Chapman theory

Gouy–Chapman theory breaks down with high charge densities and strong fields. A similar problem was already discussed for Debye–Hückel theory (Section 11.2). With a surface potential of the order of 100 mV (a bit high but not impossible), the exponential factor in Eq. (11.48) becomes $e^4 = 55$. If $c(\infty) = 0.1$ M, then the concentration of the attracted ion works out to be 5.5 M right at the surface, a value above saturation for many salts. The theory has a greater tendency to fail with multivalent ions, which interact more strongly with a charged surface, and can even bind with a high degree of specificity. Two modifications introduced by Stern lead to a theory that addresses these shortcomings (Overbeek, 1952; Aveyard and Haydon, 1973).

The first modification takes into account the finite size of an ion. An ion's center cannot actually be at a distance $x = 0$ from the surface. The closest it can get is one ionic radius away. So there is a zone right at the surface that excludes ions. This is incorporated into the theory by adding a displacement to x equal to the radius of the ion. This may seem like a small correction, but the energy can be very high at the surface of a charged membrane so the correction can make a big difference.

The second modification is to incorporate specific binding to sites on the membrane surface. To be in contact with the surface, they must be within one ionic radius, giving them a position $x = a$. The ions interacting with the surface can thus be divided into two groups. The first group, the bound ions, forms what is called the Stern layer. The second group is outside the Stern layer but within about a Debye length of the surface. These make up what is called the diffuse layer (Fig. 11.4). They are not bound like those in the Stern layer, but they are still close enough to feel the pull of the surface charge.

To maintain charge balance, the densities of these two layers must add up to the density of fixed charge in the membrane

$$\sigma = \sigma_{st} + \sigma_d \qquad (11.50)$$

The subscripts st and d denote the Stern and diffuse layers, respectively.

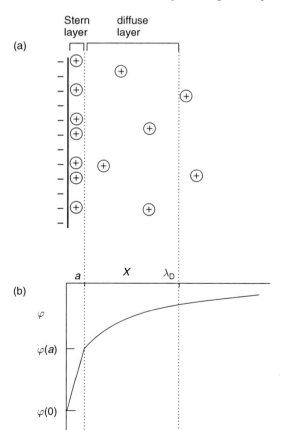

Fig. 11.4. (a) Cations in a solution adjacent to a negatively charged surface. Some are bound and form the Stern layer. Others scattered within a Debye length form the diffuse layer. The Stern layer extends to $x = a$, and the diffuse layer runs from $x = a$ to $x \sim \lambda_D$. (b) The potential has different shapes in these two regions. Within the Stern layer the potential drops linearly, and within the diffuse layer the potential is described by Eq. (11.43), but with $\varphi(a)$ in place of φ_0.

Ions in the Stern layer are described by a binding equilibrium (Chapter 4). For a cation, A^+, binding to a negatively charged site, S^-, we have (Scheme (11A))

$$A^+ + S^- \rightleftharpoons AS \tag{11A}$$

The concentrations are related according to a binding equilibrium

$$\frac{[AS]}{c(a)[S^-]} = K \tag{11.51}$$

where $c(a)$ is the concentration of A^+ at $x = a$. The binding energy is broken down into two parts, a chemical part, $RT \ln K^0$, and an electrostatic part, based on the electrostatic potential at $x = a$

$$RT \ln K = RT \ln K^0 + e\tilde{A}z\varphi(a) \tag{11.52}$$

Here K^0 is the binding constant that would be seen in the absence of electrostatic interactions with the surface.

The charge density of the Stern layer depends on the density of the fixed charges that form the binding sites, and the fraction of sites occupied. The site density is denoted as ρ_s, and the fraction of sites occupied (expressed in terms of $[S^-]$ and $[AS]$) gives σ_{st}, as follows

$$\sigma_{st} = \frac{ze\rho_s[AS]}{[S^-] + [AS]} = \frac{ze\rho_s}{1 + [S^-]/[AS]} \tag{11.53}$$

$[S^-]/[AS]$ can be replaced with the aid of Eq. (11.51)

$$\sigma_{st} = \frac{ze\rho_s}{1 + 1/(c(a)K)} = \frac{ze\rho_s}{1 + 1/(c(a)K^0 e^{e\tilde{A}z\varphi(a)/RT})} \tag{11.54}$$

The second step made use of Eq. (11.52).

To obtain the charge density of the diffuse layer we use Eq. (11.41), but with $\varphi(a)$ in place of φ_0

$$\sigma_d = -\sqrt{\frac{2\varepsilon kT\tilde{A}10^{-3}c}{\pi}} \sinh\left(\frac{-e\varphi(a)}{2kT}\right) \tag{11.55}$$

Equations (11.54) and (11.55) provide relations between the charges of the two layers and the modified surface potential, $\varphi(a)$.

Recall the analogy with a capacitor (Eq. (11.42)). The Stern layer and the diffuse layer behave as two capacitors in series. The total capacitance then becomes

$$C = \frac{C_{st}C_d}{C_{st} + C_d} \tag{11.56}$$

At high ionic strength most of the binding sites are occupied and the capacitance of the Stern layer is greatest, so C goes to C_d. At low ionic strength the capacitance of the diffuse layer is greater, so C goes to C_{st}.

Experimental analysis of the distribution of ions at a charged lipid surface has revealed both the Stern layer and the diffuse layer

(Bedzek *et al.*, 1990). The Gouy–Chapman–Stern theory gives a satisfactory account of ion-membrane interactions for a wide range of biophysical processes (McLaughlin, 1989).

From a practical point of view the most important difference between the Gouy–Chapman and Stern theories is that in Gouy–Chapman theory the charge is all that matters. Different ions will act the same if their charge is the same. Thus, one can interchange Na^+, K^+, NH_4^+, or tetraethylammonium with no change in behavior. Likewise, a series of divalent ions, Mg^{2+}, Ca^{2+}, St^{2+}, etc. should also produce indistinguishable results. This will not be the case if ions bind to the surface because the binding constant K^0 is ion specific. One often sees differences in the actions of divalent cations at a surface. This is generally interpreted as evidence for binding to sites on the membrane and the formation of a Stern layer.

11.6 | Surface charge and channel conductance

The conductance of an ion channel varies with the concentration of permeant ions. Surface charge draws ions toward the mouth of the channel, concentrating them and increasing the conductance above that expected from the bulk concentration. From Eq. (11.48) we can relate the relevant concentration of the permeant ion, c_0, to the surface potential

$$c_0 = c(\infty)e^{-e\varphi_0/kT} \tag{11.57}$$

we take c_0 as the concentration at the surface of the membrane, and presumably at the mouth of the channel as well. If there is a surface charge on the membrane and the ionic strength is altered by adding or removing an impermeant ion, the surface potential will change through Eq. (11.41), and the resulting change in c_0 will alter the channel conductance. Thus, by changing the surface potential, impermeant ions can influence the flow of permeant ions.

This effect has been demonstrated in a number of channels. For example, the conductance of a K^+ channel in heart muscle is reduced by the addition of various impermeant ions such as choline, Ca^{2+}, and Mg^{2+} (Fig. 11.5). The drop in conductance reflects the neutralization of a negative surface charge, leading to a reduction in the concentration of K^+ at the mouth of the channel. As the ionic strength increases, the effect of the existing surface charge is neutralized so the channel conductance approaches the value that would be seen in a neutral membrane. Overall, this effect is well described by Gouy–Chapman theory. Equation (11.41) gives surface charge as a function of surface potential, and this expression can be inverted to obtain the surface potential as a function of the ionic strength and surface charge (Problem 6). With Eq. (11.57), the surface potential then gives the concentration at the mouth of the channel, which is proportional to channel conductance.

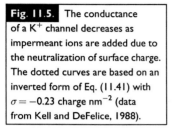

Fig. 11.5. The conductance of a K^+ channel decreases as impermeant ions are added due to the neutralization of surface charge. The dotted curves are based on an inverted form of Eq. (11.41) with $\sigma = -0.23$ charge nm^{-2} (data from Kell and DeFelice, 1988).

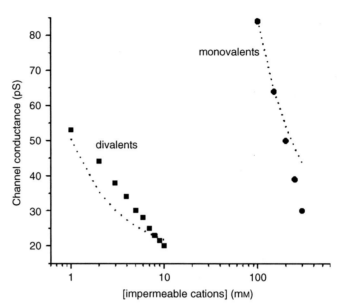

Varying the concentration of a permeant ion can produce another interesting surface charge effect. According to Eq. (11.49), c_0 at a charged surface will go to a finite limit as the bulk concentration is reduced to zero. This means that for a channel in a charged membrane, as the permeant ion concentration is reduced, the conductance can go to a nonzero limit. This seems paradoxical, because at zero ion concentration there is nothing to carry current.

In experiments with a Ca^{2+}-activated K^+ channel reconstituted into bilayers, the lowest conductance seen at low K^+ concentrations was about 30% of the highest conductance obtained with a saturating K^+ concentration. Neutralizing charge on the membrane surface by altering the lipid composition allowed the channel current to go to much lower values as the K^+ concentration was reduced (Moczydlowski *et al.*, 1985). Chemical modification of glutamates near the channel mouth removed negatively charged side chains, with a similar outcome (MacKinnon *et al.*, 1989).

It is important to distinguish whether the surface charge is on the lipid or the channel protein. Equation (11.49) depends on the charge being uniformly distributed over a planar surface. However, when the charges are on channels the distribution is not uniform. Each protein looks more or less like a point of charge at a distance that is large compared to the size of the protein. This situation was studied by Cai and Jordan (1990), who obtained a numerical solution to the Poisson–Boltzmann equation for the geometry shown in Fig. 11.6a. Treating the channel conductance as a saturating function of c_0 gave a curve that remained quite high as the concentration decreased. But as the bulk concentration fell into the low mM range, c_0 (and the channel conductance) dropped sharply (Fig. 11.6b). This behavior is sensitive to the precise nature of the charge distribution at the mouth of the

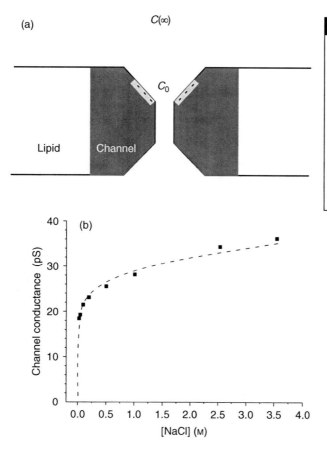

Fig. 11.6. (a) A model of a channel with fixed charge localized to two patches in the channel's vestibule. (b) A numerical solution of the Poisson–Boltzmann equation (Cai and Jordan, 1990) determined how the concentration of ions at the channel mouth, c_0, varies with the bulk concentration, $c(\infty)$. This gives a quantitative description of how channel conductance varies with Na^+ concentration (Na^+ channel data from Green et al., 1987).

channel, so one can learn something about the locations of fixed protein charges from these kinds of experiments.

11.7 | Surface charge and voltage gating

A surface potential has a direct effect on voltage gating of ion channels. This is because the voltage seen by a channel's gating charge is actually the difference between the voltages at each surface (Fig. 11.7). Changing the surface potential at one or both sides will thus change the electric field acting on the channel.

To account for this effect of surface potential on channel gating, it is clear from Fig. 11.7 that all we need to do is take the Boltzmann equation of voltage-gating (Eq. (1.28)) and add the surface potentials to ΔV. Taking $\Delta \varphi_0$ as $\varphi_{0\text{-right}} - \varphi_{0\text{-left}}$, we have

$$P_0 = \frac{1}{1 + e^{(\Delta V - \Delta V_0 + \Delta \varphi_0)/V_s}} \tag{11.58}$$

If the two sides of a membrane have different charge densities then changing the concentrations of ions will affect φ_0 differently on the two sides, and $\Delta \varphi_0$ will change. Alternatively, even if surface charge

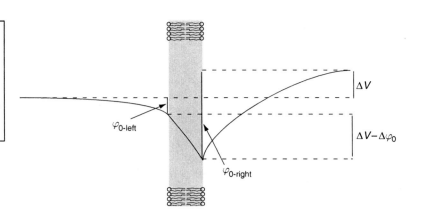

Fig. 11.7. The voltage seen by a channel in a membrane is the membrane potential, ΔV, minus the difference in surface potentials at each side. In this sketch a much larger surface potential on the right side than the left side reverses the sign of the membrane potential.

is symmetrically distributed, $\Delta\varphi_0$ can be altered by adding or removing ions on just one side of the membrane.

This kind of behavior is very common (see Chapter 17 of Hille, 1991). In general, monovalent ions produce shifts that are in reasonable agreement with Gouy–Chapman theory. In the Ca^{2+}-activated K^+ channel from skeletal muscle, the voltage dependence of channel opening followed a Boltzmann equation, and as the potassium concentration on one side of the membrane was increased from 1 to 150 mM, the voltage dependence of the open probability shifted by nearly 40 mV (MacKinnon *et al.*, 1989). The magnitude of this shift was very well described by Gouy–Chapman theory, with $\sigma = -0.23$ elementary charge nm^{-2}. On the other hand, for many channels divalent cations produce variable shifts, and this is the signature of a Stern layer formed by specific binding sites.

11.8 | Electrophoretic mobility

A charged particle migrates in an electric field, and this is termed electrophoresis. For a total charge Q, the force exerted by the field is

$$F_e = EQ \tag{11.59}$$

The electric field, E, accelerates the particle until a balance is reached with the frictional resistance to movement, at which point the particle moves with a constant velocity (Section 6.7). The friction is proportional to the velocity, v, and pushes in the opposite direction (Eq. (6.58))

$$F_f = -f V \tag{11.60}$$

This proportionality defines the frictional coefficient f.

The average velocity reflects a balance between these two forces. Setting F_e in Eq. (11.59) equal to $-F_f$ in Eq. (11.60) leads to an expression for the velocity where the forces balance

$$v = \frac{EQ}{f} \tag{11.61}$$

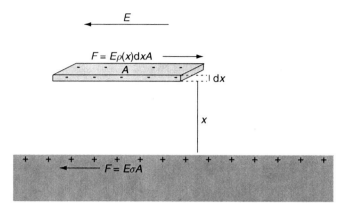

Fig. 11.8. The surface of a charged particle in an ionic solution. The field, E, pulls the positively charged surface (dark gray) one way and the negatively charged counterions the other. A volume element of solution, Adx at a distance x from the surface (light gray), is pulled by the field and by sheer forces exerted by the layers of fluid above and below.

The electrophoretic mobility, u, is defined from this equation as the ratio of the velocity to the field

$$u = \frac{v}{E} = \frac{Q}{f} \tag{11.62}$$

It is tempting at this point to take the coefficient of friction from Stokes' law (Eq. (6.65)). Then the electrophoretic mobility would be

$$u = \frac{Q}{6\pi\eta a} \tag{11.63}$$

where a is the radius of the particle (taken as a sphere) and η is the viscosity. However, when there are ions in the solution, Eq. (11.63) is incorrect. The trouble is that the counterions that are concentrated in the vicinity of the particle undergo electrophoresis in the opposite direction. The movement of these ions causes the fluid to stream along, and this moving fluid pulls the particle in the direction opposite to the direct pull exerted by the electric field (Fig. 11.8). As a result, u is much lower than that given by Eq. (11.63).

It may seem daunting to try to account for the interaction between the counterions, the fluid, and the surface charge of the particle, but a simple theory has been developed that achieves this. Consider a thin sheet-like section of solution at a distance x from the particle surface (Fig. 11.8). The field pulls the sheet because there are ions dissolved in it. The force is $E\rho(x)Adx$, where Adx is the volume of the sheet. Adjacent sheets moving at different velocities exert a sheer force on one another. The sheer force is

$$-\eta A \frac{dv}{dx}\bigg|_x \text{ at } x, \quad \text{and} \quad \eta A \frac{dv}{dx}\bigg|_{x+dx} \text{ at } x+dx$$

We now consider a steady state in which the particle moves at a constant velocity. The fluid streams along parallel to the particle surface, with layers closer to the particle matching more closely the velocity of the particle. The velocity of each layer is constant, so the forces on any layer must add up to zero. Thus, the force of the electric field is balanced by the sum of the sheer forces from above and below

$$E\rho(x)A\mathrm{d}x = \eta A\frac{\mathrm{d}v}{\mathrm{d}x}\bigg|_{x+\mathrm{d}x} - \eta A\frac{\mathrm{d}v}{\mathrm{d}x}\bigg|_{x} \tag{11.64}$$

This can be rearranged to give

$$E\rho = \eta\frac{\mathrm{d}^2 v}{\mathrm{d}x^2} \tag{11.65}$$

Using the Poisson equation (Eq. (11.2)) to replace ρ gives

$$-\frac{E\varepsilon}{4\pi}\frac{\mathrm{d}^2\varphi}{\mathrm{d}x^2} = \eta\frac{\mathrm{d}^2 v}{\mathrm{d}x^2} \tag{11.66}$$

Integration leads to a relation between the first derivatives. The derivatives of both φ and v are zero at $x = \infty$; far away from the particle the potential is flat and the fluid is stationary. Thus, we are left with the same relation between the derivatives

$$-\frac{E\varepsilon}{4\pi}\frac{\mathrm{d}\varphi}{\mathrm{d}x} = \eta\frac{\mathrm{d}v}{\mathrm{d}x} \tag{11.67}$$

Integrating again and taking advantage of $\varphi(\infty) = 0$ and $v(\infty) = 0$ gives

$$-\frac{E\varepsilon}{4\pi}\varphi(0) = \eta v(0) \tag{11.68}$$

Dividing this equation through by E and η would appear to give us an expression for the mobility u as v/E (Eq. (11.62)). But first we must realize that $\varphi(0)$ in this situation differs in a subtle way from φ_0 defined in the Gouy–Chapman theory above. There, φ_0 was the potential at the distance where an ion is in direct contact with the surface. In Eq. (11.68) $\varphi(0)$ is the potential at the distance where the fluid sticks to, and moves with, the particle. These two "zeroes" are not the same. Because of this distinction the symbol ζ is used for $\varphi(0)$ in the expression for the electrophoretic mobility, and is commonly referred to as the *zeta-potential*. Thus, Eq. (11.68) gives

$$u = \frac{v}{E} = \frac{\varepsilon\zeta}{4\pi\eta} \tag{11.69}$$

This result, originally due to Helmholtz and improved by von Smoluchowski, is referred to as the Helmholtz–Smoluchowski equation (Overbeek and Wiersema, 1967). The result is remarkable because it is independent of the particular way in which φ and ρ vary with x. The particle size and charge do not appear in Eq. (11.69), and this is mysterious because intuition tells us that the electrophoretic mobility ought to depend on these properties. This is clarified by thinking of the particle as a large uniformly charged sphere for which Eq. (11.17) can be used to express the screened potential. We take the potential at r_ζ, defined as the distance used to compute the zeta potential

$$\zeta = \frac{Q}{\varepsilon r_\zeta}\left(\frac{e^{-(r_\zeta - a)/\lambda_\mathrm{D}}}{1 + a/\lambda_\mathrm{D}}\right) \tag{11.70}$$

For a large sphere $r_\zeta \sim a$, so the exponential term is ~ 1. Substitution into Eq. (11.69) gives (Chapter 3.1.D of Benedek and Villars, 2000)

$$u = \frac{Q}{4\pi\eta a(1 + a/\lambda_D)} \tag{11.71}$$

For a particle much larger than λ_D we see how u is lower than that given by Stokes' law (Eq. (11.63)). This solves the mystery. The charge and size do affect the mobility through ζ. Equation (11.71) makes that dependence explicit.

For particles with sizes comparable to λ_D, Eq. (11.71) becomes less accurate, and as the size decreases further the electrophoretic mobility approaches that for small ions. More detailed theoretical treatments of these issues highlight the importance of the ratio a/λ_D as a critical parameter. When the particle is much larger than the thickness of the ionic double layer, Eqs. (11.69) and (11.71) are reasonably accurate (Overbeek and Wiersema, 1967).

One helpful perspective on why the electrophoretic mobility can be expressed with the simplicity of Eq. (11.69) is that the *local* interactions between the particle surface and the surrounding ions and fluid dominate in resisting the motion of the particle. By contrast, Stokes' law treats the fluid displacement by the movement of the particle (Fig. 6.7). The streaming of the fluid around the *entire* particle is then taken into account.

Although electrophoresis is an extremely widespread analytical tool in the separation and characterization of proteins and nucleic acids, quantitative theories are rarely implemented. Proteins do not have uniformly charged surfaces, and quantitative calculations of electrophoretic mobility depend on too many details such as shape and charge distribution. By contrast, the electrophoretic mobilities of cells and liposomes are well described by the Helmholtz–Smoluchowski equation. Variations in ionic strength produce the expected changes in mobility. Monovalent ions produce nonspecific effects, following the Gouy–Chapman theory. Multivalent ions produce specific effects indicative of site binding, as provided for by the Stern modifications (McLaughlin, 1989).

11.9 | Polyelectrolyte solutions I. Debye–Hückel screening

Polyelectrolytes such as DNA and RNA have charge distributed over their entire length. The density of charge is usually quite high so that the electrostatic potential near the surface is substantially larger than kT. The Boltzmann term $e^{-e\varphi/kT}$ can therefore no longer be linearized. The Poisson–Boltzmann equation can be solved exactly for a uniformly charged cylinder, but only in the restricted case where the polyelectrolyte's counterions are the only ions present, i.e. no added salt (see Oosawa, 1971). Here we will explore

Fig. 11.9. A polyelectrolyte represented as a linear chain of charges separated by a spacing, b.

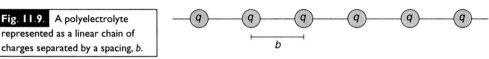

some approximate and intuitive theories that have wider applicability (Manning, 1969; Oosawa, 1971).

We assume that the repulsion between charges will straighten out the polymer and make it look like Fig. 11.9, with each charge, q, separated from its neighbor by a distance, b. The electrostatic repulsion helps give DNA its long persistence length (Section 3.4). We will now calculate the potential energy at the surface of this chain. Equation (11.17) is the starting point. The solution is taken as dilute so λ_D is large compared to the size of the individual charges on the polyelectrolyte, justifying the approximation $1 + a/\lambda_D \sim 1$. The potential for just one of the charges on the polyelectrolyte is

$$\varphi(r) = \frac{q e^{-(r-a)/\lambda_D}}{\varepsilon r} \tag{11.72}$$

Recall that this is the Coulomb potential with the term e^{-r/λ_D} reflecting the screening effect of dissolved salt.

It is important to note that λ_D is calculated using the ionic strength of the inorganic ions. Including the polyelectrolyte, with a value of z that can easily surpass 10^4, would make λ_D extremely short even with a very low concentration. However, polyelectrolytes cannot screen electrostatic interactions well because their charge is spread out over a great distance. The better screening ability of multivalent ions depends on their charge being concentrated at one point in space.

The polyelectrolyte molecules are assumed to be well separated so they do not interact with one another. That allows us to focus on the interaction between the polyelectrolyte and the surrounding salt in solution. We will now calculate the potential at the surface of one of the balls in Fig. 11.9. Number the charges on the polyelectrolyte starting from one end. The distance between charge i and j is $|i - j|b$, so from Eq. (11.72) the screened potential at the site of charge i due to charge j is $q e^{(-|i-j|b/\lambda_D)}/(\varepsilon|i - j|b)$ (a in the exponent is negligible). Summing over j, and adding in the term for charge i itself gives

$$\varphi = \frac{q}{\varepsilon a} + \frac{2q}{\varepsilon} \sum_{n=1}^{\infty} \frac{e^{-nb/\lambda_D}}{nb} \tag{11.73}$$

The first term is from the ith charge, taken from Eq. (11.72), with $r = a$. The factor of 2 in the second term reflects the interactions with charges on each side. The sum to infinity means that we are looking at a very long chain.

We now approximate the sum as an integral with $x = nb$ and with dx/b as the differential

$$\varphi = \frac{q}{\varepsilon a} + \frac{2q}{\varepsilon b} \int_b^\infty \frac{e^{-x/\lambda_D} dx}{x} \tag{11.74}$$

Transforming variables to $z = x/\lambda_D$ gives

$$\varphi = \frac{q}{\varepsilon a} + \frac{2q}{\varepsilon b} \int_{b/\lambda_D}^\infty \frac{e^{-z} dz}{z} \tag{11.75}$$

This integral cannot be evaluated analytically,[1] but when b/λ_D is small ($b << \lambda_D$ is valid for low ionic strength) it can be approximated as

$$\varphi = \frac{q}{\varepsilon a} - \frac{2q}{\varepsilon b} \ln \frac{b}{\lambda_D} \tag{11.76}$$

This approximation makes sense, because the exponential term makes the integrand small for large z. The integral blows up as $b/\lambda_D \to 0$, and this occurs in a region where the exponential is close to one. So for small b/λ_D the integral should behave like the integral of $1/z$, in which a negative logarithmic singularity dominates.

We can now calculate the work done on the dissolved ions to charge up the entire chain. This is similar to estimating the work of charging a sphere in the treatment of an ion (Eq. (11.19)). Thus, we integrate Eq. (11.76) from 0 to q

$$\begin{aligned} W &= \left(\frac{1}{\varepsilon a} - \frac{2}{\varepsilon b} \ln \frac{b}{\lambda_D} \right) \int_0^q q' \, dq' \\ &= \frac{q^2}{2\varepsilon a} - \frac{q^2}{\varepsilon b} \ln \frac{b}{\lambda_D} \\ &= \frac{q^2}{2\varepsilon a} - \frac{q^2}{\varepsilon b} \ln b + \frac{q^2}{\varepsilon b} \ln \lambda_D \end{aligned} \tag{11.77}$$

Just as Eq. (11.18) was separated into salt-dependent and salt-independent parts, we have done that again here. The last term in Eq. (11.77) contains all the salt dependence (λ_D). The other two terms are the same with or without salt. In the interest of understanding how salt influences the behavior of the polyelectrolyte, we take this

---

[1] Integrating by parts gives

$$\int \frac{e^{-z} dz}{z} = e^{-z} \ln z + \int e^{-z} \ln z \, dz$$

Integrating by parts again gives

$$e^{-z} \ln z + e^{-z}(z \ln z - z) + \int e^{-z}(z \ln z - z) dz$$

Repeating this generates terms of the form $e^{-z} z^i$ and $e^{-z} z^i \ln z$. All terms evaluated at $z = \infty$ will be zero because of the exponential. All terms evaluated at b/λ_D will by small compared to the leading term $-e^{-b/\lambda_D} \ln(b/\lambda_D) \approx -\ln(b/\lambda_D)$.

last term as the work done on the salt per charge of the polyelectrolyte to place it in solution

$$W = \frac{q^2}{\varepsilon b} \ln \lambda_D \tag{11.78}$$

This equation, due to Manning (1969), expresses the salt dependence of the free energy of dissolving a charged polymer. It is not a true limiting law such as the one from Debye–Hückel theory (Eq. (11.22)), because it has the undesirable property of becoming infinite as the salt concentration goes to zero ($\ln \lambda_D \to \infty$). This problem arises because when λ_D becomes as long as the polymer then the chain can no longer be treated as infinite. However, the salt has to be very dilute for this to happen so Eq. (11.78) is valid down to a very low salt concentrations.

Equation (11.78) can be used to calculate many fundamental properties of polyelectrolyte solutions. For example, the activity coefficient of ionic species i is influenced by polyelectrolyte, and this contribution is obtained by differentiating Eq. (11.78) with respect to the concentration of that ion

$$kT \ln \gamma_i = \frac{\partial W}{\partial c_i} = \frac{\partial}{\partial c_i} \left(\frac{q^2}{\varepsilon b} \ln \lambda_D \right) \tag{11.79}$$

With λ_D from Eq. (11.9) we can show that for a binary salt

$$\frac{\partial \lambda_D}{\partial c} = -\frac{\lambda_D}{2(c_1 + c_2 + c_c)} \tag{11.80}$$

where c_1 and c_2 refer to the anions and cations of an added salt and c_c is for the polyelectrolyte's own counterions. Using this in the differentiation of Eq. (11.79) gives (Manning, 1969)

$$kT \ln \gamma = -\frac{q^2}{2\varepsilon b(c_1 + c_2 + c_c)} \tag{11.81}$$

11.10 | Polyelectrolyte solutions II. Counterion-condensation

The foregoing analysis of Debye–Hückel screening works only for polyelectrolytes with a low charge density. When a polyelectrolyte has a high charge density it attracts ions so strongly that it gives rise to a unique form of association called *counterion-condensation* (Oosawa, 1971). This can be understood with the aid of a model in which the discrete charges of Fig. 11.9 are smeared uniformly over the length of the chain.

For very close interactions, the Debye–Hückel screening can be ignored. For the interaction with an ion carrying a charge of q_i, the Coulomb potential is integrated over interactions with a line of charge with a density of q_j/b. This is an integral of $1/r$, so we have a logarithmic function

$$\varphi(r) = -\frac{2q_i q_j}{\varepsilon b} \ln r \tag{11.82}$$

where r is small compared to the length of the polyelectrolyte.

The classical configuration integral (Eq. (1.4)) is then used to calculate the ion–polyelectrolyte contribution to the partition function

$$\Omega = \int_0^{r_{max}} e^{-\varphi(r)/kT} 2\pi r \, dr \tag{11.83}$$

The factor $2\pi r$ is due to the polar coordinate system, and r_{max} is an arbitrary large distance, which will soon be seen to be irrelevant. Substituting Eq. (11.82) into Eq. (11.83) gives

$$\Omega = \int_0^{r_{max}} r^{\frac{2q_i q_j}{\varepsilon b kT}} 2\pi r \, dr$$

$$= 2\pi \int_0^{r_{max}} r^{1 + \frac{2q_i q_j}{\varepsilon b kT}} \, dr \tag{11.84}$$

Although this is an easy integral, no effort is made to evaluate it. Instead, we focus on the qualitatively different behavior it will exhibit depending on whether the exponent of r is greater than or less than -1. For $1 + (2q_1 q_2)/(\varepsilon b kT) < -1$ the integral diverges. For values > -1, the integral converges. This defines the quantity $\xi = |(q_1 q_2)/(\varepsilon b kT)|$ as a pivotal charge-density parameter of the polyelectrolyte. For

$$\xi \geq 1 \tag{11.85}$$

the divergence of Eq. (11.84) means that the ion–polyelectrolyte system has a very high free energy, and is therefore unstable. By contrast, for

$$\xi < 1 \tag{11.86}$$

the polyelectrolyte will be screened by counterions in the standard way, and we can expect the Debye–Hückel screening of the preceding section to provide an adequate description.

Manning (1969) proposed that in the unstable situation characterized by Eq. (11.85), ions in solutions will accrete around the polyelectrolyte to reduce its effective charge density. This accumulation continues until ξ is brought down to 1. This will increase the effective mean spacing between charges to a new value b_{eff}, defined by the equation

$$\xi_{eff} = \left| \frac{q_1 q_2}{\varepsilon b_{eff} kT} \right| = 1 \tag{11.87}$$

The polyelectrolyte then behaves as though it has $b = b_{eff}$. Thus, counterion-condensation onto a polyelectrolyte reduces its charge density to a value that gives a convergent configuration integral. The nature of the bound or condensed state need not be precisely specified; counterion-condensation does not depend on the chemical nature of the counterion, only on its charge. For any choice of ions added to a solution, the same quantity will bind to the polyelectrolyte, provided that the charge is the same. Thus, like the screening interactions discussed in Debye–Hückel theory and Gouy–Chapman theory, counterion-condensation lacks chemical specificity.

The crucial spacing for which $\xi = 1$ works out to be $b = 7.1$ Å, with q_1 and q_2 taken as the unitary charge and $T = 298$ K. Many polyelectrolytes have more closely spaced charges, so counterion-condensation is quite common. For DNA $b = 1.7$ Å, so we would expect enough charge to bind to increase b to 7.1 Å. Since ξ and ξ_{eff} are proportional to the charge density before and after counterion-condensation, respectively, we can take the fraction of polyelectrolyte charge that is neutralized by counterion-condensation as $(\xi - \xi_{eff})/\xi = (\xi - 1)/\xi$ and the fraction of charge remaining as $\xi_{eff}/\xi = 1/\xi$. For DNA, $\xi = 4.2$, so the fraction of charge neutralized by monovalent counterions is 0.76. For a variety of different counterions over a wide range of concentrations, this prediction has been experimentally verified (reviewed by Manning, 1978).

Counterion-condensation tells us that the fraction of bound salt remains constant as bulk salt concentration is varied. This is not what one expects of a standard molecular association (Chapter 4). There should be a graded change in binding site occupancy as the concentration of a ligand is changed. Counterion-condensation thus violates mass action. The reason for this is the long range of the electrostatic potential of a linear array of charges. We saw that summing over all the relevant Coulomb potentials gave a potential energy that increases logarithmically with λ_D (Eq. (11.78)). Diluting the salt increases λ_D, making the attraction for ions stronger. This increase exactly cancels the increase in the entropic drive to dissociate, which is also logarithmic. The balancing of these two logarithmic terms results in a constant degree of association. Ultimately, counterions will dissociate, but only when the salt is so dilute that λ_D becomes longer than the length of the polyelectrolyte. The length of a 1000 base pair DNA molecule is 3500 Å; λ_D is this long in $\sim$1 μM NaCl.

The large amount of charge on polyelectrolytes leads to somewhat similar behavior even when a macromolecule deviates from a linear geometry. The counterions still neutralize a fraction of charge. The charge neutralized and the number of condensed counterions are not constant as in the treatment given above, but they change very slowly with salt concentration (Oosawa, 1971).

11.11 | DNA melting

The repulsion between the negatively charged phosphates is a major destabilizing factor in double-stranded DNA. Adding salt screens this repulsion, stabilizes the double helix, and raises the melting point. Experiments have shown that the melting temperature of DNA increases with the logarithm of the concentration of salt. This can be explained by a theory that incorporates the two effects, Debye–Hückel screening and counterion-condensation, developed in the preceding two sections.

Consider an equilibrium (Scheme (11B)) between double-stranded and single-stranded DNA, in which i counterions are liberated

$$D \rightleftharpoons 2S + iA^+ \tag{11B}$$

where D and S denote double and single strands. The salt is denoted as A^+. Although the helix–coil transition is not a simple two-state process (Section 3.12), we will approximate it as such and take Scheme (11B) as a two-state melting of a single cooperative unit. This is a reasonable simplification because we are mainly interested in the melting temperature and not the shape of the transition.

Scheme (11B) implies an equilibrium constant

$$K = \frac{[S]^2 c^i}{[D]} \tag{11.88}$$

for which we take the free energy change as a function of the salt concentration

$$\Delta G = \Delta G' + iRT \ln c \tag{11.89}$$

where $\Delta G'$ refers to everything that goes into the free energy other than the mass balance contribution of ion binding. It includes energies for the interaction between the DNA strands, water solvation, and Debye–Hückel screening.

First we use the results for counterion-condensation to estimate i, the difference in number of counterions bound. As noted above, the fraction of charge on the polyelectrolyte in the absence of counterion condensation is $1/\xi$. There is a corresponding number of counterions free in solution. If ξ_d and ξ_s are the charge-density parameters for double- and single-stranded DNA, respectively, then the change in the number of free counterions upon melting is

$$i = \frac{1}{\xi_s} - \frac{1}{\xi_d} \tag{11.90}$$

Now we consider the salt dependent electrostatic contribution to $\Delta G'$. This arises from the interaction between the polyelectrolyte and its ionic atmosphere, and is denoted as ΔG_{el}. The treatment of Debye–Hückel screening produced Eq. (11.78) as an expression for the work done per charge to place a polyelectrolyte in a salt solution. This result can be rewritten as $kT\xi_{eff} \ln \lambda_D$, and this is then divided by ξ to account for the fraction of unneutralized charge. Multiplying by Avogadro's number converts this to free energy per mole (replacing kT with RT). Since $\xi_{eff} = 1$, the difference between the electrostatic contributions from double- and single-stranded DNA is

$$\Delta G_{el} = \left(\frac{1}{\xi_s} - \frac{1}{\xi_d}\right) RT \ln \lambda_D \tag{11.91}$$

We now combine this with the mass action term, $iRT \ln c$, from Eq. (11.89), and take i from Eq. (11.90), to give

$$\Delta G = \Delta G'' + \left(\frac{1}{\xi_s} - \frac{1}{\xi_d}\right) RT \ln \lambda_D + \left(\frac{1}{\xi_s} - \frac{1}{\xi_d}\right) RT \ln c \tag{11.92}$$

The second and third terms on the right represent the contributions of Debye–Hückel screening and counterion-condensation, respectively, and $\Delta G''$ is what is left after subtracting ΔG_{el} from $\Delta G'$. Once again (as in Eqs. (11.18) and (11.77)), we have divided the free energy into salt-dependent and salt-independent parts so that we can focus on salt effects.

We now observe that λ_D is proportional to $1/\sqrt{c}$ (Eq. (11.9)). We can thus extract the salt-independent factor in λ_D from the middle term on the right-hand side of Eq. (11.92) and absorb it into $\Delta G''$ to create ΔG_c. Since $\ln 1/\sqrt{c} = -1/2 \ln c$, we can combine the two terms in Eq. (11.92) that depend on c as follows

$$\Delta G = \Delta G_c + \frac{1}{2}\left(\frac{1}{\xi_s} - \frac{1}{\xi_d}\right) RT \ln c \tag{11.93}$$

This is a key result. It can be used to determine how salt influences the equilibrium between single-stranded and double-stranded DNA. To look at the melting temperature we replace ΔG_c by $\Delta H_c - T\Delta S_c$, and set T equal to the melting temperature, T_m, at which point double-stranded and single-stranded DNA have equal free energies, so $\Delta G = 0$, as follows

$$0 = \Delta H_c - T_m \Delta S_c + \frac{1}{2}\left(\frac{1}{\xi_s} - \frac{1}{\xi_d}\right) RT_m \ln c \tag{11.94}$$

Solving for T_m gives

$$T_m = \frac{\Delta H_c}{\Delta S_c - \frac{1}{2}\left(\frac{1}{\xi_s} - \frac{1}{\xi_d}\right) R \ln c} \tag{11.95}$$

This equation already makes the point that raising the salt concentration raises the melting temperature, but it can be simplified by noting that ΔS_c is larger than the second term in the denominator. The fraction can therefore be expanded using $1/(1-x) \sim 1 + x$ to give

$$T_m \sim \frac{\Delta H_c}{\Delta S_c}\left(1 + \frac{1}{2\Delta S_c}\left(\frac{1}{\xi_s} - \frac{1}{\xi_d}\right) R \ln c\right) \tag{11.96}$$

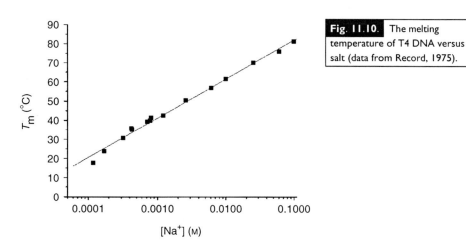

Fig. 11.10. The melting temperature of T4 DNA versus salt (data from Record, 1975).

Thus, T_m varies linearly as $\ln c$, and the slope is

$$\frac{\Delta H_c R}{2\Delta S_c^2}\left(\frac{1}{\xi_s}-\frac{1}{\xi_d}\right)=\frac{RT_{cm}}{2\Delta H_c}\left(\frac{1}{\xi_s}-\frac{1}{\xi_d}\right) \tag{11.97}$$

where we replaced ΔS_c by $\Delta H_c/T_{cm}$.[2]

Equation (11.96) describes how T_m increases linearly with the logarithm of the salt concentration. An example of such behavior is illustrated in Fig. 11.10. As noted above, $\xi_d = 4.2$. We can measure $\Delta H°$ calorimetrically or from the van't Hoff equation (Eq. (1.19)). This leaves one unknown, ξ_s, on the right-hand side of Eq. (11.97). From the slope of plots such as Fig. 11.10 one can estimate ξ_s as ~ 1.7 (Record, 1975). This is consistent with the view that single-stranded DNA has half the number of charged phosphates, and is a bit more extended than double-stranded DNA. This theory success-fully describes a large body of experimental data on how salt influ-ences DNA melting (Manning, 1978). Thus, the two effects of Debye–Hückel screening and counterion-condensation provide a good accounting of how salt affects the relative stability of single-stranded and double-stranded DNA.

Problems for Chapter 11

1. Calculate λ_D for a NaCl solution with concentrations of 25 and 50 mM. Repeat the calculation for $MgCl_2$.
2. Show that $\rho(r)$ (Eq. (11.23)) integrated over space gives a total counterion charge that perfectly balances the charge of the ion.
3. Find the distance at which the charge density is highest in Debye–Hückel theory (the maximum of Eq. (11.23)).
4. Calculate the net counterion charge within a sphere around an ion of radius λ_D.

---

[2] T_{cm} is the melting temperature of a fictitious salt-independent state so it is not equal to T_m, but the percentage difference is not large on an absolute temperature scale.

5. Use the Debye–Hückel limiting law to derive the ionic strength dependence of a simple ionization equilibrium of the form $A^+ + B^- \rightleftharpoons AB$. (Neglect the contribution of A^+ and B^- to I.)

6. Invert Eq. (11.41) to obtain an expression for φ_0 in terms of c and σ (see Appendix 5).

7. Show that the surface potential for a sphere (from Eq. (11.18)) is identical to that for a plane (Eq. (11.42)) when the radius of the sphere, $a \gg \lambda_D$. This indicates that Gouy–Chapman theory is valid for the membrane of a spherical cell.

Chapter 12

Fluctuations

Biological systems often fluctuate more noticeably than typical physical and chemical systems. This reflects the large size of many biological molecules and the small size of cells. The molecular nature of matter gives rise to fluctuations in every imaginable property. These fluctuations may or may not be easy to see, and size is a critical factor. In a system with N molecules, many measured quantities are proportional to N, but the fluctuations are proportional to $N^{1/2}$. The fluctuations relative to the mean then decrease with the size of a system as $N^{-1/2}$. When N is Avogadro's number, the task of observing these fluctuations in a conventional measurement becomes quite a challenge. Of course, there are some incredibly sensitive measurements that can be made. Signals arising from single molecules can be detected, and the fluctuations in these signals reflect the stochastic nature of molecular activity. But in many cases where the single-molecule signals are too small to see, the collective fluctuations may still be detectable. The special size scales found in biology generate a uniquely fluctuating world that merits special attention.

We have encountered fluctuations already in Chapter 3 in relation to conformations of macromolecules, and in Chapter 6 in relation to random walks. The probability of fluctuations can be calculated whenever statistical mechanics is used to develop a quantitative molecular description, and in many situations fluctuations contain important information. The study of fluctuations then becomes a powerful experimental approach by which models can be tested and molecular parameters estimated. Understanding fluctuations also has a practical value in helping one minimize the instrumental noise that limits what can be measured. Finally, since fluctuations are an intrinsic part of molecular activity, understanding them deepens one's understanding of how molecules work.

12.1 | Deviations from the mean

There are a number of ways to quantitate fluctuations. A probability distribution tells us how often different values of a number are observed. The Boltzmann distribution is a familiar example that tells us how the energy of a molecule varies. However, one would often like to view the magnitude of the fluctuations in terms of a single number. For this we naturally turn to the deviation from the mean, $\Delta x = x - \bar{x}$. But averaging this quantity gives zero. Instead, we take the mean-square deviation, or variance, $\overline{\Delta x^2} = \overline{(x - \bar{x})^2}$. Squaring before averaging avoids the problem of the positive deviations canceling the negative deviations. We can put $\overline{\Delta x^2}$ into a particularly convenient form by multiplying out the square

$$\overline{\Delta x^2} = \overline{(x - \bar{x})^2} = \overline{x^2} - \overline{2x\bar{x}} + \bar{x}^2$$
$$= \overline{x^2} - \bar{x}^2 \tag{12.1}$$

This is the most common way to represent fluctuations, and the quantity $\overline{\Delta x^2}$ is the variance of x. Taking the square root gives the familiar root-mean-square or rms deviation, which has the advantage of having the same units and dimensions as the measured quantity, x.

Note what happens if we sum two independent random variables

$$\overline{\Delta(x + y)^2} = \overline{(x + y - \overline{x + y})^2}$$
$$= \overline{x^2} - 2x\bar{x} + \bar{x}^2 + \overline{y^2} - 2y\bar{y} + \bar{y}^2 + 2\overline{xy} - 2\bar{x}\,\bar{y} \tag{12.2}$$

If x and y are uncorrelated, then we have $\overline{xy} = \bar{x}\,\bar{y}$. The resulting cancellations simplify the expression considerably

$$\overline{\Delta(x + y)^2} = \overline{x^2} - \bar{x}^2 + \overline{y^2} - \bar{y}^2$$
$$= \overline{\Delta x^2} + \overline{\Delta y^2} \tag{12.3}$$

This means that uncorrelated fluctuations add as their squares. They can thus be viewed as the edges of a right-angled triangle, with the fluctuations of the sum being the hypotenuse, as in the Pythagorean theorem. This is an important general property of fluctuations. It is a straightforward matter to generalize this form of additivity to an arbitrary number of variables as long as they are independent.

This give us a taste of the study of fluctuations, but there is quite a bit that the rms deviation cannot tell us. It tells us nothing about the shape of the distribution other than its width. For example, a distribution could be asymmetric, with larger positive deviations than negative deviations. The distribution tells us more about this kind of thing. But even if we know the complete distribution, we know nothing about the speed of the fluctuations. In fact, if a system has very slow fluctuations, then there is a serious risk that a measurement of $\overline{\Delta x^2}$ made in a short time will underestimate the true value. Techniques for quantitating the timescale of fluctuations will be developed later in this chapter.

12.2 | Number fluctuations and the Poisson distribution

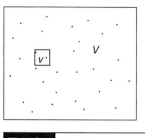

Fig. 12.1. A small volume v within a large volume V contains fluctuating numbers of molecules.

In the tiny volumes of cells and organelles, the number of molecules can be very small, and fluctuations can make this number deviate substantially from the mean. To quantitate these fluctuations we derive a probability distribution for the number of molecules in a small volume. Consider a very small volume element v within a much larger closed volume, V (Fig. 12.1). If V contains N independent molecules, then the probability of any particular molecule being in v is $p = v/V$. The probability of finding m molecules in v is then given by the binomial distribution

$$P(m) = \frac{N! p^m (1-p)^{N-m}}{(N-m)! m!} \tag{12.4}$$

We now make some approximations based on the small value of p, and the fact that expected values of m are tiny compared to N. First of all, $N!/(N-m)! = N(N-1)(N-2) \ldots (N-m+1) \sim N^m$. Thus, $N! p^m/(N-m)! \sim (Np)^m$. Furthermore, we know that $Np = \overline{m}$ (Eq. (6.37)) and this leaves us with $\overline{m}^m$. For the term $(1-p)^{N-m}$, we note that it can be rewritten as $(1-p)^{\frac{1}{p}(N-m)p}$. In this form we can recognize $\lim_{x \to 0}(1-x)^{1/x} = e^{-1}$ (closely related to the formal definition of e as $\lim_{x \to 0}(1+x)^{1/x}$), so we have $(1-p)^{N-m} \sim e^{-(N-m)p}$. Finally, we can reuse the relation $Np = \overline{m}$ for the exponent $(N-m)p \sim Np$ to obtain $e^{-\overline{m}}$. Breaking the right-hand side of Eq. (12.4) into the appropriate factors, and incorporating these approximations reduces Eq. (12.4) to

$$P(m) = \left(\frac{N! p^m}{(N-m)!} \right) ((1-p)^{N-m}) \left(\frac{1}{m!} \right)$$
$$= \frac{\overline{m}^m e^{-\overline{m}}}{m!} \tag{12.5}$$

This is the Poisson distribution. If we take $\overline{m}$ from the concentration, then it is a simple matter to calculate the probability of finding m molecules in a small volume.

The Poisson distribution is plotted for a few different values of $\overline{m}$ in Fig. 12.2. For $\overline{m} = 0.5$ the chance of finding no molecules ($m = 0$) is 0.61, and the probability falls off monotonically as m increases. For $\overline{m} = 2$ the highest probability is for $m = 1$ or 2. For $m = 10$ we have a nearly symmetrical distribution centered close to $\overline{m}$. For values of $\overline{m}$ in this range or higher, it becomes difficult to tell the difference between the Poisson distribution and the parent binomial distribution (Eq. (12.4)).

It is important to realize that information about the size of the system (the magnitude of N) is lost in the limit used to derive the Poisson distribution. Although N appears at the start (Eq. (12.4)), it

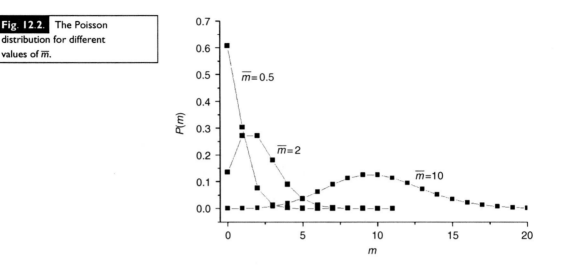

Fig. 12.2. The Poisson distribution for different values of $\overline{m}$.

drops out during the derivation. So the total number of molecules in the system is no longer relevant. If N had not been large enough, and p had not been small enough to justify the approximations leading to the Poisson limit, then an observed distribution would depend on N, and one might hope to be able to estimate N as a free parameter. In the Poisson limit this is not possible; N cannot be determined from the observed distribution. In practice it can be very difficult even to know from an observed distribution whether it is reasonable to hope to estimate N from measurements of fluctuations.

The Poisson distribution has a number of mathematical properties that aid in the analysis of fluctuations. Note that the series $\overline{m}^m/m!$ is the Taylor series for $e^{\overline{m}}$ (Eq. (A1.4))

$$\sum_{m=0}^{\infty} \frac{\overline{m}^m}{m!} = e^{\overline{m}} \tag{12.6}$$

So the sum of Eq. (12.5) from $m = 0$ to ∞ is $e^{\overline{m}}e^{-\overline{m}} = 1$. This is the expected result because a sum over the probability of all possible outcomes must be one.

We can use Eq. (12.6) to check that $\overline{m}$ is indeed the mean

$$\overline{m} = \sum_{m=0}^{\infty} mP(m) = \sum_{m=0}^{\infty} \frac{me^{-\overline{m}}\overline{m}^m}{m!} \tag{12.7}$$

The first term of the sum is zero, so we can take the sum from 1 to ∞

$$e^{-\overline{m}} \sum_{m=1}^{\infty} \frac{m\overline{m}^m}{m!} = e^{-\overline{m}}\overline{m} \sum_{m-1=0}^{\infty} \frac{\overline{m}^{m-1}}{(m-1)!} \tag{12.8}$$

By comparison with Eq. (12.6) we see that the sum is $e^{\overline{m}}$, so the right-hand side reduces to $\overline{m}$.

To evaluate the rms deviations from the mean we first determine $\overline{m^2}$, as follows

$$\overline{m^2} = \sum_m m^2 P(m) = \sum_m \frac{m^2 e^{-\overline{m}} \overline{m}^m}{m!}$$
$$= \overline{m} e^{-\overline{m}} \sum_{m=0}^{\infty} \frac{m \overline{m}^{m-1}}{(m-1)!} \qquad (12.9)$$

Again, the first term is zero, so the same procedure that led to Eq. (12.8) gives

$$\overline{m^2} = \overline{m} \, e^{-\overline{m}} \sum_{m-1=0}^{\infty} \frac{m \overline{m}^{m-1}}{(m-1)!}$$
$$= \overline{m} \, e^{-\overline{m}} \left(\sum_{m-1=0}^{\infty} \frac{(m-1) \overline{m}^{m-1}}{(m-1)!} + \sum_{m-1=0}^{\infty} \frac{\overline{m}^{m-1}}{(m-1)!} \right) \qquad (12.10)$$

Comparing with Eq. (12.8), the first sum is $\overline{m} \, e^{\overline{m}}$. According to Eq. (12.6), the second sum is $e^{\overline{m}}$. So we have

$$\overline{m^2} = \overline{m} \, e^{-\overline{m}} \left(\overline{m} \, e^{\overline{m}} + e^{\overline{m}} \right)$$
$$= \overline{m}^2 + \overline{m} \qquad (12.11)$$

Using this result together with Eq. (12.1) gives an expression for the rms deviations of the Poisson distribution

$$\sqrt{\overline{\Delta m^2}} = \sqrt{\overline{m^2} - \overline{m}^2} = \sqrt{\overline{m}^2 + \overline{m} - \overline{m}^2} = \sqrt{\overline{m}} \qquad (12.12)$$

Looking back at Fig. 12.2, we can see that the distributions do indeed get broader as $\overline{m}$ increases. However, the magnitude of the deviations relative to the mean decreases as $\sqrt{\overline{\Delta m^2}}/\overline{m} = 1/\sqrt{\overline{m}}$. This property will be used to estimate the threshold number of photons that the human eye can detect (see Section 12.3).

Before moving on, we briefly illustrate how to use the Poisson distribution for the task laid out at the beginning of this section – quantitating the fluctuations in molecule number in a small volume. Consider a 1 μm cube. The volume is 10^{-15} liters. At neutral pH, $[H^+] = 10^{-7}$ M. Multiplying the product, 10^{-22} moles, by Avogadro's number tells us that at neutral pH our 1 μm cube will contain 60 protons, on average. According to Eq. (12.12), the rms deviation around this mean will be $\sqrt{60} = 7.7$. So the number of protons in a 1 μm cube will fluctuate by ~12%. Even lower concentrations might be relevant to signaling molecules or regulatory proteins. So we could easily find ourselves in the realm where $\overline{m}$ is near one. The rms fluctuations are then similar in magnitude to the mean itself.

12.3 | The statistics of light detection by the eye

The eye can see extremely faint spots of light. The actual number of photons necessary for detection is so small that fluctuations in perception are large. These fluctuations were interpreted in terms

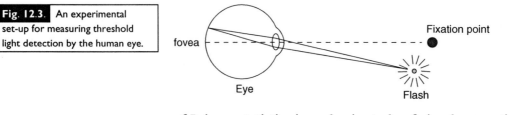

Fig. 12.3. An experimental set-up for measuring threshold light detection by the human eye.

of Poisson statistics in a classic study of visual perception (Hecht *et al.*, 1942; see Chapter 17 of Aidley, 1978, or Chapter 3.4 of Benedek and Villars, 2000).

The experiment is illustrated in Fig. 12.3. A source of light so small that it is nearly point-like was focused on the retina by the lens of the eye. All sorts of conditions were adjusted to optimize sensitivity. The light was focused on the most sensitive part of the retina (several degrees away from the fovea – a region where acuity is highest), and this required having the subject fixate on another object. Additionally, the subject was dark adapted (by sitting in a dark room for >20 min), the ideal wavelength of light was selected, and the duration of the light flash was optimized. Within the small area of retina upon which the spot of light falls, there are many photoreceptor cells, and an extremely large number of rhodopsin molecules. Absorption of a photon by any of these rhodopsin molecules can contribute to the visual response. If we take this large number of rhodopsin molecules, and accept the premise that absorption of light by just a few of them produces a threshold response, then we have a situation for which the Poisson distribution should apply – large N and low p.

The Poisson distribution is used to estimate the probability of absorbing a number of photons above the threshold, m_t. We therefore add up all the Poisson terms for $m > m_0$

$$P(m \geq m_0) = \sum_{m=m_t}^{\infty} \frac{e^{-\overline{m}} \overline{m}^m}{m!} \tag{12.13}$$

Note that the mean number of photons absorbed, $\overline{m}$, varies with the illumination strength. But for any particular $\overline{m}$, the subject may or may not see the experimental flash of light. Equation (12.13) gives this probability.

Equation (12.13) is plotted for a few different values of $\overline{m}$ in Fig. 12.4. The logarithmic scale helps visualize the change in steepness as m_0 increases. This plot shows that as the threshold rises the curve gets steeper. To see this, imagine that the number of absorption events needed for detection is large. Then there will be a very sharp threshold, with the subject seeing the light flash every time or never, depending on whether the intensity is above or below threshold. This is because the fluctuations in a large number are small relative to the mean. If the number of photons needed for detection is small enough for the fluctuations to be significant, then small changes in the light level will have a less dramatic effect on the fraction of times detection is successful.

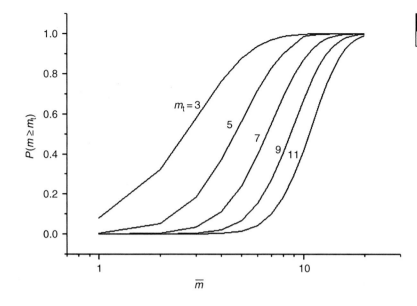

Fig. 12.4. Plots of Eq. (12.13) for various values of m_t.

In these experiments, subjects were tested repeatedly with faint spots of light to determine the probability of detection as a function of light intensity. The curves looked like those of Fig. 12.4, with the best fits obtained for m_t in the range 5–8. In the same experiments, the threshold was estimated independently by careful analysis of the many factors that determine how much light actually strikes the retina, and the fraction of this light that is absorbed. This number fell in the range 5–14, which was taken as good agreement with the estimate based on Poisson analysis.

One should wonder whether each of the 5–8 photons is absorbed by a different photoreceptor cell or whether some photoreceptors need to absorb more than one photon to register a response. It has been estimated that the area upon which the light spot is focused contains several hundred photoreceptors, so the chance of one photoreceptor absorbing two of the 5–8 photons is quite small. Therefore, a photoreceptor cell can be electrically activated by one photon. The retinal ganglion cell, which conveys the signal to the brain, receives synaptic inputs from many photoreceptor cells. The absorption events within a large pool of photoreceptors are funneled to a retinal ganglion cell, which sums these inputs and fires an action potential if the sum exceeds its threshold (Chapter 16).

12.4 | Equipartition of energy

When energy can be expressed as the square of some quantity (i.e. $E = aq^2$), then the Boltzmann distribution leads to a very simple result for the variance of that value (see Appendix 4 for the integrals)

$$\overline{q^2} = \frac{\int\limits_{-\infty}^{\infty} q^2 e^{-aq^2/kT} dq}{\int\limits_{-\infty}^{\infty} e^{-aq^2/kT} dq} = \frac{kT}{2a} \qquad (12.14)$$

So the mean energy is

$$\overline{E} = a\overline{q^2} = \frac{kT}{2} \qquad (12.15)$$

If the variable is velocity in the x direction, v_x, then the energy will be kinetic, with $a = m/2$. We then have the mean square velocity in that direction

$$\overline{v_x^2} = \frac{kT}{m} \qquad (12.16)$$

And the mean kinetic energy for motion in the x direction is $kT/2$.

If the variable is displacement within a parabolic potential energy well, then the potential energy is $\phi x^2/2$ (Eq. (2.17)), where ϕ is the force constant. Equation (12.14) now gives the mean square displacement, which tells us about fluctuations in position.

$$\overline{x^2} = \frac{kT}{\phi} \qquad (12.17)$$

This is a useful expression for estimating how much the position of an atom or molecule will vary if it is held in place by molecular forces. The various expressions for potential energy from Chapter 2 can often be used to calculate the force constants, which are obtained as half of the second derivative of the potential energy function at its minimum.

These results illustrate an important point about how thermal energy is distributed over different modes. Regardless of the mass of the molecule, or the strength of the restoring force, the energy in each mode will be $kT/2$. This is known as the principle of equipartition of energy. It is a principle rather than a law or theorem because it is not always true. The most important exceptions arise when quantum mechanics has to be considered. Quantization of energy levels invalidates the integration of a continuous variable in Eq. (12.14). For vibration energies of strong covalent bonds with light atoms, the energies of the quantum states are widely separated so the integral must be replaced by a sum. However, the weaker noncovalent forces that stabilize biological structures can be treated classically to a good approximation, so we can use the equipartition principle.

It is interesting to explore the consequences of this idea for a large molecule with many atoms. The equipartition principle states that each mode of motion will have $kT/2$ energy. If the atoms move classically (i.e. no quantum effects), then with three modes of kinetic energy, the total is $3kT/2$. Each atom lies in a potential energy well, in which displacements from the minimum are possible in the x, y, and z directions. This gives another $3kT/2$ in potential energy.[1] Thus, a molecule with N atoms will have a mean energy (kinetic plus potential) of $3NkT$. This provides the starting point for an evaluation of energy fluctuations in a large molecule.

12.5 | Energy fluctuations in a macromolecule

The ideas from the equipartition principle can be extended to gain an understanding of energy fluctuations of a large molecule in which energy is distributed over many internal modes. We first need to evaluate $\overline{E^2}$, where E^2 can be expressed as $\phi^2 q^4$ for any mode. We then obtain $\overline{q^4}$ from the Boltzmann distribution, just as $\overline{q^2}$ was earlier with Eq. (12.14) (again see Appendix 4)

$$\overline{q^4} = \frac{\int_{-\infty}^{\infty} q^4 e^{-\phi q^2/kT} dq}{\int_{-\infty}^{\infty} e^{-\phi q^2/kT} dq} = 3\left(\frac{kT}{2\phi}\right)^2 \tag{12.18}$$

The mean square energy is then

$$\overline{E^2} = \overline{\phi^2 q^4} = \frac{3}{4}(kT)^2 \tag{12.19}$$

By subtracting $\overline{E}^2 = (kT/2)^2$ (from equipartition) we can calculate the variance according to Eq. (12.1) as

$$\overline{\Delta E^2} = \frac{(kT)^2}{2} \tag{12.20}$$

Equation (12.20) applies to each internal mode of a large molecule. If a molecule has N atoms then each atom will contribute three kinetic energy modes and three potential energy modes. As in Eq. (12.3), the fluctuations add as squares, so the variance in the total energy is the sum of the variances over all $6N$ modes of the molecule. We therefore multiply Eq. (12.20) by $6N$ to obtain the energy variance for the entire molecule

$$\overline{\Delta E^2} = 3N(kT)^2 \tag{12.21}$$

Now we return to the mean energy. According to the equipartition principle, the mean energy of each mode is $kT/2$. So the mean energy of the whole molecule is $3NkT$. The heat capacity, c, which is the derivative of $\overline{E}$ with respect to T, is then $3Nk$. Combining this result with Eq. (12.21) leads to an important relation for energy fluctuations

$$\frac{\overline{\Delta E^2}}{kT^2} = c \tag{12.22}$$

---

[1] This is oversimplified, but the classical harmonic potential energy function can be transformed to a set of coordinates for which the potential energy is a sum of terms of the form γq^2 (Section 2.12). The end result is still $kT/2$ of potential energy per degree of freedom.

This equation tells us that we can determine the energy fluctuations by measuring the heat capacity. The derivation of Eq. (12.22) was not rigorous because equipartition does not hold for every internal mode of a molecule. However, this derivation is instructive because it shows how the relationship arises from the fact that both the energy fluctuations and the heat capacity reflect the number of energetic modes of a molecule.

It is possible to derive Eq. (12.22) much more rigorously, and without resorting to the equipartition principle (Chapter 2 of Hill, 1960; Chapter 25 of Kittel, 1958). We start with the mean energy of a system based on the Boltzmann distribution

$$\overline{E} = \frac{\sum_i E_i e^{-E_i/kT}}{\sum_i e^{-E_i/kT}} \tag{12.23}$$

The expression is rewritten as

$$\overline{E} \sum_i e^{-E_i/kT} = \sum_i E_i e^{-E_i/kT} \tag{12.24}$$

Each side is then differentiated with respect to temperature

$$\frac{\partial \overline{E}}{\partial T} \sum_i e^{-E_i/kT} + \overline{E} \sum_i \frac{E_i}{kT^2} e^{-E_i/kT} = \sum_i \frac{E_i^2}{kT^2} e^{-E_i/kT} \tag{12.25}$$

We can see that the derivative in the term on the left is the heat capacity, c. If we divide through by the sum $\sum_i e^{-E_i/kT}$ we see that the other two terms are $\overline{E}^2/kT^2$ and $\overline{E^2}/kT^2$. Equation (12.25) can thus be rewritten as

$$c = \frac{\overline{E^2}}{kT^2} - \frac{\overline{E}^2}{kT^2} = \frac{\overline{\Delta E^2}}{kT^2} \tag{12.26}$$

And once again we have Eq. (12.22).

This general result was discovered by Gibbs, and independently many years later by Einstein. It relates the energy fluctuations of a system to the energy absorbed during heating. It can be used in conjunction with Eq. (12.21) to illustrate how the size of a system influences its energy fluctuations. Heat capacity scales with size. If each vibrating atom absorbs thermal energy independently then the heat capacity is proportional to the number of atoms, N. Equation (12.26) then tells us that the variance in energy is also proportional to N, so the rms energy goes as $\sqrt{N}$. The energy of the system, like the heat capacity, is proportional to N, so the energy fluctuations relative to the mean vary as $1/\sqrt{N}$. This illustrates the general trend mentioned at the start of this chapter. As the system gets larger, the fluctuations become harder to see. For a mole of material, one needs a measurement sensitive to one part in 10^{12} (approximately the square root of Avogadro's number) to see the thermal energy fluctuations. We can improve our chances of observing thermal fluctuations by looking at smaller systems.

Equation (12.26) has been used to study the energy fluctuations of a protein (Cooper, 1976). The heat capacities of proteins fall in the range 0.30–0.35 cal g^{-1} K^{-1}. For a protein with a molecular weight of 25 000 Da, this gives a molar heat capacity of 7.5–8.75 kcal $mole^{-1}$ K^{-1}. With the aid of Eq. (12.22) this gives $\sqrt{\overline{\Delta E^2}} = 6.4 \times 10^{-20}$ calories $molecule^{-1}$. This works out to 38 kcal $mole^{-1}$, which is within a factor of about two of the ΔH for thermal denaturation for a protein of this size (Section 1.4). Thus, a protein undergoes energy fluctuations that are comparable to the energy that stabilizes its native state.

With such large energy fluctuations, one might think that the native state of a protein is in a precarious position. However, the fluctuations are not confined to the internal modes that unfold the protein, but are distributed over the entire molecule. In fact, the energy fluctuations calculated from Eq. (12.26) are small compared to the total thermal energy of the protein. The equipartition principle gives a mean energy per atom of $3kT \sim 10^{-21}$ cal. For a 25-kDa protein with $\sim$5000 atoms we have $\sim$5 $\times 10^{-18}$ calories. The value of $\sqrt{\overline{\Delta E^2}} = 6.4 \times 10^{-20}$ calories from Eq. (12.26) is $\sim$2 orders of magnitude smaller. The energy fluctuations of a protein reflect the independent jittering of its many atoms. Thus, a well-defined structure can still have energy fluctuations of the magnitude implied by Eq. (12.26).

12.6 | Fluctuations in protein ionization

Typical proteins have a large number of ionizable groups. Changing the pH will change the mean charge on a protein as different groups are titrated (Section 11.3). For a group with a pK equal to the pH, there will be a 50% chance of being ionized. When the pH is held constant, the charge on a protein will fluctuate as groups with pK values near the pH change their ionization state. A theory developed by Linderstrøm–Lang relates these charge fluctuations to the slope of the titration curve. The derivation has an interesting parallel with the derivation of Eq. (12.26) above.

We consider a sequence of ionization equilibria (Scheme (12A)) for a protein

$$P_0 \underset{}{\overset{H^+}{\rightleftarrows}} P_1 \underset{}{\overset{H^+}{\rightleftarrows}} P_2 \underset{}{\overset{H^+}{\rightleftarrows}} \cdots\cdots\cdots P_{n-1} \underset{}{\overset{H^+}{\rightleftarrows}} P_n \qquad (12A)$$

First we work out an expression for the state of ionization as a function of pH. This follows the analysis of binding site saturation models treated in Chapter 4. For each ionization step, an equilibrium binding relation holds

$$K_1 = \frac{[P_1]}{[P_0][H^+]}, \qquad K_2 = \frac{[P_2]}{[P_1][H^+]} \cdots K_n = \frac{[P_n]}{[P_{n-1}][H^+]} \qquad (12.27)$$

or

$$[P_1] = K_1[P_0][H^+], \quad [P_2] = K_2[P_1][H^+] \cdots [P_n] = K_n[P_{n-1}][H^+] \quad (12.28)$$

Successive substitutions with these equations lead to an expression for the concentration of a protein with m sites protonated

$$[P_m] = [P_0][H^+]^m \prod_{i=1}^{m} K_i \qquad (12.29)$$

For the sake of simplicity we replace the product of K_is by a new constant, $J_m = \prod_{i=1}^{m} K_i$, as follows

$$[P_m] = [P_0][H^+]^m J_m \qquad (12.30)$$

Now we write the equilibrium fraction of sites protonated as the number of protonated sites over the total number of sites

$$\overline{m} = \frac{\sum\limits_{m=0}^{n} m[P_m]}{\sum\limits_{m=0}^{n} [P_m]} \qquad (12.31)$$

Substituting a relation of the form of Eq. (12.30) for each P_m, and canceling out factors of P_0 from the numerator and denominator gives

$$\overline{m} = \frac{\sum\limits_{m=0}^{n} m J_m [H^+]^m}{\sum\limits_{m=0}^{n} J_m [H^+]^m} \qquad (12.32)$$

This expression tells us how the mean number of protonated sites on the protein varies as a function of $[H^+]$. To see how the value fluctuates around this mean, we multiply through by the denominator and differentiate with respect to $[H^+]$

$$\frac{\partial \overline{m}}{\partial [H^+]} \sum_{m=0}^{n} J_m [H^+]^m + \overline{m} \sum_{m=0}^{n} m J_m [H^+]^{m-1} = \sum_{m=0}^{n} m^2 J_m [H^+]^{m-1} \quad (12.33)$$

This is then multiplied on both sides by $[H^+]$. In the first term we can reexpress the derivative

$$[H^+] \frac{\partial \overline{m}}{\partial [H^+]} = \frac{\partial \overline{m}}{\partial \log[H^+]} = -\frac{\partial \overline{m}}{\partial pH} \qquad (12.34)$$

Dividing Eq. (12.33) through by the sum on the left leaves the first term on the right as $\overline{m}^2$ and the second term as $\overline{m^2}$. With the aid of Eq. (12.1) we have an expression for the fluctuations in m

$$-\frac{\partial \overline{m}}{\partial \text{pH}} = \overline{m^2} - \overline{m}^2 = \overline{\Delta m^2} \tag{12.35}$$

Thus, a protein's charge will fluctuate the most at a pH where the titration curve is steepest. This usually occurs near the isoelectric point. Equation (12.35) was used to estimate the fluctuations in charge of hemoglobin, giving an rms charge at the isoelectric point of about 1.85 (Cohn and Edsall, 1943). For ribonuclease a value of 3.65 was obtained (Tanford, 1961, p. 573).

12.7 | Fluctuations in a two-state system

The two-state model was examined in earlier chapters to understand protein conformational transitions (Scheme (12B))

$$A \underset{\beta}{\overset{\alpha}{\rightleftharpoons}} B \tag{12B}$$

At equilibrium we have

$$\frac{[B]}{[A]} = K_{eq} \tag{12.36}$$

The equilibrium constant can also be expressed in terms of the rate constants as $K_{eq} = \alpha/\beta$ (Section 7.3). If the total protein concentration is T, with $[A] + [B] = T$, then at equilibrium the number of molecules of B and A will be fractions of T. Replacing K_{eq} in Eq. (12.36) with α/β and rearranging gives

$$[A] = T\frac{\beta}{\alpha + \beta} \tag{12.37a}$$

$$[B] = T\frac{\alpha}{\alpha + \beta} \tag{12.37b}$$

Note that [A] and [B] in these two expressions are actually average concentrations. At any instant in time there will be fluctuations as small excesses of A or B appear. These fluctuations dissipate and reappear as a result of the stochastic behavior arising from the independent conformational transitions of each protein. We will first evaluate the magnitude of these fluctuations, and then in Section 12.9 we will examine their dynamics.

From Eqs. (12.37a) and (12.37b) we write the equilibrium probability of a protein molecule being in conformation A or B as

$$p_a = \frac{\beta}{\alpha + \beta} \tag{12.38a}$$

$$p_b = \frac{\alpha}{\alpha + \beta} \tag{12.38b}$$

Since each protein molecule is independent, we can use the binomial distribution to express the probabilities of finding a certain number of molecules in one conformation or the other. Take N_t as the total number of protein molecules. The probability of finding N_a molecules in conformation A is

$$P(N_a) = \frac{N_t!}{N_a!(N_t - N_a)!} p_a{}^{N_a} (1 - p_a)^{N_t - N_a} \tag{12.39}$$

A corresponding expression can be written for $P(N_b)$.

The situation is now mathematically equivalent to the random walk of Chapter 6, where the binomial distribution was used to calculate the probability of a particular number of steps to the right or left. We can thus use Eq. (6.37) to obtain the mean number of molecules of A or B. This analysis leads back to Eqs. (12.37a) and (12.37b). For A we have

$$\overline{N_a} = \sum_{N_a = 0}^{N_t} N_a \frac{N_t!}{N_a!(N_t - N_a)!} p_a{}^{N_a} (1 - p_a)^{N_t - N_a} = N_t p_a = N_t \frac{\beta}{\alpha + \beta} \tag{12.40}$$

where Eq. (12.38a) was used to replace p_a.

For the fluctuations we need to know the mean square as well, so we use Eq. (6.42) to give

$$\overline{N_a^2} = \sum_{N_a = 0}^{N_t} N_a^2 \frac{N_t!}{N_a!(N_t - N_a)!} p_a{}^{N_a} (1 - p_a)^{N_t - N_a} = (N_t p_a)^2 + N_t p_a (1 - p_a) \tag{12.41}$$

Combining Eqs. (12.40) and (12.41) gives the fluctuations in number of molecules of A around the mean

$$\overline{\Delta N_a^2} = \overline{N_a^2} - \overline{N_a}^2 = N_t p_a (1 - p_a) = N_t \frac{\alpha \beta}{(\alpha + \beta)^2} \tag{12.42}$$

Once again we see that the mean-square fluctuations scale as N. Thus, the rms fluctuations, as a fraction of the average number of molecules of A can be expressed as

$$\frac{\sqrt{\overline{\Delta N_a^2}}}{\overline{N_a}} = \sqrt{\frac{\alpha}{\beta N_p}} \tag{12.43}$$

and the fluctuations relative to the mean decrease with size, scaling as $N^{-\frac{1}{2}}$.

12.8 | Single-channel current

Fluctuation analysis has been used to study ion channels. As ion channels open and close the current through the cell membrane fluctuates. Analysis of these fluctuations provided some of the first estimates of unitary channel properties. Although this method has been largely superceded by direct measurements of single-channel currents with the patch clamp, there are still situations where the analysis of channel noise can be useful. The results from the preceding section will be used to determine the single-channel current from fluctuations in membrane current.

The gating of a channel is treated as an equilibrium between closed and open states (Scheme (12C))

$$C \underset{\beta}{\overset{\alpha}{\rightleftharpoons}} O \qquad\qquad (12C)$$

Only the open channel conducts ions. If there are N channels and N_o of them are open, then the current is simply iN_o, where i is the single-channel current. The mean number of open channels, $\overline{N_o}$, is equal to the total number of channels times the open probability p_o. We can therefore express the mean current as

$$\bar{I} = ip_o N_t \qquad\qquad (12.44)$$

with p_o taken from Eq. (12.38b) as $\alpha/(\alpha + \beta)$.

The current variance can be calculated from the variance in the number of open channels, for which Eq. (12.42) can be used

$$\overline{\Delta I^2} = i^2 \overline{\Delta N_o^2} = i^2 N_t \frac{\alpha\beta}{(\alpha + \beta)^2} \qquad\qquad (12.45)$$

Dividing the variance by the mean eliminates N_t

$$\frac{\overline{\Delta I^2}}{\bar{I}} = i(1 - p_o) = i\frac{\beta}{\alpha + \beta} \qquad\qquad (12.46)$$

where p_o varies between zero and one. If it is close to zero then the right-hand side of Eq. (12.46) reduces to i. The ratio of the variance to the mean can then be taken as an estimate of the single-channel current.

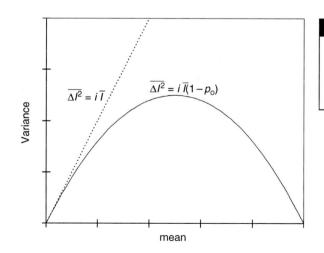

$$\overline{\Delta I^2} = i\bar{I}$$

$$\overline{\Delta I^2} = i\bar{I}(1 - p_o)$$

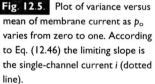

Fig. 12.5. Plot of variance versus mean of membrane current as p_o varies from zero to one. According to Eq. (12.46) the limiting slope is the single-channel current i (dotted line).

This is the basis of a basic method for studying ion channels (DeFelice, 1981; Lecar and Sachs, 1981). It is important to ascertain that p_o is in fact small. The common practice is to plot the variance versus the mean and take the limiting slope at the zero current intercept (Fig. 12.5). Note that this curve goes through a maximum and then returns to zero as the mean current increases. These zeroes at each end of the plot correspond to open probabilities of zero and one. Clearly, there must be zero variance in the limiting cases of all channels open or all closed.

12.9 | The correlation function of a two-state system

So far we have considered only the magnitudes of fluctuations and ignored the speed with which they occur. Speed is an important aspect of fluctuations. As fluctuations occur in real time, they give rise to noise in a measurement. This noise is usually a nuisance, but it does contain information about the kinetics of the underlying process that generates this noise.

One of the most important ways of analyzing noise in a signal $S(t)$ is through the correlation function, denoted as $F_c(t)$

$$
\begin{aligned}
F_c(t) &= \overline{(S(0) - \overline{S})(S(t) - \overline{S})} \\
&= \overline{S(0)S(t) - S(0)\overline{S} - S(t)\overline{S} + \overline{S}^2} \\
&= \overline{S(0)S(t)} - \overline{S}^2
\end{aligned}
$$

$$(12.47)$$

The correlation function is obtained by measuring S at time $t = 0$, and again at time t, multiplying the two measurements together, and averaging over many such measurements. If the intervening time interval is very large compared to the timescale of the fluctuations, then there will be no correlation between $S(0)$ and $S(t)$. The mean of the product $S(0)S(t)$ will then be $\overline{S}^2$ and $F_c(t)$ will be zero. If the intervening time interval is so short that S does not have time to change, then $S(0)S(t) = \overline{S^2}$, and $F_c(t)$ becomes $\overline{\Delta S^2}$. The interesting question is how $F_c(t)$ gets from $\overline{\Delta S^2}$ to zero as t increases.

We will illustrate this behavior using the example of a two-state ion channel (Scheme (12C)) (Lecar and Sachs, 1981), expressing $F_c(t)$ in terms of the opening and closing rate constants, α and β. This derivation can be seen as another twist on the analysis of the kinetics of the two-state model in Chapter 7.

Assume that we are looking at the fluctuations in the current through a single channel, so $S(t)$ becomes $I(t)$. When the channel is closed I is zero, so the only way the product $I(0)I(t)$ can be nonzero is if the channel is open at both times. The probability that the channel is open at $t = 0$ is $\alpha/(\alpha + \beta)$. The channel will be open at a later time t if there is an even number of gating transitions in the time interval between 0 and t. Thus,

$$F_c(t) = i^2 \frac{\alpha}{\alpha + \beta} P_e(t) - i^2 \left(\frac{\alpha}{\alpha + \beta}\right)^2 \tag{12.48}$$

where P_e is defined as the probability of an even number of transitions. The second term is the mean current squared, $\bar{I}^2$, corresponding to $\bar{S}^2$ in Eq. (12.47).

To obtain an expression for $P_e(t)$ we use a method similar to that used to derive the open-time distribution of a channel (Section 7.9). We write down an expression for $P_e(t + dt)$ in terms of $P_e(t)$. One way for the number of transitions to be even at $t + dt$ is for the number to be even at t, and for no transitions to occur in the intervening interval dt. The only other way to get an even number is with an odd number at t and a transition in the interval dt. For the first case we need the probability of no transition when the channel is open. The probability of closing is βdt, so the probability of not closing is $1 - \beta dt$. For the second case the channel is closed so the probability of a transition in the interval dt is αdt. Finally, we also need to note that the probability of an odd number of transitions is equal to one minus the probability of an even number of transitions. Putting this all together gives

$$P_e(t + dt) = P_e(t)(1 - \beta dt) + (1 - P_e(t))\alpha dt \tag{12.49}$$

This is rearranged into a differential equation

$$\frac{dP_e}{dt} = -(\alpha + \beta)P_e + \alpha \tag{12.50}$$

The general solution of this equation contains an exponential $e^{-(\alpha + \beta)t}$. With the initial condition that $P_e(0) = 1$, the complete solution is

$$P_e(t) = \frac{\alpha}{\alpha + \beta} + \frac{\beta e^{-(\alpha + \beta)t}}{\alpha + \beta} \tag{12.51}$$

Substituting this into Eq. (12.48) gives the correlation function

$$F_c(t) = i^2 \frac{\alpha\beta}{(\alpha + \beta)^2} e^{-(\alpha + \beta)t} = \overline{\Delta I^2} e^{-(\alpha + \beta)t} \tag{12.52}$$

where the final simplification made use of Eq. (12.45), and the assumption that the correlation function for N independent channels should scale as N.

This function has the properties expected. It starts off at $\overline{\Delta I^2}$ and decays to zero. The most important point to be seen in this expression is that the correlation function decays at the same rate as a response to a macroscopic perturbation (e.g. Eq. (7.4)). Thus, the same fundamental rate, $\alpha + \beta$, controls the dynamics at both the microscopic and macroscopic levels. This parallel between microscopic fluctuations and macroscopic kinetics is a recurring theme in dynamic processes. In the dynamics of channel noise, we will see how this can be used to investigate channel gating. But before we do, we need another tool for noise analysis.

12.10 | The Wiener–Khintchine theorem

Dynamic processes can be viewed either in real time or in terms of their characteristic frequencies. The real-time perspective is more direct; the correlation function of the preceding section is easily visualized and an exponential time constant relates directly to the time for a fluctuation to decay. In the frequency domain we imagine the signal as a random oscillation characterized by a range of frequencies. High-frequency oscillations correspond to short correlation times. These two realms are related by the Wiener–Khintchine theorem (Chapter 22 of McQuarrie, 1976; part 2 of Kittel, 1958), which will be proved here after introducing the idea of the Fourier transform of a random signal.

The frequency approach to fluctuations is based on the Fourier transform, in which a signal is decomposed into a sum of sine or cosine functions (Appendix 3). A more rapidly varying signal has higher frequency components. This is illustrated in Fig. 12.6, where noisy signals are shown with different frequencies filtered out. The top trace has contributions <200 Hz, and the noise looks slower. The traces below have more high-frequency components and the noise looks faster. The Fourier transform is a mathematical technique for determining the contribution of each frequency.

A random signal, $S(t)$, might look like one of the traces in Fig. 12.6. We can express $S(t)$ as an integral in which many sine or cosine waves of different frequencies are combined in just the right proportion. This Fourier integral takes the form

$$S(t) = \frac{1}{2\pi} \int_{-\infty}^{\infty} A(\omega)e^{i\omega t} d\omega \tag{12.53}$$

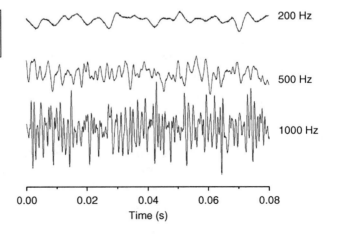

Fig. 12.6. Noise recorded by an amplifier with different high-frequency cut-offs.

200 Hz

500 Hz

1000 Hz

We use the complex expression, with $e^{i\omega t} = \cos(\omega t) + i\sin(\omega t)$ and $i = \sqrt{-1}$, because subsequent manipulations are simpler in this form. The function $A(\omega)$ is the contribution to the signal of fluctuations with a frequency of ω. For example, in the top trace of Fig. 12.6, $A(\omega)$ is very small for $f = \omega/2\pi > 200$ Hz but large for lower frequencies.[2] The function $A(\omega)$ is determined from $S(t)$ by taking its Fourier transform

$$A(\omega) = \frac{1}{T}\int_{-T}^{T} S(t)e^{-i\omega t}dt \tag{12.54}$$

where $\pm T$ is the interval in which the signal is examined.[3] We can think of $A(\omega)$ as a time average of $S(t)e^{-i\omega t}$ (times two). This interval must be longer than several periods of the slowest relevant frequency. The actual value of T often does not matter. However, it must be taken into account if you are analyzing experimental data and want to write a computer program to perform the integration numerically.

We can make use of the Fourier transform of $S(t)$ to show how noise with different frequencies adds together. Consider an integral over $A(\omega)^2$. We can replace one of the factors of $A(\omega)$ by its representation as a Fourier transform of $S(t)$ (Eq. (12.54))

$$\int_{-\infty}^{\infty} A(\omega)^2 d\omega = \frac{1}{T}\int_{-\infty}^{\infty} A(\omega) \int_{-T}^{T} S(t)e^{-i\omega t}dt\,d\omega \tag{12.55}$$

Combining the remaining $A(\omega)$ with $e^{i\omega t}$ and integrating over ω yields $2\pi S(t)$ by Eq. (12.53). The right-hand side then becomes an integral over $S(t)^2$ as follows

$$\int_{-\infty}^{\infty} A(\omega)^2 d\omega = \frac{2\pi}{T}\int_{-T}^{T} S(t)^2 dt \tag{12.56}$$

We now realize that the time average of S^2 can be written as

$$\frac{1}{2T}\int_{-T}^{T} S(t)^2 dt = \overline{S^2} \tag{12.57}$$

So Eq. (12.56) becomes

$$\overline{S^2} = \frac{1}{4\pi}\int_{-\infty}^{\infty} A(\omega)^2 d\omega \tag{12.58}$$

---

[2] Note that ω is in radian s^{-1} and f is in cycle s^{-1} so $\omega = 2\pi f$.
[3] Normally the interval for a Fourier transform is $\pm\pi$. Because the magnitude of the integral is proportional to the interval length we must divide by T.

Equation (12.58) is known as Parseval's theorem. It states the intuitively reasonable result that if we add up the noise over all frequencies we get the mean-square fluctuations in the signal. So $A(\omega)^2 d\omega$ represents the fraction of the total noise contributed by that particular frequency. It is significant that the frequency components add as squares. This is another example of the property embodied by Eq. (12.3); different frequencies act as independent sources of fluctuations. Figure 12.6 shows that including higher frequencies increases the overall amplitude of the noise.

The quantity $A(\omega)^2/(4\pi)$ is encountered a great deal in fluctuation analysis and is denoted as $P(\omega)$. It has two commonly used names, *power spectrum* and *spectral density*.

Let us now apply the Fourier transform to the correlation function of $S(t)$. To avoid the term $\bar{S}^2$ in Eq. (12.47), we subtract out the mean so that the signal fluctuates around zero. Our correlation function can now be written as a time average by taking an integral over time and dividing by the duration of the interval

$$F_c(s) = \overline{S(t)S(t+s)} = \frac{1}{2T} \int_{-T}^{T} S(t)S(t+s)dt \tag{12.59}$$

As above in Eq. (12.54), T is long enough to give a good average.

Because $F_c(s)$ has the same basic properties as $S(t)$, we can Fourier transform Eq. (12.59) in the same way as in Eq. (12.54)

$$\frac{1}{T} \int_{-T}^{T} F_c(s)e^{-i\omega s}ds = \frac{1}{2T^2} \int_{-T}^{T} \int_{-T}^{T} e^{-i\omega s}S(t)S(t+s)dtds \tag{12.60}$$

Inserting $1 = e^{-i\omega t}e^{i\omega t}$ on the right-hand side gives

$$\frac{1}{T} \int_{-T}^{T} F_c(s)e^{-i\omega s}ds = \frac{1}{2T^2} \int_{-T}^{T} \int_{-T}^{T} e^{-i\omega(t+s)}e^{i\omega t}S(t)S(t+s)dtds \tag{12.61}$$

We can now separate the right-hand side into two Fourier transforms

$$\frac{1}{T} \int_{-T}^{T} F_c(s)e^{-i\omega s}ds = \frac{1}{2T^2} \int_{-T}^{T} S(t)e^{i\omega t}dt \int_{-T}^{T} e^{-i\omega(t+s)}S(t+s)ds \tag{12.62}$$

where $1/T$ times the first integral gives $A(\omega)$ (Eq. (12.54)). The second integral is performed by replacing $t + s$ with a new variable to give another factor of $A(\omega)$. The result is

$$\frac{1}{T} \int_{-T}^{T} F_c(s)e^{-i\omega s}ds = \frac{1}{2}A(\omega)^2 = 2\pi P(\omega) \tag{12.63}$$

with $P(\omega)$ as defined above (following Eq. (12.58)).

Equation (12.63) is the Wiener–Khintchine theorem. It states that the correlation function and the power spectrum of a signal are related by Fourier transformation. With an inverse Fourier transform of the spectral density one can return to $F_c(s)$. This provides a direct route back and forth between the time and frequency representations of dynamic fluctuations. We will now use the Wiener–Khintchine theorem to look at the power spectrum of a fluctuating two-state system.

12.11 | Channel noise

Section 12.9 developed the correlation function of a two-state system, and Section 12.10 related the correlation function to the power spectrum. The natural step to take at this point is to put these two results together. We will use the Wiener–Khintchine theorem to derive the power spectrum of a two-state system. Equation (12.63) with $F_c(t)$ taken from Eq. (12.52) gives

$$P(\omega) = \frac{1}{2\pi} \int_{-\infty}^{\infty} F_c(t)e^{-i\omega t}dt = \frac{1}{2\pi} \int_{-\infty}^{\infty} \overline{\Delta I^2}e^{-(\alpha+\beta)t}e^{-i\omega t}dt \qquad (12.64)$$

Since our integral includes negative values of t we have to revise the expression for $F_c(t)$ to use the absolute value of t in the exponential. Changing the sign of t in $F_c(t)$ should not change its value since averaging over $I(s)I(s+t)$ should give the same result as averaging over $I(s)I(s-t)$. The integral above then becomes twice the integral over $[0, \infty]$

$$P(\omega) = \frac{1}{\pi} \int_{0}^{\infty} \overline{\Delta I^2}e^{-(\alpha+\beta+i\omega)t}dt = \frac{\overline{\Delta I^2}}{\pi} \frac{1}{\alpha+\beta+i\omega} \qquad (12.65)$$

This is a complex expression from which the real part must be extracted. The imaginary part can be ignored because the signal $I(t)$ is real

$$\begin{aligned} P(\omega) &= \frac{\overline{\Delta I^2}}{\pi} \frac{1}{\alpha+\beta+i\omega} \left(\frac{\alpha+\beta-i\omega}{\alpha+\beta-i\omega} \right) \\ &= \frac{\overline{\Delta I^2}}{\pi} \frac{\alpha+\beta-i\omega}{(\alpha+\beta)^2+\omega^2} \end{aligned} \qquad (12.66)$$

Throwing away the imaginary part with $i\omega$ in the numerator, and factoring $(\alpha+\beta)^2$ from the denominator, we have

$$P_{re}(\omega) = \frac{\overline{\Delta I^2}}{\pi(\alpha+\beta)} \frac{1}{1+(\omega/(\alpha+\beta))^2} \qquad (12.67)$$

Finally, we note that ω is in units of radian s^{-1}. The more familiar experimental units of cycle s^{-1} introduces a factor of 2π. We then have $2\pi f/(\alpha+\beta)$ in place of $\omega/(\alpha+\beta)$. The quantity $(\alpha+\beta)/2\pi$ is called

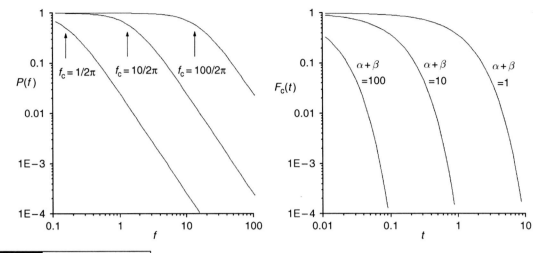

Fig. 12.7. Plots of $P(f)$ from Eq. (12.68) and $F_c(t)$ from Eq. (12.52), with $f_c = (\alpha + \beta)/2\pi$.

the corner frequency, f_c, for reasons that are easily seen when you look at a plot of this function (Fig. 12.7). In terms of f the power spectrum is then

$$P_{re}(f) = \overline{\Delta I^2}\,\frac{2}{f_c}\frac{1}{1 + (f/f_c)^2} \tag{12.68}$$

Note that a factor of 2π has been incorporated because as a density, $P(\omega)$ has an implicit $d\omega$, which must be replaced by $2\pi df$. Note also that this power spectrum is for the range of frequencies $[-\infty, \infty]$. It is common to define the power spectrum density for positive frequencies only, and then the result is multiplied by 2.

The functional form of Eq. (12.68) is known as a *Lorentzian*, and is seen in many applications. The Lorentzian is important because it is the Fourier transform of the ubiquitous exponential function. It is plotted for a few values of f_c in Fig. 12.7, along with plots of $F_c(t)$ to show how the two vary in an inverse manner: longer times go with lower frequencies. The value of f_c is marked in the plot of $P(f)$, and it can be seen to lie at the "corner" of the curve, where $P(f)$ starts to fall off rapidly with increasing f.

The curves in Fig. 12.7 can be viewed in terms of the sequence of noisy traces in Fig. 12.6. The top trace of Fig. 12.6, with a cutoff at 200 Hz, gives a power spectrum with $f_c = 200$ Hz. The lower traces in Fig. 12.6 give higher corner frequencies.

The analysis of the power spectrum in ion channel fluctuations has received wide use, primarily to measure the time constant of channel gating. DeFelice (1981) and Lecar and Sachs (1981) summarize some of the important applications. Single-channel analysis (Chapters 7 and 9) is generally more powerful than noise analysis, but when single-molecule events are too small to see, investigators can still use the power spectrum of the channel noise to estimate the rate constants. Most channels show multistate kinetic behavior and so the two-state model is not applicable. Extensions of noise theory to multi-state kinetic systems were developed by Chen and Hill (1973).

12.12 | Circuit noise

Fluctuations arise in an electrical circuit as electrons undergo random thermal motion. These fluctuations are the basis of noise in electrical measurements. This noise will now be analyzed, and the benefit will be two-fold. We will get a practical understanding of the smallest signals that can be seen with electronic instruments, and we will learn another important basic approach to the study of fluctuations.

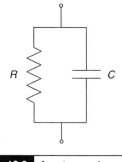

Fig. 12.8. A resistor and capacitor in parallel form a model circuit for noise calculations.

Our starting point is a circuit containing a resistor and capacitor in parallel (Fig. 12.8). If a sensitive voltmeter is used to read the voltage between the two leads at the top and the bottom, we would observe a signal that fluctuates around an average voltage of zero. These fluctuations reflect momentary excursions of the net electrostatic potential from its equilibrium value at $V = 0$. The deviations in voltage drive current through the resistor to bring the system back to equilibrium. A voltage fluctuation, V, corresponds to a charge fluctuation, q, and these are in a ratio set by the capacitance of the circuit

$$C = \frac{q}{V} \tag{12.69}$$

Moving a small amount of additional charge δq through a potential difference V changes the energy of the system by an amount $\delta q V$. The potential energy of the system with a total charge imbalance of q is then an integral reflecting the electrostatic work done to get to that point

$$E = \int_{q'=0}^{q} V dq' = \int_{q'=0}^{q} \frac{q'}{C} dq' = \frac{q^2}{2C} = \frac{CV^2}{2} \tag{12.70}$$

Since the energy goes as V^2, we can use the equipartition principle (Section 12.4) to obtain the mean energy in this mode as $kT/2$. The mean-square voltage follows directly

$$\overline{V^2} = \frac{kT}{C} \tag{12.71}$$

Note that the resistance does not appear in this expression. It will enter shortly as soon as we consider the time scale of the fluctuations.

Treating the time scale of the fluctuations requires a differential equation to express the change in voltage. The starting point is the standard analysis of the dynamic response of an RC circuit. From Eq. (12.69) we have $V = q/C$, which we differentiate with respect to time. Note that $dq/dt = -I$ because positive current through the resistor in Fig. 12.8 reduces the charge on the capacitor

$$\frac{dV}{dt} = \frac{-I}{C} \qquad (12.72)$$

Replacing I with V/R (Ohm's law) gives a first order differential equation in V

$$\frac{dV}{dt} = \frac{-V}{RC} \qquad (12.73)$$

For an initial value $V(0)$ the solution is

$$V = V(0)e^{-t/RC} \qquad (12.74)$$

Equation (12.74) does not help us understand fluctuations because it gives the response of the circuit to a macroscopic perturbation. To understand fluctuations we take Eq. (12.73) and add a noise term, Φ

$$\frac{dV}{dt} = \frac{-V}{RC} + \Phi(t) \qquad (12.75)$$

The variable $\Phi(t)$ is an electrical noise that fluctuates randomly with time and moves the voltage away from zero with small unpredictable lurches. We will now see that if it has the property that fluctuations in the positive and negative directions have equal probability, then we can derive a useful expression for the correlation function of V.

We can solve Eq. (12.75) by bringing $-V/RC$ to the left-hand side and combining these two terms as the derivative of a product

$$e^{-t/RC}\frac{dVe^{t/RC}}{dt} = \Phi(t) \qquad (12.76)$$

Multiplying Eq. (12.76) through by $e^{t/RC}$ and integrating from 0 to t leads to

$$V(t)e^{t/RC} - V(0) = \int_0^t e^{t/RC}\Phi(t)dt \qquad (12.77)$$

We then obtain

$$V(t) = V(0)e^{-t/RC} + e^{-t/RC}\int_0^t e^{t/RC}\Phi(t)dt \qquad (12.78)$$

If further progress depended on being able to evaluate the integral explicitly we would have a problem. Instead, we take an ensemble average. Because positive values of Φ are as likely as negative values, the ensemble average of the integral is zero. Thus, we can reach our goal of the correlation function of V by multiplying Eq. (12.78) by $V(0)$ and taking the ensemble average

$$F_c(t) = \overline{V(t)V(0)} = \overline{V^2}e^{-t/RC}$$

$$= \frac{kT}{C}e^{-t/RC} \tag{12.79}$$

The final step made use of Eq. (12.71).

This is an important result because it shows that microscopic fluctuations have the same dynamic character as responses to macroscopic perturbations (Eq. (12.74)). Either way, the system decays exponentially with a time constant of RC. This was also seen for the correlation function of the two-state model (Eq. (12.52), Section 12.9).

To complete the dynamic description of circuit noise we use the Wiener–Khintchine theorem to obtain the power spectrum of the noise from the correlation function. Using Eq. (12.79) for $F_c(t)$ in Eq. (12.64) gives

$$P(\omega) = \frac{kT}{2\pi C} \int_{-\infty}^{\infty} e^{-t/RC}e^{-i\omega t}dt \tag{12.80}$$

This integral can be evaluated, and by the same process that lead from Eq. (12.64) to Eq. (12.68) we have

$$P(\omega) = \frac{kT}{C\pi}RC\frac{1}{1 + (RC\omega)^2}$$

$$= \frac{kTR}{\pi}\frac{1}{1 + (RC\omega)^2}$$

$$= 2kTR\frac{1}{1 + (2\pi RCf)^2} \tag{12.81}$$

where the second step was a change of variable to $f = \omega/2\pi$ (with the implicit change to $df = d\omega/2\pi$). This has the same Lorentzian form as the two-state system result (Eq. (12.68)).

We might wonder what would happen if we simplified our circuit of Fig. 12.8 by removing the capacitor so that it contained only a resistor. The resistor would still have a small stray capacitance, so the above analysis would still be appropriate. But the consequence is that RC is tiny, allowing us to drop the $(2\pi fRC)^2$ term in the denominator. Finally, if we decide to use this expression for only positive frequencies (see comment following Eq. (12.68)), we must double the expression for $P(f)$ to obtain

$$P(f)df = 4kTRdf \tag{12.82}$$

This is a well known result known as *Nyquist's theorem* (Kittel, 1958). It makes the point that the rms voltage noise in a resistor increases as $\sqrt{R}$, and this is a well established experimental result. The thermal noise in a resistor is known as *Johnson noise*, and Nyquist's theorem is of great value in assessing noise in electrical instrumentation.

12.13 | Fluorescence correlation spectroscopy

The amplitudes of concentration fluctuations were discussed in Section 12.2. Such fluctuations also have dynamic properties. They develop and decay with a time course that reflects some underlying kinetic process. The kinetic process could include diffusion into and out of a small volume element, or a chemical reaction. By measuring the fluctuations we can probe these processes. Fluorescence measurement provides a particularly good way to study these fluctuations because lasers can be focused down to isolate a small region within a much larger volume (Fig. 12.9). Fluorescent molecules diffuse in and out of the path of the laser beam, giving rise to fluctuations. We will analyze the dynamics of these fluctuations to see how the noise in a fluorescence signal can provide useful information (Elson and Magde, 1974).

First we define the fluctuations in concentration of the fluorophore as $\delta C(\mathbf{r}, t) = C(\mathbf{r}, t) - \overline{C}$, where $\overline{C}$ is the mean. The concentration obeys the diffusion equation (Eq. (6.7)). If we replace C with δC, the derivatives of $\overline{C}$ are zero so we are left with

$$\frac{\partial \delta C}{\partial t} = D\nabla^2 \delta C \tag{12.83}$$

According to Fig. 12.9, molecules diffuse radially in and out of the laser beam, so we focus on the x–y plane perpendicular to the beam. For an initial fluctuation of magnitude $\delta C(x_0, y_0, 0)$, we can write down Eq. (6.8) for x and y, and multiply them together

$$\delta C(x, y, t) = \delta C(x_0, y_0, 0) \frac{1}{\sqrt{4\pi Dt}} e^{-(x-x_0)^2/4Dt} \frac{1}{\sqrt{4\pi Dt}} e^{-(y-y_0)^2/4Dt}$$
$$\tag{12.84}$$

Now we must relate δC to the fluorescence signal. The laser beam is not uniform but varies, decreasing in intensity as the distance from the beam center increases. This variation is approximately Gaussian so the light intensity is

$$I(x, y) = I_0 e^{-2(x^2 + y^2)/w^2} \tag{12.85}$$

where I_0 is intensity at the center and w is a parameter for the width of the beam (typically ~ 1 μm). The measured signal is the total

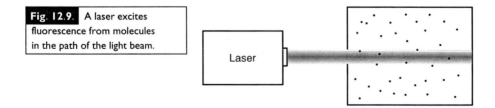

Fig. 12.9. A laser excites fluorescence from molecules in the path of the light beam.

Laser

fluorescence excited by the beam, and this can be expressed as an integral in the x–y plane

$$S(t) = A \int_0^\infty \int_0^\infty I(x,y)C(x,y)dxdy \tag{12.86}$$

where A is a factor that takes into account the experimental aspects of detection and path length as well as the extinction coefficient and quantum yield of the fluorescent molecule.

To derive the correlation function of this signal we average over the product $S(0)S(t)$

$$\langle S(0)S(t)\rangle = \left\langle A^2 \int_{-\infty}^\infty \int_{-\infty}^\infty \int_{-\infty}^\infty \int_{-\infty}^\infty I(x_0,y_0)C(x_0,y_0,0)I(x,y)C(x,y,t)\,dx_0\,dxdy_0\,dy \right\rangle \tag{12.87}$$

The bracket denotes an average over the initial fluctuation $\delta C(x_0, y_0, 0)$. With any value for $C = \overline{C} + \delta C$ the product of $C(x_0, y_0, 0)$ and $C(x, y, t)$ will produce four terms. The first is a time-independent term that gives the square of the mean signal. The second two terms are proportional to $\delta C(x_0, y_0, 0)$ and $\delta C(x, y, t)$, both of which average out to zero. The fourth term, $\delta C(x_0, y_0, 0)\,\delta C(x, y, t)$, is the key time varying part that requires attention. As defined in Eq. (12.47), the correlation function of the fluorescence signal is then the integral

$$F_e(t) = \left\langle A^2 \int_{-\infty}^\infty \int_{-\infty}^\infty \int_{-\infty}^\infty \int_{-\infty}^\infty I(x_0,y_0)\,\delta C(x_0,y_0,0)\,I(x,y)\,\delta C(x,y,t)\,dx_0\,dxdy_0 dy \right\rangle \tag{12.88}$$

Using Eq. (12.84) for $\delta C(x, y, t)$ and Eq. (12.85) for $I(x_0, y_0)$ and $I(x, y)$ gives

$$F_c(t) = A^2 I_0^{\,2} \frac{\overline{\delta C_0^{\,2}}}{4\pi Dt}$$
$$\times \int_{-\infty}^\infty \int_{-\infty}^\infty \int_{-\infty}^\infty \int_{-\infty}^\infty e^{-2(x^2 + x_0^2)/w^2} e^{-(x-x_0)^2/4Dt} e^{-2(y^2 + y_0^2)/w^2} e^{-(y-y_0)^2/4Dt}\,dx_0\,dxdy_0\,dy \tag{12.89}$$

Averaging over initial fluctuations at $t = 0$ yields a factor $\langle \delta C(x_0, y_0, 0)^2 \rangle$, which is simply denoted as $\overline{\delta C_0^{\,2}}$. It can be estimated from the Poisson distribution (Section 12.2).

The integrals in Eq. (12.89) can be factored into two identical double integrals, one over x and x_0 and the other over y and y_0. We can therefore focus on one of them

$$\vartheta = \int_{-\infty}^\infty \int_{-\infty}^\infty e^{-2(x^2 + x_0^2)/w^2} e^{-(x-x_0)^2/4Dt}\,dx_0\,dx \tag{12.90}$$

To evaluate this double integral we might try to separate the two variables into different integrals. However, this does not work. Instead, we can rearrange the exponentials to give

$$\vartheta = \int\limits_{-\infty}^{\infty} \int\limits_{-\infty}^{\infty} e^{-(x+x_0)^2/w^2} e^{-(x-x_0)^2\left(\frac{1}{w^2}+\frac{1}{4Dt}\right)} dx_0 dx \tag{12.91}$$

Now a change of variables to $x + x_0 = u$ and $x - x_0 = v$ gives

$$\vartheta = 2 \int\limits_{-\infty}^{\infty} \int\limits_{-\infty}^{\infty} e^{-u^2/w^2} e^{-v^2\left(\frac{1}{w^2}+\frac{1}{4Dt}\right)} du dv \tag{12.92}$$

where the 2 arises because the change of variables gives $dx_0 dx = 2\, du dv$. Now we have two Guassian integrals (Appendix 4). The first gives $w\sqrt{\pi}$ and the second gives $\sqrt{\pi/\left(\frac{1}{w^2}+\frac{1}{4Dt}\right)}$. Putting these results together in Eq. (12.89) squares ϑ to give

$$F_c(t) = 4A^2 I_0^2 \frac{\overline{\delta C_0^2}}{4\pi Dt} \left(w\sqrt{\pi}\sqrt{\frac{\pi}{\frac{1}{w^2}+\frac{1}{4Dt}}}\right)^2 = 4\pi A^2 I_0^2 \overline{\delta C_0^2} w^2 \left(\frac{1}{1+\frac{4Dt}{w^2}}\right) \tag{12.93}$$

This is often rewritten as

$$F_c(t) = \frac{B}{1+(t/\tau)} \tag{12.94}$$

where B lumps all the intensity parameters together, and τ is

$$\tau = w^2/4D \tag{12.95}$$

This correlation function decays to zero with time, as it should. But unlike the other correlation functions of this chapter, this decay is not exponential. In fact, the precise mathematical form of the decay depends on the shape of the laser beam. We assumed it to be Gaussian (Eq. (12.85)), and the final result, Eq. (12.94), was quite simple. If we had assumed that the laser illuminates a cylindrical volume uniformly, the mathematical result would have been more complicated. However, when this more complicated expression is plotted it looks very much like Eq. (12.94) (Elson and Magde, 1974). It should be evident from the way the integrals were factored in the derivation that each spatial dimension introduces a factor of the form $(1+t/\tau)^{-1/2}$ (Problem 2).

The special advantage of fluctuation correlation spectroscopy is that a laser beam can be focused into a cell to study kinetic processes *in vivo*. Thus, if fluorescent proteins can be introduced, their mobilities in different cellular compartments can be assessed. Figure 12.10 shows examples of fluorescent-labeled proteins injected into the squid giant axon (Terada *et al.*, 2000). One of these proteins, creatine kinase, is an enzyme, and as a globular protein it is not expected to interact with other proteins. The other protein, tubulin, is a cytoskeletal constituent, capable of oligomerizing and interacting with many other proteins. The fluorescence signals (upper part of Fig. 12.10) are noisy, and the

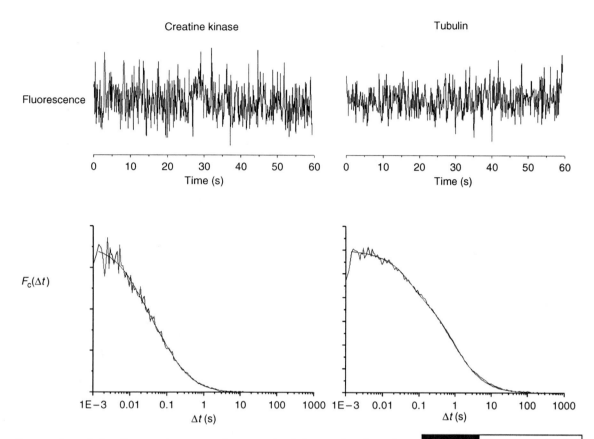

Fig. 12.10. Fluorescence signals (above) from fluorescent-labeled creatine kinase and tubulin injected into the squid giant axon. The computed correlation functions (below) yielded $\tau = 0.845$ ms for tubulin and 0.075 ms for creatine kinase (Terada *et al.*, 2000 data provided by Dr. Terada).

timescales of the fluctuations are characterized by evaluating the correlation function (lower part of Fig. 12.10) with a computer. The correlation functions are fitted to a version of Eq. (12.94) modified for a focused beam (with corrections for fluctuations due to molecules entering the triplet state during excitation). These fits yield the key parameter, τ, which in turn can be used to evaluate the diffusion coefficient with Eq. (12.95).

The diffusion constant for creatine kinase is about what one would expect from the Stokes–Einstein relation (Eq. (6.66)), except with a viscosity for cytoplasm that is 2–3 times higher than that of water. Creatine kinase does not aggregate or interact strongly with other proteins, so it is a good probe of the fluid properties of cytoplasm. By contrast, τ for tubulin was more than 10 times longer than for creatine kinase. Creatine kinase has a higher molecular weight, and thus should have a lower diffusion constant. Thus, diffusion of tubulin inside the squid axon is much slower than predicted by the Stokes–Einstein relation. To explain this result requires that tubulin forms large aggregates or is tethered to a large slow-moving complex. Fluorescence correlation spectroscopy can be applied to a wide range of interesting biological problems (Rigler and Elson, 2001). The technique is not limited to diffusion. When applied to chemical kinetics, the fluctuations in fluorescence can be used to determine the rate constants.

12.14 | Friction and the fluctuation–dissipation theorem

In liquids, molecular fluctuations are the cause of friction. The final section of this chapter will show how the study of fluctuations provides a deeper understanding of friction, and transport processes in general. One normally thinks of friction as something that resists motion and slows things down. But this oversimplifies what goes on at the molecular level in liquids. If a molecule happens to have a velocity of zero, then the molecules around it will collide with it to set it into motion. Eventually the molecule will have an rms velocity of $\sqrt{(3kT/2m)}$. On the other hand, if the molecule has a velocity greater than $\sqrt{(3kT/2m)}$, then the same collisions will slow it down. No matter what the initial velocity of a molecule is, after the collisions have done their work the rms velocity will be the same. These same fluctuations underlie Brownian motion and cause diffusion. Chapter 6 drew attention to how the diffusion and friction coefficients are related; $D = kT/f$ (Eq. (6.62)). We now return to this subject.

If we have a liquid with a high viscosity, we might think that the molecule will move less. Actually, the rms velocity is independent of viscosity. The effect of viscosity is on the time scale of the velocity fluctuations, i.e. how rapidly the velocity changes. This can be seen with the aid of the Langevin equation, which is a form of Newton's law of motion ($F = ma$) in which the molecule comes under the influence of two forces, friction and a randomly varying force uncorrelated with velocity. Since the acceleration is the derivative of velocity, we can express the law of motion as a first order differential equation in velocity

$$\frac{dv}{dt} = -\frac{fv}{m} + \Phi(t) \tag{12.96}$$

where $-fv/m$ is the effect of friction in slowing down a moving molecule. As in Eq. (6.62), f is the coefficient of friction. The function $\Phi(t)$ is a random force that bats the molecule around, and it is closely related to the noise term in Eq. (12.75). The same symbol is used because it has the same essential mathematical properties. The parallel between Eq. (12.96) and Eq. (12.75) implies a parallel between the resistance in a circuit and the friction on a moving particle.

Multiplying both sides of Eq. (12.96) by v gives

$$\frac{1}{2}\frac{dv^2}{dt} = -\frac{fv^2}{m} + \Phi(t)v \tag{12.97}$$

This equation is then averaged to give a differential equation in $\overline{v^2}$

$$\frac{d\overline{v^2}}{dt} = -\frac{2f}{m}\overline{v^2} \tag{12.98}$$

Averaging the product $\Phi(t)v$ from Eq. (12.97) gave zero because both v and Φ average to zero, and they are uncorrelated. If the molecule starts off with an initial velocity v_0 and ends up with $\overline{v^2} = (3kT/2m)$, the solution to Eq. (12.98) is

$$\overline{v^2} = \frac{3kT}{m} + (v_0^2 - \frac{3kT}{m})e^{-2ft/m} \tag{12.99}$$

So by appearing in the exponential, friction determines the time-scale for the decay of velocity fluctuations.

An explicit relation can be derived between the friction coefficient and the fluctuating force by solving Eq. (12.96) (McQuarrie, 1976, Chapter 20). We begin by writing the solution to Eq. (12.96), obtained in the same way that Eq. (12.77) was obtained as a solution of Eq. (12.75)

$$v(t) - v(0)e^{-tf/m} = e^{-tf/m} \int_0^t e^{tf/m}\Phi(t)dt \tag{12.100}$$

Squaring both sides gives

$$v(t)^2 - 2v(t)v(0)e^{-tf/m} + v(0)^2 e^{-2tf/m} = e^{-2tf/m} \int_0^t \int_0^t e^{(t+t')f/m}\Phi(t)\Phi(t')\,dtdt' \tag{12.101}$$

First we look at the right-hand side and realize that the product $\Phi(t)\Phi(t')$ is a function only of the difference $t - t'$. We transform the double integral to the variables $q = t - t'$ and $s = t + t'$

$$v(t)^2 - 2v(t)v(0)e^{-tf/m} + v(0)^2 e^{-2tf/m} = e^{-2tf/m} \int_0^{2t} e^{sf/m}ds \int_0^\infty \Phi(q)\Phi(0)dq$$

$$= (1 - e^{-2tf/m})\frac{m}{f} \int_0^\infty \Phi(q)\Phi(0)dq \tag{12.102}$$

where the first integral was evaluated as $(e^{2tf/m} - 1)(m/f)$. We next let $t \to \infty$, and take the ensemble average. The mean-square velocity is all that remains on the left, and it can be replaced by $3kT/2m$. We thus have

$$\frac{3ktf}{2m^2} = \int_0^\infty \overline{\Phi(q)\Phi(0)}dq \tag{12.103}$$

Thus, the coefficient of friction is related to the correlation function of the fluctuating force used to represent the random collisions with other molecules. Equation 12.103 is an example of the fluctuation–dissipation theorem. It is conceptually important in showing directly the relation between the friction coefficient and the random fluctuations arising from thermal motions in a liquid.

The fluctuation–dissipation theorem is a very general result. Whenever an equation such as Eq. (12.75) or Eq. (12.96) can be written down for a time varying function, the response coefficient can be expressed in terms of the integral of a correlation function of an appropriate fluctuating quantity.

Problems for Chapter 12

1. In a solution with a single fluorescent species you have measured the rms fluctuations in fluorescence in a volume of 10^{-15} liters as 3% of the mean fluorescence. What is the concentration of the fluorescent molecule?

2. Write an equation corresponding to Eq. (12.94) for a laser beam focused to a small spot with a radial width of w_b and an axial width of w_a.

3. At what value of p_o is the channel noise largest (Eq. (12.45))?

4. Calculate the rms voltage fluctuations in a 1 MΩ resistor for a frequency band defined by $f < 1000$ Hz and $T = 298$ K.

5. Calculate the rms voltage fluctuations in a 1 MΩ resistor with a 10 nF capacitor in parallel, for $f < 1000$ Hz and $T = 298$ K.

6. Calculate the power spectrum for a three-state system where the correlation function is a sum of two exponentials of the form $x_1 e^{-t/\tau_1} + x_2 e^{-t/\tau_2}$.

7. Calculate the rms fluctuations in the length of a hydrogen bond using a force constant of 30 kcal Å^{-2} mole^{-1} (Section 2.12).

Chapter 13

Ion permeation and membrane potential

A cell expends metabolic energy to transport ions, accumulating some and expelling others so that the concentrations differ between the cell's inside and outside. The cell membrane is more permeable to some of these ions than to others, so ions will move passively down their concentration gradients at different rates. The resulting separation of charge generates a voltage difference. Essentially every cell has a membrane potential arising from ion movement. The movement of different ions across the cell membrane produces a broad spectrum of electrical behavior and makes it possible for cells to communicate by electrical signaling. Electrical potentials also appear across the membranes of organelles, and across layers of cells such as epithelia. The common theme with all these phenomena is the selective permeability of a membrane to different ions. Here we will examine the relation between membrane permeability, ion flux, and membrane potential. We will not worry about the precise mechanism of ion permeation here. That is a subject for the next chapter.

13.1 | Nernst potentials

Consider two compartments separated by a membrane, with a higher concentration of salt on one side than on the other (Fig. 13.1). If only one ion is permeant, then that ion will diffuse across the membrane down its concentration gradient, but the other ions will remain behind. This produces a charge imbalance and therefore a membrane potential. The membrane potential that develops opposes the flow of that ion, and ultimately an equilibrium will be reached where the difference in electrical potential perfectly balances the difference in chemical potential.

If the concentration of the permeant ion in compartment a is c_a and the concentration in compartment b is c_b, then we have

$$\Delta G = RT \ln \frac{c_b}{c_a} \qquad (13.1)$$

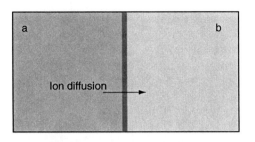

Fig. 13.1. Two compartments, a and b, with different salt concentrations are separated by a membrane that is permeable to only one ion.

for the change in free energy for transferring one mole from a to b. The free energy change per mole due to the electrostatic energy change is[1]

$$\Delta U = zF(V_b - V_a) = zF\Delta V \tag{13.2}$$

When the two solutions reach equilibrium, there is zero net change in energy for ion movement across the membrane. Setting $\Delta G + \Delta U = 0$ leads to

$$\Delta V = -\frac{RT}{zF} \ln \frac{c_b}{c_a} \tag{13.3}$$

This is the *Nernst equation*, and the potential, ΔV, defined by this equation is referred to as the Nernst potential for that particular ion. Walther Nernst derived this equation in 1890, and Katz (1966) has described it as "the best known and most frequently cited equation in the biological literature." That is somewhat overstated, but this relation between ion concentration and membrane potential is extremely important in the study of cellular physiology and membrane biophysics.

If the permeant ion is a cation, then $z = 1$. And if $c_a > c_b$, then the loss of cations from compartment a and their accumulation in compartment b makes compartment a more negative. The Nernst equation gives $V_b > V_a$, so ΔV is positive. If the permeant ion is an anion, then $z = -1$ and the situation is reversed.

The quantity RT/F that appears in Eq. (13.3) has units of volts and is equal to 25.6 mV at room temperature.[2] This quantity is of great importance and appears in most of the equations in this chapter. It plays a fundamental role in relating thermodynamic forms of free energy to electrical energy; RT/F is the increment in membrane potential for an e-fold change in concentration ratio. With $\ln 10 = 2.3$, a 10-fold change in concentration ratio produces a 58 mV

---

[1] At this point it becomes very inconvenient to continue using cgs units (Chapter 2), where charge was in esu and energy was in ergs. Electrophysiologists employ Coulombs as units of charge. The energy unit is 1 Coulomb-Volt = 1 Joule = 0.2389 calories. For monovalent ions, conversion from Coulombs to moles is achieved by multiplying by Faraday's constant (96 480 Coulomb mole^{-1}).

[2] $R = 1.98$ cal mole^{-1} K^{-1} and $T = 298$ K, so $RT = 590$ cal mole^{-1}. Also, $F = 96\ 480$ C mole$^{-1} = 96\ 480$ J V^{-1} mole$^{-1} = 23\ 050$ cal V^{-1} mole^{-1} (using 0.2389 to convert J to cal, see footnote 1 of this chapter). Therefore, $RT/F = (590$ cal mole$^{-1})/(23\ 050$ cal V^{-1} mole$^{-1}) = 0.0256$ V.

change in membrane potential. These numbers are worth remembering.

Most vertebrate cells have an intracellular $[K^+]$ of ~ 130 mM and an external $[K^+]$ of ~ 4 mM. Equation (13.3) then gives $\Delta V \sim -87$ mV as the Nernst potential for K^+. This gives a very rough accounting for the membrane potential of most excitable cells, which typically falls in the range of -60 to -90 mV. A quantitative understanding requires treatment of other ions, a subject that will be dealt with shortly.

The Nernst potential describes a thermodynamic equilibrium. The force of the concentration gradient is balanced by the voltage drop. So ions do not move from one side to the other and the system will be stable for an indefinite period of time. As will be illustrated below, with multiple permeant ions, the membrane potential represents a steady state rather than an equilibrium (except for the Donnan potential). In this steady state ions flow continually, but the fluxes balance out to give zero net current.

It is important to bear in mind that ion concentrations change very little during ion movements that change the voltage. If we look at the initial concentrations of c_a and c_b before the ions are allowed to diffuse across the membrane, and the final concentrations after the Nernst potential has been established, we would find that these concentrations are almost the same. To see this, consider a spherical cell with a radius of 100 μm. The capacitance of this cell's membrane is $4\pi \times 10^{-10}$ F (1 cm^2 of membrane has a capacitance of ~ 1 μF; see Section 15.2). It takes 12.6×10^{-11} Coulombs to change the voltage across this membrane by 100 mV. Dividing by Faraday's constant converts this charge to 1.3×10^{-15} moles of monovalent ion. The volume of the cell is 4.2×10^{-6} cm^3 or 4.2×10^{-9} liters. So the ion concentration changes by only 0.3 μM. Since concentrations are usually in the millimolar range, the change is relatively small.

Note that the capacitance goes as area and the concentration as inverse volume. Thus, in smaller cells the same voltage change will require a greater change in ion concentration. For a tiny 1 μm cell we get 1.3×10^{-19} moles and 4.2×10^{-15} liters, so the concentration increases by 0.03 mM for 100 mV. For calcium the intracellular free ion concentration is < 1 μM, so calcium currents can change the intracellular calcium concentration quite dramatically. This allows calcium to serve as powerful chemical signal. But for the more abundant ions such as Na$^+$, K$^+$, and Cl$^-$, it is usually reasonable to ignore the concentration changes that occur when their flow across the cell membrane makes the voltage change.

13.2 | Donnan potentials

For Nernst potentials, only one ion was allowed to diffuse across the membrane. The Donnan potential is the other extreme in which all of the ions but one can diffuse across the membrane. The impermeant species is typically viewed as a large polyanion, like the nucleic acids

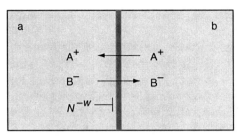

Fig. 13.2. The presence of an impermeant polyanion, N^{-w}, drives a redistribution of the permeant anions, B^-, and cations, A^+, leading to a Donnan equilibrium.

within cells. The situation is illustrated in Fig. 13.2, with N^{-w} representing a large molecule with a net charge of $-w$.

The small anions and cations, B^- and A^+, can flow across the membrane but the large anions, N^{-w} are trapped on the compartment a side. Without this polyanion the equilibrium situation would be equal concentrations of the salt AB on both sides of the membrane. However, with N^{-w} providing an excess negative charge in compartment a, the anions B^- are pushed to the right and the cations A^+ are pulled to the left (Fig. 13.2). Eventually, an equilibrium is reached in which a voltage difference is balanced by diffusive forces. The build up of A^+ in compartment a and B^- in compartment b makes $\Delta V = V_b - V_a$ negative.

The balance of diffusive and electrical forces on A^+ and B^- means that these ion concentrations satisfy the Nernst equation

$$\Delta V = -\frac{RT}{F} \ln \frac{[A^+]_b}{[A^+]_a} \tag{13.4a}$$

$$\Delta V = \frac{RT}{F} \ln \frac{[B^-]_b}{[B^-]_a} \tag{13.4b}$$

where $z = 1$ for the cations in Eq. (13.4a) and $z = -1$ for the anions in Eq. (13.4b). Equating these two gives

$$[A^+]_b[B^-]_a = [A^+]_a[B^-]_b \tag{13.5}$$

This relation can be used to test for the presence of a Donnan potential without measuring voltage. If values of $[A^+]$ and $[B^-]$ measured in both compartments obey Eq. (13.5), then it is likely to reflect a Donnan potential.

Equations (13.4a), (13.4b), and (13.5) cannot be used to solve for the concentrations and ΔV; more information is needed. We have to take into account the constraint that the total charge in each compartment is close to zero. This follows from the examples discussed at the end of the preceding section that illustrate how for typical voltage differences the excess amounts of ions are minor *compared* to their concentrations. So we can ignore the micromolar changes as small fractions of the existing millimolar concentrations. The condition of zero total charge is referred to as *electroneutrality*, and for compartment a this condition is expressed as

$$[A^+]_a - [B^-]_a - w[N^{-w}] = 0 \tag{13.6}$$

In compartment b, electroneutrality dictates that $[A^+]_b = [B^-]_b$. Since they are equal we can replace them both with salt concentration in b, c_b.

To proceed, we rearrange Eqs. (13.4a) and (13.4b), using c_b for $[A^+]_b$ and $[B^+]_b$ as follows

$$\frac{c_b}{[A^+]_a} = e^{-F\Delta V/RT} \tag{13.7a}$$

$$\frac{c_b}{[B^-]_a} = e^{F\Delta V/RT} \tag{13.7b}$$

Using these expressions to solve for $[A^+]_a$ and $[B^-]_a$, and substituting into Eq. (13.6) gives

$$c_b e^{F\Delta V/RT} - c_b e^{-F\Delta V/RT} - w[N^{-w}] = 0 \tag{13.8}$$

Multiplying by $e^{F\Delta V/RT}$ and dividing by c_b gives

$$e^{2F\Delta V/RT} - \frac{w[N^{-w}]}{c_b} e^{F\Delta V/RT} - 1 = 0 \tag{13.9}$$

This is a quadratic equation in $e^{F\Delta V/RT}$. The quadratic formula then gives the solution

$$
\begin{aligned}
e^{F\Delta V/RT} &= \frac{\frac{w[N^{-w}]}{c_b} + \sqrt{\left(\frac{w[N^{-w}]}{c_b}\right)^2 + 4}}{2} \\
&= \frac{w[N^{-w}]}{2c_b} + \sqrt{\left(\frac{w[N^{-w}]}{2c_b}\right)^2 + 1}
\end{aligned}
\tag{13.10}
$$

Of the two roots to Eq. (13.9), only the positive one has physical meaning. The membrane potential is

$$\Delta V = \frac{RT}{F} \ln\left(\frac{w[N^{-w}]}{2c_b} + \sqrt{\left(\frac{w[N^{-w}]}{2c_b}\right)^2 + 1}\right) \tag{13.11}$$

Thus, we can calculate the membrane potential in terms of the two concentrations $[N^{-w}]$ and c_b. An important difference between this situation and that described by the Nernst equation is that a substantial amount of ion flow occurs in order to reach this equilibrium. When ions flow to establish a Nernst potential, the concentrations hardly change at all. By contrast, the establishment of a Donnan potential produces substantial changes in ion concentration. Electroneutrality is preserved as this happens because the different ions maintain a charge balance as they flow.

13.3 | Membrane potentials of cells

Most excitable cells have what is referred to as a *resting potential*, a stable membrane potential seen in the absence of stimulation. Resting potentials are generally negative, that is, the inside of the cell is negative relative to the outside. The theories just presented for Nernst and Donnan potentials offer alternative ways to account for a resting potential. If a cell membrane allows only one ion (a cation) to permeate, then the cell can concentrate that ion by active transport, and then allow it to diffuse back out to create a negative Nernst potential. Pumping a permeant anion out and allowing it to diffuse back in would also work. Alternatively, a cell could make a large amount of polyanion and allow other ions to redistribute to form a Donnan potential. We will examine neurons and muscle fibers as examples of applications of these theories. Neurons use K^+ ions as the primary determinant of their resting potentials; the Nernst potential for K^+ approximates this quantity reasonably well. Skeletal muscle of vertebrates uses both K^+ and Cl^-. The two ions move together in a manner well described by a Donnan potential.

The study of resting potentials begins with calculations of the Nernst potential for each ion. Figure 13.3 provides this information for the squid axon and frog skeletal muscle. There are some qualitative similarities. Each of the ions has an unequal distribution across the membrane, with higher K^+ inside and higher Na^+ and Cl^- outside. The unequal distribution of Na^+ and K^+ is created by the pumping action of the Na^+/K^+ ATPase (Section 13.8). This ubiquitous membrane protein uses the energy released in the hydrolysis of ATP to pump Na^+ out and K^+ in. However, the membranes of the squid axon and frog muscle fiber differ in their ion permeabilities, and so the resting potential arises through different mechanisms.

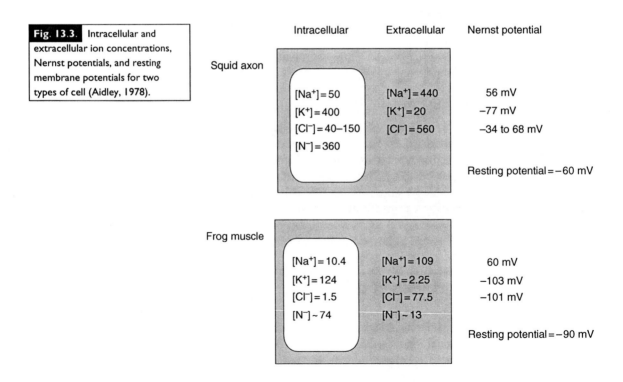

Fig. 13.3. Intracellular and extracellular ion concentrations, Nernst potentials, and resting membrane potentials for two types of cell (Aidley, 1978).

Intracellular | Extracellular | Nernst potential

Squid axon

$[Na^+] = 50$ $[Na^+] = 440$ 56 mV
$[K^+] = 400$ $[K^+] = 20$ −77 mV
$[Cl^-] = 40–150$ $[Cl^-] = 560$ −34 to 68 mV
$[N^-] = 360$

Resting potential = −60 mV

Frog muscle

$[Na^+] = 10.4$ $[Na^+] = 109$ 60 mV
$[K^+] = 124$ $[K^+] = 2.25$ −103 mV
$[Cl^-] = 1.5$ $[Cl^-] = 77.5$ −101 mV
$[N^-] \sim 74$ $[N^-] \sim 13$

Resting potential = −90 mV

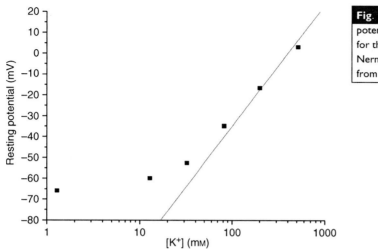

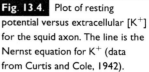

Fig. 13.4. Plot of resting potential versus extracellular $[K^+]$ for the squid axon. The line is the Nernst equation for K^+ (data from Curtis and Cole, 1942).

13.3.1 Neurons

The resting potential of the squid axon does not equal the Nernst potential for any ion, but it is close to the Nernst potentials for Cl^- and K^+, and that is an important clue. To explore this connection one must measure the membrane potential and then vary the concentrations of the different ions. If the permeability for one of those ions is important, then the membrane potential will change logarithmically, as predicted by Eq. (13.3). The result of this experiment is shown in Fig. 13.4. We see that varying extracellular $[K^+]$ produces the expected change in membrane potential. Above 30 mM, the change is nearly 58 mV per 10-fold change in $[K^+]$, and the resting potential falls close to the line drawn for the K^+ Nernst potential. Varying $[Cl^-]$ has essentially no effect. Thus, the membrane potential of the squid axon is essentially a Nernst potential for K^+ until $[K^+]$ falls below around 30 mM.

The Nernst equation fails to account for the data at low $[K^+]$. The main reason for this is that other ions permeate the membrane and make their own contribution to the membrane potential. Na^+ and Cl^- are much less permeant, but their contributions are measurable, particularly when K^+ is scarce. We therefore need to understand membrane potentials with multiple permeant ions. This will be taken up shortly, but first we look at muscle fibers.

13.3.2 Vertebrate skeletal muscle

As with the squid axon, the resting potential of a muscle fiber is close to the Nernst potentials for Cl^- and K^+ (Fig. 13.3). However, in contrast to the squid axon, the resting potential of a muscle fiber is strongly influenced by changing the extracellular concentrations of either K^+ or Cl^-. In a classical study by Boyle and Conway (1941), frog muscle was equilibrated in various solutions and the intracellular and extracellular ion concentrations were measured. $[K^+]$ and

[Cl⁻] consistently obeyed Eq. (13.5), indicating that these two ions were part of a Donnan system. The very similar Nernst potentials for these two ions, and their similarity to the resting potential (Fig. 13.3) fits with another prediction of the Donnan theory (Eqs. (13.4a) and (13.4b)).

Changing either extracellular [K⁺] or [Cl⁻] altered the membrane potential of frog skeletal muscle (Hodgkin and Horowicz, 1959). Raising extracellular [K⁺] from 2.5 mM to 10 mM produced a rapid rise in potential of 16 mV, and a slower rise taking several minutes, of another 10 mV. Reducing extracellular [Cl⁻] by a factor of 4 raised the potential by about 20 mV, but after a few minutes, the membrane potential returned to its original value. According to the Nernst equation, a 4-fold change in concentration should change the membrane potential by 35 mV, so the Nernst equation cannot apply for either ion.

The observations of Hodgkin and Horowicz can be broken into immediate and delayed changes in potential. At the start, the membrane was close to the Nernst potentials for both K⁺ and Cl⁻, as already mentioned (Fig. 13.3). Raising extracellular [K⁺] or reducing extracellular [Cl⁻] moved the Nernst potential for that ion in the positive direction, so that the two Nernst potentials were no longer equal. The immediate effect is that the voltage rose to some intermediate value between the Nernst potentials for K⁺ and Cl⁻ (how to calculate the exact value will be explained in Section 13.7). After this rapid change, the ion concentrations changed slowly to form a new Donnan potential. Raising extracellular [K⁺] caused the flux of K⁺ and Cl⁻ into the cell in exactly equal numbers (preserving electroneutrality). The Nernst potential for Cl⁻ went up much more than the Nernst potential for K⁺ because with only 1.5 mM Cl⁻, the fractional change in intracellular [Cl⁻] was much greater. The migration of KCl continued until the two Nernst potentials were equal, at which point the Donnan equilibrium had been reestablished. The response to raising extracellular [Cl⁻] follows the same logic. Once again the potential rose, but then KCl went out to reduce the Cl⁻ Nernst potential.

Skeletal muscle fibers of vertebrates generally follow this pattern (Aidley, 1978; Aickin, 1990). The membranes are very permeable to both K⁺ and Cl⁻, although the permeability of Cl⁻ is generally 2–20 times higher. Na⁺ permeability is generally not a factor because Na⁺ does not easily pass across the membrane at rest. This serves two purposes. First, the high intracellular concentration of impermeant organic anions requires an impermeant extracellular ion to maintain osmotic balance. Otherwise, water would flow into the cell and dilute the ions. High extracellular [Na⁺] keeps the osmolarity outside the cell equal to the osmolarity inside the cell. Further, the Na⁺ concentration gradient is held in reserve to drive the rapid changes in membrane potential, the action potentials, used in electrical signaling processes (Chapter 16).

13.4 | A membrane permeable to Na⁺and K⁺

When more than one ion can cross a membrane, the observed potential reflects the relative strength with which the movement of each ion pulls the voltage toward its own Nernst potential. We will now treat the case of two permeant ions of the same charge. This derivation introduces some important ideas, and the result improves the interpretation of the data in Fig. 13.4.

Consider two compartments with mixtures of NaCl and KCl in different proportions but with the same total salt concentration on each side (Fig. 13.5). Na⁺ is higher on the left side and will diffuse to the right to make the left side negative. K⁺ moving to the left will tend to make the right side more negative. Which ion wins will obviously depend on which one diffuses most easily through the membrane. But the diffusion constant does not appear in the Nernst equation. The Nernst equation applies to equilibria, and equilibria generally do not depend on transport coefficients.

The general approach to membrane potentials with multiple ions is to write down the equation for the current of each ion, add these currents together to obtain the total, set this total current equal to zero, and solve for V. The rationale for this is that if the net ion current is zero, then the membrane potential is at a stable value.

Ion flux through a membrane is treated as diffusion under the influence of the same two forces used to derive the Nernst equation: these are voltage and the ion's concentration gradient. The combined action of these two forces has come up already in the treatment of diffusion and kinetics in Chapters 6 and 7. The flux produced by the concentration gradient is $-D(dc/dx)$, where D is the diffusion coefficient of the ion within the membrane. The flux produced by voltage is $-(zFc/f)(dV/dx)$, where f is the coefficient of friction for the ion within the membrane. Adding up these responses to the two forces and using the Einstein relation, $f = kT/D$ (Eq. (6.62)) gives a total flux, as in Eq. (6.64), of

$$J = -D\left(\frac{zFc}{RT}\frac{dV}{dx} + \frac{dc}{dx}\right) \tag{13.12}$$

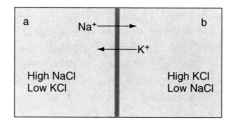

Fig. 13.5. Two compartments separated by a membrane permeable to both Na⁺ and K⁺.

The switch from kT to RT reflects the macroscopic units of moles employed here. We do not need to use partial derivatives because we are only differentiating with respect to x. This expression can be shown to be consistent with the Nernst equation (Problem 2).

The current is obtained from Eq. (13.12) by multiplying by the charge per mole. Recall that F, Faraday's constant, is the conversion factor between moles and Coulombs (see footnote 1 in this chapter). Thus, if we multiply J, in units of mole s^{-1} by zF, we have Coulomb s^{-1}, which is current in units of Amperes

$$I = -zFD\left(\frac{zFc}{RT}\frac{dV}{dx} + \frac{dc}{dx}\right)$$

(13.13)

Before proceeding with the derivation of an expression for ΔV, it is worth noting that when the concentration gradient is zero, Eq. (13.13) gives a simple linear relationship between current and voltage

$$I = -\frac{z^2F^2cD}{RT}\frac{dV}{dx}$$

(13.14)

We can integrate this equation across the membrane. The value of I is constant so the left side becomes $I\delta$, where δ is the membrane thickness. With c constant within the membrane, integration of the derivative on the right gives ΔV

$$I\delta = \frac{z^2F^2cD}{RT}\Delta V$$

(13.15)

We can now divide through by δ and replace D/δ with a new symbol, P, which will be called the permeability

$$I = \frac{z^2F^2cP}{RT}\Delta V$$

(13.16)

This is essentially Ohm's law. From the proportionality factor we obtain the following expression for the membrane conductance

$$G = \frac{z^2F^2cP}{RT}$$

(13.17)

Thus, we see that the conductance, a number that characterizes the ion current that flows in response to a voltage, is directly proportional to the permeability, a number that characterizes the ion flux driven by a concentration gradient. It makes sense that the two are related. The property of a membrane to resist ion flow induced by one of these driving forces is fundamentally the same as that to resist the other driving force.

It must be emphasized that Eqs. (13.16) and (13.17) are valid only for the case where the concentrations are very similar on both sides of the membrane (i.e. $dc/dx \sim 0$). When the concentrations are different we do not know what to use for c. We will see how this is resolved with Goldman–Hodgkin–Katz current equation (Section 13.10).

Returning to the problem of deriving the steady-state membrane potential for two permeant cations, we rearrange Eq. (13.13), take $z = 1$, and write down separate expressions for Na^+ and K^+, as follows

$$I_{Na} = -FD_{Na}e^{-FV/RT} \frac{d([Na^+]e^{FV/RT})}{dx} \qquad (13.18a)$$

$$I_K = -FD_Ke^{-FV/RT} \frac{d([K^+]e^{FV/RT})}{dx} \qquad (13.18b)$$

where D_{Na} and D_K are the diffusion constants of Na^+ and K^+, respectively. The same mathematical manipulation to get these equations from Eq. (13.13) was employed in deriving Eq. (7.28). The result is easily checked by applying the product rule to the derivative on the right-hand side.

At steady state the Na^+ and K^+ currents add up to zero

$$FD_{Na}e^{-FV/RT} \frac{d([Na^+]e^{FV/RT})}{dx} + FD_Ke^{-FV/RT} \frac{d([K^+]e^{FV/RT})}{dx} = 0 \qquad (13.19)$$

The factor $Fe^{-FV/RT}$ in front of each term can now be dropped, leaving derivatives that are easily integrated from the left side of the membrane (a) to the right side (b), as follows

$$D_{Na}([Na^+]_be^{FV_b/RT} - [Na^+]_ae^{FV_a/RT}) + D_K([K^+]_be^{FV_b/RT} - [K^+]_ae^{FV_a/RT}) = $$
$$D_{Na}([Na^+]_be^{F\Delta V/RT} - [Na^+]_a) + D_K([K^+]_be^{F\Delta V/RT} - [K^+]_a) = 0 \qquad (13.20)$$

where in the second step $e^{FV_a/RT}$ was factored out, leaving $e^{F(V_b-V_a)/RT} = e^{F\Delta V/RT}$. This expression is then solved for ΔV

$$\Delta V = -\frac{RT}{F} \ln \left(\frac{D_{Na}[Na^+]_b + D_K[K^+]_b}{D_{Na}[Na^+]_a + D_K[K^+]_a} \right) \qquad (13.21)$$

This is an example of the Goldman–Hodgkin–Katz voltage equation, which will be derived more generally for ions of different charge below (Sections 13.7 and 13.11). The derivation of Eq. (13.21) depends only on the independence of the fluxes of the two ions, and on the homogeneity of D_{Na} and D_K within the membrane (Hille, 1991). Note that if $D_{Na} \gg D_K$, ΔV goes to the Nernst potential for Na^+, and vice versa for K^+.

The Ds in Eq. (13.21) are generally replaced by permeability coefficients P (we defined P as D/δ so they are proportional). It is clear that scaling all the Ds by a common factor does not change anything. Note that ΔV depends on the ratios of the permeabilities rather than the actual values, so when we use Eq. (13.21), it does not matter whether we use D or P. The convention is to use P in applications of Eq. (13.21) to experimental membrane potential data.

13.5 | Membrane potentials of neurons again

Equation (13.21) dramatically improves our understanding of neuronal membrane potential data such as that in Fig. 13.4. First, we rewrite Eq. (13.21) in terms of the permeability ratio, $R_{K/Na} = P_K/P_{Na} = D_K/D_{Na}$, as follows

$$\Delta V = -\frac{RT}{F} \ln \left(\frac{[Na^+]_b + R_{K/Na}[K^+]_b}{[Na^+]_a + R_{K/Na}[K^+]_a} \right) \qquad (13.22)$$

In this form there is only one unknown parameter, $R_{K/Na}$. Using the numbers from Fig. 13.3 for $[Na^+]_a$, $[Na^+]_b$, and $[K^+]_b$ of the squid axon, and taking $[K^+]_a$ as the experimental variable, we can describe the variation in membrane potential with extracellular $[K^+]$ very well by setting $R_{K/Na} = 19$ (see Fig. 13.6). For many cells, Eq. (13.22) gives a good quantitative description of the membrane potential, with $R_{K/Na}$ in the range 10–100. So the deviations of the resting potential from the Nernst potential for K^+ can be accounted for by adding in a low permeability to Na^+. This analysis makes the important point that membrane permeation by another ion, in this case Na^+, substantially improves the comparison between experiment and theory. However, the actual mechanism for this variation in resting potential is more complicated and may involve a small contribution from Cl^-, as well as permeabilities that vary with membrane potential (Hille, 1977).

K$^+$ channels are the main players in determining the resting potentials of neurons. There are many types of K$^+$ channels with diverse roles in regulating excitability. Many of these exhibit a complex dependence on voltage (Chapter 16), but in frog ganglion neurons the K$^+$ channel that contributes the most K$^+$ permeability at rest has very little voltage dependence and appears to be uniquely dedicated to this function (Jones, 1989). Different K$^+$ channels are responsible for the resting potential in other cells. In cerebellar

Fig. 13.6. The data from Fig. 13.4 are well fitted by Eq. (13.22) with $R_{K/Na} = 19$.

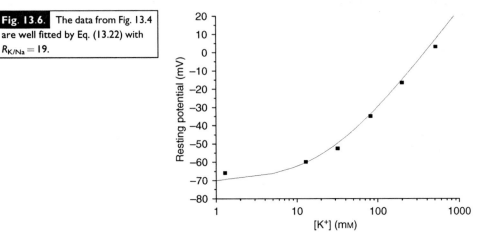

granule neurons a K^+ channel with two pore-domains is responsible for the high resting K^+ permeability (Millar *et al.*, 2000). In midbrain neurons an inward rectifier K^+ channel performs this function, and the closure of this channel by neurotransmitter partially collapses the membrane potential (Nakajima *et al.*, 1988). As noted above (Section 13.3.2), the skeletal muscle resting potential is determined by both K^+ and Cl^-, and the Cl^- channel responsible was identified as ClC-1 (Jentsch *et al.*, 2002).

13.6 | The Ussing flux ratio and active transport

How do we generalize Eq. (13.21) to handle additional ions? We need a better mathematical analysis of how fluxes are generated by the combination of electrical and chemical forces. An important element of this theory was introduced by Ussing (1949) in a classical study of the active transport of ions.

Taking Eqs. (13.18a) or (13.18b) for any arbitrary ionic species, denoted by the subscript i, we go back from current to flux (from I to J), and multiply through by the exponential factor.

$$J_i e^{z_i FV/RT} = -D_i \frac{d}{dx} c_i e^{z_i FV/RT} \tag{13.23}$$

Taking a steady state with J_i constant, and integrating both sides of Eq. (13.23) from one side of the membrane to the other gives

$$J_i \int_a^b e^{z_i FV/RT} dx = -D_i (c_{ib} e^{z_i FV_b/RT} - c_{ia} e^{z_i FV_a/RT}) \tag{13.24}$$

The integral on the left cannot be evaluated explicitly without knowing what $V(x)$ looks like within the membrane. However, that does not matter for now because the integral drops out in the following calculation.

Now we use Eq. (13.24) to write the unidirectional flux that would be seen if one added an isotope tracer to one side of the membrane and monitored its appearance on the other side. If tracer is added only to compartment a, then the tracer concentration in compartment b is zero, so Eq. (13.24) gives the flux from a to b as

$$J_{i(a \rightarrow b)} = \Theta_i c_{ia} e^{z_i FV_a/RT} \tag{13.25}$$

where the symbol Θ_i denotes D_i divided by the integral multiplying J_i on the left-hand side of Eq. (13.24). A similar equation can be derived for the tracer flux in the opposite direction

$$J_{i(b \rightarrow a)} = \Theta_i c_{ib} e^{z_i FV_b/RT} \tag{13.26}$$

where Θ_i is the same in Eqs. (13.25) and (13.26), so taking the ratio of the two fluxes leads to

$$\frac{J_{i(a \rightarrow b)}}{J_{i(b \rightarrow a)}} = \frac{c_{ia}}{c_{ib}} e^{-z_i F\Delta V/RT} \tag{13.27}$$

This is the Ussing flux ratio. It provided a valuable tool in the interpretation of tracer flux measurements in terms of passive and active transport. Experimental measurements by Ussing of tracer flux across frog skin gave a ratio that differed by a factor of several hundred from that predicted by Eq. (13.27), indicating that active transport must be at work. The dependence of ion flux on metabolic energy was verified by adding a metabolic poison. Within a few minutes the fluxes changed to values that were in agreement with Eq. (13.27).

13.7 | The Goldman–Hodgkin–Katz voltage equation

Now we will generalize Eq. (13.21) to obtain an expression for the membrane potential as a function of the concentrations and permeability ratios of all the monovalent ions (Patlak, 1960). We start by writing down the condition of zero net current (as we did for two ions in Section 13.4), but we view each net flux as the difference between the rightward and leftward unidirectional fluxes of the preceding section. The cation fluxes are perfectly balanced by the anion fluxes, so we have

$$\sum_{\text{cations}} (J_{i(b\to a)} - J_{i(a\to b)}) - \sum_{\text{anions}} (J_{i(b\to a)} - J_{i(a\to b)}) = 0 \qquad (13.28)$$

Now use the Ussing flux ratio, Eq. (13.27), to eliminate $J_{i(a\to b)}$ for cations and $J_{i(b\to a)}$ for anions

$$\sum_{\text{cations}} J_{i(b\to a)} \left(1 - \frac{c_{ia}e^{-F\Delta V/RT}}{c_{ib}}\right) - \sum_{\text{anions}} J_{i(a\to b)} \left(\frac{c_{ib}e^{-F\Delta V/RT}}{c_{ia}} - 1\right) = 0$$

$$(13.29)$$

The flux from compartment b to compartment a should be proportional to c_b, and the proportionality constant between concentration and flux is the permeability. So $J_{i(b\to a)} = P_i c_{ib}$; likewise, $J_{i(a\to b)} = P_i c_{ia}$. With these substitutions we have

$$\sum_{\text{cations}} P_i(c_{ib} - c_{ia}e^{-F\Delta V/RT}) - \sum_{\text{anions}} P_i(c_{ib}e^{-F\Delta V/RT} - c_{ia}) = 0 \qquad (13.30)$$

$$\sum_{\text{cations}} P_i c_{ib} - e^{-F\Delta V/RT} \sum_{\text{cations}} P_i c_{ia} - e^{-F\Delta V/RT} \sum_{\text{anions}} P_i c_{ib} + \sum_{\text{anions}} P_i c_{ia} = 0$$

$$(13.31)$$

We can now solve for $e^{-F\Delta V/RT}$ and take the logarithm

$$\Delta V = -\frac{RT}{F} \ln \left(\frac{\sum\limits_{\text{cations}} P_i c_{ib} + \sum\limits_{\text{anions}} P_i c_{ia}}{\sum\limits_{\text{cations}} P_i c_{ia} + \sum\limits_{\text{anions}} P_i c_{ib}}\right) \qquad (13.32)$$

This is a more general form of the Goldman–Hodgkin–Katz voltage equation (Eq. (13.21)). It expresses the membrane potential as a function of the concentrations and permeabilities of all the monovalent ions. This equation was originally derived by assuming that the electric field is constant within the membrane, so that the voltage drops linearly from one side to the other. Because of this assumption this relation has been referred to as the *constant-field equation*. The constant-field assumption will be examined shortly, but the derivation presented here shows that Eq. (13.32) is more general than originally thought. It does not depend on assuming that the field in the membrane is constant.

A more critical assumption that goes into the derivation of Eq. (13.32) is that the flux of each ion is independent of the other ion fluxes. It is easy to imagine that a channel may have ion fluxes that interact with one another and violate independence. For example, ions can saturate binding sites in channels and block the binding of other ions. The rich literature on how ion channels fail to obey the independence principle is reviewed in chapter 14 of Hille (1991), and some specific models will be studied in the following chapter. In spite of this issue, Eq. (13.32) is very widely used to determine permeability ratios from the voltage for which current is zero. Note that the ratios are the critical quantities rather than the absolute values of the permeabilities. Like Eq. (13.22), Eq. (13.32) can be rewritten in terms of permeability ratios.

It is instructive to look at how the permeabilities for various ions influence the observed value of ΔV. In most cells the monovalent ions that need to be accounted for are Na^+, K^+, and Cl^- (Fig. 13.3). Equation (13.32) then becomes

$$\Delta V = -\frac{RT}{F} \ln \left(\frac{P_{Na}[Na^+]_b + P_K[K^+]_b + P_{Cl}[Cl^-]_a}{P_{Na}[Na^+]_a + P_K[K^+]_a + P_{Cl}[Cl^-]_b} \right) \qquad (13.33)$$

Figure 13.7 shows how the membrane potential varies with $[K^+]_b$ for different values of P_{Cl}, using the concentrations given for the squid axon in Fig. 13.3. Here P_K is taken as $20 \times P_{Na}$, and a few different values of P_{Cl} are tested. When $P_K \gg P_{Cl}$ the plot is similar to Fig. 13.4, where the Nernst potential for K^+ dominates at high $[K^+]$. As P_{Cl} increases, $[K^+]$ has a weaker effect. The product of the concentration and permeability determines the impact of an ion on the potential and as P_{Cl} goes up it pulls the membrane closer to the Cl^- Nernst potential ($-88\,mV$ in this case). Mathematically, the effect is to make the $[Cl^-]$ term in the numerator and the denominator of the fraction in Eq. (13.33) larger so that the ratio becomes insensitive to variations in other ions.

It is worth commenting here about ion activities. The expression for the free energy change (Eq. (13.1)), depends on the solution being ideal. Deviations from ideality often have to be considered in quantitative studies, and this is done by replacing ion concentrations

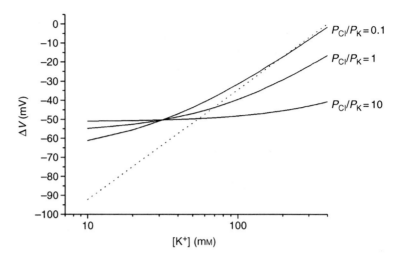

Fig. 13.7. Equation (13.33) gives ΔV versus $[K^+]_a$. Allowing Cl^- to permeate the membrane dampens the effect of varying $[K^+]$. The dashed line is the K^+ Nernst potential. Intracellular $[Cl^-]$ was taken as 75 mM.

with activities. The molar free energy is then $G^o + RT \ln a$, where a is the activity. We saw in Chapter 11 that at physiological concentrations of NaCl and KCl of $\sim$100 mM, deviations from ideality are appreciable. It is commonly assumed that the ionic strengths, and therefore the activity coefficients, are the same both inside the cell and outside the cell. This leads to a cancellation of this factor from the numerator and denominator of the logarithmic term in the Goldman–Hodgkin–Katz voltage equation. Solution nonideality then has no consequence. When the solutions have different ionic strengths, the activities will make a difference and taking this into account often improves the accuracy of this kind of analysis.

13.8 | Membrane pumps and potentials

One would think that with expressions for the membrane potential in terms of the permeabilities and concentrations of all the monovalent ions, we should be able to describe the resting membrane potential of any cell. Equation (13.33) does account for many of the effects of passive ion permeation, but there is an additional contribution made by active transport. All cell membranes contain pump proteins that harness chemical energy to drive ions against their concentration gradients. These pumps create the ion gradients that are responsible for the membrane potentials already discussed. So they are ultimately, if indirectly, responsible for the resting potential. But many active membrane pumps, most notably the Na$^+$/K$^+$ ATPase, are *electrogenic*, and generate a current by pumping more ions in one direction than the other. These pumps contribute directly to the membrane potential.

To understand the effect of pump current on membrane potential we start with the sum of all the passive ion fluxes in Eq. (13.28) and add the flux of pumped ions J_{pump} (this might be a measured flux or a measured pump current divided by zF). The same analysis that gave Eq. (13.32) now gives

$$\Delta V = -\frac{RT}{F} \ln \left(\frac{\sum\limits_{cations} P_i c_{ib} + \sum\limits_{anions} P_i c_{ia} + J_{pump}}{\sum\limits_{cations} P_i c_{ia} + \sum\limits_{anions} P_i c_{ib}} \right) \qquad (13.34)$$

In most cells, poisoning the Na^+/K^+ ATPase by an inhibitor such as ouabain depolarizes the membrane potential very quickly by 10–15 mV (see Jones, 1989). This change in ΔV results from reducing J_{pump} in Eq. (13.34) from its normal value to zero. A slower drop in potential occurs over a time course of hours. This slower change reflects the gradual collapse of the concentration gradients in the absence of pump activity.

The extent to which an electrogenic pump contributes to the resting potential of a cell can vary quite a bit. In neutrophils (a kind of white blood cell) the membrane potential appears to be generated entirely by a pump (Bashford and Pasternak, 1986). The ion gradients play a much smaller role so changing the extracellular $[K^+]$ makes little difference. In these cells, poisoning the pump reduces the membrane potential to zero. Such cells are a bit of an oddity, but the phenomenon itself is not; mitochondria also have very large membrane potentials on the order of -100 to -200 mV, and these are generated entirely by an electrogenic proton pump.

13.9 | Transporters and potentials

A large group of transport proteins couples the movement of two or more molecules across a membrane in a stoichiometric combination. For example, red blood cells have an anion exchange protein that couples the movement of Cl^- to the movement of HCO_3^- in the opposite direction. This transport process can run in either direction depending on the gradients, and in fact, the red blood cell anion exchanger reverses its direction of activity as the addition of CO_2 in the tissues and removal in the lungs leads to opposite transmembrane HCO_3^- gradients (CO_2 and HCO_3^- interconvert through the action of carbonic anhydrase; see Section 10.15). Another transporter exchanges Na^+ for Ca^{2+}. This protein harnesses the Na^+ gradient created by the Na^+/K^+ ATPase to extrude Ca^{2+} and help maintain a low cytosolic free Ca^{2+} concentration. An exchanger in neuronal membranes utilizes the driving forces for Na^+, K^+, and H^+ to recover the neurotransmitter glutamate released during synaptic transmission (Zerangue and Kavanaugh, 1996). An enormous variety of membrane transport proteins use this principle to transport metabolites in an indirect or secondary form of active transport.

They do not hydrolyze ATP themselves but they feed off the gradients created by the pump proteins that do. When the transport operation results in no net charge movement, then the membrane potential is unaffected. However, when the transport operation produces a net charge flux, then a membrane potential can influence the transport process, and the transport process can contribute to a cell's membrane potential.

Consider a transport protein that moves n molecules of an anion A^- from compartment a to compartment b, with the countertransport of m molecules of a cation B^+ from compartment b to compartment a. This exchange process can be represented by a reaction scheme (Scheme (13A))

$$nA^-_a + mB^+_b \rightleftharpoons nA^-_b + mB^+_a \tag{13A}$$

There are three forces that can drive this reaction. Two of these are the concentration gradients for A^- and B^+. The other is the membrane potential. The membrane potential will then be determined by balancing out all the forces for a system at thermodynamic equilibrium.

The concentrations specify the free energy change for the exchange process, which according to Scheme (13A) is

$$\Delta G = RT \ln \left(\frac{[A^-]_b{}^n [B^+]_a{}^m}{[A^-]_a{}^n [B^+]_b{}^m} \right) \tag{13.35}$$

Scheme (13A) also tells us how much charge is moved during the reaction. The movement of n anions one way and m cations the other way is equivalent to a net charge movement of $n+m$. And if there is a membrane potential, ΔV, then the electrostatic energy is

$$\Delta U = F \Delta V(n+m) \tag{13.36}$$

Equations (13.35) and (13.36) parallel Eqs. (13.1) and (13.2) used to derive the Nernst equation.

Equation (13.35) can be rewritten as

$$\Delta G = RT \ln \left(\frac{[A^-]_b{}^n}{[A^-]_a{}^n} \right) + RT \ln \left(\frac{[B^+]_a{}^m}{[B^+]_b{}^m} \right)$$

$$= nRT \ln \left(\frac{[A^-]_b}{[A^-]_a} \right) - mRT \ln \left(\frac{[B^+]_b}{[B^+]_a} \right) \tag{13.37}$$

These are now so close to the form of the Nernst equation (Eq. (13.3)) that we can express ΔG in terms of the Nernst potentials for A^- and B^+, which are, respectively, E_A and E_B

$$\Delta G = nF E_A + mF E_B \tag{13.38}$$

Care must be taken to keep track of the value of $z = \pm 1$ for the two ions.

Balancing out the forces at equilibrium means equating Eqs. (13.36) and (13.38)

$$\Delta V(n+m) = nE_A + mE_B \tag{13.39}$$

The membrane potential is thus the average of the two Nernst potentials, weighted according to the stoichiometric coefficients of the two ions

$$\Delta V = \frac{nE_A + mE_B}{n + m} \tag{13.40}$$

Like the Nernst potential, a membrane potential set by a counter-transport process will vary as the logarithm of the concentration ratio. But changing one of the concentrations will change the membrane potential by less than the Nernst potential because of the weighting in Eq. (13.40).

This relation was used to interpret experiments on a bacterial protein that is homologous to eukaryotic Cl^- channels (Accardi and Miller, 2004). When this protein was reconstituted into lipid bilayers it passed a Cl^- current. Imposing a Cl^- gradient moved the reversal potential away from zero, but the reversal potential was less than the Nernst potential for Cl^-. Efforts to find another permeant ion were futile. The behavior could not be explained with the Goldman–Hodgkin–Katz voltage equation (Eq. (13.32)) using any combination of Cl^- and some other ion. An important clue was found by noting that a pH gradient moved the reversal potential away from zero, even though the Cl^- concentration was the same on both sides. This suggested that the movement of Cl^- and H^+ were coupled, and systematic variations of both ions confirmed this. For variations in Cl^- concentration and pH, Eq. (13.40) predicted the reversal potential very well, with the ratio of m/n set to ~ 0.5 (assigning m to H^+ and n to Cl^-) (Fig. 13.8). This means that two Cl^- ions move with each proton.

Performing an experiment with one ion symmetrically distributed simplifies things by making one of the Nernst potentials equal to zero. Then ΔV follows the Nernst potential for the other ion, but reduced by the factor $n/(n+m)$ or $m/(n+m)$. Recall that the Nernst potential varies with the logarithm of the concentration ratio, with a 58 mV change for a 10-fold change in concentration. If we make the same plot for a countertransport protein, then the plot of ΔV versus the logarithm of the concentration will still be linear but with a lower slope. Figure 13.8 illustrates this behavior, and the plots of the Nernst equation drawn in highlight the lower slope of the experimental data. This analysis established that the bacterial protein was not a Cl^- channel, as originally surmised based on its homology with vertebrate proteins known to be Cl^- channels. Additional experiments demonstrated directly that a gradient for Cl^- drives H^+ and a gradient for H^+ drives Cl^-.

13.10 | The Goldman–Hodgkin–Katz current equation

So far we have focused on situations defined by a stable voltage where no current flows. Now we ask what happens if a voltage is

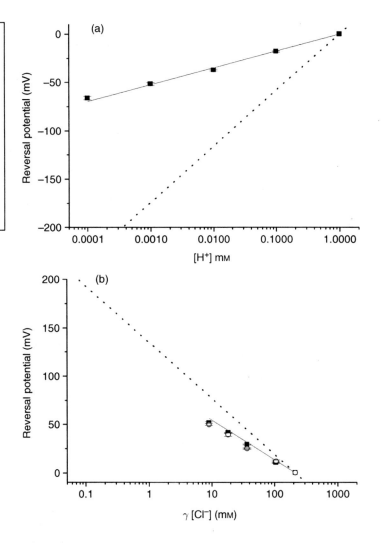

Fig. 13.8. Plots of reversal potential versus ion concentration on side a of the membrane (activity for Cl^-, i.e. $\gamma[Cl^-]$). In (a), the pH was 3 on side b, and there was 300 mM Cl^- on both sides. In (b), the Cl^- activity was 213 (300 mM) on side b. The pH was 5 for the open circles and 3 for the filled squares. The solid lines are fits of Eq. (13.40) with $m/n = 0.43$. The dotted lines are the predictions of the Nernst equation (data from Accardi and Miller, 2004).

imposed on an ion permeable membrane. This will cause current to flow and we would like to relate the magnitude of the current to the properties of the membrane. We return to Eq. (13.18a), dispense with the subscript for Na^+, and rearrange to give

$$Ie^{zFV/RT} = -zFD\frac{d(ce^{zFV/RT})}{dx} \tag{13.41}$$

As before, a steady state is assumed with constant current independent of x. We can then integrate with respect to x from one side of the membrane to the other. The integral on the right, like those evaluated in going from Eq. (13.19) to (13.20), is easy because it is already expressed as a derivative

$$-zFD\int_a^b \frac{d(ce^{zFV/RT})}{dx}dx = -zFD(c_b e^{zFV_b/RT} - c_a e^{zFV_a/RT}) \tag{13.42}$$

It was already noted in the analysis of the Ussing flux equations that in order to integrate an expression like that on the left-hand side of Eq. (13.41), something must be known about the nature of $V(x)$ within the membrane. There is very little charge in the membrane interior, so the Poisson equation (Eq. (11.2)) becomes approximately $\Delta^2\varphi(\mathbf{r}) = 0$. For one dimension the solution is simply $\varphi(x) = Ex$ (where we ignore the irrelevant constant of integration). The variable E is the electric field, and is constant. This allows us to evaluate the integral arising from the left-hand side of Eq. (13.41)

$$\int_a^b e^{zFEx/RT}\,dx = \frac{RT}{zFE}\left(e^{zFV_b/RT} - e^{zFV_a/RT}\right) \tag{13.43}$$

With these two results, the integration of Eq. (13.41) is complete, so the current can be written as the quotient of these two solved integrals

$$I = -\frac{z^2F^2DE}{RT}\left(\frac{c_b e^{zF\Delta V/RT} - c_a}{e^{zF\Delta V/RT} - 1}\right) \tag{13.44}$$

where $e_a^{zFV_a}/RT$ was factored out and $V_b - V_a$ was replaced by ΔV. Recall that $D/\delta = P$ and $E\delta = \Delta V$ (Section 13.4), so $DE = P\Delta V$, giving

$$I = -\frac{z^2F^2P\Delta V}{RT}\left(\frac{c_b e^{zF\Delta V/RT} - c_a}{e^{zF\Delta V/RT} - 1}\right) \tag{13.45}$$

This is the Goldman–Hodgkin–Katz current equation. It gives current versus voltage for a particular ion with different concentrations on each side of the membrane. It is a big improvement over Ohm's law, which states a simple linear relation, $I = G\Delta V$ (Eq. (13.16)). Equation (13.45) is not linear for $c_a \neq c_b$. Figure 13.9 shows this for $c_b = 10c_a$.

Consider what happens with either large positive or negative values of ΔV. For positive ΔV the exponential is so large it dominates, and with the resulting cancellation Eq. (13.45) goes to a line with a slope given by Eq. (13.16) with c_b. For negative ΔV the exponential becomes negligible and we get a slope from Eq. (13.16) with c_a. Thus,

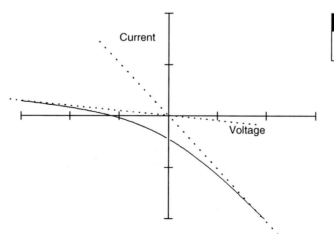

Current

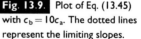

Voltage

Fig. 13.9. Plot of Eq. (13.45) with $c_b = 10c_a$. The dotted lines represent the limiting slopes.

Eq. (13.45) has two limiting slopes representing the extremes of behavior when the charge flows entirely in one direction (dotted lines in Fig. 13.9). By measuring these slopes, Eq. (13.45) can be used to determine the actual permeability of an ion. By contrast, measuring reversal potentials and using the Goldman–Hodgkin–Katz voltage equation, limits one to determining the ratios of permeabilities.

Like Eq. (13.32), Eq. (13.45) is often referred to as a constant-field equation because in the derivation we used $V(x) = Ex$, and set the field as constant. This was justified by noting that the charge density in the membrane is very low. However, there are a variety of forces acting on an ion in a membrane, and we will soon see how such forces can influence permeation (Section 13.13). Chapter 14 takes up this topic in much greater detail.

13.11 | Divalent ions

The Goldman–Hodgkin–Katz voltage equation (Eq. (13.32)) applies only to monovalent ions. There is no place in this equation for an ion with $z \neq \pm 1$. Multivalent ions make solving for ΔV a bit more difficult. The common general approach is to write down the Goldman–Hodgkin–Katz current equation for all the ions including divalents

$$\sum_i \frac{z_i^2 F^2 P_i \Delta V}{RT} \left(\frac{c_{bi} e^{z_i F \Delta V / RT} - c_{ai}}{e^{z_i F \Delta V / RT} - 1} \right) = 0 \qquad (13.46)$$

One can then use a computer to obtain the root of this equation, ΔV, for a particular set of permeability ratios (Adams *et al.*, 1980).

Analytical solutions are also known. Here we will derive an expression for the membrane potential for an arbitrary number of monovalent cations, together with a single divalent cation. All of the ionic currents are added together to give a total of zero

$$\frac{4F^2 P_{dc} \Delta V}{RT} \left(\frac{c_{bdc} e^{2F \Delta V / RT} - c_{adc}}{e^{2F \Delta V / RT} - 1} \right) + \sum_i \frac{F^2 P_i \Delta V}{RT} \left(\frac{c_{bi} e^{F \Delta V / RT} - c_{ai}}{e^{F \Delta V / RT} - 1} \right) = 0 \qquad (13.47)$$

The sum on the right contains terms of the form of Eq. (13.45), where the index i indicates each monovalent cation; $z = 1$ for each term of this sum. The other term in Eq. (13.47) is Eq. (13.45) with $z = 2$. This represents the current of the divalent cation, denoted by the subscript dc. Dividing through by common factors ($F^2 \Delta V / RT$), multiplying by $e^{2F \Delta V / RT} - 1$, and noting that $e^{2F \Delta V / RT} - 1 = (e^{F \Delta V / RT} - 1)(e^{F \Delta V / RT} + 1)$ gives

$$(e^{F \Delta V / RT} + 1) \sum_i P_i(c_{bi} e^{F \Delta V / RT} - c_{ai}) + 4P_{dc}(c_{bdc} e^{2F \Delta V / RT} - c_{adc}) = 0 \qquad (13.48)$$

This can be rewritten as a quadratic equation in $e^{F\Delta V/RT}$ as follows

$$e^{2F\Delta V/RT}\left(4P_{dc}c_{adc} + \sum_i P_i c_{ai}\right) + e^{F\Delta V/RT}\left(\sum_i P_i(c_{ai} - c_{bi})\right) - 4P_{dc}c_{bdc}$$
$$- \sum_i P_i c_{bi} = 0 \tag{13.49}$$

We now use the quadratic formula, and take the positive root,[3] as follows

$$e^{F\Delta V/RT} = \frac{-b + \sqrt{b^2 - 4ac}}{2a} \tag{13.50}$$

so $\quad \Delta V = \frac{RT}{F}\ln\left(\frac{-b + \sqrt{b^2 - 4ac}}{2a}\right) \tag{13.51}$

where $\quad a = 4P_{dc}c_{adc} + \sum_i P_i c_{ai} \tag{13.52a}$

$$b = \sum_i P_i(c_{ai} - c_{bi}) \tag{13.52b}$$

$$c = -4P_{dc}c_{bdc} - \sum_i P_i c_{bi} \tag{13.52c}$$

This expression for the membrane potential (Eq. (13.51)) is referred to as the *extended constant-field equation* (Piek, 1975). It depends on the constant field assumption used to derive the Goldman–Hodgkin–Katz current equation. The derivation can be generalized to include anions and other divalents. This equation is most often used in the analysis of Ca^{2+} permeability of ion channels, which is especially important because of the signaling functions of Ca^{2+}.

13.12 | Surface charge and membrane potentials

Surface potentials can have a strong impact on a membrane potential. Fluxes are sensitive to the surface potential, so if the surface potential affects the flux of different ions to different degrees, the reversal potential will be altered. To understand this effect we combine the results of this chapter with surface charge theory from Section 11.4.

To see a surface charge effect, we need two permeant ions with different charge, so we take Eq. (13.32) for K^+ and Cl^-

$$\Delta V = -\frac{RT}{F}\ln\left(\frac{P_K[K^+]_b + P_{Cl}[Cl^-]_a}{P_K[K^+]_a + P_{Cl}[Cl^-]_b}\right) \tag{13.53}$$

---

[3] The quantity ac in Eq. (13.50) is negative as can be seen by examining Eqs. (13.52a) and (13.52c), so the expression in the radical is $>b^2$.

Now add a surface potential φ_0 to both surfaces. The Boltzmann distribution gives the concentration of an ion at the membrane surface, c_0, as (Eq. (11.48))

$$c_0 = c(\infty)e^{-zF\varphi_0/RT} \tag{13.54}$$

where zF/RT replaced e/kT to keep with the molar units of this chapter. Normally, we calculate a reversal potential with the bulk concentration, $c(\infty)$. But c_0 is the concentration relevant to the flux across the membrane, so it belongs in Eq. (13.53). This substitution gives

$$\Delta V = -\frac{RT}{F} \ln \left(\frac{P_K[K^+]_b e^{-F\varphi_0/RT} + P_{Cl}[Cl^-]_a e^{F\varphi_0/RT}}{P_K[K^+]_a e^{-F\varphi_0/RT} + P_{Cl}[Cl^-]_b e^{F\varphi_0/RT}} \right) \tag{13.55}$$

In effect, the permeabilities are multiplied by a surface potential factor. If the permeability ratio in the absence of a surface potential is $R_{K/Cl} = P_K/P_{Cl}$, then the surface potential gives us a new effective permeability ratio of $R_{K/Cl}e^{-2F\varphi_0/RT}$.

This apparent permeability ratio can differ from the true permeability ratio by quite a bit. Surface potentials of $\sim$25 mV are common, so the surface potential factor in the permeability ratio, $R_{K/Cl}e^{-2F\varphi_0/RT}$, is $e^{-2\varphi_0/25} = 1/e^2 = 0.135$. Thus, by concentrating ions of one sign and depleting ions of another, a surface potential can alter the observed permeation properties of a membrane. Surface potential effects have been evaluated for the acetylcholine receptor channel, where increasing extracellular Ca^{2+} or Mg^{2+} shifted the reversal potential in the positive direction (Lewis, 1979). The general impact of these effects has been emphasized by Miedema (2002), who noted that the apparent selectivity of an ion channel can be switched between cation selective and anion selective as a result of the surface potential.

13.13 | Rate theory and membrane potentials

An ion's flux actually depends on how its potential energy varies inside the membrane. Up to this point, this fundamental relation was sidestepped by simply assigning values for permeabilities and deriving the membrane potential. The only mention so far of the relation between ion flux and the potential energy inside the membrane was the assumption of a constant field to derive the Goldman–Hodgkin–Katz current equation (Eq. (13.45)). The idea of a constant field can be justified with a simple electrostatics argument (Section 13.10). A quantitative treatment of the electrostatics in an ion channel (Levitt, 1978a) and an analysis that includes the distribution of ions (Jordan et al., 1989) show the approximation to be good. However, the electric field across the membrane is not the only force that affects ion movement. The drop in voltage across the membrane must be added to other forces that act on an ion. These other forces reflect the interaction between the ion and the

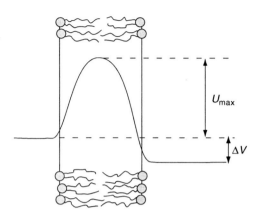

Fig. 13.10. Potential energy for the interaction between an ion and a lipid bilayer.

membrane, which varies dramatically with position. In Chapter 14 we will discuss these forces in detail, and we will make use of the method employed here of relating an ion's flux to its potential energy in a membrane. However, here the analysis is motivated by an interest in understanding how ion fluxes determined by these forces can contribute to the membrane potential.

When an ion enters the hydrophobic interior of a lipid bilayer, the potential energy will be very high because of the low dielectric constant of the hydrocarbon chains of the lipids. An image force pulls ions out of the membrane interior and into the water (Section 2.3). The potential energy function is sketched in Fig. 13.10. The peak in the middle represents the image potential, U_{max}. A voltage drop across the membrane, ΔV, is then the difference between the far right and far left limits.

The high potential energy of an ion in the middle of the membrane acts as a kinetic barrier to permeation. The rate theories of Chapter 7 were based on barrier crossing, and one of these will now be applied to the kinetics of ions moving through a membrane.

We start with the derivation of the Goldman–Hodgkin–Katz current equation, and focus on the point just before introducing the constant-field assumption. Without this assumption the integral in Eq. (13.43) must be evaluated in another way. Here, we will evaluate the integral by treating the potential energy in the exponential as a voltage drop added to a potential energy barrier. So we add a term $U(x)$ to $zFV(x)$, where U represents the peaked function sketched in Fig. 13.10. The current obtained as the ratio of two integrals (Eq. (13.44)) now has in the denominator an integral over the exponential of this energy function, instead of Eq. (13.43)

$$I = -zFD \frac{c_b e^{zF\Delta V/RT} - c_a}{\int_a^b e^{(zFV(x)+U(x))/RT} dx} \tag{13.56}$$

Since $U(x)$ peaks in the middle of the membrane, we can try the same technique used in Section 7.8, and approximate $U(x)$ as $U_{max} - w(x - x_{max})^2$. The voltage can still be taken as linear, $V(x) = x\Delta V$. Now the integral in the denominator of Eq. (13.56) is

$$\int_a^b e^{(zFx\Delta V + U_{max} - w(x-x_{max})^2)/RT}dx = e^{(U_{max}-wx_{max}^2)/RT}\int_a^b e^{(-wx^2 + (2wx_{max} + zF\Delta V)x)/RT}dx$$

(13.57)

According to Fig. 13.10 the energy should be small at $x = a$ and b, so we can let the limits of the integral go to $\pm\infty$. Now we can use Eq. (A4.7) to give

$$= e^{(U_{max} - wx_{max}^2)/RT}e^{(2wx_{max} + zF\Delta V)^2/4wRT}\sqrt{\frac{\pi RT}{w}}$$

(13.58)

Letting $x_{max} = 1/2$ (the maximum in U is at the center of the membrane) simplifies the expression to

$$= e^{U_{max}/RT}e^{(2zF\Delta V/+(zF\Delta V)^2/w)/4RT}\sqrt{\frac{\pi RT}{w}}$$

(13.59)

Finally, the quadratic term $(zF\Delta V)^2/(4wRT)$ is small and can be ignored. Using Eq. (13.59) for the integral in the denominator of Eq. (13.56) gives the current as

$$I = -zFD\sqrt{\frac{w}{\pi RT}}e^{-U_{max}/RT}e^{-zF\Delta V/2RT}(c_b e^{zF\Delta V/RT} - c_a)$$

(13.60)

We can collect the voltage and concentration independent factors together and define a permeability

$$P = \sqrt{\frac{w}{\pi RT}}De^{-U_{max}/RT}$$

(13.61)

The current can now be expressed more simply

$$I = -zFP(c_b e^{zF\Delta V/2RT} - c_a e^{-zF\Delta V/2RT})$$

(13.62)

This expression is zero at the Nernst potential for the ion (as expected). Furthermore, taking a sum of monovalent ion currents described by this equation, and setting this sum equal to zero leads to Eq. (13.32) (see Problem 12). Equation (13.62) is plotted in Fig. 13.11,

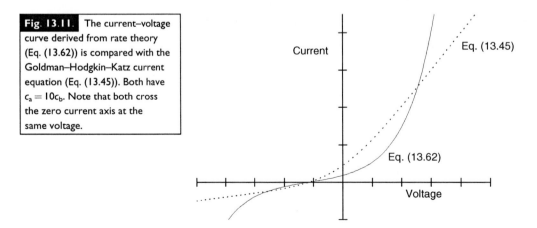

Fig. 13.11. The current–voltage curve derived from rate theory (Eq. (13.62)) is compared with the Goldman–Hodgkin–Katz current equation (Eq. (13.45)). Both have $c_a = 10c_b$. Note that both cross the zero current axis at the same voltage.

along with the Goldman–Hodgkin–Katz current equation (Eq. (13.45)). The plot becomes steeper as the driving force becomes stronger in either the positive or negative direction. This is an important hallmark of a kinetic process involving an energy barrier, and this will be examined in greater depth in the next chapter. It is actually quite common for current–voltage plots to deviate from the predictions of Eq. (13.45), and more complex permeation mechanisms involving energy barriers are invoked to explain such behavior.

Problems for Chapter 13

1. Derive the Donnan potential for N^w where the impermeant species is polycationic. Show that the result is equal to that in Eq. (13.11) for polyanions, but opposite in sign.

2. Derive the Nernst equation from the flux equation (Eq. (13.12)) by assuming that at equilibrium the flux is zero.

3. Use the Goldman–Hodgkin–Katz current equation (Eq. (13.45)) to derive the Goldman–Hodgkin–Katz voltage equation (Eq. (13.32)).

4. Extend the derivation of Eq. (13.21) to obtain a more general expression that includes any number of monovalent cations. Write the corresponding equation for anions. Compare these two equations with Eq. (13.32).

5. For the situation with NaCl on one side of a membrane and KCl on the other, at equal concentrations, use Eq. (13.33) to solve for the permeability ratio in terms of the "bionic potential" that prevails when only cations are permeant.

6. Equation (13.45) does not give zero current at $\Delta V = 0$ when $c_a \neq c_b$ (Fig. 13.10). What is the current at $\Delta V = 0$?

7. Derive the Nernst equation from Eq. (13.45) by setting $I = 0$.

8. Equation (13.16) was derived with the assumption of zero concentration gradient. This result is sometimes extended to a membrane with a concentration gradient by writing $I_i = G_i(\Delta V - E_i)$, where E_i is the Nernst potential for ion i. What is the major qualitative difference between this equation and the Goldman–Hodgkin–Katz current equation (Eq. (13.45)).

9. With the current equation from Problem 8 ($I_i = G_i(\Delta V - E_i)$) derive a general expression for the zero-current membrane potential in terms of G_i and E_i.

10. Derive Eq. (13.55) for two monovalent cations and show that surface potentials are irrelevant in this case.

11. Extend Eq. (13.55) to the case of unequal surface potentials on the different faces of the membrane (Miedema, 2002).

12. Derive Eq. (13.32) from Eq. (13.62).

13. Derive the counterpart to Eq. (13.40) for cotransport of m cations and n anions, i.e. transport of both ions in the same direction across the membrane.

14. The neuronal glutamate transporter transports $3 \, Na^+$, $1 \, H^+$, and 1 glutamate$^-$ into a cell in exchange for $1 \, K^+$ out (Zerangue and Kavanaugh, 1996). Derive the equilibrium potential for this transporter as a function of concentrations of these ions. Then express the result in terms of the four relevant Nernst potentials.

Chapter 14

Ion permeation and channel structure

Pure lipid bilayers have extremely low permeabilities to inorganic ions. Adding proteinaceous ion channels can increase the permeability by a factor of more than 10^8, allowing ions to flow across membranes and produce rapid changes in voltage. One can draw a strong analogy with enzymes. Both ion flow and the chemical reaction catalyzed by an enzyme have a favorable free energy that enables each to proceed in the absence of its respective catalyst, but at a very slow rate. Ion channels and enzymes both enhance these rates dramatically, and this enhancement is highly specific. In the case of an enzyme, small differences in the structure of a substrate can make a huge difference in catalytic efficiency. Likewise, ion channels can discriminate very effectively between different ions.

At first glance, an ion channel appears to have an easier task than an enzyme. It simply forms a water-filled pore so that ions see a continuous aqueous path through the membrane. However, a simple aqueous pore will not be specific for one particular ion. The diameter of K^+ is 1.33 Å and the diameter of Na^+ is 0.95 Å. Although this difference is small, some channels show selectivities between Na^+ and K^+ of more than 1000. Understanding this specificity is the real challenge in the study of ion channel permeation. Ion permeation depends not just on the water filling the pore but also on the detailed molecular structure of the protein that forms the channel.

14.1 | Permeation without channels

To gain an appreciation of how essential channels are to ion permeation, we first examine permeation (not necessarily of ions) through a membrane when there are no channels. The earliest insight into membrane permeation dates back to the nineteenth century. The theory of Overton relates the permeability of a substance to its tendency to partition between water and hydrophobic solvents. Since the interior of a membrane is hydrophobic,

hydrophobic substances partition into the oily interior of a membrane and cross more easily.

To develop this idea we make use of the partition coefficient, β, which is the ratio of concentrations of a dissolved substance in water and a hydrophobic solvent, when the two solvents are in contact and at equilibrium. Taking these concentrations as c_w and c_h, respectively, we have

$$\beta = \frac{c_h}{c_w} \tag{14.1}$$

Partition coefficients such as these allow one to estimate the free energy of transfer of a substance between the two environments. This was used to quantitate the hydrophobic effect (Section 2.8). If a solute dissolves in water and then equilibrates at each face of a membrane, then a concentration gradient between the two aqueous solutions will produce a proportional gradient inside the membrane between its two surfaces. Equation (14.1) thus implies that a concentration difference within the membrane is proportional to the difference in the two aqueous concentrations; $\Delta c_m = \beta \Delta c_w$. If the flux through the membrane is proportional to the internal driving force, Δc_m, then the flux will be proportional to $\beta \Delta c_w$.

Thus, a membrane's permeability to a substance is directly related to the substance's partition coefficient. Experiments have confirmed this relation for organic molecules. To take into account a molecule's mobility, a diffusion coefficient, D, can be factored in. For 16 substances a log–log plot of βD versus permeability is linear over a six-order of magnitude range, and the slope is close to one (Finkelstein, 1987).

Inorganic ions generally have extremely low values of β; they are essentially insoluble in hydrophobic solvents. Thus, we can account for the low permeability of inorganic ions through lipid bilayer membranes within the framework of Overton's simple theory. The free energy difference of an ion in water versus the membrane interior is accounted for through the self energy and the difference in dielectric constant (Section 2.2). A calculation of the image force gives the following result for the work necessary to move a charge from water, with its high dielectric constant of $\varepsilon_w \sim 80$ to the middle of a membrane, where the dielectric constant is much lower, $\varepsilon_h \sim 2$ (Eq. (2.6); see Fig. 2.3a)

$$\Delta G = \frac{q^2}{2a}\left(\frac{1}{\varepsilon_h} - \frac{1}{\varepsilon_w}\right) - \frac{q^2}{\varepsilon_h l}\ln\left(\frac{2\varepsilon_w}{\varepsilon_w + \varepsilon_h}\right) \tag{14.2}$$

where l is the thickness of the membrane, a is the radius of the ion, and q is the charge.

The variable ΔG in Eq. (14.2) is the height of an energy barrier seen by an ion as it crosses the membrane. If we envision the flux of the ion through a membrane as a barrier crossing process, then the rate will be proportional to an exponential factor (Chapter 7)

$$J \propto e^{-\Delta G/KT} \tag{14.3}$$

Anything that reduces ΔG will increase the flux across the membrane.

This theory makes two clear predictions: (1) for ions of the same charge, the larger ones cross the membrane more rapidly (by reducing the first term of Eq. (14.2)); and (2) thinner membranes are easier to cross than thicker membranes (by making the second term more negative).

Both of these predictions are testable. The first prediction was confirmed by the demonstration that large organic ions such as tetraphenylphosphonium, tetraphenylboron, and dipicrylamine cross much more rapidly than small inorganic ions. These are all monovalent ions with effective radii several times larger than those of monovalent inorganic ions such as Na^+ and Cl^-.

The second prediction of how flux changes with membrane thickness has also been tested. The thickness of an artificial bilayer membrane can be varied by forming membranes from lipids with different chain lengths. Furthermore, for reasons that are not clear, when the lipids are dissolved in alkanes with longer chains, the bilayers formed from these solutions are thicker. The thickness was determined by measuring the capacitance, which is inversely proportional to thickness. In studies of the large organic cation dipicrylamine, the barrier-crossing rate was measured from current relaxation experiments. A plot of rate versus thickness agreed well with Eqs. (14.2) and (14.3) (Fig. 14.1).

These experiments illustrate the electrostatic nature of the barrier to ion flux across the hydrophobic core of a membrane. As the radius of the ion gets smaller, the first term of Eq. (14.2) becomes very large; so the major biological ions such as Na^+, K^+, Cl^-, and Ca^{2+} see insurmountable energy barriers. One essential function of ion channels is to reduce this energy.

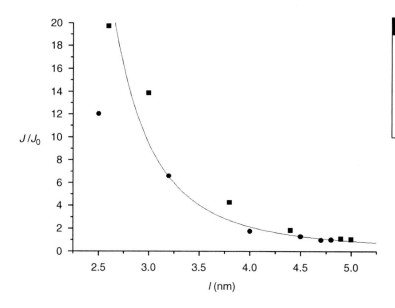

Fig. 14.1. The rate of dipicrylamine movement across a bilayer as a function of its thickness. The flux, J, was normalized to J_0, the value at $l = 5$. The solid curve is $e^{-3.69 + 17.8/l}$, based on Eqs. (14.2) and (14.3) (data from Benz and Läuger, 1977).

14.2 | The Ohmic channel

If a channel eliminates the image force then we might expect its aqueous lumen to perform like a piece of bulk electrolyte scaled down to the molecular dimensions of the channel. The bulk electrical resistivity of physiological saline, denoted as ρ, is typically about 100 Ω-cm. So a 1-cm cube has a resistance of 100 Ω for a voltage applied to two opposing faces. Whittling this cube down to the size of an ion channel would give a resistance proportional to the length, l, and inversely proportional to the area, A. This scaling gives a single-channel conductance of

$$\gamma = \frac{A}{\rho l} \tag{14.4}$$

where γ denotes the single-channel conductance and is expressed in units of Siemens (S = Ohm^{-1}). When referring to membrane conductance, the symbol G is used. In comparing these two quantities, G varies quite a bit because it scales with cell size and membrane area. By contrast, γ is a molecular property that tells us something about a channel's structure.

Take a cylindrical channel, with a length equal to the thickness of the membrane, and an area $A = \pi r^2$ (Fig. 14.2). This Ohmic channel is considered the most elementary model for an ion channel (Hille, 1991). Aside from ignoring the image force, this model assumes macroscopic behavior in a structure with microscopic dimensions. As a channel gets smaller, the water and ions interact with the channel walls, so bulk electrolyte properties become irrelevant. Both the image force and the microscopic factors become more important as channels get narrower. So Eq. (14.4) provides a test for both of these conditions.

When we know something about a channel's structure, we can estimate the conductance with Eq. (14.4) and then compare with experiments. Table 14.1 shows the results for two very large channels (the bacterial mechanosensitive channels MscL and MscS) and two intermediate sized channels (the acetylcholine receptor and gramicidin A). The smallest channels are not included because for them the Ohmic channel model is meaningless.

Comparing the theoretical and experimental conductances in Table 14.1 shows a clear trend. The larger the channel, the better the agreement with Eq. (14.4). This means that larger channels have more macroscopic character and weaker image forces. Bearing in mind that a water molecule is about 2.5 Å long, the larger channels are wide enough for several water molecules to fit side by side. So macroscopic behavior is not surprising for these channels. On the other hand, a water molecule inside an intermediate channel is likely to be in contact with the channel walls, and will behave

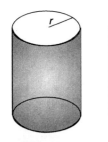

Fig. 14.2. A cylinder of electrolyte as the simplest model of an ion channel.

Channel	A (Å^2)	l(Å)	Theoretical γ_{th} (nS) from Eq. (14.4)	Experimental γ_{exp} (nS) ([KCl] (M))	$\frac{\gamma_{exp}}{\gamma_{th}}$
MscL[a]	707	40	4.4	2.5 (0.2)	0.57
MscS[b]	95	15	1.6	0.6 (0.2)	0.37
Acetylcholine receptor[c]	28	6	0.59	0.08 (0.1)	0.14
Gramicidin[d]	19	26	0.45	0.014 (0.1)	0.031

Table 14.1. *Structural parameters and conductances for various channels*

Conductance units: $1\ \text{nS} = 10^{-9}\ \text{S}$.

[a] Sukharev *et al.*, 1997, 2001.

[b] Bass *et al.*, 2002; Sukharev *et al.*, 1997.

[c] O'Mara *et al.*, 2003; Imoto *et al.*, 1988.

[d] Wallace, 1990; Andersen, 1983.

differently from bulk water. For these channels the conductance was ∼10-fold below the prediction of Eq. (14.4).

The smallest channels constitute a distinct class that was omitted from Table 14.1. Structural studies indicate that they have diameters of ∼3 Å. These channels have very high selectivity, and the conductance varies enormously depending on which ions are available in solution. Indeed, one such channel, aquaporin, only lets water through; ions of either sign are excluded and the conductance is several orders of magnitude lower than predicted by Eq. (14.4). Small channels are also sometimes referred to as "single-file channels" (Latorre *et al.*, 1992) for reasons that will become clear later in this chapter. Although the Ohmic channel only gives reasonable numbers for the largest channels, it serves as a useful gauge for the breakdown in microscopic character of intermediate and small channels.

14.3 | Energy barriers and channel properties

Energy barriers such as the image force become a major factor as ion channels get smaller. Information about a channel's energy barrier can be gained by studying how the current varies with voltage. Naturally, the Ohmic channel obeys Ohm's law and produces current that is a linear function of voltage. A barrier changes that and produces nonlinear behavior. In Section 13.13 a relevant problem of diffusion over a membrane barrier was solved, giving Eq. (13.62) (a review of this section is recommended). For the same concentration, c, of salt on each side, this equation simplifies to

$$I = zFcP(e^{zF\Delta V/2RT} - e^{-zF\Delta V/2RT}) = 2zFcP\sinh\left(\frac{zF\Delta V}{2RT}\right) \qquad (14.5)$$

where P is the permeability. The minus sign can be dropped because we no longer need to keep the ionic gradients in view. Now a positive voltage produces a positive current.

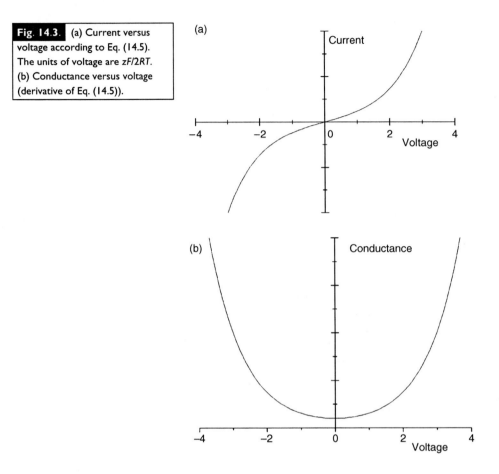

This current–voltage relation is plotted in Fig. 14.3a. The conductance is the derivative of Eq. (14.5), and is proportional to $\cosh(zF\Delta V/2RT)$ (Fig. 14.3b). This nonlinear shape is a direct consequence of the exponential dependence of the rate of barrier crossing on barrier height. Experimental current–voltage plots that resemble Fig. 14.3a are used to infer the existence of a barrier (Hall *et al.*, 1973).

When a current–voltage plot has the kind of nonlinearity shown in Fig. 14.3a, we can be reasonably sure that permeation entails surmounting an energy barrier that lies within the membrane field. We can sometimes learn a bit more, but most of the details about the barrier are absorbed into the integral in the denominator of Eq. (13.56). The barrier height thus has no direct influence on the shape of the voltage dependence. One might expect a smaller barrier to make the nonlinearity weaker. But as the barrier becomes small, the approximations used to derive this result lose their validity. When the barrier is small enough to ignore, the derivation leads to the Goldman–Hodgkin–Katz current equation (Eq. (13.45)). With the same salt concentrations on each side, this equation is linear.

One aspect of the barrier that produces an experimentally observable result is its position relative to the middle of the membrane. If a barrier is off-center, the current–voltage plot becomes asymmetric. Equation (14.5) serves as a starting point to understand this relation. Since the derivation of this equation in Section 13.13 absorbed the exponential term $e^{-U_{max}/RT}$ into the permeability, we can put this term back to give

$$I = zFcP'\left(e^{\frac{-U_{max}+zF\Delta V/2}{RT}} - e^{\frac{-U_{max}-zF\Delta V/2}{RT}}\right) \tag{14.6}$$

where $P' = Pe^{U_{max}/RT}$. In this form we see that the positive exponential is the rate of crossing in the positive direction, and the barrier seen by the ion is $U_{max} - zF\Delta V/2$. The negative exponential is the rate in the opposite direction, and this barrier is $U_{max} + zF\Delta V/2$. This equation thus has the form of a linear free energy relation (Section 7.4). In Eq. (14.6) the barrier is exactly in the middle of the membrane, so half the voltage difference is added to the barrier height for one direction of flux and subtracted for the other. Extending the logic of a linear free energy relation to the present situation, we assign the fraction of a voltage drop, δ, to the position corresponding to the top of the barrier

$$I = zFcP'\left(e^{(-U_{max}+zF\delta\Delta V)/RT} - e^{(-U_{max}-zF(1-\delta)\Delta V)/RT}\right) \tag{14.7}$$

where δ ranges from 0 to 1, and when it is 0.5, we revert back to Eq. (14.6).

Now that the analogy with the linear free energy relation has been made, we can reabsorb $e^{-U_{max}/RT}$ back into P to put the equation into the same form as Eq. (14.5)

$$I = zFcP\left(e^{zF\delta\Delta V/RT} - e^{-zF(1-\delta)\Delta V/RT}\right) \tag{14.8}$$

This equation gives asymmetric current–voltage plots when $\delta \neq 1/2$ (Fig. 14.4), and when this is observed experimentally, it can tell us the approximate position of the barrier.

Asymmetry in current–voltage plots as shown in Fig. 14.4 is referred to as rectification. In electronics, a rectifier is a device that passes current only in one direction. Channels with off-center barriers approach this behavior. This is a common property in semiconductors and serves as the starting point in the development of transistors, but semiconductor rectification arises through a fundamentally different physical mechanism. Rectification is a very common property in biological membranes, although the most common cause is voltage-dependent gating (Section 1.8) rather than barrier position. Blockade by ions binding to one side of the channel can give rise to asymmetric current-voltage plots as well (see Chapter 18 of Hille, 1991). When

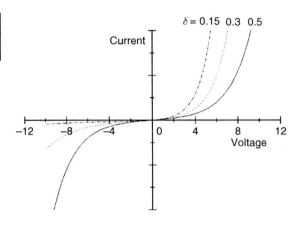

Fig. 14.4. Plots of Eq. (14.8) for the indicated values of δ, with voltage in units of zF/RT.

these alternatives can be ruled out, then rectification is likely to reflect an asymmetrically situated barrier.

14.4 | Eisenman selectivity sequences

So far nothing has been said about how channels select between different ions. There are two basic levels of selectivity. A coarse form discriminates charge, with channels preferring either cations or anions. For example, gramicidin and the acetylcholine receptor form channels selective for cations and the $GABA_A$ and glycine receptors form channels selective for anions. These channels show a weak preference between monovalent ions of the same charge. Charge selectivity is accomplished by fixed charges within the channel, or by dipoles oriented to favor one charge or the other. The second level of selectivity discriminates different ions with the same charge. Explaining this form of selectivity is comparatively difficult and involves more complicated interactions. We will now develop one of the earliest theories that attempted to explain this important property.

A popular way to classify higher order selectivity is to measure the permeabilities of several related ions and rank them in a sequence. The five commonly studied monovalent cations Li^+, Na^+, K^+, Rb^+, and Cs^+ can be combined into a total of $5! = 120$ possible sequences. However, in practice only 11 sequences are observed (Eisenman and Horn, 1983). Eisenman developed a simple explanation for the pattern of observed sequences. Although the physical model it is based on is not very sophisticated, it provides some useful qualitative insights into how channels can preferentially pass different ions.

The Eisenman theory begins by assuming that permeation is controlled by a *selectivity filter*, which is here envisioned as a site within the channel with a charge opposite to that of the permeating ions. The energy of an ion's interaction with this site determines the

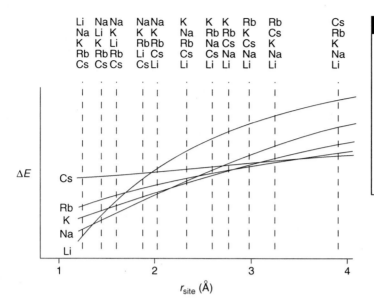

Fig. 14.5. The 11 Eisenman sequences for selectivity among 5 monovalent cations. A sequence is formed by the energies of the different ions along a vertical dashed line. The permeability sequences above are the inverse of the energies. The smooth curves are roughly based on Eq. (14.10). Plots of this equation tend to make the sequences cluster in a way that makes them hard to distinguish.

flux of that ion. The Coulomb potential (Eq. (2.1)) gives that inter-action energy, E_{site}, for the distance of closest approach

$$E_{site} = \frac{q_{site} q_{ion}}{\varepsilon(r_{site} + r_{ion})} \tag{14.9}$$

where r represents radius and q represents charge, with the sub-scripts indicating the site or ion. Add to this a hydration energy, E_{hyd} (which we note also depends on the ionic radius, Section 2.2), and take this sum as the energy change for moving an ion from the bulk water to the selectivity filter

$$\Delta E = E_{hyd} - E_{site} = E_{hyd} - \frac{q_{site} q_{ion}}{\varepsilon(r_{site} + r_{ion})} \tag{14.10}$$

Since the flux will be proportional to $e^{-\Delta E/RT}$, permeabilities for different ions should follow the inverse of the ΔE sequence. Figure 14.5 shows how ΔE for each ion varies with r_{site}. The depend-ences vary so that at different values of r_{site} the energies will fall into a different sequence. A selectivity sequence is defined by the order of energies along a vertical line for a particular value of r_{site}.

Each plot in Fig. 14.5 crosses every other plot only once. Each crossing exchanges two ions in their rank order of energy. The total number of crossings is 10 (this is fixed by topology), so the total number of sequences is 11. It is remarkable that the 11 experimentally observed sequences were generated by varying a single channel-specific parameter, r_{site}. This represents a striking success for the theory. Similar agreement has been found for anion selective channels (Eisenman and Horn, 1983).

Figure 14.5 makes some useful points about how channels work. The right-most sequence represents the case where the interaction between the ion and the channel is small and the hydration energies dominate. Hydration energies fall in the sequence $Li^+ > Na^+ > K^+ > Rb^+ > Cs^+$, with the smaller ions being the most difficult to dehydrate. Thus, when a selectivity sequence is the inverse of the hydration energy sequence, the implication is that the electrostatic interactions between the channel and the ion are less important and the primary obstacle to permeation is dehydration. On the other hand, when electrostatic interactions with the channel dominate, the smaller ions, which can get closer to the site, will be favored. This corresponds to the sequence on the left side of Fig. 14.5. The intermediate sequences reflect the interplay between these two forces.

There is no doubt that the physical model of the Eisenman theory is an oversimplification. Interactions between ions and channels are more complicated, and deeper examination of these interactions will occupy much of this chapter. A paradox in the theory is that dehydrating an ion requires a great deal of energy, which means that the ion must interact very strongly with the channel. Yet the theory gives the sequence as the inverse of the hydration energies only when the term representing interactions with the selectivity filter is small (large r_{site} in Eq. (14.10)).

Nevertheless, the generation of the correct selectivity sequences indicates that this simple picture has an important message. In general, ion permeation reflects a balance between a number of large energies. When two of these energies vary monotonically with ionic radius, the crossovers generate 11 sequences for 5 ions. The hydration and electrostatic interactions of the Eisenman theory are just two such contributions, and the sections that follow will examine these and others.

14.5 | Forces inside an ion channel

As discussed earlier, an ion in a channel is subject to a large image force due to the low polarizability of the membrane compared with that of water. The magnitude of this force increases as the channel radius gets smaller, and this is a major factor in the breakdown of the Ohmic channel model (Section 14.2). A theoretical calculation of this force can be made using electrostatics, treating the water and membrane as continuous media with different dielectric constants. This sort of problem was discussed in Section 2.3. The potential energy is obtained by solving the Laplace equation of electrostatics for an ion positioned within the channel. In the simplest case, the channel is assumed to be a cylindrical hole through a planar slab (Fig. 14.6).

This problem was first considered for an infinite cylinder (Parsegian, 1969). For this case the energy difference between an ion in bulk water and a channel takes the form

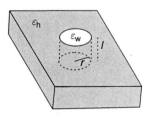

Fig. 14.6. A cylindrical hole containing water ($\varepsilon_w = 80$) embedded in a hydrophobic slab ($\varepsilon_h = 2$).

$$\Delta E = \frac{q^2}{2\varepsilon_h r} F\left(\frac{\varepsilon_h}{\varepsilon_w}\right) \qquad (14.11)$$

where r is the radius of the channel, and F is a mathematical function that must be evaluated with a computer. For $\varepsilon_w = 80$ and $\varepsilon_h = 2$, Eq. (14.11) works out to

$$\Delta E = \frac{28.4}{r} \, kcal \, mole^{-1} \tag{14.12}$$

when r is in angströms. For $r = 3$ Å, we get $9.8 \, kcal \, mole^{-1}$ or $16 \, RT$.

A barrier of this magnitude is far too high to allow ion flux to occur at the rates observed for most channels. More realistic calculations performed on channels with a finite length indicated that the energy is very sensitive to channel length and that for lengths comparable to the thickness of a membrane the energy could be reduced by a factor of 2 or 3 (Levitt, 1978a). With barrier heights so sensitive to channel dimensions a critical test of the role of image forces in ion permeation requires working with specific channels of known structure.

Many ion channels bind ions. This cannot be explained by the image force, which is purely repulsive. Thus, an additional attractive force is necessary to draw ions into the channel. The attractive force is often pictured as a short-range interaction between the ion and pore-lining amino acids of the channel. Since it is short range it acts on ions as soon as they enter and is roughly constant through the length of the channel protein. Such an attractive force is drawn in Fig. 14.7. When this attractive force is added to the image force, potential energy wells, which can serve as ion binding sites, appear at each end. The central energy barrier limits the rate of ion flux through the channel. This turns out to be a generic feature of ion channel potential energy profiles and a great deal of effort has gone into the interpretation of permeation and conductance data in terms of barriers and binding sites. A potential energy function with the same features captures the essential elements of ion permeation in gramicidin A channels (Section 14.6).

Rate theory provides a method for calculating the rate of ion movement between the two energy wells of an ion channel (Levitt, 1978b). We take Eq. (13.56) as the starting point, with $U(x)$

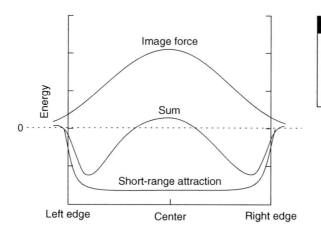

Left edge   Center   Right edge

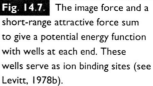

Fig. 14.7. The image force and a short-range attractive force sum to give a potential energy function with wells at each end. These wells serve as ion binding sites (see Levitt, 1978b).

representing the summed energy of the ion interaction with the channel and membrane (Fig. 14.7) and $V(x)$ representing the membrane potential

$$I = -zFD \frac{p_b e^{zF\Delta V/RT} - p_a}{\int_a^b e^{(zFV(x)+U(x))/RT} dx} \tag{14.13}$$

Here, the points a and b refer to the two energy minima on either side of the channel, and p_a and p_b are the probabilities of an ion being at those points. Taking symmetrical ionic conditions ($p_a = p_b$) and small voltage differences, we can expand the exponential in the numerator as $1 + zF\Delta V/RT$. We then have

$$I = -\frac{Dz^2 F^2 \Delta V p_a}{RT} \frac{1}{\int_a^b e^{(zFV(x)+U(x))/RT} dx} \tag{14.14}$$

The conductance of the channel varies with ion concentration. We assume that at the concentration where the conductance is maximal, the channel contains one ion. The probability of that ion being at a position x is given by the Boltzmann distribution, so $p_a = e^{-(zFV(a)+U(a))/RT} / \int_a^b e^{-(zFV(x)+U(x))/RT} dx$. Offsetting the energy to make it zero at $x = a$ makes the numerator equal to 1. Using this expression for p_a in Eq. (14.14) and dividing through by ΔV gives the maximal conductance

$$\gamma_{max} = \frac{Dz^2 F^2}{RT} \frac{1}{\int_a^b e^{(zFV(x)+U(x))/RT} dx \int_a^b e^{-(zFV(x)+U(x))/RT} dx} \tag{14.15}$$

Evaluating the integrals in Eq. (14.15) yields the maximum single-channel conductance, providing a way to connect a theoretical energy profile to a readily measurable quantity.

14.6 | Gramicidin A

Gramicidin A is an antibiotic that forms a cation selective channel in lipid bilayers. This 15-amino acid peptide has an unusual composition of alternating D and L amino acids. Its small size makes it amenable to structural studies (Wallace, 1990). The peptide rolls up into an unusual barrel-shaped helix. Two such barrels dimerize in a bilayer to form a channel ~26 Å long and 4–5 Å in diameter. With its structural simplicity, gramicidin provides an important testing ground for physical models of ion permeation.

Studies with NMR have shown that cations bind at the ends of the gramicidin channel (Tian and Cross, 1999). These sites fill as the ion concentration is raised. The conductance reaches a maximum when there is on average one ion in a channel, and as both sites start to fill the conductance begins to decline. The attractive interaction that drives this binding is an example of the kind of interaction depicted in Fig. 14.7. In gramicidin A, peptide backbone carbonyls line the pore with the electronegative oxygen atoms facing inward. These oxygen atoms attract cations and this attraction is invoked to account for the selectivity of gramicidin for cations over anions (Jordan, 1990; Roux *et al.*, 2000; Kuyucak *et al.*, 2001).

Analysis of realistic potential energy functions for gramicidin A yielded a large central energy barrier and conductances a few orders of magnitude below those observed, raising serious questions about the energetics of ion permeation in this channel (Jordan, 1990, Kuyucak *et al.*, 2001). A large part of the difficulty arises from the sensitivity to errors in two large opposing contributions. The image force is large and repulsive and the ion–peptide interaction energy is large and attractive. The small difference is especially sensitive to errors.

Although quantitative calculations of these energies have been difficult, qualitative ideas about the ion selectivity of the gramicidin channel are still instructive (Edwards *et al.*, 2002). The energy from the image force is proportional to z^2 (Eq. (14.2)), where z is the valence of the ion. The ion–dipole interaction energy is proportional to z (Eq. (2.9)). Thus, we can express the energy barrier as a sum, $z^2 U_{im} + z U_{di}$, representing the image force and ion–dipole energies, respectively. If we let $U_{im} = U$ and $U_{di} = -U$, so they are approximately equal in magnitude and opposite in sign, then for a monovalent cation with $z = 1$, the barrier is $z^2 U - z U = 0$. Monovalent cations see no energy barrier and thus permeate readily. On the other hand, for monovalent anions $z = -1$. The terms now add instead of subtract to give $2U$. This large energy accounts for gramicidin's low anion permeability. Finally, divalent cations are also impermeant, and this reflects the dependence on z^2. For a divalent cation, $z = 2$ and we have $z^2 U - z U = 2U$. The image force thus wins out to produce a barrier that is comparable to that seen by anions.

The difficulty in achieving quantitative agreement between theory and experiment with gramicidin A has been examined in depth (Edwards *et al.*, 2002). Reproducing the observed conductance and concentration dependence requires 8 kT energy wells at the channel ends and a 5 kT barrier in the middle. Such an energy profile cannot be obtained by adding a detailed ion-peptide potential energy function and an image force. The problem arises from treating the water in the gramicidin channel as a continuous medium. Gramicidin contains ~6 water molecules lined up end-to-end (Finkelstein and Andersen, 1981). The polarization and orientation of these water molecules is not a simple linear response to the field of an ion, so quantitative calculations of ion interactions in gramicidin A require using a molecular theory for the water. Recent progress along those lines indicated that the linear arrangement of waters in gramicidin A is surprisingly effective in stabilizing the ion within the channel (Allen *et al.*, 2004). This study developed a detailed potential energy function, which when used in an equation closely related to

Eq. (14.15) yielded a maximum conductance within a factor of 25 of the experimental value.

14.7 | Rate theory for multibarrier channels

Binding sites separated by an energy barrier are a generic feature of the potential energy function of an ion in a channel (Fig. 14.7). These basic features have been inferred again and again in studies of ion permeation. We can thus envision ion movement within a channel as a series of hops over barriers (Fig. 14.8). A voltage drop across the membrane can then be superimposed to bias the jumps in one direction or the other.

For now we will take a simple approach and assume that the barriers and sites are all the same. Focusing on one barrier and the two adjacent sites, the rate of crossing to the right is taken as α and the rate of crossing to the left is taken as β. Rate theory gives these rates as exponential functions of the barrier height

$$\alpha = \omega e^{-(E + \frac{zF\Delta V}{2n})/RT} \tag{14.16a}$$

$$\beta = \omega e^{-(E - \frac{zF\Delta V}{2n})/RT} \tag{14.16b}$$

Note that ω is a preexponential factor, which can be calculated from the potential energy function around the minimum (Section 7.8), but need not be written out here; and E is the height of the energy barrier when $\Delta V = 0$. Figure 14.8 indicates that a fraction of the voltage drop must be included in each energy barrier. The voltage difference between two adjacent sites is given by $\Delta V/n$, and the top of the barrier is half way in between, so the voltage drop between a site and an adjacent barrier is $\Delta V/2n$. As in Section 14.3, this quantity is added to the energy barrier for transitions to the right and subtracted for transitions to the left.

The ion concentration on each side of the membrane is under experimental control, but the probability of an ion occupying a site

Fig. 14.8. A series of energy barriers and wells encountered by an ion passing through a channel. The variables α and β represent the forward and reverse barrier crossing rates. The voltage drop across the membrane is ΔV and the voltage drop across one barrier is $\Delta V/n$.

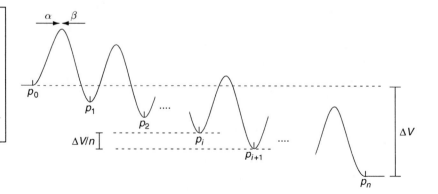

within the channel is unknown and must be derived from the model. We denote these probabilities as p_i and use them to express the net flux over a barrier as the difference in frequency of rightward and leftward transitions

$$J_i = \alpha p_i - \beta p_{i+1}$$
$$= \omega e^{-E/RT}\left(p_i e^{-zF\Delta V/2nRT} - p_{i+1} e^{zF\Delta V/2nRT}\right) \quad (14.17)$$

where the second step made use of Eqs. (14.16a) and (14.16b).

We take the system at steady state, so the p_is are time independent. That makes J_i the same for each barrier, so we can drop the subscript and use the same J everywhere. For small voltage drops across the barrier, $F\Delta V/n < RT$, we can expand the exponentials

$$J = \omega e^{-E/RT}\left(p_i\left(1 - \frac{zF\Delta V}{2nRT}\right) - p_{i+1}\left(1 + \frac{zF\Delta V}{2nRT}\right)\right)$$
$$= \omega e^{-E/RT}\left((p_i - p_{i+1}) - (p_i + p_{i+1})\frac{zF\Delta V}{2nRT}\right) \quad (14.18)$$

A simple result can be obtained for the case of equal concentration on both sides of the membrane because p_i and p_{i+1} will then also be equal. Multiplying J by zF to convert flux to current, and dividing by ΔV gives the channel conductance

$$\gamma = \frac{\omega\, zF p_i}{nRT} e^{-E/RT} \quad (14.19)$$

This result indicates that the conductance is inversely proportional to the number of barriers. The barriers thus act like resistors in series, with the resistance of each barrier summing to give the resistance of the entire transmembrane pathway.

The above result was for small voltage changes where current varies linearly. For larger voltages the expansion of the exponential to get Eq. (14.18) is no longer valid; current ceases to be a linear function of voltage when ΔV approaches nRT/zF. We saw earlier how a single barrier model gives a highly nonlinear voltage dependence (Fig. 14.3). Now we will explore this nonlinearity for a multibarrier model.

The procedure is to start from the left and write the flux over each barrier. For the first barrier we use Eq. (14.17), with $i = 0$

$$J = \alpha\, vc_a - \beta p_1 \quad (14.20)$$

where vc_a replaces p_0. The probability of being at the foot of the first barrier at the channel entrance is proportional to c_a, the bulk concentration on that side. The variable v is a proportionality constant with dimensions of volume. It can be envisioned as a small volume element at the entry to the channel. Solving Eq. (14.20) for p_1 gives

$$p_1 = \phi vc_a - \frac{J}{\beta} \quad (14.21)$$

where ϕ was introduced for α/β. Note that by Eqs. (14.16a) and (14.16b) $\phi = e^{-zF\Delta V/nRT}$.

The flux over the second barrier is

$$J = \alpha p_1 - \beta p_2 \tag{14.22}$$

As noted above, J is the same for each barrier due to steady state. Solving for p_2 gives

$$p_2 = \phi p_1 - \frac{J}{\beta} \tag{14.23}$$

Substituting Eq. (14.21) eliminates p_1, as follows

$$p_2 = \phi^2 \, vc_a - (1 + \phi)\frac{J}{\beta} \tag{14.24}$$

Repeating this process for the third barrier gives

$$p_3 = \phi^3 \, vc_a - (1 + \phi + \phi^2)\frac{J}{\beta} \tag{14.25}$$

The trend is now clear, so we can write the general result

$$p_i = \phi^i \, vc_a - \frac{J}{\beta}\sum_{j=0}^{i-1} \phi^j \tag{14.26}$$

The sum on the right is a geometric series, which is easily evaluated (Eq. (A1.9))

$$p_i = \phi^i \, vc_a - \frac{J(\phi^i - 1)}{\beta(\phi - 1)} \tag{14.27}$$

At the right side of the channel $p_n = vc_b$ (just as $p_0 = vc_a$ in Eq. (14.20)). We then take the expression for p_n from Eq. (14.27) and equate it with vc_b as follows

$$vc_b = \phi^n \, vc_a - \frac{J(\phi^n - 1)}{\beta(\phi - 1)} \tag{14.28}$$

This is solved for J. We then use $\phi = e^{zF\Delta V/nRT}$, $\phi^n = e^{zF\Delta V/RT}$, and Eq. (14.16b) to express β

$$J = v\omega e^{\frac{-(E - (zF\Delta V/2n))}{RT}} \frac{(c_a e^{zF\Delta V/RT} - c_b)(e^{zF\Delta V/nRT} - 1)}{e^{zF\Delta V/RT} - 1} \tag{14.29}$$

When $c_a = c_b = c$ the result simplifies to

$$I = zFv\omega \, ce^{-E/RT}(e^{zF\Delta V/2nRT} - e^{-zF\Delta V/2nRT}) = zFv\omega \, ce^{-E/RT} \sinh\left(\frac{zF\Delta V}{2nRT}\right) \tag{14.30}$$

where multiplication by zF converted flux to current.

Thus, we have a hyperbolic sine function like Eq. (14.5). When $n = 1$ this equation reverts to Eq. (14.5) for a single barrier, as it should. Recall that the plot of Eq. (14.5) showed a steeply rising conductance with voltage (Fig. 14.3). When there are n barriers

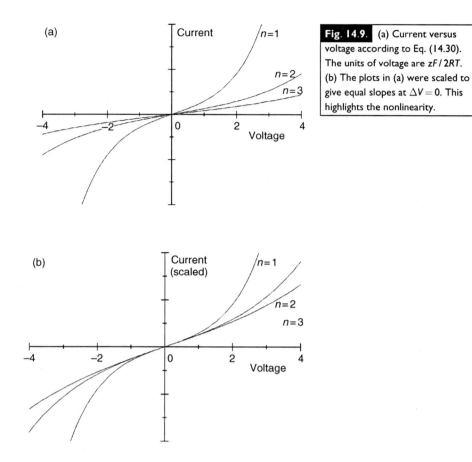

Fig. 14.9. (a) Current versus voltage according to Eq. (14.30). The units of voltage are $zF/2RT$. (b) The plots in (a) were scaled to give equal slopes at $\Delta V = 0$. This highlights the nonlinearity.

instead of one, the nonlinearity becomes noticeable as ΔV approaches $2nRT/zF$ rather than $2RT/zF$. Thus, adding barriers spreads out the current–voltage curve, pushing the nonlinearity out to larger and less accessible voltages (Fig. 14.9a, b). This illustrates a fundamental relation between the energy profile for an ion in a channel and the nonlinearity of the current–voltage plot. The more barriers, the weaker the nonlinearity. Fig. 14.9 also illustrates the point made above with Eq. (14.19a) that adding barriers reduces the conductance.

The derivation here was for a channel with identical barriers but it can be extended to nonidentical barriers, as will be seen in the following section. A full treatment of barrier models by Läuger (1973) illuminated many interesting properties of ion channels. In addition to the nature of the nonlinearity in the current–voltage plot, this study showed that when the barriers at the channel entrances are larger than those within the channel, the nonlinearity has the opposite form. The conductance decreases as the voltage increases.

The multi-barrier model can also be used to derive the flux in response to a concentration gradient and thus give us the channel permeability. Start with Eq. (14.29), and let $\Delta V \to 0$. The limit of the ratio $(e^{zF\Delta V/nRT} - 1)/(e^{zF\Delta V/RT} - 1)$ in Eq. (14.29) is $1/n$ after taking the first order expansions of the exponentials, so we end up with

$$J = \frac{v\omega e^{-E/RT}}{n}(c_a - c_b) \tag{14.31}$$

The permeability is the proportionality constant between J and $c_a - c_b$

$$P = \frac{v\omega e^{-E/RT}}{n} \tag{14.32}$$

The conductance can also be obtained from Eq. (14.30). Let $\Delta V \to 0$ and multiply by zF to turn flux into current. The conductance is the proportionality constant between current and voltage

$$\gamma = \frac{v\omega\,cz^2\,F^2 e^{-E/RT}}{nRT} \tag{14.33}$$

Comparing Eqs. (14.32) and (14.33) leads directly to a relation between permeability and conductance identical to Eq. (13.17).

14.8 | Single-ion channels

The Ohmic channel gives a conductance that is inversely proportional to the medium resistivity, ρ (Eq. 14.4). The value of ρ decreases with ion concentration and so the conductance will increase. However, channels often reach a maximum conductance as the ion concentration increases, and in fact many channels have an optimal concentration, above which conductance starts to fall. Gramicidin A provided an example of this behavior (Section 14.6). This is clearly inconsistent with an Ohmic channel, but barrier models can explain it.

One way to incorporate saturation into a model is to assume that only one ion can occupy the channel at a time. This is based on the idea that electrostatic repulsion prevents ions of the same charge from being near each other. Once one ion is in the channel, it keeps others out. Clearly, this assumption is violated by gramicidin; ions can occupy each end. But for some other channels the single-ion assumption holds, and theoretical predictions of single-ion models have proven useful in this assessment.

Here, a model for a single-ion channel will be developed for an arbitrary collection of sites and barriers with different heights and depths (Fig. 14.10). This makes the present analysis more general than that of the preceding section, where all sites and barriers were identical. This section will thus build on the preceding section both with the single-ion occupancy assumption and the generalization to heterogeneous structure.

The single-ion assumption is incorporated by modifying Eq. (14.20). The flux over the first barrier, the rate of entry into the channel, is reduced by the factor $1 - p$ (Läuger, 1973), where p is the

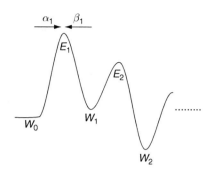

probability that an ion resides in any of the sites within the channel $(p = \sum_{i=1}^{n-1} p_i)$. Thus, the entry step can only occur when the channel is empty

$$J = \alpha_1 v c_a (1 - p) - \beta_1 p_1 \tag{14.34}$$

For the second barrier we have

$$J = \alpha_2 p_1 - \beta_2 p_2 \tag{14.35}$$

The general expression for p_i corresponding to Eq. (14.26) can be developed by successive solutions for p_1, p_2, etc. The result is a more general version of Eq. (14.26)

$$p_i = c_a v (1 - p) \prod_{j=1}^{i} \phi_j - J \left(\frac{\prod_{j=2}^{i} \phi_j}{\beta_1} + \frac{\prod_{j=3}^{i} \phi_j}{\beta_2} + \cdots + \frac{\phi_i}{\beta_{i-1}} + \frac{1}{\beta_i} \right) \tag{14.36}$$

where $\phi_i = \alpha_i / \beta_i$.

This daunting expression simplifies quite nicely when we use rate theory

$$\alpha_i = w e^{-(E_i - W_{i-1})/RT} \tag{14.37a}$$

$$\beta_i = w e^{-(E_i - W_i)/RT} \tag{14.37b}$$

The Es and Ws are as indicated in Fig. 14.10. Equations (14.37a) and (14.37b) are generalizations of Eqs. (14.16a) and (14.16b). This gives

$$\phi_i = \frac{\alpha_i}{\beta_i} = e^{-(W_i - W_{i-1})/RT} \tag{14.38}$$

The barrier energies in these equations (the Es) have the form $E + \delta z F \Delta V$, where E is the energy in the absence of a membrane potential and δ is the fraction of membrane potential traversed to reach the top of the barrier. The well energies depend on voltage in the same way; $W + \delta z F \Delta V$.

When the ϕs are multiplied together, the Ws in successive terms cancel out. This gives a general expression for the products

$$\prod_{k=j}^{i} \phi_k = e^{-(W_i - W_{j-1})/RT} \tag{14.39}$$

With the energy on the left side of the channel taken as a reference value of zero ($W_0 = 0$), use of these expressions simplifies Eq. (14.36) to

$$p_i = c_a v(1-p)e^{-W_i/RT} - Je^{-W_i/RT} \sum_{j=1}^{i} \frac{e^{W_j/RT}}{\beta_j}$$

$$= c_a v(1-p)e^{-W_i/RT} - \frac{J}{\omega}e^{-W_i/RT} \sum_{j=1}^{i} e^{E_j/RT} \tag{14.40}$$

where the second step made use of Eq. (14.37b) in the second term on the right.

The flux over the last barrier on the right side of the channel is

$$J = \alpha_n p_{n-1} - \beta_n c_b v(1-p) \tag{14.41}$$

Including $1-p$ here reflects the inability of an ion to enter an occupied channel from the right side of the membrane. Equation (14.41) is in a sense a mirror image of Eq. (14.34). Using Eq. (14.40) to replace p_{n-1} in Eq. (14.41) and dividing through by β_n gives

$$\frac{J}{\beta_n} = c_a v(1-p)\phi_n e^{-W_{n-1}/RT} - \frac{J}{\omega}\phi_n e^{-W_{n-1}/RT} \sum_{j=1}^{n-1} e^{E_j/RT} - c_b v(1-p) \tag{14.42}$$

This can be simplified by noting that $\phi_n e^{-W_{n-1}/RT} = e^{-(W_n - W_{n-1})/RT}e^{-w_{n-1}/RT} = e^{-w_n/RT}$. Here W_n reflects the energy after the final barrier on the right side of the channel. It is thus the energy of an ion that has passed through the membrane potential, or $-zF\Delta V$. Taking J/β_n over to the right side, and using Eq. (14.37b) for β_n gives a result that can be recognized as an nth term to be incorporated into the sum in Eq. (14.42). Now the sum extends to n instead of $n-1$, as follows

$$0 = c_a v(1-p)e^{zF\Delta V/RT} - \frac{J}{\omega}e^{zF\Delta V/RT} \sum_{j=1}^{n} e^{E_i/RT} - c_b v(1-p) \tag{14.43}$$

Solving for J gives

$$J = \frac{\omega v(1-p)(c_a - c_b e^{-zF\Delta V/RT})}{\sum_{j=1}^{n} e^{E_j/RT}} \tag{14.44}$$

Although something sensible is starting to take shape, it is not a complete result yet because we do not know p. However, Eq. (14.44) still provides an important insight into reversal potentials. Take two permeant ions A and B, with fluxes J_A and J_B. At the reversal potential the two fluxes must add up to zero. When this is written out, the factor $\omega v(1-p)$, which is the same for the two fluxes, can be dropped to leave (Läuger, 1973)

$$\frac{c_{Aa} - c_{Ab}e^{-zF\Delta V/RT}}{\sum\limits_{j=1}^{n} e^{E_{jA}/RT}} + \frac{c_{Ba} - c_{Bb}e^{-zF\Delta V/RT}}{\sum\limits_{j=1}^{n} e^{E_{jB}/RT}} = 0 \qquad (14.45)$$

This equation cannot be solved for ΔV because the barrier energies, the E_{jA}s and E_{jB}s, depend on voltage. Even if a linear dependence of these energies on voltage is assumed, as in Fig. 14.8, Eq. (14.45) cannot be solved for arbitrary barriers. Without solving for the voltage, we can still see that if the concentrations of both ions are multiplied by the same factor, this will not alter the value of ΔV that satisfies this equation. The scaling factor for concentration cancels out to leave the equation unchanged. The linear dependence on concentration thus insures that the reversal potential is independent of concentrations as long as they remain in a fixed ratio. This statement is valid in spite of the fact that we cannot explicitly solve for ΔV. Thus, Eq. (14.45) provides us with an important and experimentally testable prediction that is demonstrated in Fig. 14.11 below.

The Goldman–Hodgkin–Katz voltage equation (Eq. (13.32)) gave the reversal potential in terms of concentrations and permeabilities, but the permeabilities were taken as fixed quantities that do not depend on voltage. In fact, if the sums in the denominator on each side of Eq. (14.45) are assigned symbols corresponding to the reciprocals of the permeabilities, and their voltage dependence is ignored, the Goldman–Hodgkin–Katz equation can be derived (Problem 3). Thus, these sums are effective permeabilities, even though they depend on voltage. If concentrations are changed asymmetrically, the reversal potential will change, but so will the permeability ratio. This represents a clear departure from the classical Goldman–Hodgkin–Katz theory of Chapter 13. There are many examples of channels showing this kind of behavior, and barrier models have proven very useful in interpreting such results, providing

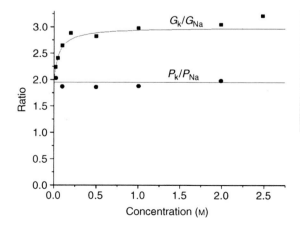

Fig. 14.11. The permeability ratio (P_K/P_{Na}) is independent of concentration, and is equal to the conductance ratio (G_K/G_{Na}) in the limit of zero concentration. The curve for G_K/G_{Na} is based on Eq. (14.49) with $K = 60$ mM (from Coronado et al., 1980).

insight into the energy profile for an ion in a channel (Eisenman and Horn, 1983).

The constancy of the reversal potential and permeability ratio when all of the concentrations are scaled together can be contrasted with other properties that change. In particular, the channel conductance varies with ion concentration, and the ratio of the conductances of two different ions can change as the concentrations are scaled up or down in parallel. Thus, the permeability, which is the coefficient relating ion flux to the concentration gradient, shows fundamentally different behavior from the conductance, which is the coefficient relating ion flux to voltage.

To understand this different behavior of conductance ratios we must pick up where we left off after Eq. (14.44), and derive an expression for p, the probability that a channel is occupied. We can get a reasonable picture of how p behaves by taking $J = 0$ and $c_a = c_b = c$. Equation (14.36) then simplifies to

$$p_i = vc(1 - p) \prod_{j=1}^{i} \phi_j = vc(1 - p)e^{-(W_i - W_0)/RT} \tag{14.46}$$

where the second step made use of Eq. (14.38). Summing terms from Eq. (14.46) gives p as

$$p = \sum_{i=1}^{n-1} p_i = vc(1 - p) \sum_{i=1}^{n-1} e^{-W_i/RT} \tag{14.47}$$

where $e^{W_0/RT} = 1$ because $W_0 = 0$. Now we can solve for p

$$p = \frac{vc \sum_{i=1}^{n-1} e^{-W_i/RT}}{1 + vc \sum_{i=1}^{n-1} e^{-W_i/RT}} \tag{14.48}$$

This expresses the saturation of the channel with increasing ion concentration. The value of p reaches a plateau as c exceeds $1/\left(v \sum_{i=1}^{n-1} e^{-W_i/RT}\right)$. In fact, $\sum_{i=1}^{n-1} e^{-W_i/RT}$ is actually a binding constant, so we will call it $1/K$. (The resemblance between this sum and a partition function is significant.) Keep in mind as we continue that K depends on the W_is, the energies of the wells in the channel (Fig. 14.10).

We can now obtain the current by substituting p from Eq. (14.48) into Eq. (14.44). Again, taking $c_a = c_b = c$, and multiplying by zF leads to

$$I = \frac{zF\omega vc}{\sum_{j=1}^{n} e^{E_j/RT}} \left(\frac{1 - e^{-zF\Delta V/RT}}{1 + (c/K)} \right) \tag{14.49}$$

The concentration dependence has the standard single-site saturation behavior discussed in Chapter 4. At low concentrations, c/K is small and can be omitted from the denominator, to give

$$I = \frac{zF\omega vc}{\sum\limits_{j=1}^{n} e^{E_j/RT}} (1 - e^{-zF\Delta V/RT}) \tag{14.50}$$

Note that K, which depends on the energies of the wells, is absent. Thus, at low concentrations I does not depend on the wells, only on the barriers.

For small voltages we can take $1 - e^{-zF\Delta V/RT} \sim zF\Delta V/RT$. Equation (14.50) then becomes

$$I = \frac{z^2 F^2 \omega vc \Delta V}{RT \sum\limits_{j=1}^{n} e^{E_j/RT}} \tag{14.51}$$

The conductance is the proportionality constant between I and ΔV. A ratio of the conductances for two different ions would depend on the sums in the denominator. These sums depend only on the barriers, and not on the wells that form the binding sites. The permeabilities (the denominator in Eq. (14.45)) have the same dependence on barriers. Thus, the conductance ratios and permeability ratios are equal at low ion concentrations.

As c increases, I reaches a limit where the channel is always occupied. In this limit of large c and small ΔV, Eq. (14.49) becomes

$$I = \frac{z^2 F^2 \omega vK \Delta V}{RT \sum\limits_{j=1}^{n} e^{E_j/RT}} \tag{14.52}$$

Now the conductance depends on K, and thus on the W_is that represent well energies. The situation at low concentrations was very different (Eq. (14.50)). Symmetric increases in the concentrations do not alter the permeabilities determined from the reversal potential (Eq. (14.45)), but do change the conductances. This means that when one measures the conductance of a series of ions, the ratios of these conductances can vary. Indeed, the conductance sequence can be switched by changing concentration.

These ideas have been nicely illustrated with a cation selective channel of sarcoplasmic reticulum (Coronado et al., 1980). The permeability ratio was measured from the reversal potential with Na^+ on one side and K^+ on the other. Single channel conductance was measured for symmetrical K^+ and Na^+. The permeability ratio P_K/P_{Na} was close to 2, and there was no change over a wide range of concentrations (Fig. 14.11), as predicted by Eq. (14.45). The conductance ratio went to the permeability ratio at low concentrations, as predicted by Eq. (14.50). But at higher concentrations the conductance ratio increased to about 3. Thus, the energy wells come into play as the channel becomes saturated. The wells would appear to select K^+ to a greater degree than the barriers, and so the conductance ratio is higher than the permeability ratio.

The single-channel conductance saturated as the concentrations increased, and this behavior was similar to that of the conductance ratio plotted in Fig. 14.11. This saturation behavior fits with the basic assumption that no more than one ion can occupy the channel at a time. Relatively few channels show this behavior, and most channels hold more than one ion. It takes a different kind of model to address this feature, and one important example is the single-file model to be examined next.

14.9 | Single-file channels

The term single file describes a class of models in which ions form a linear chain through a channel that is so narrow that ions cannot get past one another. Since one ion cannot move without forcing others to move, they move together. If they do not move as a perfect unit, then they move in a sequence of hops, with each ion moving into a vacancy created when its neighboring ion moves. Single-file models are especially well suited for highly selective channels that have strong and specific ion binding sites.

The single-file model originated in a classical study by Hodgkin and Keynes (1955). It is a remarkable outcome of recent structural advances that the key features of the single-file model have been verified. Hodgkin and Keynes examined K^+ flux in the squid axon after blocking active transport. With the pumps shut down K^+ only flowed down its electrochemical gradient. The inward and outward fluxes were measured with isotope tracers to determine the Ussing flux ratio (Section 13.6). According to this theory the unidirectional flux ratio is given by (Eq. (13.27))

$$\frac{J_{i(a \to b)}}{J_{i(b \to a)}} = \frac{c_{ia}}{c_{ib}} e^{-z_i F \Delta V / RT} \tag{14.53}$$

The flux data of Hodgkin and Keynes did not obey this relation. Although the fluxes were still equal at the Nernst potential (where the right-hand side of Eq. (14.53) equals one), Hodgkin and Keynes found that to fit the rest of the data they had to raise the right side of the equation to a higher power than one. The exponent they obtained was 3.5. This result did not negate the basic premise of passive transport, because the fluxes were balanced at the Nernst potential. However, the derivation of the Ussing flux ratio was also based on the assumption that inward and outward fluxes are independent, so Hodgkin and Keynes sought to explain their result with a theory in which inward and outward fluxes are strongly coupled. This led to the single-file model.

Figure (14.12) shows a single-file channel filled with ions. There are K^+ ions on both sides of the membrane. The ones on the left are labeled A and the ones on the right are labeled B. A channel has n sites that are occupied by either As or Bs. A sharp boundary separates the As and Bs in the channel because a sequence of the

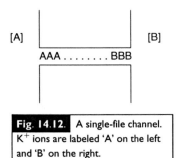

[A] AAA . . . . BBB [B]

Fig. 14.12. A single-file channel. K^+ ions are labeled 'A' on the left and 'B' on the right.

form ABA or BAB requires an ion to enter from a side where all of the ions have the other label. This cannot happen because a radioactive tracer is instantly diluted after it reaches the other side. So all of the occupancy states of a single-file channel have the form AA ... AB ... BB.

At this point a distinction is drawn between isotope tracer fluxes, where the labels are key, and net ion flux or current, where the labels do not matter. A tracer transfer event occurs after there have been $n+1$ more single-file hops in one direction than in the other direction, but a charge transfer event occurs with just one hop. We denote the rate of right and left charge transfer events as r and s, respectively. The Nernst potential indicates that these rates are in the ratio

$$\frac{r}{s} = \frac{[A]}{[B]} e^{-zF\Delta V/RT} \tag{14.54}$$

This way $r = s$ when $\Delta V = -\dfrac{RT}{zF} \ln \dfrac{[B]}{[A]}$ (Eq. (13.3)).

Each charge movement event, or hop, shifts the single-file line of ions in the channel by one position. If we start with a channel with all n sites occupied by As, and denote this as species A_n, it will be converted to a single-file line of the form $A_{n-1}B$ with a rate s. The reverse process occurs with a rate r. This specifies a differential equation for the rate of change of $[A_n]$

$$\frac{d[A_n]}{dt} = -s[A_n] + r[A_{n-1}B] \tag{14.55}$$

At steady state the rate of change is zero, so

$$\frac{r}{s} = \frac{[A_n]}{[A_{n-1}B]} \tag{14.56}$$

The single-file complex $[A_{n-1}B]$ can go to either $[A_n]$ or $[A_{n-2}B_2]$, and can be created from these species by the reverse processes. Thus, the rate of change of $[A_{n-1}B]$ is

$$\frac{d[A_{n-1}B]}{dt} = -(s+r)[A_{n-1}B] + r[A_{n-2}B_2] + s[A_n] \tag{14.57}$$

Again with a steady state, we set this expression equal to zero, rearrange, and make use of Eq. (14.56) to obtain

$$\left(\frac{r}{s}\right)^2 = \frac{[A_n]}{[A_{n-2}B_2]} \tag{14.58}$$

This can be continued for $[A_{n-2}B_2]$ and $[A_{n-3}B_3]$, etc., so we have a general relation

$$\left(\frac{r}{s}\right)^j = \frac{[A_n]}{[A_{n-j}B_j]} \tag{14.59}$$

Tracer flux events only occur from the states A_n and B_n. Movement of an A into the solution on the right in Fig. 14.12 occurs with a rate $r[A_n]$ and movement of a B into the solution on the left occurs with a rate $s[B_n]$. Dividing these two gives the tracer flux ratio as

$$\frac{J_{i(a \to b)}}{J_{i(b \to a)}} = \frac{r[A_n]}{s[B_n]} = \left(\frac{r}{s}\right)^{n+1} \tag{14.60}$$

where the second step made use of Eq. (14.59) with $j = n$. Combining this with Eq. (14.54) gives the final result

$$\frac{J_{i(a \to b)}}{J_{i(b \to a)}} = \left(\frac{[A]}{[B]} e^{-\frac{zF\Delta V}{RT}}\right)^{n+1} \tag{14.61}$$

As noted above, Hodgkin and Keynes fitted this equation to their experimental data on tracer flux and obtained an exponent of 3.5, so $n = 2.5$. They therefore concluded that the single-file line of K^+ ions in the channel is 2–3 ions long.

A key feature of this model, the concerted hopping of all the ions together, is not very realistic. A more plausible picture of single-file movement has been developed with multiple barriers and sites. When all but one of the sites are occupied, then there is a single vacancy. As an ion hops to fill the vacancy, the vacancy appears to move in the opposite direction. We can picture a single-vacancy model, which might be described by very similar equations as the single-ion model of the preceding section. The mapping of one of these problems to the other has been formally carried out by Schumaker and MacKinnon (1990), who showed how to obtain a hyperbolic sine function for the current–voltage relation that is essentially the same as that for a multi-barrier model (Eq. (14.30)).

Barrier models for single-file channels tend to be mathematically complex because multiple ion occupancy creates a proliferation of states. The kinetic equations can be solved with a computer or by the general methods employed in Chapter 9. A number of experimental observations are recovered from these models (Hille and Schwarz, 1978), including the exponential form for the flux ratio (Eq. (14.61)). Single-file models also give rise to a conductance–concentration plot with a peak above which conductance declines. This happens when the concentration is so high that a vacancy reaching the edge of a channel is filled more rapidly by ions from solution than from the neighboring site. Additional predictions of single-file models include steeply voltage-dependent block by certain impermeant ions and steeply voltage-dependent changes in conductance. The observation of similar properties in many different K^+ channels indicates that the single-file mechanism is widespread.

There are a number of interesting examples of single-file behavior in membrane transport. A single-file model explained how Ca^{2+} channels exclude monovalent cations such as Na^+ so effectively. The sites in a Ca^{2+} channel bind Ca^{2+} with a high affinity; the dissociation constant for filling the first site is $\sim 1\,\mu M$ and the other sites are occupied at somewhat higher concentrations. Na^+ can also bind these sites, but weakly. At physiological Ca^{2+} concentrations of ~1 mM, Ca^{2+} channels are always filled with Ca^{2+}, which blocks the entry of monovalent cations. If by chance a Na^+ enters, its lower charge provides a relatively weak electrical push, which is insufficient to drive the neighboring Ca^{2+} ion into a deeper site and engage the single-file mechanism. When a Ca^{2+} ion enters, its electrostatic repulsion is stronger, and forces a single-file motion of other Ca^{2+} ions. Reducing the Ca^{2+} concentration below ~100 μM leaves only one of the sites occupied. Conduction requires multiple ion occupancy so the current goes to zero. Further reduction of the Ca^{2+} concentration below 1 μM empties out the last remaining Ca^{2+} from the channel, at which point Na^+ readily flows through. Thus, complete removal of Ca^{2+} converts the Ca^{2+} channel into a channel that is selective for Na^+ (Almers and McCleskey, 1984; Hess and Tsien, 1984).

Single-file models have been applied to the gramicidin channel to analyze the flux of water (Finkelstein and Andersen, 1981). The water flux is coupled to the ion flux, and an analysis of this coupling produced an estimate of the number of water molecules in the channel as ~6. Computer simulations confirmed the single-file nature of water and ion movement, showing that the ions and water molecules never passed one another inside the gramicidin A channel (Edwards *et al.*, 2002). Ion flux measurements showed that for ion concentrations below ~100 mM (depending on choice of ion) the ions flow through the channel one at a time, with no single-file character. However, as the concentration is raised the flux ratio exceeds one, indicating that both of the sites in Fig. 14.7 can be occupied. When this happens, an ion at one site must wait for the other site to empty before it can cross to the other side. Tracer flux then becomes a two-ion event so the exponent in Eq. (14.61) exceeds one.

Another interesting example of single-file behavior was found for the protein that transports the neurotransmitter serotonin across the membrane (Adams and DeFelice, 2002). The flux of labeled serotonin varied as unlabeled serotonin was added, indicating that molecules of serotonin interact as they cross the membrane. Additional experiments suggested that the serotonin transporter is a channel filled with both inorganic ions and serotonin. Ion gradients can thus drive serotonin across the membrane by pushing the serotonin in a single-file line. This is likely to be a common mechanism by which electrochemical gradients are harnessed to drive diverse substrates across membranes (DeFelice, 2004).

The various models discussed in this and the two preceding sections have enjoyed considerable success in accounting for a wide range of ion permeation phenomena. The barriers and binding sites of these models represent real features of channels. But the energies obtained as parameters by applying barrier models to experimental data have a rather fictional character. They generally cannot be related explicitly to the channel structure. The problem is not in the models but in the paucity of structural information available when the studies were carried out. Recent advances in channel structure have dramatically changed this situation, providing proof for many of the features inferred from barrier models. We will now see how crystal structure has provided clear answers to some long-standing questions about ion permeation and has spawned a transition from barrier models to detailed computational theories.

14.10 | The KcsA channel

The first selective ion channel for which a high resolution structure was determined was the KcsA K$^+$ channel (Doyle *et al.*, 1998). The sequence of this bacterial channel contains many features that are almost universal in K$^+$ channels, so the insights derived from this structure can be applied broadly.

The KcsA channel consists of four subunits coming together in a clover-leaf arrangement to form an aqueous pore down the middle (Fig. 14.13a). The pore is lined partly by a membrane spanning α-helix, and partly by a sequence of five amino acids in an extended configuration (Fig. 14.13b). This threonine–valine–glycine–tyrosine–glycine

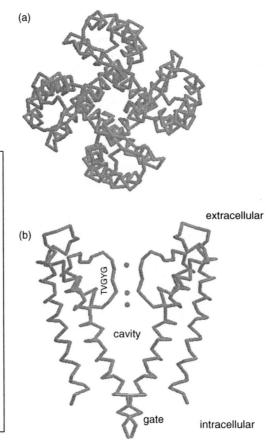

(a)

(b)

extracellular

TVGYG

cavity

gate   intracellular

Fig. 14.13. The KcsA K$^+$ channel (backbone structures). (a) Looking into the plane of the membrane shows the four-fold symmetry of the channel and its central pore. (b) Viewed from the side (a plane perpendicular to that in (a)), the two major membrane spanning helices of two opposing subunits form the channel walls. The TVGYG segment forms the selectivity filter at the extracellular face. The cavity in the center is filled with water. The gate of the channel lies on the intracellular face. Three ions are present, two in the selectivity filter and one at the interface between the selectivity filter and the cavity (Doyle *et al.*, 1998).

(TVGYG) sequence is a signature of K^+ channels. These segments from the four subunits come together to form the selectivity filter, which is shaped like a 3 Å wide pipe. A large water-filled cavity lies between the selectivity filter and the channel gate. The solved structure contained three ions, and this represented a remarkable confirmation of the single-file model just discussed. The selectivity filter thus forms the physical substrate for the classical single-file mechanism.

Many of these structural features are tailored for ion permeation. The peptide backbone carbonyls of the selectivity filter extend into the channel to form a series of binding sites that fit K^+ ions with remarkable precision. The large cavity provides a means for hydrated ions to approach the selectivity filter at minimal energy cost (Problem 5). A short α-helix connecting one of the membrane spanning helices to the TVGYG segment is oriented so that its dipole moment favors cation entry and retards anion entry (Roux and MacKinnon, 1999; Fig. 2.12). Additional polar groups form dipoles that favor cation entry. Some of these are acidic amino acids that can form negative charges, but the state of ionization of these residues has been difficult to establish.

The selectivity filter is at the heart of the permeation process. It contains four binding sites, which fill with two ions as the ion concentration is raised (Morais-Cabral et al., 2001). The ions distribute roughly evenly among the four sites, suggesting that the two occupancy states, K^+–W–K^+–W and W–K^+–W–K^+ (where W is a water molecule) are present in equal proportions. Ion flow can be pictured as transitions between these two configurations. Using a theoretical force field based on the channel structure, and evaluating the energies of various configurations, it was found that the K^+–W–K^+–W and W–K^+–W–K^+ configurations were the two most stable occupancy states (Åqvist and Luzhkov, 2000). With roughly equal energies, they interconvert easily. Molecular dynamics simulations show that water–K^+ complexes hop in just a few nanoseconds in a concerted transition together with a third K^+ in the cavity (Berniche and Roux, 2000).

The various energetic contributions to permeation can be estimated by distilling the key electrostatic features from the KcsA structure (Fig. 14.14a; Chung et al., 1999). These include a low dielectric region surrounding the water-filled pore, selectivity filter dipoles, α-helix dipoles, and dipoles at the channel mouth on each side. The shape of the channel determines the image force, which can be calculated for this particular structure by extending the methods discussed in Section 14.6. For a single ion moving down the axis of the channel, this energy is plotted as trace a in Fig. 14.14b. It reaches a peak of about 20 kT near the center of the selectivity filter.

The selectivity filter dipoles (Fig. 14.14b, trace b) and helix dipoles (trace c) together with the mouth dipoles attract ions into the channel. These attractive forces overcome the image force so that the summed energy profile has a deep minimum in the

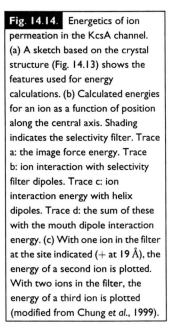

Fig. 14.14. Energetics of ion permeation in the KcsA channel. (a) A sketch based on the crystal structure (Fig. 14.13) shows the features used for energy calculations. (b) Calculated energies for an ion as a function of position along the central axis. Shading indicates the selectivity filter. Trace a: the image force energy. Trace b: ion interaction with selectivity filter dipoles. Trace c: ion interaction energy with helix dipoles. Trace d: the sum of these with the mouth dipole interaction energy. (c) With one ion in the filter at the site indicated (+ at 19 Å), the energy of a second ion is plotted. With two ions in the filter, the energy of a third ion is plotted (modified from Chung et al., 1999).

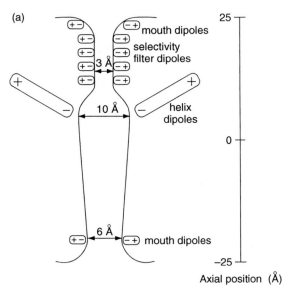

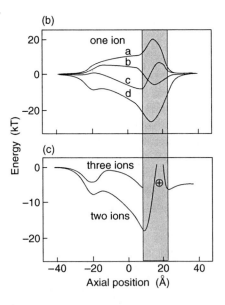

selectivity filter (trace d). With a depth of more than 25 kT a single ion will be bound very tightly. If this were the whole story then there would be very little permeation. An ion would enter the selectivity filter and stay there for a long time.

However, more ions enter the channel. A second ion sees all the forces that the first ion sees, as well as a direct repulsion arising from the first ion. The second ion also sees the medium polarization induced by the first ion, and this part is attractive. The result is the two-ion trace in Fig. 14.14c. The second ion falls into an energy minimum near the end of the selectivity filter close to the cavity. Even a third ion is drawn into the channel, but now the attractive forces are balanced by the repulsion from the first two ions. This third ion sees two energy minima, both in the cavity.

To relate these energy plots to permeation, the computational method of Brownian dynamics was used to simulate the movement of ions in the channel (Chung *et al.*, 1999; Kuyucek *et al.*, 2001). This method starts with the equations of motion for the ions in a three-dimensional version of the force field plotted in Fig. 14.14. A random fluctuating force is added to represent thermal collisions with the surrounding medium. For a system with many ions, the force on the *i*th ion generates an acceleration according to Newton's equation of motion.

$$m_i \frac{dv_i}{dt} = fv_i - \nabla U + \phi(t) \qquad (14.62)$$

where ∇U is the force field that includes the interactions enumerated above, and $\phi(t)$ is the fluctuating random force reflecting thermal activity. This is the Langevin equation (Eq. (12.96)) in the presence of an external force.

A computer then solves these equations numerically by calculating the changes in positions for steps of 10^{-13} s. Trajectories are run for several microseconds. When a potential difference across the membrane is incorporated into the equation by modifying the potential energy function (U in Eq. (14.62)), a significant number of ions pass through the channel in the time computed so that the basic permeation process is effectively simulated. This is important because it permits detailed comparisons between theory and experiment.

An analysis of the ion trajectories from these simulations illuminated the basic nature of the permeation process (Chung *et al.*, 2002). Contrary to expectations, the rate-limiting step was not movement through the selectivity filter. For K^+ ions moving to the right (out of a cell), the slowest step was the crossing of the barrier within the cavity in the three-ion trace of Fig. 14.14c. Once this third ion reaches the entrance to the selectivity filter, the ion lodged in the opposite side of the selectivity filter is rapidly expelled to the right by the electrostatic repulsion. The other ion in the selectivity filter assumes the position previously occupied by the expelled ion and the ion in the cavity enters the selectivity filter. These steps are quite rapid compared to the approach of ions through the cavity. This whole process repeats itself to generate a single-file permeation process.

This indicates that understanding the actual permeation rate requires a shift in attention from the selectivity filter to the cavity. The height of the cavity energy barrier was found to be very sensitive to the cavity radius, so that variations in this region produce dramatic changes in the channel conductance. The many K^+ channels found in nature have conductances running from about

1 to 250 pS (10^{-12} S). Variations in cavity dimensions over a reasonable range of 4–10 Å can account for this diversity.

It is notable that the selectivity of the channel and the magnitude of the current are controlled by two different parts of the protein. The selectivity filter forms sites that bind K^+ preferentially over Na^+, and this selectivity is destroyed by mutations in the TVGYG motif. Nature cannot tamper with this region without sacrificing selectivity, so instead the cavity is molded to achieve a wide range of structures without affecting selectivity. Likewise, channel opening and closing do not involve the selectivity filter. Instead, gating is accomplished by movements at the other end of the cavity toward the intracellular side of the membrane (Fig. 14.13b).

The secret of the ion selectivity of K^+ channels lies in the physical performance of the selectivity filter and its TVGYG motif. Two contrasting physical pictures have been developed. On the one hand, a very rigid structure can discriminate. When the oxygen atoms of the backbone carbonyls are held in place, a 1.33 Å K^+ fits perfectly into the sites of the selectivity filter (Garofoli and Jordan, 2003). Putting in a smaller 0.95 Å Na^+ stretches these bonds. If the structure is rigid then the stretching requires quite a bit of work because the force constants of covalent bonds are strong. Recall that stretching bonds by as little as 0.1 Å raises the energy by $\sim kT$ (Section 2.12). Conversely, the filter excludes a larger Cs^+ (1.69 Å) because that ion distorts bonds in the other direction.

A soft selectivity filter cannot discriminate by this mechanism, because then the carbonyl bonds move to accommodate ions of different size. Stretching is no longer necessary. Doubts about the rigidity of this part of the channel motivated an effort to test another mechanism for selectivity (Noskov et al., 2004). The carbonyl dipoles of the filter that coordinate the permeating ions repel one another. Binding an ion then involves the favorable ion–dipole interactions and the unfavorable dipole–dipole interactions. For the geometry of the K^+ channel's selectivity filter, the interplay between these two interactions gives a lower energy minimum for K^+ than for the smaller Na^+. This theory explains selectivity in terms of the specifics of the interactions between a permeating ion and the ligands within the channel, much as the Eisenman theory did with a simple Coulombic interaction (Section 14.4).

At the outset of this chapter a parallel was drawn between ion channels and enzymes. It is thus interesting to compare the progress in quantitative theoretical modeling of ion channel permeation with that of enzyme catalysis. One advantage of ion channels is that ions get through very rapidly so that simulations only need to be on the order of a microsecond. For most enzymes simulations of ~ 1000 times longer are needed to capture catalytic events. During ion permeation, there are no chemical reactions. Covalent bonds do not form and break. The standard types of potential energy functions suffice to describe the trajectory of an ion. Modeling enzyme catalysis requires quantitative representation of the forces that come into play as chemical bonds form, and potential energy functions for these processes are not readily accessible (Section 10.10). Furthermore, the motions of the ions are not intimately dependent on the motions of the protein, so Brownian dynamics in a relatively static environment simulates ionic motions well. By neglecting protein dynamics the equations are much simpler, so computation is faster and simulations can run for times long enough to generate an appreciable number of barrier crossings. In enzymes the motions of the protein play an important role, and this requires the method of molecular dynamics, which is computationally much more demanding.

Problems for Chapter 14

1. Use the Ohmic channel model to calculate the conductance of the selectivity filter with the dimensions indicated in Fig. 14.14.

2. Show that when all of the barriers in a single-ion channel (Fig. 14.10) for two different ions differ by the same amount, ΔE, the permeability ratio is $e^{-\Delta E/RT}$ (Coronado et al., 1980).

3. Assign permeabilities, P_A and P_B to the appropriate sums in Eq. (14.45), ignore their voltage dependence, and derive the Goldman–Hodgkin–Katz voltage equation (Eq. (13.22)).

4. Use the results from the single-ion channel model (Section 14.8) to show that the ratio KP/γ_{max} is a constant independent of choice of ion (P is permeability) (Coronado et al., 1980). For P use the relevant sums in Eq. (14.45). Use Eq. (14.52) to derive γ_{max}.

5. Use Eq. (2.7) to calculate the energy of a K^+ ion at the center of the water-filled cavity of the KcsA channel, assuming the cavity is spherical with a radius of 5 Å (Roux and MacKinnon, 1999; Roux et al., 2000).

6. Consider a channel with a single site at $\delta = 1/2$, and two barriers with $\delta = 1/2 \pm \psi$, where ψ can range from 0 to 1/2. The site can be empty or contain one ion. The energy barriers at zero voltage are ART, the well is 0, and the preexponential factor is one. Derive the current as a function of voltage and ion concentration. For symmetrical ions and low concentrations, plot the I-V curve for $A = 10$ and $\psi = 0.1$, 0.25, and 0.4.

7. Consider a single barrier to ion flow across a membrane. At $\Delta V = 0$ the peak is at the center of the membrane, and in the neighborhood of the peak the barrier energy has the form $E - \phi(x - x_0)^2$. Add a linear voltage drop to this energy and determine the position and height of the barrier for an arbitrary value of ΔV. The result is a Marcus relation (Section 7.6). Write out the current–voltage relation.

8. When n in Eq. (14.61) equals one, the model can no longer be called single file because there is only one ion. Are the fluxes now independent? Why does the flux ratio deviate from the Ussing equation (Eq. (13.27))?

Chapter 15

Cable theory

Cells can have very complex geometries, and when they do the voltage can vary dramatically between different regions. If ionic current flows through a restricted part of a cell's membrane, then the membrane potential at that location will change rapidly, but the membrane potential at distant locations will change more slowly and the change will be smaller. Voltage changes spreading through a cell act as signals to change membrane properties and trigger cellular events such as exocytosis and muscle contraction. Electrical signaling allows the nervous system to control and organize behavior.

Electrical signals in cells fall into two general classes. If the membrane conductance is independent of voltage, then the spread of voltage is passive. This type of signal, also referred to as electrotonic, travels a limited distance. On the other hand, when voltage alters the membrane conductance, then a voltage signal can regenerate itself and propagate without decrement over unlimited distances. This chapter will examine passive electrical signaling and the following chapter will treat active propagation.

The study of passive voltage changes serves a number of purposes. (1) Some biologically important voltage changes spread passively; passive spread is especially important when voltage changes are small. (2) Passive voltage changes are of technical importance in the design and interpretation of electrophysiological experiments. (3) Passive signaling serves as a baseline from which one goes on to study active propagation.

The principles of passive signaling derive from the basic rules of electrical circuits. Voltage drives current through resistors. Current flow changes the voltage by charging a capacitor. In cable theory these processes are generalized to continuous geometries with circuit elements distributed through cell membranes and cytoplasm. This generalization leads to the cable equation. For any particular geometry of a cell, we can solve the cable equation to obtain a description of the relevant passive voltage changes.

15.1 | Current through membranes and cytoplasm

Ionic current flows through both membrane and cytoplasm. Current through the membrane changes the voltage, and current through the cytoplasm makes these voltage changes spread. The relative magnitudes of the membrane and cytoplasmic currents determine the degree to which voltage spreads. To compare these two currents, we need the resistances of equivalent pieces of membrane and cytoplasm. A sheet of biological membrane with area $1\,\mathrm{cm}^2$ has a resistance of about $10^4\,\Omega$ (the resistance of a unit area is the membrane resistivity, ρ_m, with units of $\Omega\,\mathrm{cm}^2$). A cube of cytoplasm of volume $1\,\mathrm{cm}^3$ has a resistance of about $100\,\Omega$ (this is the cytoplasmic resistivity, ρ_c with units of $\Omega\,\mathrm{cm}$). Scaling down the cube of cytoplasm to a unit square sheet with the thickness of a membrane ($\sim5\times10^{-7}\,\mathrm{cm}$) gives $5\times10^{-5}\,\Omega$. Thus, cytoplasm conducts about eight orders of magnitude better than cell membrane. This gives us a qualitative perspective on voltage spread in a cell. For a voltage difference between two points, one inside a cell and one outside a cell, nearly the entire voltage drop occurs across the membrane.

At this level of analysis, the voltage within a cell tends to be uniform. If a cell is spherical, the resistance between any two points on the inside is negligibly small compared to the resistance across the cell membrane. The situation changes only when extreme geometries provide very long cytoplasmic pathways. Then substantial voltage gradients can occur in cytoplasm. This condition arises in cells with long, thin, fiber-like extensions.

Since spatial variations in voltage within a cell require fiber-like processes, we use a cylindrical cable with a radius of a as our basic geometric model (Fig. 15.1). In this model the cytoplasmic current flows almost entirely along the cylindrical axis. This axial current is denoted as i_a. The current that traverses the membrane is denoted as i_m.

We can now estimate the relative resistances of these two pathways in the context of the cable illustrated in Fig. 15.1. The resistances to axial and transverse current are defined in terms of axial and transverse resistivities, r_a and r_m, of a unit length of cable. The axial resistivity is ρ_c divided by the cross-sectional area

$$r_a = \frac{\rho_c}{\pi a^2} \tag{15.1}$$

It has units of $\Omega\,\mathrm{cm}^{-1}$ so multiplying by the length gives the resistance to current flow in the axial direction of a specified length of cable.

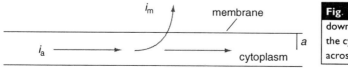

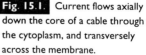

Fig. 15.1. Current flows axially down the core of a cable through the cytoplasm, and transversely across the membrane.

The transverse resistivity is ρ_m divided by the membrane circumference

$$r_m = \frac{\rho_m}{2\pi a} \tag{15.2}$$

The units here are $\Omega\,cm$, so we *divide* by length to get the resistance of a segment of cable to current flow across the membrane. Each of these quantities decreases with radius; $r_m \propto 1/a$, and $r_a \propto 1/a^2$. So as a increases the axial resistance goes down more rapidly than the transverse resistance. As a result, voltage spreads further in wider cables. The ratio of r_a to r_m is the natural variable with which to evaluate voltage nonuniformity. This ratio is generally low, so that voltage can spread quite far in the axial direction.

It is also instructive to compare the time scale for the relaxation of a spatial voltage gradient within cytoplasm versus relaxation of a voltage difference across a membrane. For each we use the model of a resistor and capacitor in parallel, where voltage decays exponentially as $e^{-t/RC}$, with the product RC as the time constant. For the first case, picture a unit cube of solution. Two of its opposing faces can be thought of as a capacitor, with capacitance $C = \varepsilon_w A/l$, where ε_w is the dielectric constant of water, and A and l are the area and thickness. If the capacitor carries a charge, current will flow between the faces, and this current sees a resistance $R = \rho_c l/A$. The product RC works out to be independent of the details of the cubic volume element because A and l cancel. We are left with the product $\varepsilon_w \rho_c$, which is known as the Maxwell time constant.[1] For a typical physiological saline $\varepsilon_w \rho_c$ is of the order of 1 ns.

For the decay of a voltage difference across a membrane we use the membrane resistance $R = \rho_m/A$ and capacitance $C = c_m A$, where c_m is the specific membrane capacitance. Again, A cancels when we compute $RC = \rho_m c_m$. For a typical membrane $\rho_m c_m \sim 10$ ms. So spatial voltage variations within a solution relax about 10^7 times faster than voltage drops across a membrane. This constitutes the dynamic counterpart to the above comparison of spatial extent of spread of voltages.

---

[1] For a more rigorous derivation of the Maxwell time constant, start with the Poisson equation.

$$\frac{d^2V}{dx^2} + \frac{d^2V}{dy^2} + \frac{d^2V}{dz^2} = -\frac{q}{\varepsilon}$$

Since $i = (1/\rho)(dV/dx)$, the first derivatives in V can be replaced by $i_x\rho$, $i_y\rho$, or $i_z\rho$, where i_x, i_y and i_z are the current densities in the direction of each voltage derivative. We then have the divergence of the current density, which gives the total charge flowing into a volume element minus the total charge flowing out. This must equal the rate of change of the charge density at that site (conservation of charge) so we can replace the left-hand side of the Poisson equation with $\rho(dq/dt)$ as follows

$$\rho\frac{dq}{dt} = -\frac{q}{\varepsilon}$$

The solution is $q = q_0 e^{-t/\rho\varepsilon}$, where q_0 is the initial value, so we recover the Maxwell time constant $\rho\varepsilon$.

These qualitative calculations tell us where we should look for spatial variations in voltage and where we need not bother. For example, within a typical spherical cell of diameter 25 μm we might ask what will happen when a bunch of channels are concentrated in one small raft of membrane. Opening all of those channels suddenly would change the voltage at that spot but the voltage change would spread through a roughly spherical cell within the Maxwell time constant of $\sim 10^{-9}$ s. This is much faster than any relevant biological signaling process. On the other hand, in an axon or dendrite of length ~ 1 mm, cytoplasm would produce an accumulating resistance while the large area of membrane would provide a significant exit pathway. The voltage along this process will then vary quite a bit.

15.2 | The cable equation

To proceed from these qualitative order-of-magnitude arguments to a quantitative understanding requires a more precise mathematical representation. The cable equation gives us just that. Consider a cable of radius a, broken up into infinitesimally thin slices (Fig. 15.2). The tendency toward uniformity over short distances means that the voltage within a slice can be taken as constant.

We assume that the voltage is zero everywhere outside the cable and that the voltage in the cable is a function only of axial distance, $V(x)$. Ohm's law states that the axial current through each slice will be equal to the voltage difference divided by the resistance, which is r_a (Eq. (15.1)) times the distance between slices, dx. For the two slices in Fig. 15.2 we have

$$i_a(x + \tfrac{1}{2}dx) = -\frac{V(x + dx) - V(x)}{r_a dx} = \frac{-1}{r_a} \frac{\partial V(x + \tfrac{1}{2}dx)}{\partial x} \tag{15.3a}$$

$$i_a(x - \tfrac{1}{2}dx) = -\frac{V(x) - V(x - dx)}{r_a dx} = \frac{-1}{r_a} \frac{\partial V(x - \tfrac{1}{2}dx)}{\partial x} \tag{15.3b}$$

The position for evaluating the derivative is taken as half way between the two surfaces, giving $x + dx/2$ and $x - dx/2$.

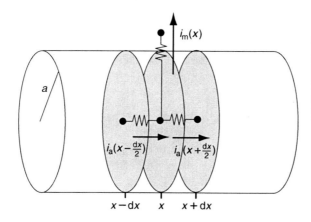

Fig. 15.2. Axial current, $i_a(x)$, flows between two thin slices and membrane current, $i_m(x)$, flows across the cell membrane. One slice is bounded by $x - dx$ and x. The adjacent slice is bounded by x and $x + dx$.

The transverse current through the membrane around the central slice is also given by Ohm's law. The resistance will be r_m/dx (Eq. (15.2)), so we have

$$i_m(x) = dx\, V(x)/r_m \tag{15.4}$$

By Kirchhoff's law, the net current into a slice centered at x must produce a voltage change at x proportional to the capacitance of the membrane. The net current is obtained by summing the axial and transverse contributions, into and out of the central point in Fig. 15.2

$$i_a(x - \tfrac{1}{2}dx) - i_a(x + \tfrac{1}{2}dx) - i_m(x) = dx\, c_{m'} \frac{\partial V(x)}{\partial t} \tag{15.5}$$

where $c_{m'}$ is the capacitance of the same area of membrane for which r_m is the resistance (the subscript m' is necessary because c_m is used for the capacitance of a unit area of membrane).

We now use Eqs. (15.3a), (15.3b), and (15.4) to replace all the current terms in Eq. (15.5) and get an equation in which voltage is the only variable

$$-\frac{1}{r_a}\frac{\partial V(x - \tfrac{1}{2}dx)}{\partial x} + \frac{1}{r_a}\frac{\partial V(x + \tfrac{1}{2}dx)}{\partial x} - \frac{dx}{r_m}V(x) = dx\, c_{m'} \frac{\partial V(x)}{\partial t} \tag{15.6}$$

The difference between derivatives can be converted to a second derivative after division by dx. Multiplying by r_m then produces

$$\frac{r_m}{r_a}\frac{\partial^2 V}{\partial x^2} - V = r_m c_{m'} \frac{\partial V}{\partial t} \tag{15.7}$$

Equation (15.7) is called the cable equation. It is a partial differential equation with essentially the same mathematical properties as the diffusion equation of Chapter 6. This equation tells us how an initial spatial variation in V evolves through time. Given an initial condition $V(x, 0)$, and boundary conditions for what happens at the cable ends, the solution to the cable equation gives us $V(x, t)$, the distribution at any later time. The cable equation has a long history going back to the study of telegraph lines. It was introduced into biophysics by Hodgkin and Rushton (1946), and is the starting point for the analysis of a wide range of electrical processes in cellular physiology (Rall, 1977; Jack et al., 1983).

The cable equation can be expressed more compactly by changing variables to $T = t/(r_m c_{m'})$ and $X = x/\sqrt{r_m/r_a}$ as follows

$$\frac{\partial^2 V}{\partial X^2} - V = \frac{\partial V}{\partial T} \tag{15.8}$$

Now the equation has no constants because they have been absorbed into the variables.

This transformation confers importance on the quantities $r_m c_{m'}$ and $\sqrt{r_m/r_a}$, which define the fundamental units of time and length. The quantity

$$\tau_m = r_m c_{m'} \tag{15.9}$$

is the *membrane time constant*, already examined at the end of the preceding section. Because resistance and capacitance have the opposite proportionality with area, the dimensions of the membrane cancel, leaving τ_m with units of time.

The quantity

$$\lambda = \sqrt{r_m/r_a} \tag{15.10}$$

has units of length and is defined as the *cable length constant* (also referred to as the *space constant*). The variables τ_m and λ are the basic units of the cable equation, and one often finds oneself thinking with reference to these basic measures of length and time.

Note that λ is the square root of that important ratio of resistivities introduced in the qualitative analysis of voltage spread in the preceding section. To see what λ means, and also to get a more quantitative sense for the distance over which voltage spreads, let us take the steady-state case of Eq. (15.8), setting $\partial V/\partial T = 0$

$$\frac{\partial^2 V}{\partial X^2} = V \tag{15.11}$$

If the cable is infinite, and we let $V = V_0$ at $X = 0$ and $V = 0$ at $X = \infty$ (boundary conditions), the solution for positive X is

$$V = V_0 \, e^{-X} = V_0 e^{-x/\lambda} \tag{15.12}$$

Thus, λ has a clear physical meaning as the length over which the voltage changes e-fold. That is why it is called the length constant. It is the natural unit for thinking about spatial variations in the voltage in a cell.

Equation (15.10) makes the point that λ increases with r_m, and decreases with r_a. This shows how voltage spread is influenced by these two parameters, and these dependences are intuitively reasonable if one uses Fig. 15.1 to envision how spread of voltage reflects a division in the flow of current between the two pathways provided by the membrane and cytoplasm.

Since r_m and r_a depend on the radius, so does λ. This becomes explicit when we use Eqs. (15.1) and (15.2) to express λ in Eq. (15.10) in terms of more fundamental quantities

$$\lambda = \sqrt{\frac{a\rho_m}{2\rho_c}} \tag{15.13}$$

The fact that λ increases with a illustrates another qualitative point of the preceding section that voltage spreads further in a wider cable.

The cable equation brings into focus a few key electrical parameters that influence voltage spread in cells. Among these are the resistivities, ρ_m and ρ_c, and the specific membrane capacitance, c_m (for unit area; $c_{m'}$ was for unit length of cable). Some experimentally measured values are given in Table 15.1.

	Squid axon[a]	Lobster axon[b]	Hippocampal neuron[c]
$c_m(\mu F\, cm^{-2})$	0.9	1.3	0.9
$\rho_m(\Omega\, cm^2)$	7600	2300	14 000
$\rho_c\ (\Omega\, cm)$	61	60	300

Table 15.1. *Cable parameters*

[a] Hodgkin *et al.* (1952).
[b] Hodgkin and Rushton (1946).
[c] Gentet *et al.* (2000); Meyer *et al.* (1997).

The value of c_m is fairly constant from one cell to the next and is very close to what is computed from the formula $c = \varepsilon/l$ with the dielectric constant $\varepsilon \sim 5$ and membrane thickness $l \sim 50$ Å. The value of ρ of the extracellular saline is simply its bulk resistivity. More rigorous analysis that treats voltage changes outside the cell includes this quantity, but the effects are usually small. Note that ρ_c is hard to measure but tends to be about two to three times greater than ρ of isotonic saline, with the difference reflecting the higher viscosity of cytoplasm and the large amount of charge attached to high molecular weight molecules with lower mobility; ρ_m for a resting membrane varies from cell to cell. When a cell is stimulated with chemicals or voltage, channels can open, causing ρ_m to change by factors of greater than 100. This is the basis for the active propagation of voltage signals discussed in Chapter 16.

15.3 | Steady state in a finite cable

The steady-state cable equation provides a description of standing voltage gradients within a cell. Voltage signals are often rapid and dynamic, and in these cases the steady state is not very relevant. However, it is much easier to solve the steady-state equation, and when voltage changes last for times significantly longer than the membrane time constant, τ_m, these solutions are very useful.

The steady-state cable equation (Eq. (15.11)) has the general solution

$$V = Ae^{-x} + Be^{x} \qquad (15.14)$$

A specific solution is found by imposing boundary conditions to determine A and B. The infinite cable was one example (Eq. (15.12)). It was easy to solve so we did not need to deal formally with the boundary conditions. Now we will take a finite cable of length L. An electrode positioned at one end imposes a voltage of V_0 at $X = 0$. The other end of the cable at $X = L$ is open so the voltage is zero. (This is an absorbing boundary condition (Section 6.2.4).)

The expression $V(0) = V_0$ constitutes the boundary condition at $X = 0$. This gives

$$A + B = V_0 \tag{15.15}$$

The other boundary condition, $V(L) = 0$, gives

$$Ae^{-L} + Be^{L} = 0 \tag{15.16}$$

Solving Eqs. (15.15) and (15.16) for A and B gives

$$A = \frac{-V_0 e^{2L}}{1 - e^{2L}} = \frac{-V_0 e^{L}}{e^{-L} - e^{L}} \tag{15.17a}$$

and

$$B = \frac{V_0}{1 - e^{2L}} = \frac{V_0 e^{-L}}{e^{-L} - e^{L}} \tag{15.17b}$$

The second steps in each of these entailed multiplying by e^{-L}/e^{-L}. Using these expressions in Eq. (15.14) completes the determination of the specific solution

$$
\begin{aligned}
V &= \frac{-V_0 e^{L}}{e^{-L} - e^{L}} e^{-X} + \frac{V_0 e^{-L}}{e^{-L} - e^{L}} e^{X} \\
&= \frac{V_0}{e^{L} - e^{-L}} (e^{L-X} - e^{-(L-X)})
\end{aligned}
\tag{15.18}
$$

Hyperbolic sines simplify this result to

$$V = V_0 \frac{\sinh(L - X)}{\sinh(L)} \tag{15.19}$$

You can check that this solves Eq. (15.14) and satisfies the boundary conditions.

If the cable is sealed at $X = L$, then the boundary condition is different. There is no axial current, making $i_a = -(1/r_a)(\partial V/\partial x) = -(1/\lambda r_a)(\partial V/\partial X) = 0$ (see Eqs. (15.3a) and (15.3b)). So instead of setting Eq. (15.14) equal to zero, as we did when the end was open, we set its derivative equal to zero (a reflecting boundary (Section 6.2.4))

$$-Ae^{-L} + Be^{L} = 0 \tag{15.20}$$

Combining this with Eq. (15.15) to solve for A and B leads to

$$V = \frac{V_0}{e^{L} + e^{-L}} (e^{L-X} + e^{-(L-X)}) = V_0 \frac{\cosh(L - X)}{\cosh(L)} \tag{15.21}$$

These two solutions (Eqs. (15.19) and (15.21)) are plotted in Fig. 15.3, along with the solution for the infinite cable (Eq. (15.12)). These results illustrate how different boundary conditions alter the solution. This point will be explored further with the dynamic behavior of the cable equation in Sections 15.4 and 15.5.

The sealed-end case is especially useful as a model for cells with processes. In Fig. 15.3 we see that if $L = 1$, the voltage at the end of the process is about 65% of V_0. The value of L is a useful number to know because it enables us to estimate the voltage nonuniformity.

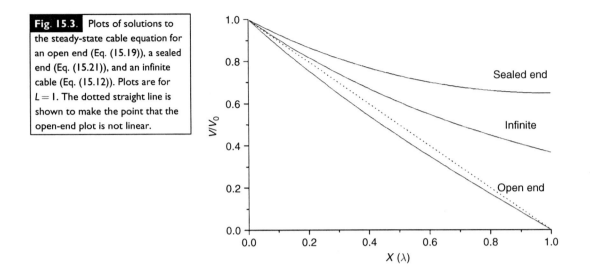

Fig. 15.3. Plots of solutions to the steady-state cable equation for an open end (Eq. (15.19)), a sealed end (Eq. (15.21)), and an infinite cable (Eq. (15.12)). Plots are for $L = 1$. The dotted straight line is shown to make the point that the open-end plot is not linear.

15.4 | Voltage steps in a finite cable

We will now proceed from the steady-state equation to explore situations where voltage varies with time. This means using the full partial differential equation, Eq. (15.8). The method of separation of variables (Section 6.2.4, footnote 2) produces the following general solution

$$V(X, T) = (A \sin(\alpha X) + B \cos(\alpha X))e^{-(1+\alpha^2)T} \tag{15.22}$$

This can be verified by direct substitution back into Eq. (15.8). The only difference between this and the general solution to the diffusion equation (Eq. (6.16)) is that the exponential has $1 + \alpha^2$ in place of α^2. The general strategy for going from this general solution to a particular solution is the same as for the diffusion equation in Chapter 6. One determines boundary conditions at the ends of the cable and uses them to solve for α. There are usually an infinite number of solutions, and they must be indexed with a subscript, α_i. This gives an infinite number of cosine and sine functions that can be combined as a Fourier sum to satisfy a particular initial condition

$$V(X, 0) = \sum_i (A_i \sin(\alpha_i X) + B_i \cos(\alpha_i X)) \tag{15.23}$$

(Note that $e^{-(1+\alpha^2)T} = 1$ at $T = 0$.) We can determine A_i and B_i by taking a Fourier transform (Appendix 3) of the initial condition. Once they are known, we compose the solution as

$$V(X, T) = \sum_i (A_i \sin(\alpha_i X) + B_i \cos(\alpha_i X)e^{-(1+\alpha_i^2)T}) \tag{15.24}$$

The different values of α_i correspond to different frequencies of spatial variation in the initial voltage distribution. The way that the α_i appear in both the spatial and temporal part of the equation means that the spatial features with higher frequencies (variations over shorter distances) decay more rapidly in time. Diffusion has the same property.

We will now solve the time-dependent equation for a cable of length L with a voltage clamp at $X = 0$ to maintain $V(0,T) = 0$. At the other end, where $X = L$, a sealed end means that the derivative with respect to X must be equal to zero. We apply the condition at $X = L$ first, differentiating Eq. (15.24) to obtain

$$\sum_i (A_i \alpha_i \cos(\alpha_i L) - B_i \alpha_i \sin(\alpha_i L)) e^{-(1+\alpha_i{}^2)T} = 0 \tag{15.25}$$

Since this holds for all T it must apply to each term

$$A_i \alpha_i \cos(\alpha_i L) - B_i \alpha_i \sin(\alpha_i L) = 0 \tag{15.26}$$

The boundary condition $V(0, T) = 0$ is applied to Eq. (15.24), and after using $\sin(0) = 0$ and $\cos(0) = 1$ we have

$$\sum_i B_i e^{-(1+\alpha_i{}^2)T} = 0 \tag{15.27}$$

Again, this holds for all T, so it applies to each term and makes all the B_i zero. This reduces Eq. (15.26) to

$$A_i \alpha_i \cos(\alpha_i L) = 0 \tag{15.28}$$

This equation defines a set of values for α. One of these is zero, which we designate as $\alpha_0 = 0$. However, this value contributes nothing to the solution because $\sin(0) = 0$ and $B_0 = 0$. Other values of α_i that solve Eq. (15.28) are $\pi/2$, $3\pi/2$, $5\pi/2$, etc. The general expression is

$$\alpha_i = (2i - 1)\frac{\pi}{2L} \tag{15.29}$$

where i is a positive integer. The exponentials in the solution then have the form $e^{-\left(1 + \left((2i-1)(\pi/2L)\right)^2\right)T}$.

It is significant that the boundary conditions are sufficient to determine the values of α_i. The exponential functions thus defined can then be combined in various ways to satisfy different initial conditions.

The time constants of a response to a voltage step can provide important information about a cell, regardless of the initial condition. Setting $i = 1$ and 2 in Eq. (15.29) gives the time constants of the two slowest exponentials as

$$\tau_1 = \frac{\tau_m}{1 + \alpha_1{}^2} = \frac{\tau_m}{1 + (\pi/2L)^2} \tag{15.30}$$

and

$$\tau_2 = \frac{\tau_m}{1 + \alpha_2{}^2} = \frac{\tau_m}{1 + (3\pi/2L)^2} \tag{15.31}$$

where τ_m reappears because units were converted back to real time, $t = \tau_m T$. We can use these two relations to solve for L as follows (Rall, 1969)

$$L = \frac{\pi}{2}\sqrt{\frac{9\tau_2 - \tau_1}{\tau_1 - \tau_2}} \qquad (15.32)$$

Thus, L can be calculated from the time constants extracted from an experiment, and τ_m is then calculated from either τ_1 or τ_2 with Eqs. (15.30) or (15.31).

The time constants are an important part of the solution, but to solve the equation completely one must use an initial condition to determine the A_i. The value of V was taken as zero at $X = 0$, but this value was imposed by the voltage clamp at $T = 0$. Before $T = 0$ the voltage clamp was at another voltage, which we will call V_0. We will assume that the prior voltage was held long enough for a steady state to be reached. That means that $V(X)$ immediately before the step is given by Eq. (15.21). The combination of sine functions that sums to this particular function is found by Fourier transformation

$$A_i = \frac{2}{L\cosh(L)} \int_0^L \cosh(L - X)\sin\left((2i - 1)\frac{\pi X}{2L}\right) dX$$
$$= \frac{4\pi(2i - 1)}{4L^2 + (2i - 1)^2 \pi^2} \qquad (15.33)$$

From this we can reconstruct the initial condition as a sum of sine functions

$$V_0 \frac{\cosh(L - X)}{\cosh(L)} = V_0 \sum_{i=1}^{\infty} \frac{4\pi(2i - 1)}{4L^2 + (2i - 1)^2 \pi^2} \sin\left((2i - 1)\frac{\pi X}{2L}\right) \qquad (15.34)$$

Since this sum of sine functions satisfies the initial condition, the sum of exponentials

$$V(X, T) = V_0 \sum_{i=1}^{\infty} \frac{4\pi(2i - 1)}{4L^2 + (2i - 1)^2 \pi^2} \sin\left(\frac{(2i - 1)\pi X}{2L}\right) e^{-\left(1 + \left((2i - 1)\pi/2L\right)^2\right)T} \qquad (15.35)$$

satisfies both the initial condition and the cable equation, and is thus a complete solution. Each term in Eq. (15.35) has the form of the general solution (Eq. (15.22)).

With this solution in hand it is apparent that if we go the other way and step the voltage from 0 to V_0 at $T = 0$, we have the solution

$$V(X, T) = V_0 \frac{\cosh(L - X)}{\cosh(L)}$$
$$- V_0 \sum_{i=1}^{\infty} \frac{4\pi(2i - 1)}{4L^2 + (2i - 1)^2 \pi^2} \sin\left(\frac{(2i - 1)\pi X}{2L}\right) e^{-\left(1 + \left((2i - 1)\pi/2L\right)^2\right)T} \qquad (15.36)$$

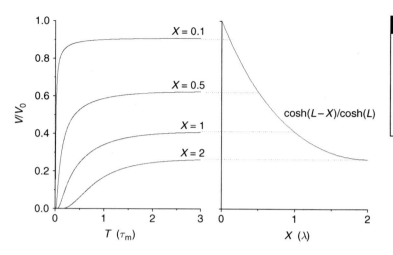

Fig. 15.4. Plots of Eq. (15.36) as a function of T (left) for $L = 2$ and for the indicated values of X. Note that V rises with varying speeds to reach steady-state values (right), determined by Eq. (15.21). The plot on the right is like Fig. 15.3, but with $L = 2$.

At $T = 0$ the sum cancels out the steady-state $\cosh(L - X)/\cosh(L)$ term leaving the voltage zero everywhere. As the exponentials decay, the terms in the sum gradually disappear and we arrive at the final steady state.

Equation (15.36) is plotted for $L = 2$ and for $X = 0.1$, 0.5, 1, and 2 (Fig. 15.4). This graph shows how the voltage at different positions responds to the voltage step applied at $X = 0$. The locations closest to the site of the applied voltage step respond with the largest change, and with the greatest speed. This illustrates at the most basic level how a cable shapes a passive voltage signal.

15.5 | Current steps in a finite cable

The particular experimental method used to record an electrical signal is connected in a very interesting way with the mathematics of solving the cable equation. This connection will be illustrated by looking at the current clamp and comparing it with the results of the preceding section on the voltage clamp. With the voltage clamp, a voltage is imposed at a particular location and one measures the current. A current clamp imposes a current while measuring voltage. These experimental conditions are expressed mathematically as boundary conditions. Clamping the voltage to V_0 at $X = 0$ prescribes the boundary condition $V(0) = V_0$. Clamping the current to I_0 at $X = 0$ specifies the derivative of the voltage at that point. We will simplify this to $I_0 = 0$ so that

$$\frac{\partial V}{\partial X} = 0 \tag{15.37}$$

The derivative of the general solution (Eq. (15.22)) is a sum of sine and cosine terms. Since $\cos(0) = 1$, the A_i, which become the coefficients of the cosine terms after the derivative is taken, must all be zero. The mathematical solution to this problem must therefore be composed entirely from the terms of the form $B_i\cos(\alpha_i X)e^{-(1+\alpha_i^2)T}$. Compare this with the voltage clamp scenario of the preceding section, where the Bs were forced to zero and we were left with terms of the form $A_i\sin(\alpha_i X)e^{-(1+\alpha_i^2)T}$.

The derivative at a sealed end at $X = L$ is zero. Applying this condition to the remaining terms of the form $B_i\cos(\alpha_i X)e^{-(1+\alpha_i^2)T}$ gives

$$\alpha_i B_i \sin(\alpha_i L) = 0 \tag{15.38}$$

Since the sine function is zero for 0, π, 2π, 3π, etc., we have

$$\alpha_i = \frac{i\pi}{L} \tag{15.39}$$

where i is zero or a positive integer (compare with Eq. (15.29)). The exponentials will then have the form $e^{-\left(1+(i\pi/L)^2\right)T}$, with time constants

$$\tau_i = \frac{\tau_m}{1 + (i\pi/L)^2} \tag{15.40}$$

The slowest time constant is simply τ_m, for $i = 0$. By contrast, the slowest time constant under voltage clamp (Eq. (15.30)), is always smaller than τ_m.

This theoretical result suggests a simple experimental way to determine τ_m: take the time constant of the slowest component of the response to a current step. However, this approach can lead to errors for more realistic models of neurons (Section 15.7).

The way in which the difference between voltage clamp and current clamp comes about can be summarized as follows. Fixing $V = 0$ removes the cosine terms, and when the remaining sine terms are used they specify the α_i as half-integral multiples of π/L that do not include zero. Fixing $I = 0$ removes the sine terms, and when the remaining cosine functions are used they specify the α_i as integral multiples of π/L that do include zero.

15.6 | Branches and equivalent cylinder representations

Cable theory is clearly appropriate for axons and muscle fibers because they are shaped like cables, but what about cells with other geometries? Many neurons have extraordinarily complex dendrites (Fig. 15.5). Their extensive branching looks like a hopeless obstacle to mathematical modeling. However, in 1959 W. Rall carried out an important analysis of branching dendrites to identify conditions under which a complex dendritic tree can be expected to behave like a single cylindrical process. The key conditions were as

(a)

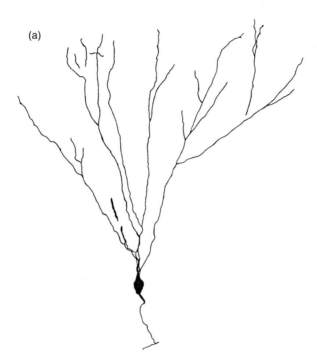

Fig. 15.5. (a) A granule cell from the dentate gyrus (Courtesy of Dr. Helen Scharfman). (b) An interneuron from the lateral geniculate nucleus (Courtesy of Dr. Dan Uhlrich). Both are from rat.

follows. (1) At a branch the radius of the parent segment, a_p, and the daughter branches, a_{b1} and a_{b2}, obey the relation

$$a_p^{3/2} = a_{b1}^{3/2} + a_{b2}^{3/2} \tag{15.41}$$

This is known as the 3/2 power law. (2) All branches terminate with the same electrotonic length. So for two branches

$$L_{b1} = L_{b2} \tag{15.42}$$

Each L is in units of its own length constant, λ_{b1} or λ_{b2}. Rall's work raised the hope that a cylinder with the right dimensions could model the passive voltage changes in a complex dendrite. The idea of an equivalent cylinder representation of a complex dendrite has had a considerable impact on electrophysiological studies of neurons.

15.6.1 Steady state

We first use a steady-state solution of the cable equation to show how the Rall branching conditions make the dendritic resistance equal to that of an equivalent cylinder. The input resistance of a cable of length L with a sealed end is the voltage, V_0, applied at $X = 0$ divided by the total membrane (transverse) current. Current flows across the membrane all along the cable. At each point the current is $V(x)dx/r_m$, so the total current is the integral

$$I = \frac{1}{r_m} \int_0^{\lambda L} V(x)dx = \frac{\lambda}{r_m} \int_0^L V(X)dx \tag{15.43}$$

Fig. 15.5. (cont.)

(b)

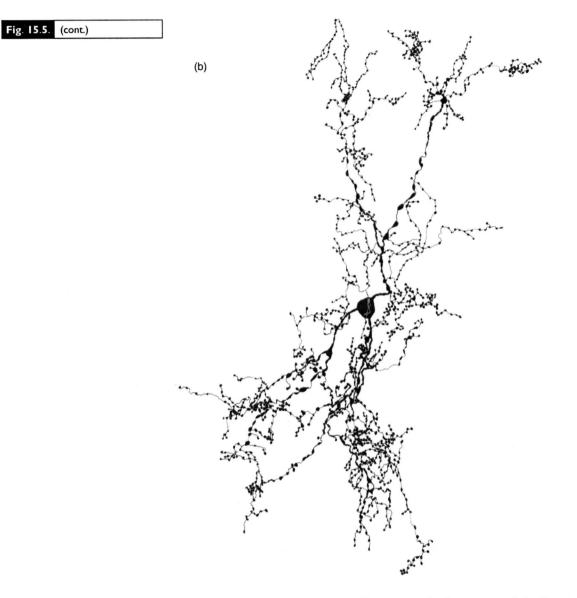

The integral was first expressed in terms of x, because r_m is in $\Omega\,\text{cm}$, and then converted to units of L. Taking $V(X)$ from Eq. (15.21) gives

$$I = \frac{V_0 \lambda}{r_m} \int\limits_0^L \frac{\cosh(L - X)}{\cosh(L)}\, dX \tag{15.44}$$

The integral of $\cosh(L - X)$ is $-\sinh(L - X)$ (Appendix 5), so the resistance, R_L, is

$$R_L = \frac{V_0}{I} = \frac{r_m \cosh(L)}{\lambda \, \sinh(L)} = \frac{r_m}{\lambda \, \tanh(L)} \tag{15.45}$$

Recall how λ depends on the radius, a, of the cable. With the aid of Eqs. (15.2) and (15.13) we obtain

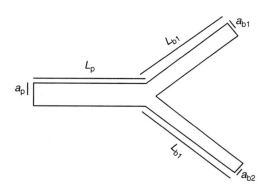

$$R_L = \frac{\rho_m}{2\pi a \, \tanh(L)} \sqrt{\frac{2\,\rho_c}{a\,\rho_m}} = \frac{a^{-3/2}}{\sqrt{2}\,\pi \, \tanh(L)} \sqrt{\rho_c \rho_m} \qquad (15.46)$$

Now consider a cylinder that branches (Fig. 15.6). According to Eq. (15.46), the resistance of each branch is

$$R_{b1} = \frac{a_{b1}^{-3/2}}{\sqrt{2}\,\pi \, \tanh(L_{b1})} \sqrt{\rho_c \rho_m} \qquad (15.47a)$$

$$R_{b2} = \frac{a_{b2}^{-3/2}}{\sqrt{2}\,\pi \, \tanh(L_{b2})} \sqrt{\rho_c \rho_m} \qquad (15.47b)$$

These two resistors are in parallel, so their combined resistance is

$$R_b = \frac{1}{(1/R_{b1}) + (1/R_{b2})} = \frac{\sqrt{\rho_c \rho_m}}{\pi\sqrt{2}} \left(\frac{1}{\tanh(L_{b1})a_{b1}^{3/2} + \tanh(L_{b2})a_{b2}^{3/2}} \right)$$

$$(15.48)$$

Now applying the Rall conditions (Eqs. (15.41) and (15.42)) simplifies Eq. (15.48) to

$$R_b = \frac{a_p^{-3/2}}{\sqrt{2}\,\pi \, \tanh(L_b)} \sqrt{\rho_c \rho_m} \qquad (15.49)$$

which is identical to Eq. (15.46).

Thus, the two branches behave just like an extension of the parent segment. We can therefore expect the resistance of the entire structure shown in Fig. 15.6 to be (Problem 4)

$$R = \frac{a_p^{-3/2}}{\pi \, \tanh(L_p + L_b)} \sqrt{\rho_c \rho_m / 2} \qquad (15.50)$$

The parent segment with its two branches has the resistance of a single cylindrical segment with a length $L_p + L_b$. This calculation indicates that the steady-state solution of a branching process will be the same as that for an equivalent unbranched cylinder if the segments obey the two branching rules of Rall.

15.6.2 Time constants
The time course of voltage spread in a cable with branches also reduces to the equivalent cylinder result when the Rall conditions are met. This will now be demonstrated for the time constants.

The voltage in each segment is a sum of terms of the form

$$V_p(X_p, T) = (A_p\sin(\alpha(L_p - X_p)) + B_p\cos(\alpha(L_p - X_p)))e^{-(1+\alpha^2)T}$$

$$\text{(15.51a)}$$

$$V_{b1}(X_{b1}, T) = (A_{b1}\sin(\alpha(L_{b1} - X_{b1})) + B_{b1}\cos(\alpha(L_{b1} - X_{b1})))e^{-(1+\alpha^2)T}$$

$$\text{(15.51b)}$$

$$V_{b2}(X_{b2}, T) = (A_{b2}\sin(\alpha(L_{b2} - X_{b2})) + B_{b2}\cos(\alpha(L_{b2} - X_{b2})))e^{-(1+\alpha^2)T}$$

$$\text{(15.51c)}$$

These expressions differ from Eq. (15.22) in the replacement of X with $L - X$. This still satisfies Eq. (15.8) and aids in the present analysis.

The ends are sealed so the boundary conditions at $X_{b1} = L_{b1}$ and $X_{b2} = L_{b2}$ are that the derivative with respect to X is equal to zero. The terms $A\alpha\cos(\alpha(L - X))$ (derivatives of sine terms) cannot be zero at $X = L$ because $\cos(0) = 1$, so A_{b1} and A_{b2} must be zero.

The branch point, where $X_p = L_p$ and $X_{b1} = X_{b2} = 0$, provides two additional conditions. All three voltage functions must be equal at this point; $V_p(L_p, T) = V_{b1}(0, T) = V_{b2}(0, T)$. So $V_p(L_p)$ is simply $B_p e^{-(1+\alpha^2)T}$ because $\cos(0) = 1$ and $\sin(0) = 0$, and A_{b1} and A_{b2} are zero owing to the condition just stated for the branch ends. After canceling out the time dependent factors, which are the same in all three equations, the voltage equivalence at the branch point gives us

$$B_p = B_{b1}\cos(\alpha L_{b1}) = B_{b2}\cos(\alpha L_{b2}) \tag{15.52}$$

We also have a constraint on the current flow at the branch point. The current out of the parent segment is equal to the sum of the currents flowing into the two branches. Each current is given as the spatial derivative divided by the axial resistivity (from Eq. (15.1))

$$\frac{\pi a_p^2}{\rho_c \lambda_p}\frac{\partial V_p}{\partial X_p} = \frac{\pi a_{b1}^2}{\rho_c \lambda_{b1}}\frac{\partial V_{b1}}{\partial X_{b1}} + \frac{\pi a_{b2}^2}{\rho_c \lambda_{b2}}\frac{\partial V_{b2}}{\partial X_{b2}} \tag{15.53}$$

The length constants are there because we need units of cm to be compatible with the units used for ρ_c. If we assume that ρ_c and ρ_m are the same for all segments, then all of the factors multiplying each derivative cancel except for $a^{3/2}$, as follows

$$a_p^{3/2}\frac{\partial V_p}{\partial X_p} = a_{b1}^{3/2}\frac{\partial V_{b1}}{\partial X_{b1}} + a_{b2}^{3/2}\frac{\partial V_{b2}}{\partial X_{b2}} \tag{15.54}$$

Inserting the expressions for V_p, V_{b1}, and V_{b2} from Eqs. (15.51a)–(15.51c) into Eq. (15.54) gives

$$-a_p^{3/2}A_p = a_{b1}^{3/2}B_{b1}\sin(\alpha L_{b1}) + a_{b2}^{3/2}B_{b2}\sin(\alpha L_{b2}) \tag{15.55}$$

The final constraint of a sealed end at $X_p = 0$ gives a derivative of Eq. (15.51a) equal to zero

$$\alpha(A_p \cos(\alpha L_p) - B_p \sin(\alpha L_p)) = 0 \qquad (15.56)$$

Equations (15.52), (15.55), and (15.56) can now be used to solve for α. Equation (15.52) gives B_{b1} and B_{b2} as $B_p/\cos(\alpha L_{b1})$ and $B_p/\cos(\alpha L_{b2})$, respectively. Equation (15.56) gives A_p as $B_p \sin(\alpha L_p)/\cos(\alpha L_p)$. Making these substitutions in Eq. (15.55) gives

$$-a_p{}^{3/2}B_p \frac{\sin(\alpha L_p)}{\cos(\alpha L_p)} = a_{b1}{}^{3/2}B_p \frac{\sin(\alpha L_{b1})}{\cos(\alpha L_{b1})} + a_{b2}{}^{3/2}B_p \frac{\sin(\alpha L_{b2})}{\cos(\alpha L_{b2})}$$

$$(15.57)$$

This simplifies to

$$a_p{}^{3/2}\tan(\alpha L_p) + a_{b1}{}^{3/2}\tan(\alpha L_{b1}) + a_{b2}{}^{3/2}\tan(\alpha L_{b2}) = 0 \qquad (15.58)$$

This equation can in principle be solved for α, but for arbitrary values of the as and Ls these results will not show a simple pattern. Invoking the Rall rules changes that. First, using Eq. (15.42), $L_{b1} = L_{b2} = L_b$, leads to

$$a_p{}^{3/2}\tan(\alpha L_p) + (a_{b1}{}^{3/2} + a_{b2}{}^{3/2})\tan(\alpha L_b) = 0 \qquad (15.59)$$

Now if the 3/2 power rule (Eq. (15.41)) is obeyed we have

$$\tan(\alpha L_p) + \tan(\alpha L_b) = 0 \qquad (15.60)$$

Finally, examination of the trigonometric identity for the tangent of a sum

$$\tan(\alpha + \beta) = \frac{\tan(\alpha) + \tan(\beta)}{1 - \tan(\alpha)\tan(\beta)} \qquad (15.61)$$

makes it apparent that Eq. (15.60) will have the same roots as $\tan(\alpha L_p + \alpha L_b)$, so

$$\tan(\alpha(L_p + L_b)) = 0 \qquad (15.62)$$

This defines α as integral multiples of $\pi/(L_p + L_b)$, just as in Eq. (15.39). So the transient response of a branched cable will have the same time constants as a single cylinder with a length of $L_p + L_b$.

This analysis can be extended to a very complicated dendritic arbor. Working one's way back from the ends, each pair of branches can be replaced by an equivalent segment to reduce the entire dendrite to a single equivalent cylinder. Rall (1959, 1977) has shown that the complete dynamic response characteristics are reproduced completely by this equivalent cylinder representation.

Of course, the big question raised by this mathematical analysis is whether real dendrites obey the Rall rules for branching. A substantial amount of effort has been invested in anatomical examinations of neurons such as those shown in Fig. 15.5 to try to determine whether dendritic branches obey these two rules. The diversity of neuronal morphologies is so vast that it is impossible to draw general conclusions. A study of two types of relay neurons in the lateral geniculate nucleus suggested that the Rall rules are obeyed

(Bloomfield *et al.*, 1987). On the other hand, in motoneurons the dendrites get thinner as they branch so that the sum of $a^{3/2}$ is less than that of the parent segment (Barrett and Crill, 1974; Clements and Redman, 1989). This suggests that the dendrite should be represented by a cylinder that tapers with distance from the cell body.

Ways of testing the equivalent cylinder model are discussed near the end of the following section. If the Rall branching rules are violated, or if one is uncertain about their applicability, morphological information about a cell can be used to develop a compartmental model that can be solved with a computer to simulate passive voltage changes (Section 15.9).

15.7 | Cable analysis of a neuron

If the dendritic tree of a neuron can be represented by a single cylinder, then we can use the model shown in Fig. 15.7. The simple branching pattern of the neuron in Fig. 15.5a makes it a good candidate for such a representation. The cell body is represented by a sphere and the dendrite by a cylinder. The axon is so narrow that it can be ignored. This physical representation of a neuron is known as the Rall model (Rall, 1969). Here we will consider the Rall model with a patch electrode in the "whole-cell configuration" (Jackson, 1992). The tip of a patch pipette provides a direct electrical link from the amplifier to the interior of the cell. Experimentally, a voltage is imposed at the point denoted by V_c in Fig. 15.7.

Before examining the complete model shown in Fig. 15.7 it is worth a brief comment on how this system performs when there is no dendrite, just a voltage-clamped spherical cell (Marty and Neher, 1995). If the cell resistance R_{cb} is very high compared to the

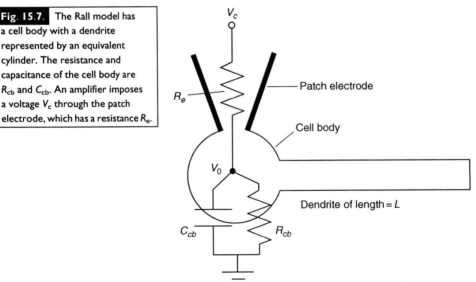

Fig. 15.7. The Rall model has a cell body with a dendrite represented by an equivalent cylinder. The resistance and capacitance of the cell body are R_{cb} and C_{cb}. An amplifier imposes a voltage V_c through the patch electrode, which has a resistance R_e.

resistance of the electrode, R_e, then the only relevant current is through the patch electrode, which by Ohm's law is $(V_c - V_0)/R_e$. The capacitance of the cell body, C_{cb}, is charged by this current, so V_0 changes at a rate dictated by the current and capacitance

$$\frac{V_c - V_0}{R_e} = C_{cb} \frac{dV_0}{dt} \tag{15.63}$$

A term V_0/R_{cb} was omitted because R_{cb} is very large. According to this equation, the step response will be a single exponential with a time constant equal to $R_e C_{cb}$.

Now we add the dendrite. This allows current to leave the cell body through the cable. A term accounting for this axial current must be added to Eq. (15.63). This current is given by the derivative of the voltage at the entrance to the cylinder divided by the axial resistivity, r_a. Incorporating this term on the left-hand side of Eq. (15.63) gives

$$\frac{V_c - V_0}{R_e} + \frac{1}{r_a} \frac{\partial V_0}{\partial x} = C_{cb} \frac{\partial V_0}{\partial t} \tag{15.64}$$

To apply the cable equation we need to convert to units of X and T as follows

$$\frac{V_c - V_0}{R_e} + \frac{1}{\lambda r_a} \frac{\partial V_0}{\partial X} = \frac{1}{R_{cb}} \frac{\partial V_0}{\partial T} \tag{15.65}$$

The factor R_{cb} appears because after replacing t by $\tau_m T$, we realize that C_{cb} cancels if we take $\tau_m = R_{cb} C_{cb}$. At $X = L$ we have the usual sealed-end condition

$$\frac{\partial V}{\partial X} = 0 \tag{15.66}$$

As with the step response of a finite cable, the general solution, Eq. (15.22), is inserted into each of the boundary conditions. We use Eq. (15.65) first, set $V_c = 0$, cancel out the exponential factor present in every term, and after noting that $\sin(0) = 0$ and $\cos(0) = 1$, we obtain

$$-\frac{B}{R_e} + \frac{A\alpha}{\lambda r_a} = -\frac{B(1 + \alpha^2)}{R_{cb}} \tag{15.67}$$

This can be rearranged to

$$\frac{A}{B} = \frac{\lambda r_a}{\alpha} \left(\frac{1}{R_e} - \frac{1 + \alpha^2}{R_{cb}} \right) \tag{15.68}$$

Inserting Eq. (15.22) into the other boundary condition (Eq. (15.66)) gives

$$A\cos(\alpha L) - B\sin(\alpha L) = 0 \tag{15.69}$$

which can be rearranged to

$$\frac{A}{B} = \tan(\alpha L) \tag{15.70}$$

Now A and B can be eliminated by equating Eqs. (15.68) and (15.70) to obtain an equation for α

$$\tan(\alpha L) = \frac{\lambda r_a}{\alpha}\left(\frac{1}{R_e} - \frac{1+\alpha^2}{R_{cb}}\right) \tag{15.71}$$

The values for α that solve this equation define the exponential time constants, just as Eqs. (15.28) and (15.38) defined α for other conditions.

Although Eq. (15.71) is a transcendental equation that cannot be solved analytically, its behavior can be understood by rearranging to give

$$\frac{R_{cb}}{\lambda r_a}\alpha\tan(\alpha L) - \frac{R_{cb}}{R_e} + 1 + \alpha^2 = 0 \tag{15.72}$$

The left-hand side is plotted versus α in Figure 15.8 to show the roots.

Figure 15.8 shows that Eq. (15.72) is satisfied by values of α very near $\pi/2$, $3\pi/2$, etc. These roots arise because the tangent function approaches ∞ at half-integral multiples of π. Since the ratio R_{cb}/R_e is large the tangent part of the equation has to be very large to cancel it. For low values of α, the roots are determined by the interplay between these two terms. It is important to recognize that these roots are the same as for a finite cable without a cell body (Eq. (15.29)). Thus, Eqs. (15.30)–(15.32) can be used to calculate L and τ_m.

It is significant that as α becomes large a solution to the equation can be found that does not depend on the singularity in the tangent function. At this point the equation crosses the x-axis at $\alpha^2 \sim R_{cb}/R_e$, because R_{cb}/R_e is of the order of 100. This means that one of the exponentials will have a time constant given by

$$\tau \sim \tau_m \frac{R_e}{R_{cb}} = C_{cb}R_e \tag{15.73}$$

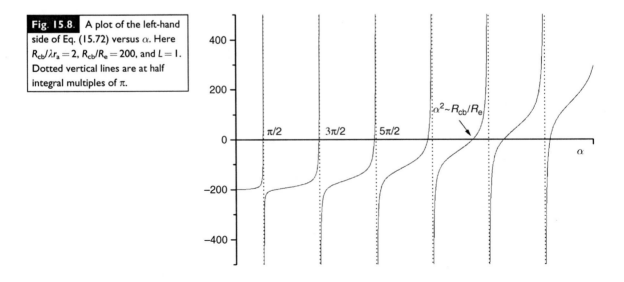

Fig. 15.8. A plot of the left-hand side of Eq. (15.72) versus α. Here $R_{cb}/\lambda r_a = 2$, $R_{cb}/R_e = 200$, and $L = 1$. Dotted vertical lines are at half integral multiples of π.

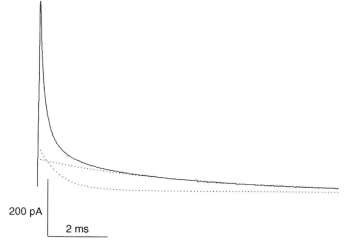

Fig. 15.9. A passive current transient evoked by a 10 mV voltage step in a cell with a process (a granule cell of the dentate gyrus like Fig. 15.5a). The transient was fitted to three exponentials $I(t) = 3.24e^{-t/.013} + 1.41e^{-t/0.7} + 1.09e^{-t/4.8}$; the dotted curves indicate the two slowest components.

200 pA

2 ms

which is identical to the time constant of the single exponential function obtained by solving Eq. (15.63). Thus, this value of α is fundamentally different from the others in corresponding to the charging of the cell body. The other exponential time constants reflect the charging of the dendrite. In patch clamp recordings from neurons, this part of the charging transient is quite prominent and easily distinguished from slower dendrite charging terms. In spite of the presence of the dendrite, this term provides a measurement of the cell body capacitance, C_{cb}.

An example is shown in Fig. 15.9 of an experimental recording of a passive charging transient in a neuron elicited by a voltage step. The current is fitted to a sum of three exponentials, and τ_1 and τ_2 can be used to determine L and τ_m with Eqs. (15.30)–(15.32).

The charging transient also can be used to evaluate the validity of the equivalent cylinder representation. For example, Eq. (15.32) tells us that if $\tau_1 > 9\tau_2$ then L is imaginary. No cylinder can give a transient with such charging dynamics so the dendrite cannot have an equivalent cylinder representation. This makes it relatively easy to reject the Rall model. However, the converse does not hold. A transient with $\tau_1 < 9\tau_2$ will allow one to calculate L and τ_m, but that does not prove that the dendrite being charged has a cylindrical representation.

If three or more exponentials can be resolved in a step response then L and τ_m are overdetermined and one can assess consistency with an equivalent cylinder representation. After calculating L and τ_m from the two slowest time constants of a transient response (Eqs. (15.30)–(15.32)), one can check the amplitudes and time constants of the other components. Additional criteria for consistency with an equivalent cylinder representation have been developed (Jackson, 1992, 1993a), and satisfying several theoretical predictions would increase the likelihood that this model is valid. However, it should be emphasized that the charging transient of a neuron has nowhere near enough information for a detailed reconstruction of a cell's morphology. The Rall model fails quite often, and the equivalent cylinder representation cannot be used in these cases. More complicated analytical methods (Major *et al.*, 1993) and computer models (Section 15.9) must then be used.

It is significant that in many respects a voltage clamped model neuron behaves similarly to a voltage clamped cylindrical segment (Section 15.4). The slower components have the same time constants, and Eqs. (15.30)–(15.32) can be used to make quantitative determinations of τ_m and L. The situation is less fortunate for a current clamped neuron. For this case the time constants are determined by the formula (Problem 9; Rall, 1969)

$$\alpha \cot(\alpha L) = -\frac{R_{cb}}{R_d \tanh(L)} \qquad (15.74)$$

where R_d is the resistance of the dendrite. This equation does not have simple solutions because the right-hand side is neither very small nor very large (Problem 10). Depending on the ratio of R_{cb}/R_d, the α_i, and therefore the time constants, can vary between the two extremes of Eqs. (15.29) and (15.39). This makes it very risky to take the slowest exponential time constant from a current clamp experiment as τ_m (Section 15.5). It is necessary to use additional theoretical expressions to determine the ratio of the cell body and dendrite resistances and then determine the appropriate roots of Eq. (15.74) (Rall, 1969; Jack and Redman, 1971).

The voltage clamp is a much better technique for cable analysis than the current clamp. Equations (15.30)–(15.32) provide a straightforward path from voltage clamp data to the accurate determination of L and τ_m. In spite of this, the voltage clamp has not been widely exploited in the analysis of cable properties. Rall's 1969 analysis clearly demonstrated the advantages of the voltage clamp. However, at that time there were serious technical problems with voltage clamping neurons. Voltage clamping then required placing two microelectrodes in a cell, but this was quite difficult. The microelectrodes used in those days had high resistances, so that with only one, the electrode and cell were resistors in series. They divided an applied voltage in a manner that made the voltage of the cell difficult to control. Current was easier to control because with two resistors in series the same current passes through both. For these reasons experimenters used the current clamp for cable analysis. Much later, around 1990, electrophysiologists learned how to use low resistance patch electrodes to record from neurons in brain slices. These electrodes ($< 10\,M\Omega$) made voltage clamping easy, but investigators long accustomed to doing cable analysis under current clamp have been slow to change their habits.

15.8 | Synaptic integration in dendrites: analytical models

Neurons use synapses to communicate. Although virtually every part of a neuron can form a synapse, the generic case involves an

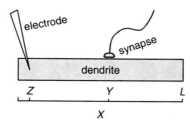

Fig. 15.10. A synapse on a cylindrical dendrite. The X-axis marks distance along the dendritic cable. The recording electrode is near one end at $X = Z$. The synapse is at $X = Y$. The electrotonic length of the dendrite is L.

axon terminal forming a synapse with a dendrite. Neurotransmitter released by a nerve terminal opens channels and elicits current flow through a small spot of membrane on a dendrite. An electrode in the cell body can measure the effect of this current originating at a distant site, and cable theory can describe how the voltage change produced at the site of the synaptic input spreads through the rest of the cell (Rall, 1967; Rall *et al.*, 1967; Jack and Redman, 1971). A complete analysis of this problem using the Rall model is quite complicated. However, the qualitative features are well represented by a simpler model consisting of a cylindrical segment with two sealed ends (Fig. 15.10).

15.8.1 Impulse responses

We begin with the notion that a synaptic potential starts off as an instantaneous impulse of voltage, which takes the form of a delta function. The results will later be extended to a more realistic input. The analysis developed in Section 15.5 provides the starting point. With both ends sealed, the α_i are defined in Eq. (15.39) as integral multiples of π/L, and the coefficients of the sine terms, the A_i, are all zero. We must use the initial condition to determine the B_i that multiply the cosine terms. With an instantaneous impulse as the synaptic input, the initial voltage at $T = 0$ is a delta function

$$V(X,0) = S_0 \delta(X - Y) \tag{15.75}$$

where Y is the site of the input indicated in Fig. 15.10. Here S_0 might be thought of as the charge that enters during the synaptic impulse divided by the capacitance of subsynaptic membrane at Y.

We must now use a Fourier integral to find the B_i in order to compose a delta function from cosines

$$B_i = \frac{2S_0}{L} \int_0^L \delta(X - Y) \cos\left(\frac{i\pi Y}{L}\right) dX \tag{15.76}$$

The delta function is one of the easiest functions to Fourier transform. Since it is large at $X = Y$ and zero everywhere else, the integral extracts the value of the cosine function at $X = Y$

$$B_i = \frac{2S_0}{L} \cos\left(\frac{i\pi Y}{L}\right) \tag{15.77}$$

The determination of B_0 is a little different, being half as great as Eq. (15.76) with $i = 0$ (Eq. A3.2)

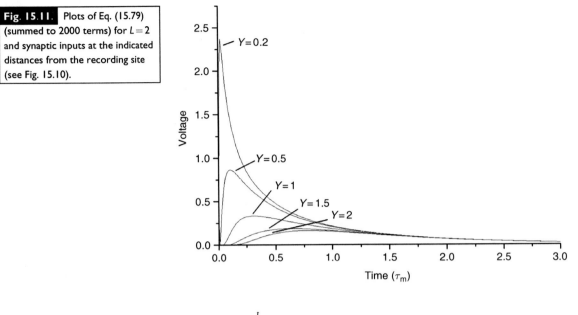

Fig. 15.11. Plots of Eq. (15.79) (summed to 2000 terms) for $L = 2$ and synaptic inputs at the indicated distances from the recording site (see Fig. 15.10).

$$B_0 = \frac{S_0}{L} \int_0^L \delta(X - Y)dX = \frac{S_0}{L} \tag{15.78}$$

Now we can put these pieces together in Eq. (15.24) to obtain the complete solution

$$V(X, T) = \frac{S_0}{L} \left(e^{-T} + 2 \sum_{i=1}^{\infty} \cos\left(\frac{\pi i Y}{L}\right) \cos\left(\frac{\pi i X}{L}\right) e^{-\left(1 + (i\pi/L)^2\right)T} \right) \tag{15.79}$$

This result is plotted in Fig. 15.11 for a cable with $L = 2$, setting X equal to the recording site ($Z = 0.01$ for recording near the left end as indicated in Fig. 15.10) and taking the input at various sites (values of Y). The plots show how a very brief transient signal is shaped as it spreads.

Figure 15.11 makes a number of important points about the passive spread of voltage signals. A proximal input, one near the recording site, rises almost instantly. It decays rapidly at first and slowly later on. A distal input, one far from the recording site, rises after a delay and decays slowly. The decay of the distal input has the same time constant as the slowest component of decay of the proximal input. In fact, all of the various signals end up decaying with the same final time constant, equal to τ_m. This corresponds to $\alpha_0 = 0$. This is a direct consequence of the fact that according to Eq. (15.79), all responses are composed as sums of the same set of exponential functions.

Another significant feature in Fig. 15.11 is that the peaks of the responses decrease very strongly with increasing distance from the recording site. The most extreme case would arise if the response were plotted for an input exactly at the recording site. The initial voltage would be infinite, but that is an artificial result due to the delta function. That issue will be dealt with shortly, but for now it is still useful to note that the peak decreases from 0.86 to 0.33 to 0.18 to 0.16 for $Y = 0.5$, 1.0, 1.5, and 2, respectively. This decrease is far greater than that predicted for the solution to the relevant steady-state problem (Eq. (15.21)). These values of 0.62, 0.41, 0.3, and 0.27 span a factor of 2.3 compared to a factor of more than 5 for the impulse responses. This illustrates a general aspect of cable theory. Transient signals do not spread as far as steady signals. The reason for this is that the membrane capacitance acts as a low-pass filter to remove rapidly varying voltages.

15.8.2 Realistic synaptic inputs

The impulse response illustrates some important qualitative features about how synaptic inputs are processed by the dendrite of a neuron, but we are left wondering how the time course of a real synaptic conductance will be shaped by a dendrite. The impulse response provides a tool to address this question. Envisage an input with a particular time dependence as an envelop of impulses. S_0 of Eq. (15.75) can be replaced by this function of time. Because time here refers to the time course of the input, we represent it with a different symbol, S, and call the function $\varphi(S)$.

If we denote Eq. (15.79) as $V_{imp}(X, T)$, then the response to a particular part of the input at time S is $\varphi(S)V_{imp}(X, T - S)$. The complete response is then obtained by integrating this product from 0 to T as follows

$$V(X, T) = \int_0^T V_{imp}(X, T - S)\varphi(S)dS \qquad (15.80)$$

This is called a convolution integral and it appears in many different problems in physics. Here we are expressing the synaptic response as a convolution of the impulse response and the synaptic input function.

We will evaluate this integral (Eq. (15.80)) using the function[2]

$$\varphi(S) = \sigma^2 Se^{-\sigma S} \qquad (15.81)$$

This form of $\varphi(S)$ rises, peaks at $S = 1/\sigma$, and then decays exponentially. It does a pretty good job of reproducing the actual time course of the conductance change seen at a real synapse (Finkel and Redman, 1983). The σ^2 in front normalizes the area so that the time course can be stretched or contracted without changing the amount of charge entry.

Putting Eq. (15.81) and Eq. (15.79) together in Eq. (15.80) gives the response to an alpha function synaptic input.

---

[2] This function is commonly referred to as an "alpha" function, with the symbol α in place of σ. Using α here would create confusion because this symbol appears in the time constants in the solution of the cable equation.

$$V(X,T) = \frac{\sigma^2}{L} \int_0^T \left(e^{-(T-S)} + 2 \sum_{i=1}^{\infty} \cos\left(\frac{\pi i Y}{L}\right) \cos\left(\frac{\pi i X}{L}\right) e^{-\left(1+(i\pi/L)^2\right)(T-S)} \right) S e^{-\sigma S} \, dS$$

(15.82)

We can rewrite this expression as

$$V(X,T) = \frac{\sigma^2}{L} \left(e^{-T} \int_0^T S e^{S(1-\sigma)} \, dS \right.$$

$$\left. + 2 \sum_{i=1}^{\infty} \cos\left(\frac{\pi i Y}{L}\right) \cos\left(\frac{\pi i X}{L}\right) e^{-\left(1+(i\pi/L)^2\right)T} \int_0^T S e^{\left(1+(i\pi/L)^2 - \sigma\right)S} \, dS \right) \quad (15.83)$$

There are now two integrals to evaluate, both of the form $\int_0^T S e^{cS} dS = \frac{1}{c^2}\left(1 + e^{cT}(cT - 1)\right)$, so

$$V(X,T) = \frac{\sigma^2}{L} \left(\frac{e^{-T}}{(1-\sigma)^2} \left(1 + (T(1-\sigma) - 1) e^{(1-\sigma)T} \right) \right.$$

$$+ 2 \sum_{i=1}^{\infty} \cos\left(\frac{\pi i Y}{L}\right) \cos\left(\frac{\pi i X}{L}\right) \frac{e^{-\left(1+(i\pi/L)^2\right)T}}{\left(1 + (i\pi/L)^2 - \sigma\right)^2}$$

$$\left. \times \left(1 + \left(T\left(1 + \left(\frac{i\pi}{L}\right)^2 - \sigma \right) - 1 \right) e^{\left(1+(i\pi/L)^2 - \sigma\right)T} \right) \right) \quad (15.84)$$

Figure 15.12 plots this result for the same values of Y as in Fig. 15.11. A rapidly varying input is used for Fig. 15.12a. This was created by taking $\sigma = 5$. Since T is in units of τ_m, this means that the synaptic input peaks at $\tau_m/5$. The other case (Fig. 15.12b) shows a slowly varying input, created by taking $\sigma = 0.2$. This synaptic input peaks at $5\tau_m$.

The responses to a brief input in Fig. 15.12a can be compared with Fig. 15.11. The qualitative trend is similar. The more distal inputs are smaller in amplitude and more spread out over time. The effect is not as dramatic as for the impulse responses in Fig. 15.11 because an impulse is briefer and more sensitive to the high frequency filtering action of the cable. It is significant that for $Y \geq 1$ the responses are very similar for both the impulse input and rapid version of $\varphi(S)$. This means that an experimental recording of a distal synaptic potential contains very little information about the actual time course of a fast synaptic input. The cable filters out this information. It is also significant that the shape of the response in Fig. 15.12a does not change much as Y gets closer to L. This makes it harder to evaluate the site of the input as it approaches the distal end of the dendrite (Jack and Redman, 1971).

In contrast to the fast input, Fig. 15.12b shows that the slow input is not distorted as much by the cable. The amplitudes are reduced, and they scale roughly according to the steady-state expression (Eq. (15.21)). These inputs are slow enough for the voltage to approach a steady-state.

(a)

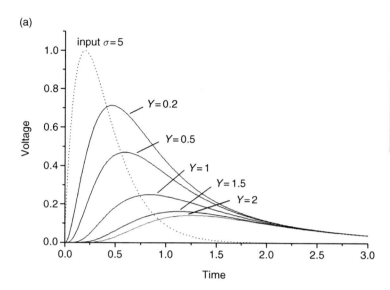

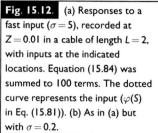

Fig. 15.12. (a) Responses to a fast input ($\sigma = 5$), recorded at $Z = 0.01$ in a cable of length $L = 2$, with inputs at the indicated locations. Equation (15.84) was summed to 100 terms. The dotted curve represents the input ($\varphi(S)$ in Eq. (15.81)). (b) As in (a) but with $\sigma = 0.2$.

(b)

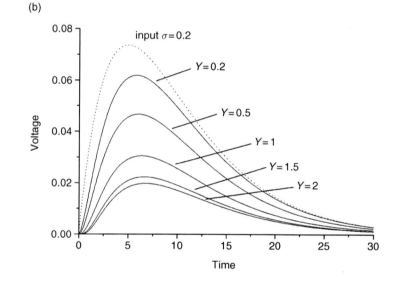

Plots such as those in Figs. 15.11 and 15.12a have been used to characterize how the shape of a response depends on the location of a synaptic input. By comparing the synaptic recordings with these kinds of theoretical curves it was possible to estimate the distance of the synaptic input from the recording site (Rall *et al.*, 1967; Jack and Redman, 1971). When this analysis was followed by anatomical examination to locate the synapse (in experiments on spinal motoneurons), the two methods were in good agreement (Redman and Walmsley, 1983).

One interesting experimental approach to the question of location is to apply voltage steps to the cell body and activate a synaptic input at various times after the voltage step. The amplitude of the synaptic response is then plotted versus the time interval. A proximal input settles rapidly to the value specified by the new voltage, but a distal input takes a longer time (Hestrin *et al.*, 1990). Imaging techniques have permitted the detection of highly localized rises in intracellular Ca^{2+} induced by synaptic inputs. This method localizes an input far more precisely than is possible with cable analysis. As yet, there has been little effort to combine cable analysis and Ca^{2+} imaging to gain a deeper understanding of synaptic integration.

15.9 | Compartmental models and cable theory

Most of the analytical results of cable theory are limited to cylinders. Extending cable analysis to the complicated geometries routinely encountered in the nervous system requires a general approach that does not depend on the mathematical convenience of cylindrical geometry. The most useful general approach is computational, in which a computer solves the problem numerically for an arbitrary shape (Segev *et al.*, 1989).

Recall that to derive the cable equation (Section 15.2) we sliced a cylinder into discrete sections (Fig. 15.2), and then wrote down the equations for current flow and voltage change. Taking the limit of infinitely many infinitely thin sections led to the cable equation. For the more general approach here we will again slice or subdivide our cell into discrete sections. But instead of going to a limit of infinitely many infinitesimal slices, we will try to find a suitable size in which the voltage can be taken as essentially uniform. Recall that in Section 15.1 qualitative reasoning indicated that the voltage within a cell varies only over rather large distances. This means that we do not need to go to an extreme limit to obtain a useful model. We can take the compartments as small enough to be nearly isopotential (uniform in voltage), but not so small that there are too many for the computer to handle.

Consider a neuron with a cell body and a dendrite (Fig. 15.13). We might divide it up into compartments as shown. As a rough and tentative guide we can calculate the length constant locally and

Figure 15.13. A cell with a cell body (1), primary dendrite (2–4), and a branch point (5) with two tapering segments (6, 7 and 8, 9). Each compartment has its own membrane resistance and capacitance. Adjacent compartments are connected by resistors.

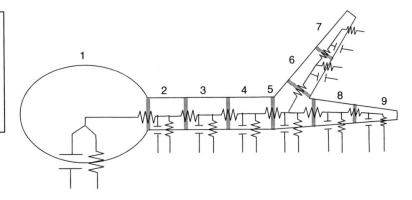

make the length of the compartment much smaller, say less than 5%, of the length constant. The voltage will vary within this section roughly by a fraction of $\sim e^{-0.05} = 0.95$, so the change is only 5%.

Figure 15.13 shows the resistors and capacitors that represent the relevant circuit elements of each compartment. Borders were drawn in an arbitrary manner, and it does not really matter how they are drawn as long as the compartments are small enough to be isopotential. The membrane surrounding each compartment has its own capacitance, which is proportional to the surface area, and its own resistance, which is inversely proportional to the surface area. These numbers can all be calculated using the unitary membrane quantities ρ_m and c_m. Between each compartment is a resistor for the cytoplasm. The cross-sectional area at the border between the two compartments is used to calculate the connecting resistance using ρ_c.

The voltage of each compartment will change with the charging of its membrane capacitance as current flows through the various resistors connected to that compartment. We can write down the basic circuit equation for each compartment. Compartment 1 is the cell body, with a membrane resistance of R_{m1}. Its voltage, V_1, changes as its membrane capacitance, C_1, is charged by the membrane and axial current

$$C_1 \frac{dV_1}{dt} = -\frac{V_1}{R_{m1}} - \frac{V_1 - V_2}{R_{a1-2}} \tag{15.85}$$

The first term on the right is the current through the membrane of the cell body and the second is the axial current through the cytoplasmic connection to compartment 2. This connection is represented by the resistance, R_{a1-2}.

For compartment 2 we have

$$C_2 \frac{dV_2}{dt} = -\frac{V_2}{R_{m2}} + \frac{V_1 - V_2}{R_{a1-2}} - \frac{V_2 - V_3}{R_{a2-3}} \tag{15.86}$$

where the resistances and capacitance are analogous to those of Eq. (15.85). Here there are two axial current terms, one for current between compartments 1 and 2 and another for current between compartments 2 and 3. The equation for compartment 4 would have three such terms because it contains the branch. When we get to the end of a branch there is only one axial term.

Each compartment has its own equation of the form of Eqs. (15.85) and (15.86). The full description of a cell with N compartments is then N such equations. They form a system of coupled linear first order differential equations of exactly the same form that arises in multistate kinetics (Chapter 9). The matrix method developed for those problems is fully applicable (Perkel et al., 1981), but in practice the analytical solutions are rarely used. Instead, computers are used to integrate the system of equations numerically, computing V_i as a function of time for all the compartments.

Neuroscientists have developed powerful computer modeling programs based on compartmental models. Use of these programs requires detailed knowledge of a cell's geometry. Photomicrographs of cells (Fig. 15.5) are used to render a system of compartments, in some cases numbering in the thousands. With the membrane area of each compartment, as well as the local cross-sectional area, the parameters C_i, R_{mi}, and $R_{ai, i+1}$ can be calculated. With each parameter specified the system of equations can then be integrated and the passive voltage changes simulated with excellent accuracy. It is common practice in work with such models to carefully vary the compartmentalization scheme to make certain that compartments are small enough to be isopotential. For example, if each compartment is divided in half to generate twice as many compartments, the computer time needed to perform the integration will be greater but if this check produces the same behavior then results with the original number of compartments can be trusted as accurate.

15.10 | Synaptic integration in dendrites: compartmental models

To illustrate compartmental modeling we return to the subject of synaptic integration, and use a hippocampal pyramidal neuron (Fig. 15.14a) to expand on the principles from Section 15.8. The extensive branching of its dendrites makes it unlikely that an equivalent cylinder representation will work for this cell. Furthermore, we might also wonder about how voltage signals in one branch spread to other branches. The equivalent cylinder model ignores this question.

Figure 15.14b shows how the voltage in the cell body responds to synaptic inputs at different locations. These traces were generated with the computer program NEURON (Hines and Carnevale, 1997) using a compartmental model based on the cell in Fig. 15.14a. The model has nearly 200 compartments connected by the simple rules of Section 15.9. NEURON then integrated these ~200 coupled equations, with the conductance of one of the compartments changing to simulate a synapse at the various sites indicated (input function was Eq. (15.81) with $\sigma = 5$).

As in Figs. 15.11 and 15.12a, the synapses at more distant inputs produce responses at the cell body that are smaller and more spread out in time. All of the responses coalesce to the same slow exponential. Again, we can see how passive voltage changes are sums of the same exponentials, weighted differently according to the conditions of each simulation. The equivalent cylinder captures these qualitative effects, but the differences between proximal and distal synapses are far greater in this real neuron. The distal synapses attenuate to the point where they have almost no discernable impact on the cell body. This implies that many distal synaptic inputs must be activated simultaneously to generate a noticeable response. The pyramidal cells of the hippocampus compensate for

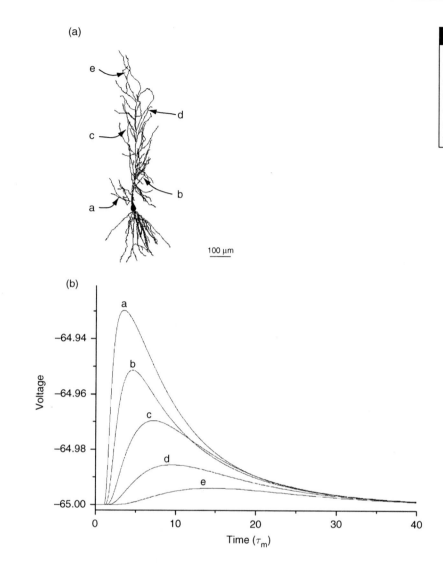

(a)

100 μm

(b)

Fig. 15.14. (a) A picture of a hippocampal pyramidal cell with five sites (labeled a–e) selected for synaptic inputs (neuron n170 from http:\\neuron.duke.edu). (b) Computed responses in the cell body to synaptic inputs at the sites indicated in (a).

this attenuation by having higher receptor densities at distal locations so that activating synapses at those sites produces larger currents (Magee and Cook, 2000).

The compartmental model can be used to illustrate a few more interesting features of synaptic integration. First, although the dendrite attenuates and retards responses at the cell body, the local response at the location of the synapse is much larger, and quite rapid (Fig. 15.15a). The synapse depolarizes the membrane quite a bit in the segment of the dendrite it contacts. The dendrite is very narrow so the capacitance of that region of membrane is small. The current that enters the cell at that site can thus exert a large effect on such a small capacitance. As this current spreads through the dendrite, the large surface area of the many branches acts as a sink to absorb more charge in the production of local voltage changes. That leaves little charge to reach the cell body, hence the large disparity.

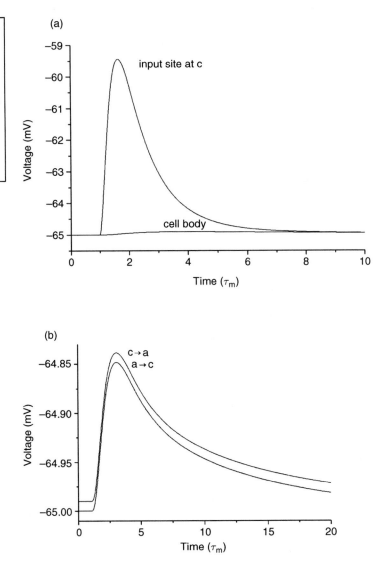

Fig. 15.15. (a) A synaptic input at site c (Fig. 15.14a) depolarizes that site far more rapidly and to a far greater degree than the cell body. Note that the cell body trace is trace c in Fig. 15.14b. (b) A synaptic input at site d produces a response at site b. A synaptic input at site b produces a response at site d. The two are displaced slightly to make the point that they are identical.

It is also interesting to look at how synaptic inputs influence other parts of the dendrite. In Fig. 15.15b we see that a synaptic input at one site depolarizes another site with exactly the same amplitude and time course as an equivalent synaptic input at the second site depolarizes the first. In fact, this is a general reciprocity relation that can be proven mathematically for any pair of sites in a structure of arbitrarily branching cylinders (Major *et al.*, 1993).

The location of a synaptic input in a dendritic arbor profoundly influences how it functions in a neural circuit. The complexities of synaptic integration make neurons very interesting and versatile electrical devices. Different cell morphologies in various brain regions serve specialized functions, allowing neural circuits to handle information in different ways. The functional diversity is amplified enormously by the presence of voltage-activated channels in dendrites (Stuart *et al.*, 2000). This makes the spread of all but the smallest voltage signals active, and the passive models fail to capture these important features (Section 16.11).

Problems for Chapter 15

1. Show that the transformation $V = Ue^{-T}$ converts the cable equation into a dimensionless form of the diffusion equation.

2. Determine the resistance of a semi-infinite cable and of a finite cable with an open end at $X = L$, and then with a sealed end at $X = L$.

3. Solve the cable equation in the steady-state for a finite length segment with $V = V_0$ at $X = 0$. The end at $X = L$ terminates with a resistance R_{term} (hint: the boundary condition at $X = L$ is $1/r_a\lambda(dV/dX) = (V/R_{\text{term}})$). Calculate the resistance in terms of L, R_{term}, ρ_c and ρ_m, and a.

4. In the preceding problem, place a sealed-end cylinder of length L_B of the same diameter at the end of a segment of length L_A (L of the preceding problem). Show that the combined resistance equals that of a finite sealed-end cylinder of length $L_A + L_B$.

5. Solve the cable equation for a cable extending to $\pm\infty$, with $V(X,0) = \delta(X)$ as the initial condition.

6. Show that two branches that satisfy the Rall conditions for an equivalent cylinder have the same surface area (not including ends) as the equivalent cylinder.

7. Determine the time constants for a current clamped cable with an open end at $X = L$.

8. Determine the time constants of the branched structure in Fig. 15.6 with a voltage clamp at $X_p = 0$, and with the Rall branching criteria fulfilled. This requires repeating the analysis of Section 15.6.2 with $V_p(0) = 0$ instead of with the sealed end condition.

9. Derive Eq. (15.74).

10. Plot Eq. (15.74) to visualize the roots, with the right-hand side equal to 0.2, 1, and 5. Compare the two smallest roots indicated from this plot with the integral and half-integral multiples of π.

11. Calculate L and τ_m from the time constants given in the legend of Fig. 15.9. Is τ_3 consistent with this result? What does this tell you about the validity of the equivalent cylinder representation for recordings from this cell?

Chapter 16

Action potentials

Many types of cells, including neurons, muscle fibers, and endocrine cells, have the capacity to generate electrical impulses. These impulses, known as action potentials, play an important role in the regulation of cell function, and constitute a biological mechanism for the digitization of information. Action potentials arise from a very special combination of cell membrane properties. The selective permeability of ion channels (Chapter 14) and voltage induced transitions in membrane proteins (Chapter 1) join forces to generate rapid voltage changes with unique features. Action potentials are not the passive decremental processes of Chapter 15, but wavelike events that propagate over great distances. This chapter will bring together many ideas developed earlier in this book to provide a quantitative description of this fundamental form of bioelectric signaling.

16.1 | The action potential

An action potential presents a striking contrast with the passive behavior of the preceding chapter. The basic phenomenon is most easily visualized in a compact, spherical cell with no internal voltage gradients. This avoids the complication of propagation, which can be incorporated into the picture later.

First consider the passive response of a cell with only the channels that generate a resting potential, $V_r = \sim -70\,\text{mV}$ (Section 13.3). These channels are always open so the cell responds to a stimulus current injected through an electrode according to the differential equation

$$C\frac{dV}{dt} = I_{\text{stim}} + (V_r - V)/R \tag{16.1}$$

where C is the cell's capacitance, I_{stim} is the current applied through an electrode, and R is the resistance, which for now is taken as constant. The term $(V_r - V)/R$ represents the current through the channels that set the resting potential. The linear expression simplifies things quite a bit compared to the Goldman–Hodgkin–Katz

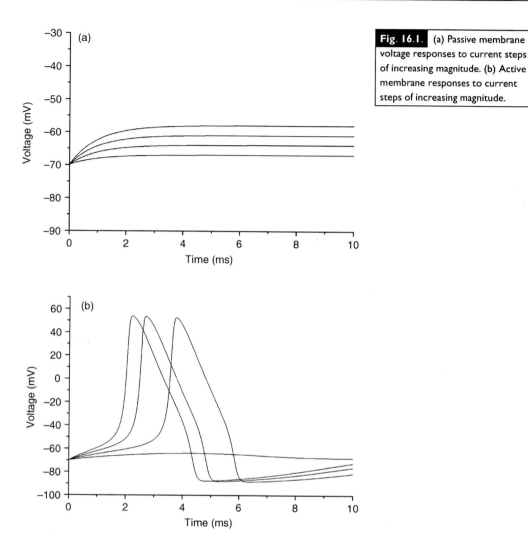

Fig. 16.1. (a) Passive membrane voltage responses to current steps of increasing magnitude. (b) Active membrane responses to current steps of increasing magnitude.

current equation (Eq. (13.45)) or a rate theory expression (Eq. (13.62)). This is entirely adequate for the present purpose, which is to compare passive and active responses.

Equation (16.1) has the solution

$$V = V_r + I_{stim}R(1 - e^{-t/RC}) \tag{16.2}$$

Note that $V = V_r$ at $t = 0$, and the final value of $V_r + I_{stim}R$ is reached after the exponential term has decayed. This response is plotted in Fig. 16.1a. We see that increasing the stimulus current moves the voltage in the positive direction, to *depolarize* the membrane. The final steady-state voltage increases linearly with current, and the time constant, RC, is the same for each trace.

An excitable cell generates a response that is strikingly different from the passive responses in Fig. 16.1a. These responses are shown in Fig. 16.1b. For the smallest stimulus current the response roughly resembles the smallest passive response in Fig. 16.1a, but for larger stimulus currents the response changes dramatically. Once the stimulus current exceeds a certain value, the *threshold*, the voltage shoots up to a value close to the Na^+ Nernst potential (~50 mV), and then returns to a negative value, undershooting the initial voltage before settling back to the resting potential. This brief impulse of voltage is highly characteristic of excitable cells and is known as an action potential. Because of their shape action potentials are often referred to simply as "spikes."

Action potentials cannot be explained with a static membrane conductance such as the one that generates the resting potential. Qualitatively, the action potential can be understood in terms of the gating of ion channels, which makes the voltage swing back and forth between the Nernst potentials for K^+ and Na^+ (E_K and E_{Na}). Recall from Chapter 13 that the membranes of neurons rest at a negative voltage a bit above E_K, because the permeability to K^+ is much higher than to other ions. By contrast, E_{Na} is around 50 mV, so when enough Na^+ channels open to enable Na^+ permeability to dominate, the voltage approaches E_{Na}.

The action potential is initiated by the opening of voltage-dependent Na^+ channels. Making the voltage more positive opens these channels by inducing a conformational transition in the channel protein (Section 1.8). The Na^+ channels are nearly all closed at the resting potential of −70 mV. The Na^+ channel has a midpoint for activation at about −45 mV, and if a stimulus raises the membrane potential, then the Na^+ channels open. Na^+ flows into the cell through these channels, and when enough Na^+ channels are open so that the inward Na^+ current exceeds the outward K^+ current, the voltage will start to move toward E_{Na}. Once this starts to happen the process has become regenerative. Even if the stimulus current is turned off, Na^+ will keep moving into the cell, the voltage will keep moving in the positive direction, and more Na^+ channels will open. Eventually a potential is reached just below E_{Na}, reflecting the new dominance of Na^+ (Fig. 16.2a).

This movement of Na^+ through voltage-dependent Na^+ channels accounts for the rising phase of the action potential. If nothing else happened then an action potential would be a one-way event. However, there are two other voltage-activated processes that follow Na^+ channel activation and limit the time spent near E_{Na}. One of these is Na^+ channel inactivation. This is a separate voltage-induced transition in the Na^+ channel protein that plugs or closes the channel. The other process is the activation of additional K^+ channels (different from those that produce the resting potential). These two processes work together to terminate the action potential and repolarize the membrane to a value near E_K. The opening of additional K^+ channels raises the K^+ to Na^+ permeability ratio to a value much higher than the resting value of ~20 (Section 13.5). This makes the voltage undershoot the resting potential and move closer to E_K. Returning to a negative membrane potential closes the voltage-gated K^+ channels so the membrane potential can finally return to its resting state (Fig. 16.2a).

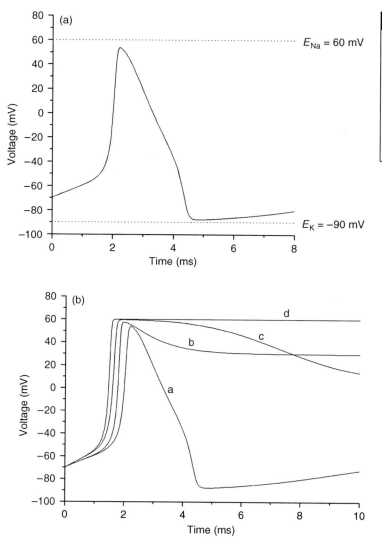

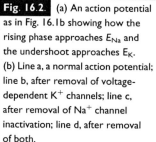

Fig. 16.2. (a) An action potential as in Fig. 16.1b showing how the rising phase approaches E_{Na} and the undershoot approaches E_K. (b) Line a, a normal action potential; line b, after removal of voltage-dependent K^+ channels; line c, after removal of Na^+ channel inactivation; line d, after removal of both.

There are a total of three distinct voltage-dependent channel transitions in the action potential, Na^+ channel activation, Na^+ channel inactivation, and K^+ channel activation. Their impacts are illustrated in Fig. 16.2b. A complete action potential is shown, as in Fig. 16.2a, together with the voltage wave-forms that are computed without Na^+ channel inactivation, without K^+ channel inactivation, and with only Na^+ channel activation.

The next section will examine these voltage-dependent conductances in detail, but now we continue the qualitative discussion by visualizing how these membrane mechanisms can make the action potential propagate along a cable. If current is injected at one end of a cable to depolarize it, the voltage will rise above the threshold. An action potential will then be initiated near where the current is injected, but current entering through the membrane will spread axially, away from the site of excitation. Adjacent membrane will

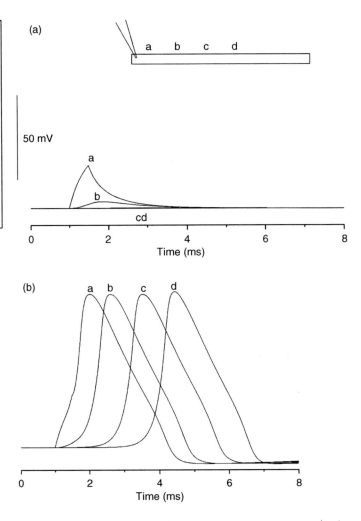

Fig. 16.3. Spread of voltage in passive (a) and active (b) cables, in response to stimulation (indicated on the left in the sketch above). The simulation was performed in a cable of length 2 mm, and diameter 2 μm (sketch in (a)), with and without the voltage-gated channels responsible for an action potential. Voltage is shown from the indicated locations, which are separated by 300 μm. The delays to peak in (b) reflect the time for the action potential to propagate to those points.

then be depolarized sufficiently to open new Na$^+$ channels and initiate a new action potential. In this way the action potential can propagate all the way down the cable to its far end. The regenerative nature of the action potential supports propagation over unlimited distances.

The responses of passive and active cables are compared in Fig. 16.3. In the passive case the response decreases with distance from the stimulus site (like Fig. 15.4). In the active case the peak amplitude is the same over the entire cable. The shape is also essentially the same wherever it is observed, but the distant sites have a delay reflecting the time it takes for the action potential to propagate there from its site of initiation. Action potentials generally propagate with a constant velocity.

Action potentials have a number of essential properties. Most importantly, they are all-or-none events, with a sharp threshold. It is almost impossible to produce a half-amplitude action potential. As the stimulus is increased above threshold, the action potential's peak amplitude remains virtually constant, and indistinguishable from those evoked by weaker stimulus currents. But when the stimulus drops below the threshold, the response is small and resembles the passive responses of Figs. 16.1a and 16.3a. Another important feature of an action potential is its refractory period, a period immediately after the action potential during which a supra-threshold stimulus will fail to evoke another action potential (discussed in Section 16.4). Two action potentials propagating toward one another from opposite ends of an axon will collide and annihilate one another due to this refractory period.

The voltage-dependent transitions of Na$^+$ and K$^+$ channels give a full account of the action potential. The following sections will develop a detailed quantitative analysis of the relation between these ion currents and the action potential.

16.2 | The voltage clamp and the properties of Na$^+$ and K$^+$ channels

To understand the action potential we need to study the Na$^+$ and K$^+$ channels and see how their activity varies with voltage and time. This is experimentally challenging because the complex geometry typical of excitable cells makes the voltage across the membrane hard to control. Controlling the voltage at one region of the cell membrane is not good enough because the voltage elsewhere will vary. The currents through different parts of membrane will then add together to create a confusing mixture. It is necessary to "clamp" the voltage of an entire cell so that a current measurement reflects only the intended voltage. K.S. Cole and G. Marmont performed the first voltage clamp experiments in 1949. A.L. Hodgkin and A.F. Huxley then used the technique in a landmark study in 1952 that elucidated the basic properties of neuronal Na$^+$ and K$^+$ channels. They went on to show how these channels generate action potentials.

These experiments were performed on a particularly large axon, called the giant axon, located in the mantel of the squid. In the typical animals used for experiments the axon was $\sim$20 cm long and $\sim$0.5 mm wide. The unusually large diameter of the squid giant axon was exploited by inserting a long thin silver wire axially as shown in Fig. 16.4. An electrical stimulus applied to the wire is then transmitted nearly uniformly along the entire length of the axon. The axial wire effectively reduces the axon's cylindrical geometry to a circular geometry.

The axial wire eliminates spatial nonuniformity, but even with this "space-clamp" the currents responsible for the action potential

Fig. 16.4. The axial wire distributes an applied voltage (V) uniformly through an axon.

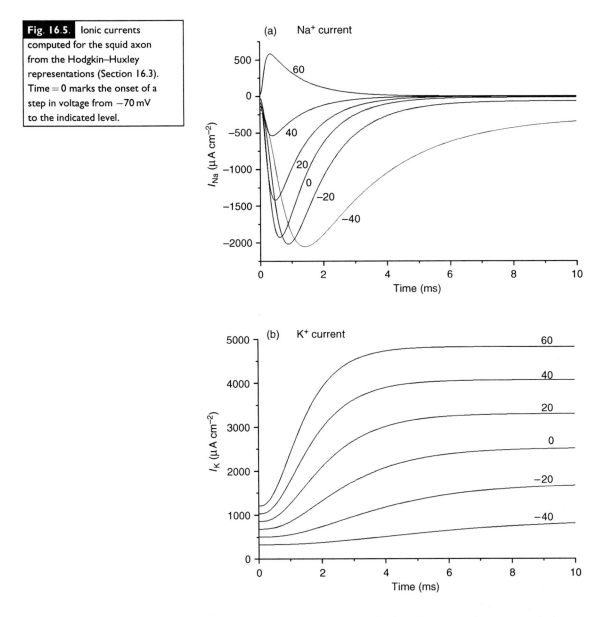

Fig. 16.5. Ionic currents computed for the squid axon from the Hodgkin–Huxley representations (Section 16.3). Time = 0 marks the onset of a step in voltage from −70 mV to the indicated level.

are still very difficult to study because membrane conductances vary with voltage. As channels open and close to produce changes in current, the voltage will change. To solve this problem, a circuit is needed to hold the voltage constant as the current varies. A voltage clamp amplifier is designed to control the voltage on the axial wire, and measure the current as the conductance of the membrane changes. The standard protocol is to change the voltage in a step-wise fashion. This gates the ion channels and changes the current according to how the channels open and close.

Examples of Na⁺ current and K⁺ current activated by a series of voltage steps are shown in Fig. 16.5. Both Na⁺ and K⁺ channels are mostly closed at the resting potential and open as the voltage steps are made more positive. With K$^+$ current (Fig. 16.5b) the voltage is always above E_K (-77 mV) so the current gets larger and larger as the voltage increases. The value of E_{Na} is 50 mV for these calculations so the Na$^+$ current is negative as long as the voltage is below 50 mV (following the convention that positive charge leaving a cell produces positive current). Once the voltage steps are large enough to open all the Na$^+$ channels, further increases make the Na$^+$ current smaller as the voltage gets closer to E_{Na}. For the one pulse shown above E_{Na} the current is positive.

The Na$^+$ and K$^+$ currents are shown separately in Fig. 16.5. The actual current recorded when the squid giant axon was bathed in normal seawater was a sum of these two ionic components. To separate these two currents seawater was prepared in which choline replaced the Na$^+$. Choline does not pass through Na$^+$ channels, so with this solution only K$^+$ current was seen. Subtracting the isolated K$^+$ current from the total current recorded in normal seawater gave the Na$^+$ current. In later years, much better methods were developed for separating different ionic components of membrane current. In addition to improved ion substitution methods, pharmacological agents were found that bind to specific channel proteins and block them. For example, tetrodotoxin blocks most Na$^+$ channels and tetraethylammonium blocks most K$^+$ channels. Currently, there is a huge arsenal of highly specific channel blockers available for use in these kinds of experiments.

The Na$^+$ and K$^+$ currents shown in Fig. 16.5 have exactly the right properties to account for the action potential. These currents reveal directly the three basic voltage-dependent processes invoked in the preceding qualitative discussion of the action potential's rising and falling phases. Figure 16.5a shows that depolarization makes Na$^+$ channels open rapidly. The actual rate depends on voltage but for pulses above -20 mV the peak is reached in less than one millisecond. By contrast, the K$^+$ currents are activated more slowly. This leaves a brief window in time during which the Na$^+$ channels are the dominant permeation pathway in the axonal membrane. So the Na$^+$ channels are initially unopposed and Na$^+$ can push the voltage toward E_{Na}.

After peaking, the Na$^+$ current in Fig. 16.5a decays. The decay is due to the inactivation process discussed in the preceding section. It is distinct from activation and represents a separate effect of voltage on the Na$^+$ channel. Both the inactivation process in Fig. 16.5a and the activation of K$^+$ current in Fig. 16.5b occur after Na$^+$ channel activation. These two later processes contribute jointly to the repolarization of the action potential, and as illustrated in Fig. 16.2, both are necessary. Without the voltage-dependent K$^+$ current, Na$^+$ channel inactivation leaves only the resting K$^+$ channels and the current through this pathway is too small to change the voltage rapidly. Without Na$^+$ channel inactivation, the repolarizing action of the voltage-activated K$^+$ current meets too much resistance from open Na$^+$ channels so the voltage settles at a level closer to E_{Na} than to E_K.

16.3 | The Hodgkin–Huxley equations

The current recorded in a voltage clamp experiment provides the essential information for a quantitative understanding of action potentials. We will now see how quantitative descriptions of the kinetics of channel gating can be used to reconstruct the basic action potential wave-form. Section 16.5 will show how these ideas can be extended to understand propagation.

The quantitative description of Na^+ and K^+ channels introduced by Hodgkin and Huxley (1952) is still in wide use today. They fitted the separate components of Na^+ and K^+ current in Fig. 16.5 to appropriate empirical functions. For example, the decay of the Na^+ current of the squid axon looks like a single exponential (Fig. 16.5a), and an exponential function fits this part of the data very well. An exponential decay indicates that we have a two-state kinetic process (Section 7.1). We therefore define inactivation as a transition from an activatable/open state, A, to an inactivated closed state, I.[1]

$$A \overset{\alpha_h}{\underset{\beta_h}{\rightleftarrows}} I \qquad (16A)$$

In Scheme (16A) α_h and β_h are the forward and reverse rate constants. The subscript for these rate constants, h, is also a variable that denotes the fraction of channels in A ($h = [A]/([A] + [I])$). This scheme is a simple two-state process, so we can use Eq. (7.4) for the time dependence of h. In the present notation we have

$$h = (h_i - h_\infty)e^{-t/\tau_h} + h_\infty \qquad (16.3)$$

where h_∞ is the equilibrium value of h, equal to $\beta_h/(\alpha_h + \beta_h)$, and τ_h is the time constant, equal to $1/(\alpha_h + \beta_h)$. Following a voltage step, h relaxes from its initial value, h_i, defined by the voltage prior to the step, to a new value, h_∞, defined by the voltage stepped to.

In contrast to inactivation, the activation of Na^+ channels is not exponential but sigmoidal (Fig. 16.5a). The rising phase looks like an exponential raised to the third power. This suggests that there are three independent subunits within the Na^+ channel protein, all of which must be turned on to open a channel. The term "gating particle" is widely used for these voltage sensing units.

---

[1] The letter A actually includes both the closed state that can open in response to voltage and the open state. The inactive state is a distinct closed state that remains closed at positive voltage. Thus, the kinetic process of inactivation includes transitions from either the closed or open states to the inactive state. This idea is imbedded in the expression below of the conductance as proportional to $m^3 h$ (Eq. (16.5)). The mechanistic basis for this model is generally recognized as inadequate for a full and detailed kinetic description of the Na^+ channel, but it is entirely adequate for a quantitative modeling of the Na^+ current for the purpose of understanding action potentials.

With a two-state mechanism such as Scheme (16A) for each of these gating particles, we use the variable m to denote the fraction of these particles in the activated state, and write an expression corresponding to Eq. (16.3)

$$m = (m_i - m_\infty)e^{-t/\tau_m} + m_\infty \tag{16.4}$$

We can raise m to the third power to obtain the fraction of channels with all three gating particles activated. This is then the fraction of activated channels.

To obtain the fraction of channels that have undergone activation but have not yet inactivated, we multiply m^3 by h. The Na^+ current is then equal to the number of open channels times the driving force $V - E_{Na}$. If the number of Na^+ channels is N_{Na}, and each has a single-channel conductance of γ_{Na}, then we can write the Na^+ current, I_{Na}, as

$$
\begin{aligned}
I_{Na} &= (V - E_{Na})N_{Na}\gamma_{Na}m^3h \\
&= (V - E_{Na})G_{Na\text{-}max}((m_i - m_\infty)e^{-t/\tau_m} + m_\infty)^3((h_i - h_\infty)e^{-t/\tau_h} + h_\infty)
\end{aligned}
\tag{16.5}
$$

where the second step used Eqs. (16.3) and (16.4) to replace m and h. Note that $N_{Na}\gamma_{Na}$ was replaced by $G_{Na\text{-}max}$ because this product is the maximum possible conductance of the membrane to Na^+, realized when all of the channels are simultaneously open. In the squid giant axon $G_{Na\text{-}max} = 120\,mS\,cm^{-2}$. With a single channel conductance of 4 pS, we can estimate the channel density as $300\,\mu m^{-2}$ (Hille, 1991).

The representation of K^+ current is simpler since we only have an activation process to deal with. Like Na^+ channel activation, K^+ channel activation is sigmoidal (Fig. 16.5b). It is well described by raising the two-state kinetics expression to the fourth power, as though this channel protein contains four gating particles. Using n to represent the fraction of K^+ channel gating particles in the activated state, the same logic that lead to Eq. (16.5) gives the K^+ current as

$$
\begin{aligned}
I_K &= (V - E_K)N_K\gamma_K n^4 \\
&= (V - E_K)G_{K\text{-}max}((n_i - n_\infty)e^{-t/\tau_n} + n_\infty)^4
\end{aligned}
\tag{16.6}
$$

In the squid giant axon $G_{K\text{-}max} = 36\,mS\,cm^{-2}$. With a single channel conductance of 20 pS the density of these channels is $18\,\mu m^{-2}$ (Hille, 1991).

Current traces recorded under voltage clamp are readily fitted to Eqs. (16.5) and (16.6), yielding values for m_∞, h_∞, and n_∞, as well as values for the time constants, τ_m, τ_h, and τ_n. These are plotted as a function of voltage in Fig. 16.6. The m_∞, h_∞, and n_∞ plots resemble the plot of the Boltzmann equation (Fig. 1.5), as expected for a voltage-dependent two-state equilibrium (Section 1.8). Likewise, the τ_m, τ_h, and τ_n plots resemble Fig. 7.9 (Section 7.5).

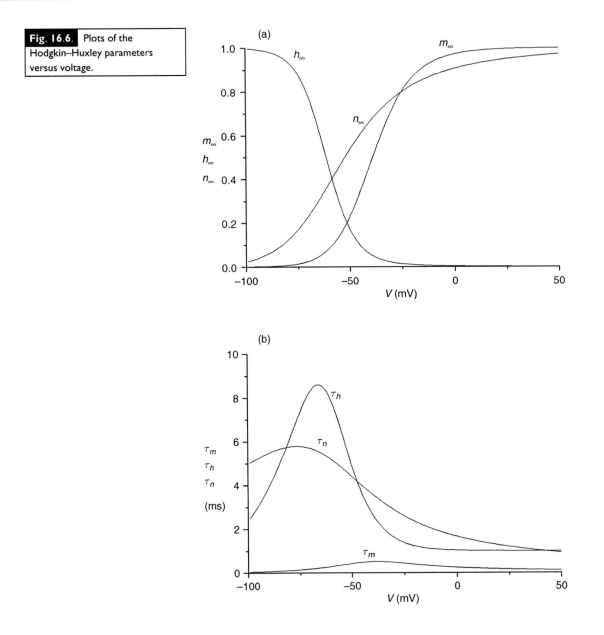

Fig. 16.6. Plots of the Hodgkin–Huxley parameters versus voltage.

The plots of the Hodgkin–Huxley parameters in Fig. 16.6 provide another tour through the basic features of the action potential. Near the resting potential, h_∞ is high and m_∞ is low. The product $m_\infty^3 h_\infty$ is therefore low so Na^+ channels are closed. A sudden change in the voltage to the neighborhood of $-50\,mV$ raises m, and the small value for τ_m means that the change is rapid. This starts the cycle of Na^+ channels opening and upward movement in voltage. The changes in h and n would oppose this process, but the longer values of τ_h and τ_n compared to τ_m (Fig. 16.6b) insure that the initial opening of Na^+ channels can dominate at the earliest time. When h and n catch up with m, their processes of Na^+

channel inactivation and K^+ channel activation, respectively, initiate the recovery of the action potential and the return of the voltage toward its resting value.

For any voltage, we can take the measured values of τ_m and m_∞, and using the simple expressions written in the text just below Eq. (16.3), solve for α_m and β_m. The same can be done for h and n. Hodgkin and Huxley (1952) did this and found that their αs and βs were empirically fitted by the following expressions

$$\alpha_m(V) = \frac{0.1(V+40)}{1 - e^{-(V+40)/10}} \quad \beta_m(V) = 0.108e^{-V/18} \tag{16.7a}$$

$$\alpha_h(V) = 0.0027e^{-V/20} \quad \beta_h(V) = \frac{1}{1 + e^{-(V+35)/10}} \tag{16.7b}$$

$$\alpha_n(V) = \frac{0.01(V+55)}{1 - e^{-(V+55)/10}} \quad \beta_n(V) = 0.055e^{-V/80} \tag{16.7c}$$

It is important to realize that these equations were not derived from physical principles, but are empirical representations of the voltage-dependent gating processes. Although a satisfactory physical interpretation for the expressions for α_m and α_n has been developed in terms of Kramers' theory (Goychuk and Hänggi, 2002), the utility of these expressions derives from their role in accounting for electrical impulses in terms of these conductances, rather than explaining the actual conductance mechanisms.

These voltage-dependent rate constants can now be used to calculate the precise trajectory of an action potential. For each of the different gating particle transitions, we can write the rate of change of the state variable in terms of these rate constants. With reference to the differential equation for this kind of process (see Eq. (7.2)), we have the rate of change of each gating variable as the sum of the forward and reverse velocities

$$\frac{dm}{dt} = \alpha_m(1-m) - \beta_m m \tag{16.8a}$$

$$\frac{dh}{dt} = \alpha_h(1-h) - \beta_h h \tag{16.8b}$$

$$\frac{dn}{dt} = \alpha_n(1-n) - \beta_n n \tag{16.8c}$$

Finally, we need an equation that describes the rate of change of V. The change in voltage depends on the current through the channels, so we can combine Eqs. (16.5) and (16.6)

$$C\frac{dV}{dt} = G_{\text{Na-max}}m^3h(V - E_{\text{Na}}) + G_{\text{K-max}}n^4(V - E_K) + G_L(V - E_L) + I_{\text{stim}} \tag{16.9}$$

The additional "leakage" current, $G_L(V - E_L)$, represents current through channels that maintain the resting potential. It is essentially like the linear term in Eq. (16.1). In the squid giant axon the measurements gave $G_L = 0.3$ mS cm^{-2} and $E_L = -54$ mV. Here, I_{stim} is a current provided by an electrode under experimental control. If I_{stim} is small, the voltage remains near rest where m^3h and n^4 are small. The channels do not open so nothing happens. Making I_{stim} large and positive moves V to a level where m^3h becomes large. The large value of CdV/dt means that the voltage will change.

Equations (16.8a)–(16.8c) and (16.9) are a system of coupled differential equations that can be solved with a computer. To visualize how a computer solves this problem, take an initial condition in which n, m, h, and V are specified. For a very small increment in time, δt, the changes δn, δm, δh, and δV can be calculated from the right-hand sides of the relevant equation by simply multiplying by δt. The slightly changed values of the αs and βs are then calculated for this new voltage, $V + \delta V$, with Eqs. (16.7a)–(16.7c). This completes one time step. Now the calculation can be repeated for another time step and so on. The computer code for this iterative stepping through time is actually quite short and simple. A computer is essential for practical work with these equations as there is no analytical solution.

The action potentials shown in Figs. 16.1 and 16.2 were generated by this process of integrating Eqs. (16.8a)–(16.8c) and (16.9). The close resemblance of these computed action potentials to experimental recordings from the squid axon represents an impressive success of the theory. It is important to appreciate that the Hodgkin–Huxley equations were all based on voltage clamp analysis, so the generation of an action potential is a *bona fide* theoretical prediction. Section 16.5 describes the additional success of calculating the action potential propagation velocity when the Hodgkin–Huxley equations are combined with the cable equation.

Before moving on it is worth pointing out that the solution to these equations not only yields the voltage as a function of time, it also yields the values of m, h, and n as a function of time. Thus, we can reconstruct not only the action potential, but also the time course of the Na$^+$ and K$^+$ currents. Figure 16.7a shows an action potential, and Fig. 16.7b plots the time course of the normalized conductances for Na$^+$ and K$^+$ computed as the action potential progresses: $G_{Na}/G_{Na\text{-max}}$ is m^3h and $G_K/G_{K\text{-max}}$ is n. We see that $G_{Na}/G_{Na\text{-max}}$ rises rapidly, and this coincides with the upstroke of the action potential; $G_K/G_{K\text{-max}}$ rises later, and this coincides with the repolarization and undershoot.

Values of m and h are plotted in Fig. 16.7c to show how the rise in m corresponds well with the rise in $G_{Na}/G_{Na\text{-max}}$ and the fall in h corresponds well with the decay in $G_{Na}/G_{Na\text{-max}}$. Figure 16.7 shows how the Na$^+$ and K$^+$ channels change their activity during an action potential and how the m and h gating processes relate to the Na$^+$ current.

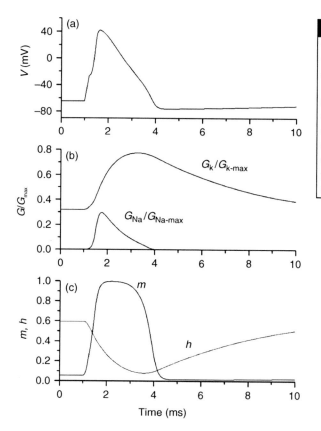

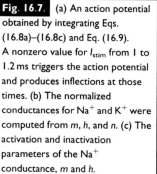

Fig. 16.7. (a) An action potential obtained by integrating Eqs. (16.8a)–(16.8c) and Eq. (16.9). A nonzero value for I_{stim} from 1 to 1.2 ms triggers the action potential and produces inflections at those times. (b) The normalized conductances for Na$^+$ and K$^+$ were computed from m, h, and n. (c) The activation and inactivation parameters of the Na$^+$ conductance, m and h.

16.4 | Current–voltage curves and thresholds

Two special limiting cases of the Hodgkin–Huxley equations help to illustrate how the threshold for action potential generation arises. When a constant current is applied to an axonal membrane for a long time, all the kinetic processes will eventually settle to an end point. To describe this situation, we take Eq. (16.9), set $dV/dt = 0$, and set m, h, and n equal to m_∞, h_∞, and n_∞. The result is a steady-state current–voltage relation. Calling this voltage $V_{s\text{-}s}$, we have

$$-I_{stim} = (V_{s\text{-}s} - E_{Na})G_{Na\text{-}max}m_\infty(V_{s\text{-}s})^3 h_\infty(V_{s\text{-}s})$$
$$+ (V_{s\text{-}s} - E_K)G_{K\text{-}max}n_\infty(V_{s\text{-}s})^4 + G_L(V_{s\text{-}s} - E_1) \qquad (16.10)$$

On the other hand, immediately after turning I_{stim} on, m, with the fastest time constant, will change before h and n. If the initial voltage is V_0, then we will have h and n still at their initial values $h_\infty(V_0)$ and $n_\infty(V_0)$. The peak current will be close to that described by $m_\infty(V)$ for a new voltage, V, and $h_\infty(V_0)$ and $n_\infty(V_0)$ for the prior voltage. We call this voltage V_{fast}

$$-I_{stim} = (V_{fast} - E_{Na})G_{Na\text{-}max}m_\infty(V_{fast})^3 h_\infty(V_0)$$
$$+ (V_{fast} - E_K)G_{K\text{-}max}n_\infty(V_0)^4 + G_L(V_{fast} - E_1) \qquad (16.11)$$

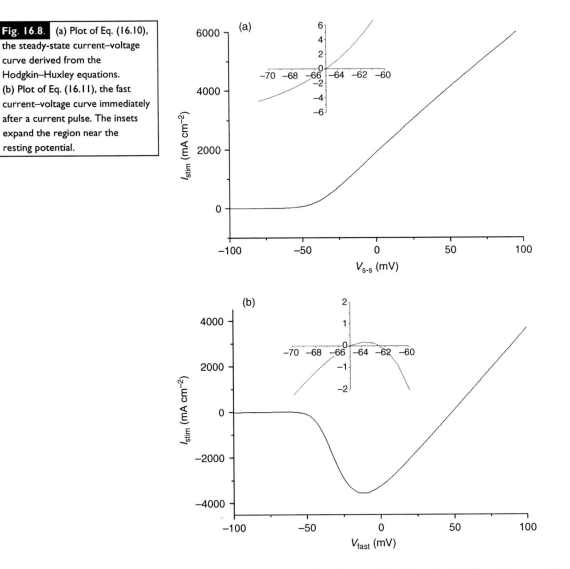

Fig. 16.8. (a) Plot of Eq. (16.10), the steady-state current–voltage curve derived from the Hodgkin–Huxley equations. (b) Plot of Eq. (16.11), the fast current–voltage curve immediately after a current pulse. The insets expand the region near the resting potential.

Although Eqs. (16.10) and (16.11) look quite similar, the use of V_0 for n and h makes a big difference. Equation (16.11) applies to a time soon after the stimulus. We do not know exactly when, but it is generally <1 ms, because τ_m is always <1 ms (Fig. 16.6b). The important thing is that it approximates the voltage that gives the peak current reasonably well.

These two equations define current–voltage relations with fundamentally different properties. A plot of Eq. (16.10) in Fig. 16.8a shows that the curve crosses the zero current axis only at the resting potential of −65 mV (see inset). This is the only voltage for which the current is zero, so no matter where the voltage is moved to initially, it ultimately returns to the resting potential after I_{stim} is turned off. The rapid voltage-dependent channel gating transitions do not change the resting potential. Because Na$^+$ channels inactivate, the K$^+$ channels dominate in the long term at any voltage.

Now we turn to the fast current–voltage curve (plot of Eq. (16.11) in Fig. 16.8b). Expanding this plot around the resting potential (inset of Fig. 16.8b) shows that for small displacements away from the resting potential, the positive, outward current will pull the voltage back. However, slightly above −64 mV the current peaks and curves down. At ~ −62.5 mV the current is zero again and above this point the current is negative. This marks the threshold for action potential generation. If the voltage is displaced past this point, even slightly, the negative, inward current through newly opened Na^+ channels will move the voltage in the positive direction, further away from the resting potential. This will open more Na^+ channels and initiate an action potential. Figure 16.8b is very useful for this purpose, as it gives a good picture of what happens at the threshold.

The precise voltage at the threshold is not an absolute property of the axon. This curve in Fig. 16.8b is based on the assumption used to derive Eq. (16.11) that m changes instantaneously as the voltage changes while n and h do not change at all. If the voltage is changed slowly, then n and h will have time to keep up with m and this will alter the current. In fact, if the voltage is changed very slowly, then the curve will follow Eq. (16.10) and Fig. 16.8a. There will be no action potential. A very slow depolarization, by inactivating Na^+ channels, renders the axon inexcitable. This is called *accommodation* and is a well-known property of excitable membranes. The concept of a voltage threshold is thus somewhat fluid. The actual threshold voltage depends on the nature of the stimulus as well as the initial voltage.

The shape of the current–voltage curve in Fig. 16.8b is one of the hallmarks of an excitable membrane. Taking the electrical resistance as the slope of the current–voltage plot, a negative slope means a negative resistance. In membrane biophysics, this only occurs when voltage gates a channel. When the slope is positive everywhere (Fig. 16.8a), then current always pulls the voltage back to the resting potential. The membrane is stable and returns to its resting state in response to any perturbation. By contrast, a region of negative slope can be unstable. When the voltage is displaced to such a point, then the current will move the voltage further away rather than returning it to its original state. That is what makes the downward crossing of the V-axis in the inset of Fig. 16.8b so critical.

In addition to the threshold just discussed, the Hodgkin–Huxley equations provide a clear explanation for the refractory period. This is an increase in threshold following an action potential. With Na^+ channels inactivated and more K^+ channels open, it takes more stimulus current to produce an action potential. The recovery of Na^+ channels from inactivation (Fig. 16.7c) and the closure of voltage-gated K^+ channels (Fig. 16.7b) jointly describe the decay of the refractory period.

After the refractory period an axon is *supernormal* (Swadlow et al., 1980). During this time the axon has a lower threshold than it did before the action potential. This period varies quite a bit between different animals and types of axons but is usually maximal between 7 and 20 ms after an action potential. One can imagine either a reduction in Na^+ channel inactivation (increase in h) or a reduction in K^+ conductance (decrease in n) giving rise to an increase in excitability. An analysis of their relative contributions indicated that the decrease in n has a greater impact and can account for most of the reduction in threshold of the supernormal period (Stockbridge, 1988).

16.5 | Propagation

The analysis of membrane currents in Sections 16.2–16.4 did not treat spatial variations in voltage. In fact, the most important property of the action potential is that it propagates (Fig. 16.3). Now that we have seen how the voltage-dependent gating of Na^+ and K^+ channels can generate an action potential we can use these properties to achieve a quantitative understanding of propagation. The starting point for spatial variations in voltage is the cable equation from the preceding chapter. We take Eq. (15.7) and divide through by r_m as follows

$$c_{m'} \frac{\partial V}{\partial t} = \frac{1}{r_a} \frac{\partial^2 V}{\partial x^2} - \frac{V}{r_m} \tag{16.12}$$

In this form we see that the change in voltage with time at a particular position depends on two terms. The first is the axial term, reflecting an imbalance of axial current flowing in from the left and out to the right. The second term is the current through the membrane. To incorporate current through the channels into the cable equation, we replace this second term, V/r_m, with the sum of the membrane current terms from Eq. (16.9)

$$c_{m'} \frac{\partial V}{\partial t} = \frac{1}{r_a} \frac{\partial^2 V}{\partial x^2} - (V - E_{Na})G_{Na\text{-}max}m^3 h - (V - E_K)G_{K\text{-}max}n^4 - G_L(V - E_L) \tag{16.13}$$

The variables m, h, and n now vary with x, the position along the cable, in addition to being functions of voltage and time. The numerical procedure for computing how V, m, h, and n evolve through time must be extended to include the spatial dependence. The axon must be subdivided into many tiny segments of length Δx (Fig. 15.2). The increments in V, m, h, and n in each segment must be calculated as outlined in Section 16.3. But an additional contribution arising from neighboring segments must be computed by evaluating $(1/r_a)(\partial^2 V/\partial x^2)$. Computer programs can readily perform this integration by extending the compartmental model methods of Section 15.9.

Hodgkin and Huxley had a calculating machine but no computer when they studied axons around 1950 so they modified Eq. (16.13) to facilitate hand calculations. Their method also provides an important insight into the velocity of action potential propagation so it will be examined closely here.

If one accepts the fact that an action potential preserves its shape as it propagates, then the solution to Eq. (16.13) must have the form

$$V(x, t) = V(x - \theta t) \tag{16.14}$$

where θ is the velocity. This is a very general mathematical expression that arises in the physics of waves. If $V(x, 0)$ has some complex shape centered around x_0, then Eq. (16.14) says that at some later time V will have the same shape, but centered at a new value of $x = x_0 + \theta t$. Equation (16.14) is a general solution to a partial differential equation known as the wave equation

$$\frac{\partial^2 V}{\partial x^2} = \frac{1}{\theta^2} \frac{\partial^2 V}{\partial t^2} \tag{16.15}$$

as is easily checked by substitution.

Equation (16.13), along with expressions for m, h, and n (Eqs. (16.8a)–(16.8c)), form a system of partial differential equations in space and time. Using Eq. (16.15) allows us to eliminate the second derivative with respect to x, replacing it with the second derivative with respect to time. In this way we reduce this differential equation to a dependence on only one independent variable

$$c_{m'} \frac{dV}{dt} = \frac{1}{r_a \theta^2} \frac{d^2 V}{dt^2} + (V - E_{Na})G_{Na\text{-}max}m^3h + (V - E_K)G_{K\text{-}max}n^4 + G_L(V - E_L) \tag{16.16}$$

With time as the only independent variable, performing the numerical integration by hand becomes practical. In a few hours one can integrate in small steps (as small as 0.01 ms) for a few milliseconds.

The only number not known in Eq. (16.16) is θ, the conduction velocity. Hodgkin and Huxley found that for most choices of θ the numerical solution goes to $V = \pm\infty$ as time evolves. If θ is too large or too small V blows up. After repeated attempts with different values of θ, they found that one particular value resulted in a $V(t)$ that returns to the resting potential after producing an impulse that looks like the experimentally recorded action potential; $\theta = 18.8 \text{ m s}^{-1}$ was the value that produced a stable result. This compares well with the experimental propagation velocity of 21.2 m s^{-1}.

This determination of the propagation velocity from an analysis of Eq. (16.16) represents a remarkable success of the ionic theory of membrane excitability. It must be emphasized that the parameters used as input in the calculation were based on voltage clamp measurements. Although the use of Eq. (16.15) in the derivation of Eq. (16.16) assumes a wavelike solution, the velocity of the wave was not known. Thus, the similarity between the experimental and theoretical values of θ provides a stringent test of the theory.

Equation (16.16) cannot be solved analytically to obtain the propagation velocity in terms of the parameters that represent the channel properties. However, the dependence of conduction velocity on the cable parameters can be explicitly derived. We collect the basic parameters of Eq. (16.16) together by dividing through by $c_{m'}$. This yields $1/(\theta^2 r_a c_{m'})$ as the factor multiplying $\partial^2 V/\partial t^2$. The factors multiplying the specific channel current terms are now all of the form $G_x/c_{m'}$. Since the densities of all the channels are fixed, then each G_x will scale with the area; $c_{m'}$ scales with area as well so the ratios $G_x/c_{m'}$ are constant. This makes the constant $1/(\theta^2 r_a c_{m'})$ invariant with respect to axon diameter. Once the value of this factor that gives a stable solution to Eq. (16.16) is found, it will work for any other axonal diameter. Denoting this factor as k gives

$$\theta = \sqrt{\frac{k}{r_a c_{m'}}} \tag{16.17}$$

Recall from Chapter 15 that $r_a = \rho_c/\pi a^2$ (Eq. (15.1)), where a is the radius. Note that $c_{m'}$ is the capacitance of the membrane of a unit length of axon; $c_{m'} = 2\pi a c_m$. These substitutions give

$$\theta = \sqrt{\frac{ka}{2\rho_c c_m}} \tag{16.18}$$

This is an important result because it tells us that the velocity increases with the square root of the axon radius. In fact, this is an old experimental result, as shown in a plot of conduction velocity versus diameter for many different squid and cuttlefish giant axons (Fig. 16.9).

Equation (16.18) and Fig. 16.9 illustrate that, for axons, bigger is faster. Since speed helps an organism in many ways, evolution will favor large axons. The giant axon of the squid is a perfect example. This axon runs through the mantel, triggering a contraction to generate a rapid burst of locomotion. This motion is part of

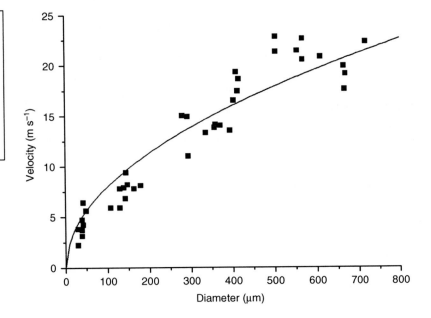

Fig. 16.9. Action potential conduction velocity is plotted versus axon diameter. Measurements were made in a wide range of axons from squid and cuttlefish of different sizes. The curve drawn is 0.8 $\sqrt{\text{diameter}}$. (Data from Table 1 of Pumphrey and Young, 1938; temperature range 19.5–22.1°C)

an escape reflex of the animal to avoid predators, and the conduction velocity of this axon contributes to the speed of this response.

16.6 | Myelin

The advantage of speed provided by a large axon weighs against the disadvantage of size and energy cost. In a complex nervous system with billions of neurons, too many large axons would make the nervous system unmanageably huge and consume an inordinate amount of metabolic energy. Invertebrates such as the squid generally have fewer neurons, but vertebrates have much more elaborate nervous systems and need another strategy to speed up axonal conduction. They have achieved this with myelin, a thick insulating sheath that wraps around axons to increase their membrane resistance and reduce their capacitance (Fig. 16.10a). This increases the conduction velocity. With myelin, a 10-μm diameter nerve fiber of a frog conducts at about $20 \, \text{m} \, \text{s}^{-1}$ (Hodgkin, 1964), which is roughly three times faster than a 100-μm diameter axon in the squid (Fig. 16.9).

Myelin sheaths do not cover the entire axon surface. There are bare spots, called *nodes of Ranvier*, spaced at regular intervals (Fig. 16.10b). At the node there is no myelin, so ions have unobstructed access to the membrane. Electrical recordings from myelinated axons showed that the nodes are the sites of action potential generation. Extracellular fields recorded near a node showed that action potentials are accompanied by active inward currents (Huxley and Stämpfli, 1949). Recordings at other sites detected only passive outward currents. The Na^+ channels, which are essential to the initiation and propagation of action potentials, are concentrated at the nodes of Ranvier.

The high resistance of the myelin allows the depolarization at one node to spread over a great distance without much current loss. One way to think about this is that increasing the membrane resistance increases the length constant, λ. Myelin also reduces the capacitance so that charge entering a node can change the voltage of a greater area of axonal membrane. Both the increased resistance and reduced capacitance work together to promote rapid spread of an action potential from one node to the next. Because this entails

(a) (b)

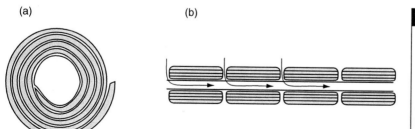

Fig. 16.10. (a) Myelin wraps itself around an axon in a spiral form. (b) Between the sheaths of myelin are nodes of exposed axonal membrane. The nodes are spaced at regular intervals. Arrows indicate ion entry. Action potentials propagate by jumping from node to node.

jumping from node to node, this type of conduction in myelinated axons is referred to as *saltatory*.

The geometry of myelinated axons shows a remarkable scaling property in which the inner diameter, outer diameter, and internode distance are in the same proportion for nerve fibers of different sizes. This scaling has been shown to optimize conduction velocity (Rushton, 1951). This can be understood in terms of the length constant and time constant of the myelinated axon. First, consider the length constant $\lambda = \sqrt{r_m/r_a}$ (Eq. (15.10)). A longer λ means the action potential will spread a greater distance, so we will consider how this number can be maximized for a given outer diameter, d_o, of the myelin. As the inner diameter, d_i, is reduced the layer of myelin will become thicker, so r_m will increase. The value of r_m can be calculated as the resistance of many concentric cylindrical resistors in series. The membrane resistance of one of these cylinders of diameter s is $\rho_m/(\pi s)$ for a unit length of axon (Eq. (15.2)). We can therefore take r_m as an integral from the inner diameter to the outer diameter

$$r_m = \frac{\rho_m}{\pi} \int_{s=d_i}^{d_o} \frac{1}{s} ds = \frac{\rho_m}{\pi} \ln(d_o/d_i) \qquad (16.19)$$

However, as d_i gets smaller r_a decreases because the area of the axial cytoplasmic core goes down. With $r_a = \rho_c/(4\pi d_i^2)$ (Eq. (15.1)) we use Eq. (15.10) to write

$$\lambda = \sqrt{\frac{r_m}{r_a}} = \sqrt{\frac{(\rho_m/\pi)\ln(d_o/d_i)}{\rho_c/(4\pi d_i^2)}} = 2d_i\sqrt{\frac{\rho_m}{\rho_c}\ln(d_o/d_i)} \qquad (16.20)$$

The value of d_i that maximizes λ can be found by differentiating with respect to d_i and setting the derivative equal to zero. The result is

$$\frac{d_i}{d_o} = e^{-1/2} = 0.607 \qquad (16.21)$$

Thus, scaling the inner and outer diameters of myelin so they keep to this ratio maximizes λ. Furthermore, fixing the ratio d_i/d_o in Eq. (16.20) makes λ proportional to d_i (or d_o). So scaling the internodal distance, l, with d_i or d_o, means that l will have the same electrotonic length for different axonal diameters.

Now we can use the cable equation to extend this scaling rule to the velocity of propagation. First, note that adding myelin leaves the membrane time constant, $\tau_m = r_m c_m$, unchanged. Just as $r_m \propto \ln(d_o/d_i)$ (Eq. (16.19)), we have $c_m \propto 1/(\ln(d_o/d_i))$ by the same kind of reasoning. So the product remains the same. Recall from Chapter 15 that the cable equation becomes dimensionless when expressed in units of λ and τ_m. Thus, the myelinated section between two nodes is equivalent for any axonal diameter because τ_m is the same and the scaling rule makes l the same when expressed in units of λ. An action potential at a node produces a voltage that spreads through the myelinated portion of the axon. Because the spread is passive, the cable equation applies. Even though the time course of the voltage change at the node is complicated and cannot be expressed mathematically, the solution of the dimensionless cable equation describing the spread through the myelinated region will be the same for different sized axons of equivalent electrotonic lengths. The time it takes for a neighboring node to reach threshold will therefore be the same. Denoting this time as t_l, we can write the conduction velocity as l/t_l. With the scaling of $l \propto d_i$ and d_o, we have the result that the conduction velocity increases linearly with diameter in myelinated axons. This linear relationship has been demonstrated experimentally (Waxman and Swadlow, 1977). The theoretical basis was pointed out by Huxley and Stämpfli (1949) and the general principles of scaling were elaborated by Rushton (1951).

Quantitative descriptions of Na^+ and K^+ currents in the node membrane follow the same basic form as the Hodgkin–Huxley equations (Eqs. (16.7a)–(16.7c)), although in some instances the exponents of m and n are lower (Hille, 1977). Computations of action potentials with these representations gave reasonable velocities, which increased linearly with diameter, provided that l and d_i were scaled as specified by Rushton (Goldman and Albus, 1968).

Ultrastructural studies determined that the ratio d_i/d_o ranges between 0.64 and 0.87, with a mean of 0.77 (Waxman and Swadlow, 1977). This may be significantly greater than the optimal value of 0.607 obtained above (Eq. (16.21)). However, the functional dependence on d_i in Eq. (16.20) actually has a broad maximum so that a deviation of this magnitude will hardly change λ. The variable l is also somewhat larger than optimal for conduction velocity, and it has been argued that increasing l may have another advantage in conserving energy. Each node is a site where energy is dissipated so that spacing them out over greater distances will reduce the energy cost of an action potential (Rushton, 1951).

Tradeoffs between speed and processing power bring up some very interesting biological questions. Rushton (1951) pointed out that increasing the velocity from 30 to 90 m s^{-1} might gain a rabbit about 2 or 3 ms. This is not a significant advantage in a complex behavior involving sequential activation of many neurons, because the accumulated synaptic delays may consume much more time, say ~50 ms. On the other hand, a 3-fold reduction in diameter allows an animal to put 9 times as many axons into the same space, providing $2^9 = 512$ possible on–off combinations. Thus, myelination will be selected for axons carrying a sensory input that requires minimal interpretation. When subtle differences must be discerned, speed will be sacrificed in favor of more information.

16.7 | Axon geometry and conduction

Although long cables are a fundamental structural unit of axons, there are many variations on this geometry. Axons branch

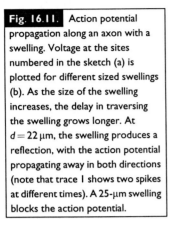

Fig. 16.11. Action potential propagation along an axon with a swelling. Voltage at the sites numbered in the sketch (a) is plotted for different sized swellings (b). As the size of the swelling increases, the delay in traversing the swelling grows longer. At $d = 22\,\mu m$, the swelling produces a reflection, with the action potential propagating away in both directions (note that trace 1 shows two spikes at different times). A 25-μm swelling blocks the action potential.

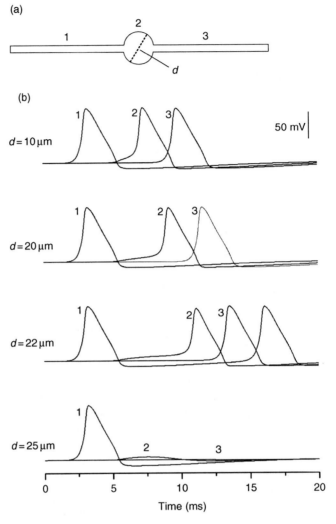

extensively, change diameters either abruptly or gradually, and have swellings or varicosities. As an action potential approaches a point of increased diameter, the axial current encounters a larger membrane capacitance. This mismatch will reduce the voltage change. A region of axon enlargement thus acts as an obstacle to action potential propagation and a number of interesting things can occur (Goldstein and Rall, 1974; Swadlow *et al.*, 1980). The various possibilities are illustrated in Fig. 16.11. This figure shows simulated action potentials in an unmyelinated axon with a diameter of $0.5\,\mu m$ and with a single spherical varicosity of varying diameter.

The traces for swellings of 10 and 20 μm illustrate how the action potential can be delayed by the obstacle (Fig. 16.11). Thus, we see a longer time delay between the peaks at sites 1 and 2 compared to the delay between the peaks at sites 2 and 3. Increasing the varicosity diameter increases the delay. Charging of the varicosity membrane is slowed by its greater capacitance. The voltage at the swelling (site 2 in the sketch (Fig. 16.11a)) clearly shows the slower onset of the action potential. When this delay outlasts the refractory period, the continuing depolarization in the swelling initiates an action potential both in the forward and reverse directions ($d = 22\,\mu m$). This process is known as reflection. Finally, a varicosity can be so large that the axial current can no longer drive the voltage to threshold. Propagation then fails. So depending on the size of the varicosity, the action potential will either be delayed, or reflected, or blocked.

Action potential failure has been documented at axonal enlargements, branch points, and varicosities (Swadlow et al., 1980). In most cases, it is a close call so that small changes in excitability will alter the outcome. Repeated action potentials can reduce excitability by mechanisms such as Na^+ channel inactivation or accumulation of extracellular K^+. This will then make action potentials fail at the obstacles. When swellings are borderline for action potential propagation, changes in membrane properties can tip the balance toward success (Obaid and Salzberg, 1996), or failure (Segev, 1990; Jackson and Zhang, 1995).

The delays of a few milliseconds seen with $10–20\,\mu m$ varicosities may be important for neural functions that involve precise timing. These delays will vary if a neurotransmitter activates receptors on the varicosity membrane. It is interesting that swellings actually slow propagation since a uniformly larger diameter will conduct more rapidly (Section 16.5).

Reflections are quite strange and one can imagine a dumbbell-like structure with an action potential reverberating back and forth indefinitely. Reflection has been demonstrated directly in recordings from molluscan neurons. Action potentials generated in the axon propagate into the large cell body and bounce back (Tauc, 1962). Reflection is a rare occurrence in normal biological function, but this form of recurring electrical activity is a major factor in cardiac arrhythmias (Antzelevitch, 2001).

16.8 | Channel diversity

The voltage-gated Na^+ and K^+ channels that generate the action potential belong to an enormous superfamily of related proteins (Ashcroft, 2000; Caterall et al., 2002). There are at least nine mammalian Na^+ channels that show qualitatively similar behavior but differ quantitatively in the details of their voltage dependence and kinetics. There are at least 50 K^+ channels and they are divided into four distinct families. There are also at least ten voltage-gated Ca^{2+} channels. They are related to the Na^+ channels and in some types of cells they generate an inward current that produces impulses, which can propagate in the same way as an action potential.

The threshold, duration, shape, refractory period, velocity, and other properties of an action potential vary according to the specific properties of the ion channels present in the cell membrane. It is easy to see that a positive shift in the voltage dependence of activation of a Na$^+$ channel (the m_∞ plot in Fig. 16.6a) will increase the threshold. Slower kinetics of K$^+$ channel closure at negative voltages will prolong the refractory period and delay the onset of the supernormal period. In the case of Ca^{2+} channels, inactivation is slow or in some cases almost nonexistent, and Ca^{2+} spikes generally have a much longer duration. The diversity of the voltage-gated ion channels translates directly into a rich diversity of electrical signaling processes in biology.

Many of these channels have been subjected to voltage-clamp analysis and in some cases quantitative representations of their gating have been rendered in the same form used by Hodgkin and Huxley. The methods developed in Sections 16.3–16.6 readily incorporate these different channels. So we have a very powerful general method to relate the biophysical properties of these proteins to their biological function. The biophysical properties in turn can be understood in terms of their molecular structure, and ion channels were used to illustrate many principles in earlier chapters of this book. In the following sections we will see how the biophysical representations of membrane conductance illuminate aspects of complex electrical activity in neurons and muscle fibers.

16.9 | Repetitive activity and the A-current

A single pulse of current will usually evoke one action potential, but when the stimulus current is sustained, action potentials can keep firing repetitively. As the level of constant current increases, the squid axon jumps from firing only one spike to firing more than 50 spikes per second. This frequency does not increase much with further increases in current (Fig. 16.12). So the squid axon loses the information about how strong the stimulus current was. That may not matter for the escape reflex triggered by this giant axon, but other functions may by improved be smarter neurons.

Just as the ion channels of the squid axon determine the shape of the action potential, they also determine the frequency response. Na$^+$ channel inactivation and K$^+$ channel closure determine the refractory period, which limits how high the frequency can get. Na$^+$ channel inactivation also makes the axon accommodate, or lose its sensitivity, when a small stimulus current induces a slow depolarization. That is what limited the response to a single spike when the stimulus current was 5 μA (Fig. 16.12). Other excitable cells have very different frequency response characteristics, due to differences in the kinetic properties of their voltage-gated channels. Various forms of repetitive activity and the theories used to understand them are reviewed in chapter 11 of Jack et al. (1983). Here we will develop one important example to illustrate how adding a different kind of K$^+$ channel, originally named the "A-current," can dramatically alter the electrical behavior of a neuron. This

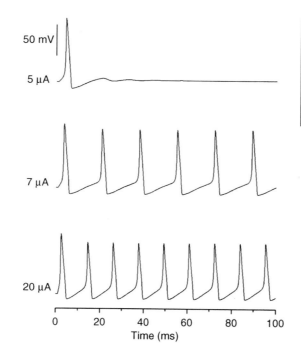

Fig. 16.12. Simulations of repetitive activity in the squid axon. The Hodgkin-Huxley equations were integrated with a stimulus current (indicated on the left) starting at 1 ms and continuing during the entire simulation.

channel enables neurons to vary their frequency of repetitive activity linearly over a wide range.

Voltage clamp experiments in the cell body of the *Anisidoris* snail revealed three ionic currents (Connor and Stevens, 1971a, b). There was a Na$^+$ current with qualitatively similar behavior to the squid axon Na$^+$ current, and two K$^+$ currents. One of the K$^+$ currents, I_K, bore some resemblance to that of the squid axon, but the other, I_A (the A-current), was very different. Figure 16.13 reproduces the calculated currents under voltage clamp. A voltage step from −80 to 40 mV opens the I_K channels with sigmoidal activation kinetics. Note that I_A is also activated with sigmoidal kinetics, but this current inactivates with a time constant of about 200 ms. The inactivation increases as the voltage goes from −80 to −40 mV. That means that when the starting potential is −40 mV (before the step to +40 mV), I_A is already inactivated and the pulse to +40 mV cannot open a significant number of these K$^+$ channels.

The A-current accounts for some very distinctive forms of electrical activity seen in the *Anisidoris* snail neuron. The consequences of these channel properties can be explored using quantitative expressions for the rate constants (the αs and βs) for each activation and inactivation process. Such expressions with the same general form as Eqs. (16.7a)–(16.7c) were developed by de Schutter (1986). These expressions can be substituted in place of the Hodgkin-Huxley expressions for Na$^+$ and K$^+$ channels. The solutions to these equations for sustained current (Fig. 16.14) look quite different from those for the squid axon (Fig. 16.12). Now, increases in the stimulus current produce a proportional increase in action

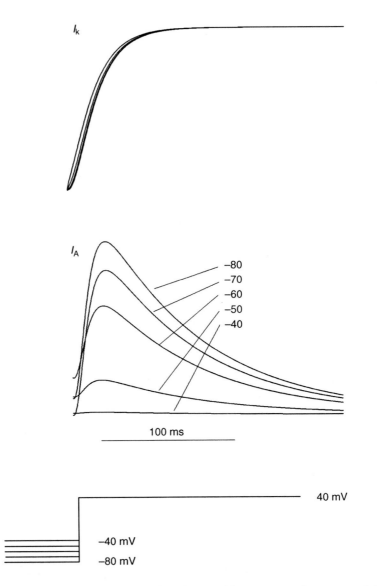

Fig. 16.13. A comparison of the kinetics of activation of I_K, the conventional K$^+$ current, and I_A, the transient K$^+$ current of a snail neuron, plotted using the equations of de Schutter (1986). The voltage is held at levels ranging from −80 to −40 mV (indicated for each I_A trace), and then stepped to 40 mV. Voltage steps are sketched below.

potential frequency. The time interval between action potentials is determined largely by the inactivation kinetics of the A-current. Stronger stimulus currents move the voltage through the inactivation range (−80 to −40 mV, Fig. 16.13) more rapidly, so the A-current inactivates faster and another action potential can fire sooner. The relation between current and frequency is almost linear (Connor and Stevens, 1971b; Jack et al., 1983), although there is no simple mathematical analysis that explains this result.

The A-current is widespread and confers neurons with a number of interesting characteristics (Rogawski, 1985). In addition to the capacity for repetitive firing over a wide frequency range, A-currents make the response of a cell very sensitive to small voltage changes around rest. For a cell near −70 or −80 mV, there is little inactivation so the A-current will be fully primed. It will then serve as

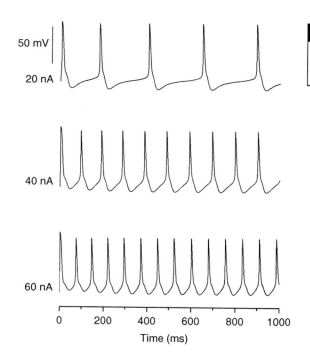

Fig. 16.14. Simulations of repetitive activity in the *Anisidoris* snail neuron with increasing stimulus current.

a powerful brake to sudden excitatory impulses. A sustained stimulus will only fire an action potential after a long delay during which the A-current inactivates. If a cell has been above −50 mV for more than ~200 ms, then the A-current will have largely inactivated, so the cell will be much more sensitive to stimulation. In addition, the A-current, and other K$^+$ channels with similar properties, can be inactivated by repetitive activity. Once these channels have inactivated, there is less K$^+$ current to repolarize the action potential. Action potentials then become broader, (Aldrich *et al.*, 1979), and when this happens in a nerve terminal, more Ca^{2+} enters during a spike to trigger more neurotransmitter release (Jackson *et al.*, 1991).

Although the term A-current is still widely used, it has become somewhat obsolete. A number of distinct K$^+$ channel proteins show the requisite transient behavior. These include the product of the *Drosophila Shaker* gene and its mammalian homologue Kv1.4. The mammalian K$^+$ channels Kv3.2, Kv3.3, and Kv4.1 show similar behavior and auxiliary K$^+$ channel subunits (β subunits) can combine with the channel forming α subunits to convert a noninactivating form into an inactivating form.

16.10 | Oscillations

Oscillations in voltage regulate a wide range of biological processes. Biologically realistic oscillations can be generated by a minimal complement of two channels with very different rates of activation.

If the channels conduct ions with very different Nernst potentials, then they will compete for control and the voltage will swing back and forth. To illustrate this we will look at a model of Morris and Lecar (1981), which was developed to explain voltage oscillations in barnacle muscle fibers.

This model is based on two voltage-activated channels, one permeable to Ca^{2+} and the other permeable to K^+. The Nernst potential for K^+ in the barnacle muscle fiber is $-70\,mV$ so opening K^+ channels moves the voltage in the negative direction. The Nernst potential for Ca^{2+} is $100\,mV$ so opening Ca^{2+} channels moves the voltage in the positive direction. Both channels open as the voltage becomes more positive, but the Ca^{2+} channels open rapidly and at slightly more negative voltages so that a stimulus that moves the voltage in the positive direction opens the Ca^{2+} channels first. The voltage is then pulled toward $E_{Ca} = 100\,mV$. The K^+ channels open more slowly, but when they do they produce a larger conductance that pulls the voltage back down. Both channels then close. A constant stimulus current will then move the voltage up to repeat the cycle.

With Ca^{2+} channels, K^+ channels, and a leakage pathway, we can write a differential equation to describe how currents make the voltage change.

$$C\frac{dV}{dt} = I_{stim} + G_{Ca}(V - E_{Ca}) + G_K(V + E_K) + G_L(V + E_L) \tag{16.22}$$

The change in voltage reflects the sum of the various currents. Note the resemblance to Eq. (16.9). Here $E_L = -50\,mV$.

One of the attractive features of the Morris–Lecar model is that it builds on the simplest picture of voltage gating of ion channels. Recall that the curves for m_∞, h_∞, and n_∞ (Fig. 16.6a) resemble plots of the Boltzmann equation used to describe voltage gating in Section 1.8 (Fig. 1.5), and the curves for τ_m, τ_h, and τ_n (Fig. 16.6b) resemble plots for voltage dependent rates in Section 7.5 (Fig. 7.9). It is therefore appealing to replace the purely phenomenological equations that describe m, h, and n (Eqs. (16.7a)–(16.7c)) with simpler equations based on elementary models of channel gating.

A Boltzmann function (Eq. (1.28)) is used to express the equilibrium voltage dependence of the Ca^{2+} channel open probability, m_∞, as follows

$$m_\infty = \frac{1}{1 + e^{-V/7.5}} \tag{16.23}$$

The K^+ channels open at a more positive voltage, so the Boltzmann equation for n_∞, the K^+ channel open probability, is offset by $10\,mV$

$$n_\infty = \frac{1}{1 + e^{-(V-10)/7.5}} \tag{16.24}$$

As already noted, an essential feature of this model is that the Ca^{2+} channels gate much more rapidly than the K^+ channels.

We can therefore assume that the open probability for Ca^{2+} channels tracks the voltage perfectly with no delay (in the spirit of the instantaneous response of m in Eq. (16.10)). The time constant for m is thus zero.

The K^+ channel responds more slowly to a change in voltage, so we have

$$\frac{dn}{dt} = (n_\infty - n)/\tau_n \tag{16.25}$$

This is like Eqs. (16.8a)–(16.8c) only in a slightly different form (Problem 1). This tells us that n, the K^+ channel gating parameter, changes with exponential kinetics, and τ_n is the time constant. So τ_n will depend on the voltage, and we can use the model for barrier crossing transitions in a membrane. With the open probability in Eq. (16.24), inspection of Eq. (7.17) gives us

$$\tau = \frac{\alpha_n}{e^{(V-10)/15} + e^{-(V-10)/15}} \tag{16.26}$$

where α_n is a parameter reflecting the height of the energy barrier of the transition in the absence of a voltage. Here a value of $\alpha_n = 0.05\ ms^{-1}$ is chosen to give oscillation frequencies in the desired range.

Equations (16.22)–(16.26) are all we need to determine how the voltage will vary with time. This set of equations is like the Hodgkin–Huxley equations, only it is simpler because there are only two independent variables. The same numerical method can be used to solve these equations. We calculate the derivatives of n and V with respect to time for an initial condition and then calculate new values of n and V for a tiny time increment. The new n and V values are used to calculate the derivative again and the cycle is repeated for as long a time as necessary to simulate the behavior of interest. The results are shown in Fig. 16.15 for four different stimulus currents.

A subthreshold stimulus current, 35 nA, simply displaces the voltage to a new value (one of the upper traces of Fig. 16.15; rest is $-50\ mV$ and the new steady-state value is $-27\ mV$). A suprathreshold stimulus current of 40 nA sends the system into a regular oscillation, and a larger stimulus current, 80 nA, makes the system oscillate faster. Finally, when the stimulus exceeds a critical value near 115 nA, the oscillation starts but then dies out, settling to a new stable voltage of $\sim 7\ mV$.

There are very powerful mathematical theories for oscillations, so we will study the equations a bit more to touch on some of these ideas. One can set the time derivatives of V and n equal to zero to obtain two conditions of stability. Setting Eq. (16.22) equal to zero and solving for n gives

$$n = \frac{-I_{stim} - G_L(V - V_L) - G_{Ca\text{-}max}m(V)(V - V_{Ca})}{G_{K\text{-}max}(V - V_K)}$$

$$= \frac{-I_{stim} - G_L(V - V_L) - (G_{Ca\text{-}max}/(1 + e^{-V/7.5}))(V - V_{Ca})}{G_{K\text{-}max}(V - V_K)} \tag{16.27}$$

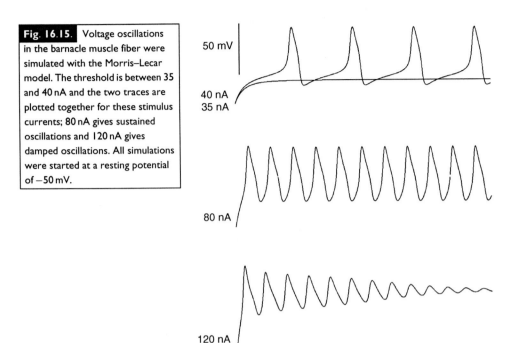

Fig. 16.15. Voltage oscillations in the barnacle muscle fiber were simulated with the Morris–Lecar model. The threshold is between 35 and 40 nA and the two traces are plotted together for these stimulus currents; 80 nA gives sustained oscillations and 120 nA gives damped oscillations. All simulations were started at a resting potential of −50 mV.

where we have taken $G_K = G_{K\text{-max}}n$, $G_{Ca} = G_{Ca\text{-max}}m(V)$, and used Eq. (16.23) for $m(V)$ in the second step. Setting Eq. (16.25) equal to zero gives

$$n = n_\infty = \frac{1}{1 + e^{-(V-10)/7.5}} \tag{16.28}$$

where Eq. (16.24) was used for n_∞.

These two relations between n and V are called *nullclines*. They are plotted in Fig. 16.16. The stimulus current was varied, using some of the same values used to generate the traces in Fig. 16.15. The nullclines for dV/dt (Eq. 16.27) climb upward as I_{stim} is increased. There is only one nullcline drawn for dn/dt because I_{stim} does not affect it.

We can hope to find a stable point where both the nullclines are satisfied. Satisfying just one is clearly not good enough because the other quantity will change. But the intersection of the two null-clines is a place where both variables can remain constant. For $I = 35$ nA the nullclines intersect in three places. The lowest is −27 mV, which is the stable endpoint of the corresponding simulation in Fig. 16.15. The other intersections are not stable even though the time derivatives are both zero. The reason is that when infinitesimal displacements from these points in n–V space make the derivatives nonzero, the signs of these derivatives are such that n and V will always move away from those intersection

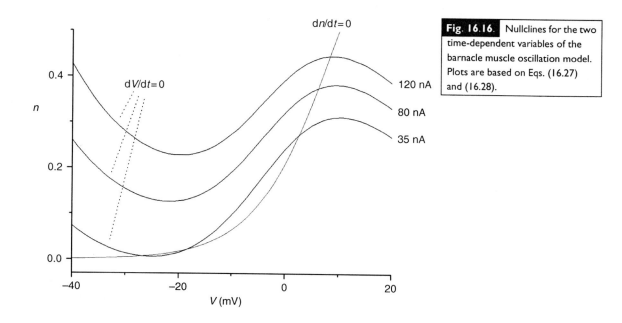

Fig. 16.16. Nullclines for the two time-dependent variables of the barnacle muscle oscillation model. Plots are based on Eqs. (16.27) and (16.28).

points. These points are like a chair balanced on one leg. The slightest disturbance will tip it over.

It is worth mentioning that the threshold voltage for generation of an action potential has the same unstable property. Look at the point in the inset of Fig. 16.8b where the current crosses zero near −62 mV. There is zero current so the voltage will not change. But the smallest perturbation to the left returns the membrane to the resting potential and the smallest perturbation to the right fires an action potential.

Reducing I_{stim} below 35 nA moves the dV/dt nullcline down, so the stable intersection point moves to the left. However, increasing I_{stim} moves the curve to the right and eventually removes this intersection point. For $I_{stim} = 80$ nA there is only one intersection point and it is unstable. The oscillations for this stimulus current reflect the absence of a stable point in n–V space. There is no pair of values of n and V that holds still for this value of I_{stim}. Increasing I_{stim} further moves the one intersection point to the right, and as the two curves intersect at or past the local peak in the dV/dt nullcline, stability becomes possible. The final settling voltage of the 120 nA trace in Fig. 16.15 corresponds with the intersection of the 120 nA nullclines in Fig. 16.16.

The concept of n–V space provides a useful method of visualizing oscillations. Figure 16.17 plots n versus V together with the null-clines. The oscillating voltage for $I_{stim} = 80$ nA settles into a closed loop that cycles endlessly in the counterclockwise direction. The direction is evident from the way the loop is approached from the initial point at −50 mV. Note that the loop encloses the intersection of the two nullclines.

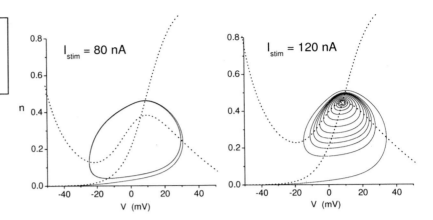

Fig. 16.17. Plots of *n–V* trajectories for two of the traces in Fig. 16.15, together with the corresponding nullclines (dotted curves) from Fig.16.16.

The damped oscillation seen with $I_{stim} = 120$ nA gives an n–V plot that spirals inward to the nullcline intersection point. This relates directly to the damping of the oscillations in the lowest trace of Fig. 16.15.

Much more can be done in the mathematical analysis of oscillations. For any set of parameters controlling the voltage dependence of m, n, and τ_n, one can partition parameter space into regions with qualitatively different dynamic behavior. Morris and Lecar (1981) illustrate this and provide references to additional work in this area.

16.11 | Dendritic integration

The dendrites of a neuron shape its synaptic inputs. This topic commanded quite a bit of attention in Chapter 15, where it was assumed that the cell membrane was passive. Although investigators had serious reservations about this assumption, it was widely used because rigorous tests were difficult. In 1994 Stuart and Sakmann used patch recording to demonstrate directly that dendritic membranes have voltage gated Na$^+$ channels. This changed the way neurophysiologists study dendritic integration. Now the passive theory has been replaced by more realistic models with voltage-gated channels. Passive theory still explains the integration of very small subthreshold inputs, and it still serves as a useful baseline for the development of active theories.

Now we will revisit the analysis of the preceding chapter, and add channels to the membrane to see what changes. Experiments such as those of Stuart and Sakmann showed that while the dendritic membrane contains voltage-gated ion channels, their density is quite low, just a few percent of the density in the squid giant axon membrane. The low density makes it difficult to initiate action potentials by injecting current into the dendrite. In fact, when large currents are injected into the dendrite, the voltage change spreads to the cell body and to the initial segment of the axon. The action potential is then triggered in the axon and propagates back out into the cell body and dendrites.

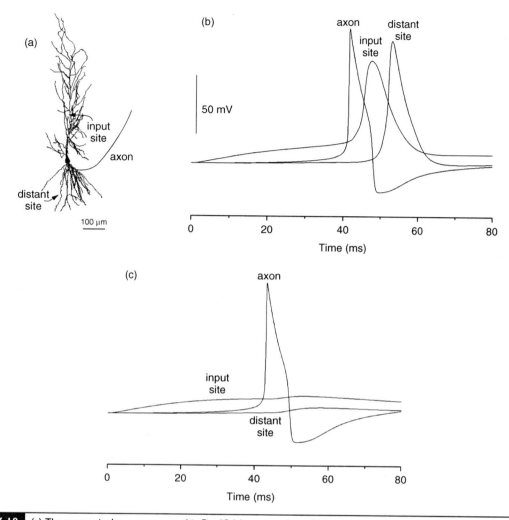

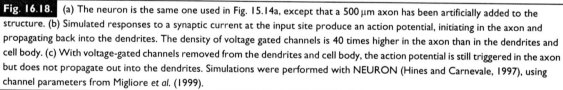

Fig. 16.18. (a) The neuron is the same one used in Fig. 15.14a, except that a 500 μm axon has been artificially added to the structure. (b) Simulated responses to a synaptic current at the input site produce an action potential, initiating in the axon and propagating back into the dendrites. The density of voltage gated channels is 40 times higher in the axon than in the dendrites and cell body. (c) With voltage-gated channels removed from the dendrites and cell body, the action potential is still triggered in the axon but does not propagate out into the dendrites. Simulations were performed with NEURON (Hines and Carnevale, 1997), using channel parameters from Migliore et al. (1999).

The simulations in Fig. 16.18b illustrate these points, using the same neuronal morphology employed in the analysis of passive synaptic integration (Fig. 15.14a), but with an added stretch of axon. These simulations were produced using a compartmental model (Section 15.9) to which voltage-gated ion channels were added (Mainen et al., 1995; Rapp et al., 1996). The dendrites and soma contained a dendritic Na^+ channel and the axon contained Na^+ channels with a lower threshold and slower kinetics at a 40-fold higher density compared to dendritic channels. K^+ channels were present at 30% of the density of Na^+ channels. A very slow synaptic input was applied to a site in the dendritic shaft indicated in Fig. 16.18a. This produced a slow depolarization that spread to the axon where the threshold was low enough for a spike to be triggered. The action potential then spread throughout the dendrite.

Figure 16.18c shows the results of a simulation in the same cell, but with the channels removed from the dendrites. A synaptic input can still generate an action potential in the axon but there is almost no spread into the dendrites.

Simulations such as these defined the biophysical properties that are essential for the replication of the experimentally observed behavior (Mainen et al., 1995; Rapp et al., 1996). The low density of channels in the dendrites and the high density in the axon creates a large disparity in the thresholds in these two regions. If this difference exceeds the attenuation of voltages as they spread from the dendrites to the axon, then the threshold will be surpassed in the axon before it is surpassed in the dendrite (Rapp et al., 1996). The threshold difference is further amplified by differences in the voltage dependence and kinetics of the Na^+ channels in the two regions.

Although the low channel density in the dendrites prevents action potential initiation, there are enough channels to enable action potentials that start elsewhere to propagate into the dendrites. Action potential spread through dendrites is especially important because these voltage changes are sufficient to open voltage-gated Ca^{2+} channels. The Ca^{2+} can then enter to initiate signaling cascades that alter synaptic strength. Such activity-related changes in synaptic function are what make it possible for the brain to learn and to modify an organism's response to outside influences.

This back-propagation of action potentials into dendrites can be regulated by voltage-gated channels. The A-current, just discussed in the context of repetitive activity, is present in dendritic membranes. Recall that the degree of inactivation of these channels varies with the resting voltage. The A-current can also be regulated by phosphorylation. By keeping the membrane below threshold, the A-current can limit how far an action potential will propagate back into the dendrite, reducing the voltage change in remote regions. But if the A-current has been inactivated, simulations suggest that back-propagation can be quite pronounced (Migliore et al., 1999). By varying the Ca^{2+} entry during electrical activity, this mechanism can regulate how much the synapses of a neuron will be modified by conditions and experience.

Problems for Chapter 16

1. Show that Eq. (16.8a) can be expressed equivalently as $dm/dt = (m_\infty - m)/\tau_m$. (Of course, the other equations of this group can be put in this form too.)
2. Why will the threshold by somewhat higher than the second zero-crossing in the inset in Fig. 16.18b?

3. Use Eqs. (16.7a)–(16.7c) and (16.9) to write a complete expression that can be solved (numerically) to determine the resting potential.

4. Plot Eq. 16.11 using an initial voltage of $-75\,mV$ instead of $-65\,mV$ and use this plot to estimate the threshold. How does the value differ from that indicated in Fig. 16.18b? First determine the holding current to add to I_{stim} to obtain $-75\,mV$. The plotted current must be corrected for this holding current.

5. Use any computer language or a mathematical modeling program such as MATLAB, MATHCAD, or MATHEMATICA to write code that integrates the Hodgkin–Huxley system of equations with a stimulus current to generate an action potential (many simulations here used MATHCAD). Do not use a modeling program such as NEURON or GENESIS because these programs already contain the code you should write.

6. For the case where Na^+ channels do not inactivate (Fig. 16.2b, trace c), use the Hodgkin–Huxley equations to calculate the final value to which the voltage settles after the action potential.

7. Consider a squid axon where the Na^+ channels have been completely blocked. A depolarizing current of $20\,mA\;cm^{-2}$ is applied for $50\,ms$. Sketch the voltage response and use the Hodgkin–Huxley equations to estimate voltage at key time points both during the stimulus and after it is turned off.

Appendix 1
Expansions and series

A1.1 | Taylor series

Any function can be approximated in the vicinity of a particular point by a tangent line through that point. This approximation takes the form

$$F(x + \delta x) \approx F(x) + \delta x \frac{dF(x)}{dx} \tag{A1.1}$$

This is illustrated graphically in Fig. A1.1. The deterioration of this approximation with distance is clear. As δx increases the curvature pulls the function away from the line. The approximation can be improved by incorporating the second derivative

$$F(x + \delta x) \approx F(x) + \delta x \frac{dF(x)}{dx} + \frac{\delta x^2}{2} \frac{d^2 F(x)}{dx^2} \tag{A1.2}$$

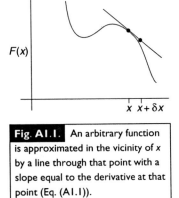

Fig. A1.1. An arbitrary function is approximated in the vicinity of x by a line through that point with a slope equal to the derivative at that point (Eq. (A1.1)).

Now we are using a parabola through the point at x instead of a line. This works better but it too fails as δx grows.

This idea is generalized with the Taylor expansion, which can approximate any continuous function at arbitrary distances from x to any desired degree of accuracy, provided that all the derivatives are defined at x, as follows

$$F(x + \delta x) = F(x) + \delta x \frac{dF(x)}{dx} + \frac{\delta x^2}{2} \frac{d^2 F(x)}{dx^2} + \frac{\delta x^3}{6} \frac{d^3 F(x)}{dx^3} \cdots$$
$$= F(x) + \sum_{n=1}^{\infty} \frac{\delta x^n}{n!} \frac{d^n F(x)}{dx^n} \tag{A1.3}$$

Taylor expansions of functions to first or second order are encountered very often and are extremely useful in theoretical analysis. The expansions of the following functions are especially common. They were calculated directly from Eq. (A1.3) to third order. They are generally valid for small values of x

$$e^x = 1 + x + \frac{x^2}{2} + \frac{x^3}{6} \cdots \tag{A1.4}$$

$$\ln(1 + x) = x - \frac{x^2}{2} + \frac{x^3}{3} \cdots \tag{A1.5}$$

$$\sqrt{1 + x} = 1 + \frac{x}{2} - \frac{x^2}{8} + \frac{3x^3}{48} \cdots \tag{A1.6}$$

AI.2 | The binomial expansion

This takes the form

$$(\alpha + \beta)^N = \sum_{i=0}^{N} \frac{N!}{(N-i)!i!} \alpha^i \beta^{N-i} \tag{A1.7}$$

This is easily checked by trying out small values of N. For arbitrarily large N the combinatoric term $N!/(N-i)!i!$ represents the number of ways of selecting α i times from N factors.

AI.3 | Geometric series

It is easy to check by multiplying out the product that

$$(1-\alpha)(1+\alpha+\alpha^2+\alpha^3 \cdots \alpha^n) = 1 - \alpha^{n+1} \tag{A1.8}$$

The sum of a geometric series is the second factor on the left, so

$$\sum_{i=0}^{n} \alpha^i = \frac{1-\alpha^{n+1}}{1-\alpha} \tag{A1.9}$$

Differentiating this expression with respect to α and then multiplying by α gives

$$\sum_{i=0}^{n} i\alpha^i = \frac{\alpha(1-\alpha^{n+1})}{(1-\alpha)^2} - \frac{(n+1)\alpha^{n+1}}{1-\alpha} \tag{A1.10}$$

For $n \to \infty$ and $0 < \alpha < 1$, we have

$$\sum_{i=0}^{\infty} \alpha^i = \frac{1}{1-\alpha} \tag{A1.11}$$

$$\sum_{i=0}^{\infty} i\alpha^i = \frac{\alpha}{(1-\alpha)^2} \tag{A1.12}$$

from Eq. (A1.9) and Eq. (A1.10), respectively.

Appendix 2
Matrix algebra

A2.1 | Linear transforms

Matrices simplify the mathematical analysis of problems with multiple linear equations and variables. For example, four linear equations with four unknowns take the form

$$a_{11}x_1 + a_{12}x_2 + a_{13}x_3 + a_{14}x_4 = y_1 \tag{A2.1a}$$

$$a_{21}x_1 + a_{22}x_2 + a_{23}x_3 + a_{24}x_4 = y_2 \tag{A2.1b}$$

$$a_{31}x_1 + a_{32}x_2 + a_{33}x_3 + a_{34}x_4 = y_3 \tag{A2.1c}$$

$$a_{41}x_1 + a_{42}x_2 + a_{43}x_3 + a_{44}x_4 = y_4 \tag{A2.1d}$$

Using matrices and vectors, we can compress this into one equation

$$\mathbf{Ax} = \mathbf{y} \tag{A2.2}$$

where $\mathbf{x}$ and $\mathbf{y}$ are vectors of the form (x_1, x_2, x_3, x_4) and (y_1, y_2, y_3, y_4), and $\mathbf{A}$ is a matrix

$$\mathbf{A} = \begin{pmatrix} a_{11} & a_{12} & a_{13} & a_{14} \\ a_{21} & a_{22} & a_{23} & a_{24} \\ a_{31} & a_{32} & a_{33} & a_{34} \\ a_{41} & a_{42} & a_{43} & a_{44} \end{pmatrix} \tag{A2.3}$$

The vectors are expressed in column form so that Eq. (A2.2) is

$$\begin{pmatrix} a_{11} & a_{12} & a_{13} & a_{14} \\ a_{21} & a_{22} & a_{23} & a_{24} \\ a_{31} & a_{32} & a_{33} & a_{34} \\ a_{41} & a_{42} & a_{43} & a_{44} \end{pmatrix} \begin{pmatrix} x_1 \\ x_2 \\ x_3 \\ x_4 \end{pmatrix} = \begin{pmatrix} y_1 \\ y_2 \\ y_3 \\ y_4 \end{pmatrix} \tag{A2.4}$$

where the multiplication of a matrix times a column vector means treating each row of the matrix as a vector and then taking vector dot products of each row with the column vector. This gives each entry of $\mathbf{y}$ as the sums in Eqs. (A2.1a)–(A2.1d).

The compactness of this notation is maintained when we consider a series of successive transformations. After taking $\mathbf{Ax} = \mathbf{y}$, in which $\mathbf{x}$ is transformed to $\mathbf{y}$, we can perform a second transform $\mathbf{By} = \mathbf{z}$. If we kept the notation of Eqs. (A2.1a)–(A2.1d) and wrote out z_1, z_2, z_3, and z_4 in terms of x_1, x_2, x_3, and x_4, the result would be exceedingly complicated. But matrix form makes it simple.

$$\mathbf{BAx} = \mathbf{z} \tag{A2.5}$$

This works because multiplication of matrices follows the rule that each element of the product $\mathbf{BA}$ is given by the vector dot product of the appropriate row of **B** and the appropriate column of **A**. For the element at the intersection of the ith row and jth column of **BA**

$$(\mathbf{BA})_{ij} = \sum_{k=1}^{4} b_{ik} a_{kj} \tag{A2.6}$$

A2.2 | Determinants

The determinant is defined for a square matrix as the sum of all the products that can be formed by multiplying together elements, where each column and each row index is used only once in each product. The sign of each term is negative if an odd number of index swaps makes all elements have the form ii; otherwise the sign is positive. The determinant of **A** is denoted as $|\mathbf{A}|$. In the case of a 2×2 matrix, we have

$$|\mathbf{A}| = \begin{vmatrix} a_{11} & a_{12} \\ a_{21} & a_{22} \end{vmatrix} = a_{11}a_{22} - a_{12}a_{21} \tag{A2.7}$$

For a 3×3 matrix,

$$|\mathbf{A}| = \begin{vmatrix} a_{11} & a_{12} & a_{13} \\ a_{21} & a_{22} & a_{23} \\ a_{31} & a_{32} & a_{33} \end{vmatrix}$$
$$= a_{11}a_{22}a_{33} - a_{11}a_{32}a_{23} - a_{21}a_{12}a_{33} + a_{21}a_{13}a_{32} + a_{31}a_{12}a_{23} - a_{31}a_{22}a_{13}$$
$$\tag{A2.8}$$

If one row (or column) can be formed by a linear combination of the other rows (or columns) then the determinant is zero. Thus, the condition of a nonzero determinant tells us whether the rows (and columns) are linearly independent. For example, consider a matrix **A** with the fourth column all zeroes. Now, make a new matrix **A'**, and replace the zeroes in the fourth column with **y** obtained as **Ax**. The determinant of **A'** would be zero because the fourth column is a linear combination of the other three. This makes the fourth column redundant. It can be absorbed into the other three columns and the transformation can be expressed with a matrix consisting of only three columns.

If there is redundancy in a matrix then the corresponding linear equations are also redundant. That means that when we transformed the four elements of **x** to four elements in **y**, one of the four elements of **x** was unnecessary, and we could have calculated the four elements of **y** with just three xs. That means we cannot use **y** to recover the four values in **x** because that would entail solving for four unknowns with only three equations.

If the matrix is not redundant, then we can return from **y** back to **x**. This reverse transformation defines the inverse matrix.

$$\mathbf{A}^{-1}\mathbf{y} = \mathbf{x} \tag{A2.9}$$

This has the same basic form as Eq. (A2.2), and $\mathbf{A}^{-1}$ is a square matrix of the same dimensions as $\mathbf{A}$. A matrix can be inverted if and only if $|\mathbf{A}| \neq 0$.

If we multiply Eq. (A2.9) on the right by $\mathbf{A}$

$$\mathbf{A}\mathbf{A}^{-1}\mathbf{y} = \mathbf{A}\mathbf{x} \tag{A2.10}$$

The product $\mathbf{A}\mathbf{A}^{-1} = \mathbf{I}$, where $\mathbf{I}$ is the identity matrix (such that $\mathbf{I}\mathbf{x} = \mathbf{x}$). The matrix $\mathbf{I}$ has ones along the diagonal, and zeros everywhere else

$$\mathbf{I} = \begin{pmatrix} 1 & 0 & 0 & 0 \\ 0 & 1 & 0 & 0 \\ 0 & 0 & 1 & 0 \\ 0 & 0 & 0 & 1 \end{pmatrix} \tag{A2.11}$$

An important general theorem for matrices with nonzero determinants is that the determinant of a product is equal to the product of the determinants.

$$|\mathbf{AB}| = |\mathbf{A}||\mathbf{B}| \tag{A2.12}$$

For 2×2 matrices

$$|\mathbf{A}||\mathbf{B}| = \begin{vmatrix} a_{11} & a_{12} \\ a_{21} & a_{22} \end{vmatrix} \begin{vmatrix} b_{11} & b_{12} \\ b_{21} & b_{22} \end{vmatrix} = (a_{11}a_{22} - a_{12}a_{21})(b_{11}b_{22} - b_{12}b_{21})$$
$$= a_{11}a_{22}b_{11}b_{22} - a_{11}a_{22}b_{12}b_{21} - a_{12}a_{21}b_{11}b_{22} + a_{12}a_{21}b_{12}b_{21} \tag{A2.13}$$

and

$$|\mathbf{AB}| = \begin{vmatrix} a_{11}b_{11} + a_{12}b_{21} & a_{11}b_{12} + a_{12}b_{22} \\ a_{21}b_{11} + a_{22}b_{21} & a_{21}b_{12} + a_{22}b_{22} \end{vmatrix}$$
$$= (a_{11}b_{11} + a_{12}b_{21})(a_{21}b_{12} + a_{22}b_{22}) - (a_{11}b_{12} + a_{12}b_{22})$$
$$\times (a_{21}b_{11} + a_{22}b_{21})$$
$$= a_{11}b_{11}a_{21}b_{12} + a_{11}b_{11}a_{22}b_{22} + a_{12}b_{21}a_{21}b_{12} + a_{12}b_{21}a_{22}b_{22}$$
$$- a_{11}b_{12}a_{21}b_{11} - a_{11}b_{12}a_{22}b_{21} - a_{12}b_{22}a_{21}b_{11} - a_{12}b_{22}a_{22}b_{21}$$
$$= a_{11}b_{11}a_{22}b_{22} + a_{12}b_{21}a_{21}b_{12} - a_{11}b_{12}a_{22}b_{21} - a_{12}b_{22}a_{21}b_{11} \tag{A2.14}$$

The final results of Eqs. (A2.13) and (A2.14) are equal. Equation (A2.12) can be proven for square matrices of any dimension.

A2.3 | Eigenvalues, eigenvectors, and diagonalization

Consider an equation where a matrix times a vector equals a scalar times the same vector

$$\mathbf{A}\mathbf{x} = \lambda\mathbf{x} \tag{A2.15}$$

In this equation λ is known as an *eigenvalue* (or characteristic value) of matrix $\mathbf{A}$, and $\mathbf{x}$ is known as an *eigenvector* (or characteristic vector) of matrix $\mathbf{A}$. This kind of equation appears quite often so we will outline how to solve for the eigenvalues, and explore some properties of the eigenvectors.

Since $\lambda\mathbf{x} = \lambda\mathbf{I}\mathbf{x}$, we can rearrange Eq. (A2.15) to give

$$(\mathbf{A} - \lambda\mathbf{I})\mathbf{x} = 0 \tag{A2.16}$$

Because this is equal to zero, we could express one of the columns of the matrix $\mathbf{A} - \lambda\mathbf{I}$ as minus the sum of the others.[1] So the columns of this matrix are not linearly independent, and its determinant must be zero

$$|\mathbf{A} - \lambda\mathbf{I}| = 0 \tag{A2.17}$$

This is called the *characteristic equation* of $\mathbf{A}$. It is a polynomial in λ of order equal to the dimension, n, of the matrix. This can be seen using the above comments on determinants. A product of n elements taken from different rows and columns can have up to n diagonal elements. The determinant includes the product of the diagonal elements $\prod_{i=1}^{n}(a_{ii} - \lambda)$, and this is an nth order polynomial in λ. Other terms of the determinant will be lower order polynomials in λ. There will also be a term without λ (it equals $|\mathbf{A}|$). So the determinant will be a sum of powers of λ with the highest order term being λ^n.

The eigenvalues of a matrix can thus be determined as the roots of Eq. (A2.17). An nth order polynomial has n roots, so an $n \times n$ matrix has n eigenvalues. Sometimes, some of the eigenvalues are identical, and they are called degenerate. Eigenvalues can also be complex.

Each eigenvalue is assumed to have an associated eigenvector which satisfies Eq. (A2.15). We give the eigenvalues and eigenvectors indices to keep track of them. So there are then n equations like Eq. (A2.15)

$$\mathbf{A}\mathbf{x}_i = \lambda_i\mathbf{x}_i \tag{A2.18}$$

Now construct an $n \times n$ matrix, the columns of which are the eigenvectors

$$\mathbf{X} = (\mathbf{x}_1, \mathbf{x}_2, \ldots \mathbf{x}_n) \tag{A2.19}$$

The transpose of this matrix, defined by switching the rows into columns or, equivalently, the columns into rows, can be written as

$$\mathbf{X}^t = \begin{pmatrix} \mathbf{x}_1^t \\ \mathbf{x}_2^t \\ \cdot \\ \cdot \\ \cdot \\ \mathbf{x}_n^t \end{pmatrix} \tag{A2.20}$$

where each vector $\mathbf{x}_i^t$ is $\mathbf{x}_i$ transposed from a column to a row.

---

[1] Note that $\mathbf{x} = 0$ also solves Eq. (A2.14), but this result is of no interest.

With no loss of generality, we can stipulate that each $\mathbf{x}$ has a length of one in an n-dimensional space; $\mathbf{x}_i^2 = 1$ (an eigenvector scaled by a factor still satisfies Eq. (A2.15)). Furthermore, it can be shown that the $\mathbf{x}_i$ form an orthogonal set of vectors such that the product of any two is zero; $\mathbf{x}_i^t \mathbf{x}_j = 0$, if $i \neq j$. Note that multiplying a row vector by a column vector follows the convention for matrix multiplication but is equivalent to taking a vector dot product. Because of the way the products of the vectors work out to zero or one, $\mathbf{X}^t$ and $\mathbf{X}$ are inverses; $\mathbf{X}^t \mathbf{X} = \mathbf{I}$.

Multiplying $\mathbf{A}$ by $\mathbf{X}$ and using Eq. (A2.18) gives

$$\mathbf{AX} = (\lambda_1 \mathbf{x}_1, \lambda_2 \mathbf{x}_2, \ldots \lambda_\nu \mathbf{x}_n) \tag{A2.21}$$

Multiplying this equation on the left by $\mathbf{X}^t$ gives a matrix on the right-hand side of the equation for which each element has the form $\mathbf{x}_i^t \lambda_j \mathbf{x}_j$. The off-diagonal elements on the right-hand side are all zero because $\mathbf{x}_i^t \mathbf{x}_j = 0$, and the diagonal elements are the eigenvalues, because $\mathbf{x}_i^2 = 1$. So we have

$$\mathbf{X}^t \mathbf{AX} = \Lambda \tag{A2.22}$$

where Λ is a diagonal matrix with $\Lambda_{ii} = \lambda_i$.

The operation shown in Eq. (A2.22) of multiplying on the left and right by matrices to obtain a matrix with the eigenvalues of the original matrix along the diagonal and zeroes everywhere else is called *diagonalization*. Recall that $\mathbf{X}^t \mathbf{X} = \mathbf{I}$. Because of the product rule (Eq. (A2.12)), that means that diagonalizing a matrix does not change the determinant. The determinant of Λ is just the product of its elements. since it is the only nonzero product that can be formed from elements from different rows and columns. Since the determinant of $\mathbf{A}$ and Λ are the same, we have

$$|\mathbf{A}| = |\Lambda| = \prod_{i=1}^{n} \lambda_i \tag{A2.23}$$

Thus, once the eigenvalues of a matrix have been found, the determinant is readily calculated.

Appendix 3
Fourier analysis

Fourier analysis enables one to express functions as sums of the trigonometric sine and cosine functions. We will first show how this works with an example. Take the periodic function that flips between $+1$ and -1 at regular intervals of π (Fig. A3.1). This function is perfectly represented by the sum

$$F(x) = \frac{4}{\pi}\left(\sum_{i=1}^{\infty} \frac{\sin((2i-1)x)}{2i-1}\right) \tag{A3.1}$$

The first four terms of this series are plotted on the left in Fig. A3.2 and the sums produced by successive addition of each term are plotted on the right. With the addition of each higher frequency term the sum looks more and more like the function plotted in Fig. A3.1.

Equation (A3.1) is an example of a Fourier series. In general, we can express any function in the interval $[-\pi, \pi]$ as a sum of the form

$$F(x) = \frac{A_0}{2} + \sum_{i=1}^{\infty}(A_i\sin(ix) + B_i\cos(ix)) \tag{A3.2}$$

To put this representation to use we need to know the values for A_i and B_i. These are derived by multiplying Eq. (A3.2) by $\sin(jx)$ or $\cos(jx)$ and integrating over the interval (where j is a positive integer). The right-hand side then becomes a sum of integrals of the form

$$\int_{-\pi}^{\pi} \sin(jx)\sin(ix)dx = 0 \text{ for } i \neq j$$

$$= \pi \text{ for } i = j \tag{A3.3a}$$

$$\int_{-\pi}^{\pi} \cos(jx)\cos(ix)dx = 0 \text{ for } i \neq j$$

$$= \pi \text{ for } i = j \tag{A3.3b}$$

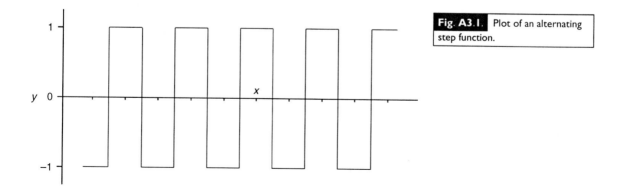

Fig. A3.1. Plot of an alternating step function.

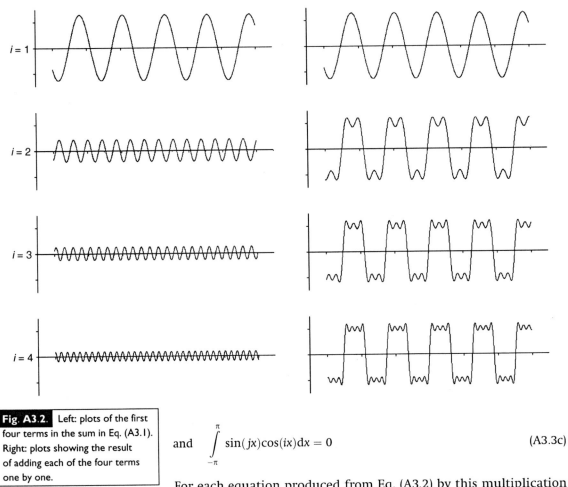

Fig. A3.2. Left: plots of the first four terms in the sum in Eq. (A3.1). Right: plots showing the result of adding each of the four terms one by one.

and
$$\int_{-\pi}^{\pi} \sin(jx)\cos(ix)dx = 0 \qquad (A3.3c)$$

For each equation produced from Eq. (A3.2) by this multiplication and integration, only one of the terms is nonzero. For the equations produced by multiplying Eq. (A3.2) by $\sin(jx)$, the only nonzero term gives

$$A_j = \frac{1}{\pi} \int_{-\pi}^{\pi} F(x)\sin(jx)dx \qquad (A3.4)$$

For the equations produced by multiplying by $\cos(jx)$, we have

$$B_j = \frac{1}{\pi} \int_{-\pi}^{\pi} F(x)\cos(jx)dx \qquad (A3.5)$$

To obtain A_0, integrate Eq. (A3.2) as is. The integrals of $\sin(ix)$ and $\cos(ix)$ in this interval are all zero, so we are left with the integral of A_0 from $-\pi$ to π. This gives 2π, so

$$A_0 = \frac{1}{\pi} \int_{-\pi}^{\pi} F(x)dx \qquad (A3.6)$$

Equations (A3.4)–(A3.6) provide a general method for deriving explicit forms for the coefficients in Eq. (A3.2). Virtually any function can be expressed in this way. Fourier coefficients for a large number of commonly encountered functions have been derived, and the results can be found in mathematical tables.

It is worth noting that it is usually easy to see when a Fourier series should consist of either sines only or cosines only. If $F(x)$ is even, with $F(x) = F(-x)$, then it will be constructed from cosines, which are also even. If $F(x)$ is odd, with $F(x) = -F(-x)$, then it will be constructed from sines, which are odd.

Now look at the closely related Fourier transform. Instead of using a sum of trigonometric functions, we can go to the limit of infinitesimal spacing between the frequencies of the sine and cosine waves. This leads to the representation of our function as an integral

$$F(x) = \int_0^\infty A(s)\sin(sx)ds + \int_0^\infty B(s)\cos(sx)ds \tag{A3.7}$$

This is like the sums in Eq. (A3.2), but the continuous variable s has replaced our summation index i.

For Fourier transforms it is common to take advantage of Euler's formula

$$e^{ix} = \cos(x) + i\sin(x) \tag{A3.8}$$

Now we express our function $F(x)$ as an integral

$$F(x) = \int_{-\infty}^\infty G(s)e^{ixs}ds \tag{A3.9}$$

The cosine integral in Eq. (A3.7) is the real part of this expression, except that the limits of integration have been extended to $\pm\infty$. The $-\infty$ limit facilitates the analysis without abandoning the spirit of adding together wave-like trigonometric functions.

An expression for $G(s)$ is derived by multiplying by $e^{-ixs'}$ and integrating over x

$$\int_{-\infty}^\infty F(x)e^{-ixs'}dx = \int_{-\infty}^\infty \int_{-\infty}^\infty G(s)e^{ixs}e^{-ixs'}dsdx$$

$$= \int_{-\infty}^\infty G(s)\left(\int_{-\infty}^\infty e^{ix(s-s')}dx\right)ds \tag{A3.9}$$

The second integral in brackets over x is $1/2\pi$ times a well-known representation of the delta function, $\delta(s - s')$.

$$\int_{-\infty}^\infty F(x)e^{-ixs'}dx = \frac{1}{2\pi}\int_{-\infty}^\infty G(s)\delta(s - s')ds \tag{A3.10}$$

Integrating over s on the right-hand side pulls out the value of $G(s)$ for which $s = s'$

$$\int_{-\infty}^{\infty} F(x)e^{-ixs'}\,dx = \frac{1}{2\pi}G(s')$$

(A3.11)

This is the inverse Fourier transform. Note the symmetry between this expression and the Fourier transform (Eq. (A3.9)). We can thus go back and forth between representations in terms of the two continuous variables x and s. In practice, these two variables are often time and frequency.

Fourier transforms are particularly useful in solving differential equations. This is because of the simple results obtained when taking the Fourier transformation of a derivative. Use the notation $\phi(G)$ to denote the Fourier transform of G. In this notation, the Fourier transform of the derivative of $G(s)$ is

$$\phi\left(\frac{dG}{ds}\right) = \int_{-\infty}^{\infty} \frac{dG}{ds}e^{ixs}\,ds$$

(A3.12)

Integrating by parts gives

$$\phi\left(\frac{dG}{ds}\right) = G(s)e^{ixs}\Big|_{-\infty}^{\infty} - ix\int_{-\infty}^{\infty} G(s)e^{ixs}\,ds$$

(A3.13)

One often knows that the value of $G(s)$ at $\pm\infty$ is zero, so

$$\phi\left(\frac{dG}{ds}\right) = -ix\int_{-\infty}^{\infty} G(s)e^{ixs}\,ds = -ix\phi(G)$$

(A3.14)

In words, the Fourier transform of the derivative of a function is $-ix$ times the Fourier transform of the function. It is easy to extend this to higher order derivatives. Multiplying by $-ix$ again gives

$$\phi\left(\frac{d^2G}{ds^2}\right) = -x^2\phi(G)$$

(A3.15)

Appendix 4
Gaussian integrals

The integral of a Gaussian function cannot be evaluated for arbitrary limits

$$\vartheta = \int_a^b e^{-\alpha x^2} x \tag{A4.1}$$

However, it can be evaluated for the limits $-\infty$ to ∞. This is accomplished by taking the product of two such integrals

$$\vartheta^2 = \int_{-\infty}^{\infty} e^{-\alpha x^2} dx \int_{-\infty}^{\infty} e^{-\alpha y^2} dy = \int_{-\infty}^{\infty} \int_{-\infty}^{\infty} e^{-\alpha x^2} e^{-\alpha y^2} dxdy \tag{A4.2}$$

We now transform to polar coordinates, with $x^2 + y^2 = r^2$, $dxdy = rdrd\theta$

$$\vartheta^2 = \int_{\theta=0}^{2\pi} \int_{r=0}^{\infty} e^{-\alpha r^2} rdrd\theta \tag{A4.3}$$

The integral over θ is simply 2π. The integral over r is solved by transforming to $u = r^2$ to give $1/2\alpha$. So $\vartheta^2 = \pi/\alpha$. Taking the square root gives

$$\vartheta = \int_{-\infty}^{\infty} e^{-\alpha x^2} dx = \sqrt{\frac{\pi}{\alpha}} \tag{A4.4}$$

Of course, the integral from $x = 0$ to ∞ is just half this because the integrand is symmetrical around $x = 0$.

Additional useful integrals can be generated by differentiation of Eq. (A4.4) with respect to α. Differentiating once gives the first of these

$$\int_{-\infty}^{\infty} x^2 e^{-\alpha x^2} dx = \frac{\sqrt{\pi}}{2\alpha^{3/2}} \tag{A4.5}$$

This can be continued to get the integral of any even power of x times a Gaussian function.

Integrals of the form $\int_{-\infty}^{\infty} e^{-\alpha x^2 - \beta x} dx$ are often encountered. They are converted to the form of Eq. (A4.4) by multiplying by $1 = e^{-\beta 2/4\alpha} e^{\beta 2/4\alpha}$. This completes the square in the exponent

$$\int_{-\infty}^{\infty} e^{-\alpha x^2 - \beta x} dx = e^{\beta^2/4\alpha} \int_{-\infty}^{\infty} e^{-\alpha(x^2 + \beta x/\alpha + \beta^2/4\alpha^2)} dx$$

$$= e^{\beta^2/4\alpha} \int_{-\infty}^{\infty} e^{-\alpha(x + \beta/2\alpha)^2} dx \tag{A4.6}$$

Now with a new variable $u = x + \beta/2\alpha$ the integral is solved

$$\int_{-\infty}^{\infty} e^{-\alpha x^2 - \beta x}\,dx = e^{\beta^2/4\alpha} \int_{-\infty}^{\infty} e^{-\alpha u^2}\,du = e^{\beta^2/4\alpha}\sqrt{\frac{\pi}{\alpha}} \tag{A4.7}$$

Appendix 5
Hyperbolic functions

The hyperbolic cosine is defined as

$$\cosh(x) = \frac{e^x + e^{-x}}{2} \qquad (A5.1)$$

The hyperbolic sine is

$$\sinh(x) = \frac{e^x - e^{-x}}{2} \qquad (A5.2)$$

It is easy to see that the two are interconverted by differentiation. The hyperbolic tangent is defined as the ratio of the two, just as the trigonometric tangent is defined as the ratio of the sine to the cosine

$$\tanh(x) = \frac{\sinh(x)}{\cosh(x)} = \frac{e^x - e^{-x}}{e^x + e^{-x}} \qquad (A5.3)$$

In contrast to trigonometric functions hyperbolic functions can be inverted. Multiplying Eq. (A5.1) by e^x and rearranging gives

$$e^{2x} - 2\cosh(x)e^x + 1 = 0 \qquad (A5.4)$$

This is a quadratic equation in e^x, so the solution is

$$e^x = \cosh(x) \pm \sqrt{\cosh^2(x) - 1} \qquad (A5.5)$$

$$\text{and} \quad x = \ln(\cosh(x) \pm \sqrt{\cosh^2(x) - 1}) \qquad (A5.6)$$

Thus, we have inverted the hyberbolic cosine. Replacing $\cosh(x)$ by u gives

$$x = \cosh^{-1}(u) = \ln(u \pm \sqrt{u^2 - 1}) \qquad (A5.7)$$

Likewise,

$$\sinh^{-1}(u) = \ln(u \pm \sqrt{u^2 + 1}) \qquad (A5.8)$$

Appendix 6
Polar and spherical coordinates

In many problems everything revolves around a central point and the key variable is the distance, r, to this center. In these cases one uses polar or spherical coordinates, and writes an equation as a function of r. The standard operations in calculus must then be modified. When everything lies in one plane, then we go from x and y to the polar coordinates r and θ. In a three-dimensional space, we go from x, y, and z to spherical coordinates r, θ, and ϕ (Fig. A6.1).

If a function $f(x, y)$ is transformed to polar coordinates, it is expressed as $f(r, \theta)$. A function $f(x, y, z)$ transformed to spherical coordinates becomes $f(r, \theta, \phi)$. This book deals with isotropic situations where the function depends only on r and does not vary with the angles. So we are left with functions of r only, $f(r)$.

To integrate the function over all space we must transform the differential to the new coordinate system. In polar coordinates, dxdy becomes $r dr d\theta$, so we can write

$$\int_{x=-\infty}^{\infty} \int_{y=-\infty}^{\infty} f(x,y)dxdy = \int_{r=0}^{\infty} \int_{\theta=0}^{2\pi} f(r,\theta)rdrd\theta \qquad (A6.1)$$

When f depends only on r, the integral over θ can be performed to give

$$\int_{r=0}^{\infty} \int_{\theta=0}^{2\pi} f(r,\theta)rdrd\theta = \int_{r=0}^{\infty} f(r)2\pi rdr \qquad (A6.2)$$

Note that the factor $2\pi r dr$ is the area of a circular shell of radius r and thickness dr.

In spherical coordinates the differential dxdydz becomes $r^2\sin\theta\, drd\theta d\phi$, so the integral is

$$\int_{x=-\infty}^{\infty} \int_{y=-\infty}^{\infty} \int_{z=-\infty}^{\infty} f(x,y,z)dxdydz = \int_{r=0}^{\infty} \int_{\theta=0}^{\pi} \int_{\phi=0}^{2\pi} f(r,\theta)\sin\theta\, r^2 drd\theta d\phi$$

$$(A6.3)$$

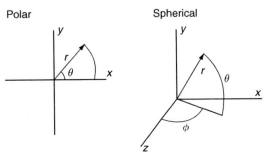

Fig. A6.1. Polar and spherical coordinate systems are shown on a cartesian background.

Polar

Spherical

Again in an isotropic situation we can integrate over the angles

$$\int_{r=0}^{\infty}\int_{\theta=0}^{\pi}\int_{\phi=0}^{2\pi} f(r,\theta,\phi)\sin\theta\, r^2 dr d\theta d\phi = \int_{r=0}^{\infty} f(r)4\pi r^2 dr \qquad (A6.4)$$

Now note that the factor $4\pi r^2 dr$ is the volume of a spherical shell of radius r and thickness dr.

A coordinate transformation also has an impact on a derivative. Of particular importance is the Laplacian differential operator, which appears in many equations of this book, including the diffusion equation, the Poisson and Poisson–Boltzmann equations, and the cable equation. In spherical coordinates the full Laplacian is a complicated expression, but in an isotropic system the derivatives with respect to θ and ϕ are zero so we have

$$\nabla^2 f = \frac{\partial^2 f}{\partial x^2} + \frac{\partial^2 f}{\partial y^2} + \frac{\partial^2 f}{\partial z^2} = \frac{1}{r^2}\frac{\partial}{\partial r}\left(r^2\frac{\partial f}{\partial r}\right) \qquad (A6.5)$$

An alternative version is

$$\nabla^2 f = \frac{1}{r}\frac{\partial^2 (rf)}{\partial r^2} \qquad (A6.6)$$

The two expressions can be shown to be equivalent by applying the product rule to the derivatives.

References

Abbott, A. J. and Nelsestuen, G. L. (1988). The collisional limit: an important consideration for membrane-associated enzymes and receptors. *Faseb J.*, **2**, 2858–2866.

Accardi, A. and Miller, C. (2004). Secondary active transport mediated by a prokaryotic homologue of ClC Cl⁻ channels. *Nature*, **427**, 803–807.

Adam, G. and Delbrück, M. (1968). Reduction of dimensionality in biological diffusion processes. In *Structural Chemistry and Molecular Biology*, ed. A. Rich and N. Davidson. San Francisco: Freeman, pp. 198–215.

Adams, D. J., Dwyer, T. M. and Hille, B. (1980). The permeability of endplate channels to monovalent and divalent metal cations. *J. Gen. Physiol.*, **75**, 493–510.

Adams, S. A. and DeFelice, L. J. (2002). Flux coupling in the human serotonin transporter. *Biophys. J.*, **83**, 3268–3282.

Ahern, C. A. and Horn, R. (2004). Stirring up controversy with a voltage sensor paddle. *TINS*, **27**, 303–307.

Aicken, C. C. (1990). Chloride transport across the sarcolemma of vertebrate smooth and skeletal muscle. In *Chloride Channels and Carriers in Nerve, Muscle, and Glial Cells*, ed. F. J. Alvarez-Leefmans and J. M. Russell. New York: Plenum Press, pp. 209–249.

Aidley, D. J. (1978). *The Physiology of Excitable Cells*. Cambridge: Cambridge University Press.

Alber, T., Dao-pin, S., Wilson, K., Wozniak, J. A., Cook, S. P. and Matthews, B. W. (1987). Contributions of hydrogen bonds of Thr 157 to the thermodynamic stability of phage T4 lysozyme. *Nature*, **330**, 41–46.

Albery, W. J. and Knowles, J. R. (1976). Evolution of enzyme function and the development of catalytic efficiency. *Biochemistry*, **15**, 5631–5640.

Aldrich, R. W., Getting, P. A. and Thompson, S. H. (1979). Mechanism of frequency-dependent broadening of molluscan neurone soma spikes. *J. Physiol.*, **291**, 531–544.

Allen, T. W., Andersen, O. S. and Roux, B. (2004). Energetics of ion conduction through the gramicidin channel. *Proc. Natl Acad. Sci.*, **101**, 117–122.

Almers, W. and McCleskey, E. W. (1984). The non-selective conductance in calcium channels of frog muscle: calcium selectivity in a single-file pore. *J. Physiol.*, **353**, 585–608.

Andersen, O. S. (1983). Ion movement through gramicidin A channels. Single-channel measurements at very high potentials. *Biophys. J.*, **41**, 119–133.

Anderson, C. R. and Stevens, C. F. (1973). Voltage-clamp analysis of acetylcholine produced end plate current fluctuations at frog neuromuscular junction. *J. Physiol.*, **235**, 655–691.

Antzelevitch, C. (2001). Basic mechanisms of reentrant arrhythmias. *Curr. Opin. Cardiol.*, **16**, 1–7.

Åqvist, J. and Luzhkov, V. (2000). Ion permeation mechanism of the potassium channel. *Nature*, **404**, 881–884.

Ashcroft, F. M. (2000). *Ion Channels and Disease*. San Diego: Academic Press.

Aurora, R., Creamer, T. P., Srinivasan, R. and Rose, G. D. (1997). Local interactions in protein folding: lessons from the α-helix. *J. Biol. Chem.*, **272**, 1413–1416.

Aveyard, R. and Haydon, D. A. (1973). *An Introduction to the Principles of Surface Chemistry*. Cambridge: Cambridge University Press, p. 231.

Axe, D. D., Foster, N. W. and Fersht, A. R. (1996). Active barnase variants with completely random hydrophobic cores. *Proc. Natl Acad. Sci.*, **93**, 5590–5594.

Baldwin, R. L. (1996). How Hofmeister ion interactions affect protein stability. *Biophys. J.*, **71**, 2056–2063.

Barrett, J. N. and Crill, W. E. (1974). Specific membrane properties of cat motoneurons. *J. Physiol.*, **239**, 301–324.

Bashford, C. L. and Pasternak, C. A. (1986). Plasma membrane potential of some animal cells is generated by ion pumping, not by gradients. *Trends Biochem. Sci.*, **11**, 113–116.

Bass, R. B., Strop, P., Barclay, M. and Rees, D. C. (2002). Crystal structure of *Escherichia coli* MscS, a voltage-modulated and mechanosensitive channel. *Science*, **298**, 1582–1587.

Bedzek, M. J., Bommarito, G. M., Caffrey, M. and Penner, T. L. (1990). Diffuse-double layer at a membrane-aqueous interface measured with X-ray standing waves. *Science*, **248**, 52–56.

Beece, D., Eisenstein, L., Frauenfelder, H. *et al.* (1980). Solvent viscosity and protein dynamics. *Biochemistry*, **19**, 5147–5157.

Ben-Shaul, A., Ben-Tal, N. and Honig, B. (1996). Statistical thermodynamic analysis of peptide and protein insertion into lipid membranes. *Biophys. J.*, **71**, 130–137.

Benedek, G. B. and Villars, F. M. H. (2000). *Physics with Illustrative Examples from Medicine and Biology*. New York: Springer-Verlag.

Benz, R. and Läuger, P. (1977). Transport kinetics of dipicrylamine through lipid bilayer membranes. *Biochem. Biophys. Acta*, **468**, 245–258.

Berezin, I. V., Kazanskaya, N. F. and Klyosov, A. A. (1971). Determination of the individual rate constants of α-chymotrypsin-catalyzed hydrolysis with the added nucleophilic agent 1,4-butanediol. *FEBS Lett.*, **15**, 121–124.

Berg, H. C. (1983). *Random Walks in Biology*. Princeton: Princeton University Press. [Cited in Chapters 6 and 8. This is an accessible and clear introduction to Brownian motion and diffusion with good biological examples.]

Berg, H. C. and Purcell, E. M. (1977). Physics of chemoreception. *Biophys. J.*, **20**, 193–219. [Cited in Chapter 8. A seminal paper that introduces many important concepts in diffusion-limited encounters.]

Berg, O. G. and von Hippel, P. H. (1985). Diffusion-controlled macromolecular interactions. *Ann. Rev. Biophys. Biophys. Chem.*, **14**, 131–160.

Berniche, S. and Roux, B. (2000). Molecular dynamics of the KcsA K^+ channel in a bilayer membrane. *Biophys. J.*, **78**, 2900–2917.

Betz, S. F., Bryson, J. W. and DeGrado, W. F. (1995). Native-like and structurally characterized designed α-helical bundles. *Curr. Opin. Struct. Biol.*, **5**, 457–463.

Bezanilla, F. (2000). The voltage sensor in voltage-dependent ion channels. *Physiol. Rev.*, **80**, 555–592.

Blacklow, S. C., Raines, R. T., Lim, W. A., Zamore, P. D. and Knowles, J. R. (1988). Triosephosphate isomerase catalysis is diffusion controlled. *Biochemistry*, **27**, 1158–1167.

Blangy, D., Buc, H. and Monod, J. (1968). Kinetics of the allosteric interactions of phosphofructokinase from *Escherichia coli*. *J. Mol. Biol.*, **31**, 13–35. [Cited in Chapter 5. This provides an example of how the MWC theory can synthesize a broad range of observations.]

Bloomfield, S. A., Hamos, J. E. and Sherman, S. M. (1987). Passive cable properties and morphological correlates of neurones in the lateral geniculate nucleus of the cat. *J. Physiol.*, **383**, 653–692.

Bloomfield, V. A., Crothers, D. M. and Tinoco, I. (1974). *The Physical Chemistry of Nucleic Acids*. New York: Harper and Row, Publishers, Inc.

Boyle, P. J. and Conway, E. J. (1941). Potassium accumulation in muscle and associated changes. *J. Physiol.*, **100**, 1–63.

Breslow, E. and Gurd, F. R. N. (1962). Reactivity of sperm whale metmyoglobin towards hydrogen ions and p-nitrophenyl acetate. *J. Biol. Chem.*, **237**, 371–381.

Brooks, B. R., Bruccoleri, R. E., Olafson, B. D., States, D. J., Swaminathan, S. and Karplus, M. (1983). CHARMM: A program for macromolecular energy, minimization, and dynamics calculations. *J. Comput. Chem.*, **4**, 187–217.

Brouwer, A. C. and Kirsch, J. F. (1982). Investigation of diffusion-limited rates of chymotrypsin reactions by viscosity variation. *Biochemistry*, **21**, 1302–1307.

Bruice, T. C. (1970). Proximity effects and enzyme catalysis. In *The Enzymes II*, ed. P. D. Boyer. New York: Academic Press, pp. 217–279.

Bryant, R. G. (1996). The dynamics of water–protein interactions. *Ann. Rev. Biophys. Biomol. Struct.*, **25**, 29–53.

Cai, M. and Jordan, P. C. (1990). How does vestibule surface charge affect ion conduction and toxin binding in a sodium channel. *Biophys. J.*, **57**, 883–891.

Camacho, C. J. and Thirumalai, D. (1993). Minimum energy compact structures of random sequences of heteropolymers. *Phys. Rev. Lett.*, **71**, 2505–2508.

Cantor, P. R. and Schimmel, P. R. (1980). *Biophysical Chemistry*. San Francisco: W. H. Freeman and Co. [Cited in Chapter 3. This thorough three-volume set has excellent chapters on conformational statistics of polymers and the helix–coil transition.]

Cardinale, G. J. and Abeles, R. H. (1968). Purification and mechanism of action of proline racemase. *Biochemistry*, **7**, 3970–3978.

Carra, J. H., Murphy, E. C. and Privalov, P. L. (1996). Thermodynamic effects of mutations on the denaturation of T4 lysozyme. *Biophys. J.*, **71**, 1994–2001.

Carslaw, H. S. and Jaeger, J. C. (1959). *Conduction of Heat in Solids*. Oxford: Oxford University Press.

Cassidy, C. S., Lin, J. and Frey, P. A. (1997). A new concept for the mechanism of action of chymotrypsin: the role of the low-barrier hydrogen bond. *Biochemistry*, **36**, 4576–4584.

Caterall, W. A., Chandy, G. K. and Gutman, G. A. (eds.) (2002). *The IUPHAR Compendium of Voltage-Gated Ion Channels*. Leeds: IUPHAR Media.

Chan, H. S. and Dill, K. A. (1991). Polymer principles in protein structure and stability. *Ann. Rev. Biophys. Biophys. Chem.*, **20**, 447–490. [Cited in Chapter 3. This article draws many interesting connections between basic polymer theory and protein structure.]

Chandrasekhar, S. (1943). Stochastic problems in physics and astronomy. *Rev. Mod. Phys.*, **15**, 1–89. [Cited in Chapter 6. This is a clear account of Brownian motion, and is also included in an excellent selection of related papers, edited by N. Wax and published by Dover.]

Changeux, J. P. (1984). Acetylcholine receptor: an allosteric protein. *Science*, **225**, 1335–1345.

Chapman, M. L., Van Donger, H. M. A. and Van Donger, A. M. J. (1997). Activation-dependent subconductance levels in the drk 1 K channel suggest a subunit basis for ion permeation and gating. *Biophys. J.*, **72**, 708–719.

Checover, S., Nachliel, E., Dencher, N. A. and Gutman, M. (1997). Mechanism of proton entry into the cytoplasmic section of the proton-conducting channel of bacteriorhodopsin. *Biochemistry*, **36**, 13 919–13 928.

Chen, Y.-D. and Hill, T. L. (1973). Fluctuations and noise in kinetic systems: Applications to K^+ channels in the squid axon. *Biophys. J.*, **13**, 1276–1295.

Chung, S.-H., Allen, T. W., Hoyles, M. and Kuyucak, S. (1999). Permeation of ions across the potassium channel: Brownian dynamics studies. *Biophys. J.*, **77**, 2517–2533. [Cited in Chapter 14. A lucid treatment of K^+ channel permeation with modern computational methods.]

Chung, S.-H., Allen, T. W. and Kuyucak, S. (2002). Conducting-state properties of the KcsA potassium channel from molecular and Brownian dynamics simulations. *Biophys. J.*, **82**, 628–645.

Cleland, W. W. (1970). Steady state kinetics. In *The Enzymes*, ed. P. D. Boyer. New York: Academic Press, Vol. ii, pp. 1–65.

Cleland, W. W., Frey, P. A. and Gerlt, J. A. (1998). The low barrier hydrogen bond in enzymatic catalysis. *J. Biol. Chem.*, **273**, 25 529–25 532.

Clements, J. D. and Redman, S. J. (1989). Cable properties of cat spinal motoneurons measured by combining voltage clamp, current clamp and intracellular staining. *J. Physiol.*, **409**, 63–87.

Cohn, E. J. and Edsall, J. T. (1943). *Proteins, Amino Acids and Peptides as Ions and Dipolar Ions*. New York: Reinhold Publishing Corporation.

Colquhoun, D. and Hawkes, A. G. (1982). On the stochastic properties of bursts of single ion channel openings and of clusters of bursts. *Phil. Trans. R. Soc. Lond.*, B**300**, 1–59.

Colquhoun, D. and Hawkes, A. G. (1995). A Q-matrix cookbook. In *Single-Channel Recording*, ed. B. Sakmann and E. Neher. New York: Plenum, pp. 589–633.

Connelly, P., Ghosaini, L., Hu, C.-Q., Kitamura, S., Tanaka, A. and Sturtevant, J. M. (1991). A differential scanning calorimetric study of the thermal unfolding of seven mutant forms of phage T4 lysozyme. *Biochem.*, **30**, 1887–1891.

Connor, J. A. and Stevens, C. F. (1971a). Inward and delayed outward membrane current in isolated neural somata under voltage clamp. *J. Physiol.*, **213**, 1–19.

(1971b). Prediction of repetitive firing behavior from voltage clamp data on an isolated neurone soma. *J. Physiol.*, **213**, 31–53.

Cooper, A. (1976). Thermodynamic fluctuations in protein molecules. *Proc. Natl Acad. Sci.*, **73**, 2740–2741.

Coronado, R., Rosenberg, R. L. and Miller, C. (1980). Ionic selectivity, saturation, and block in a K^+-selective channel from sarcoplasmic reticulum. *J. Gen. Physiol.*, **76**, 425–446.

Cox, D. R. and Miller, H. D. (1965). *The Theory of Stochastic Processes*. New York: Chapman and Hall.

Crank, J. (1975). *The Mathematics of Diffusion*. Oxford: Oxford University Press.

Crothers, D. M., Drak, J., Kahn, J. D. and Levene, S. D. (1992). DNA bending, flexibility, and helical repeat by cyclization kinetics. *Methods Enzymol.*, **212**, 3–29.

Cubero, E., Luque, F. J. and Orozco, M. (1998). Is polarization important in cation–π interactions? *Proc. Natl Acad. Sci.*, **95**, 5976–5980.

Curtis, H. J. and Cole, K. S. (1942). Membrane resting and action potentials from the squid giant axon. *J. Cell. Comp. Physiol.*, **19**, 135–144.

Daggett, V. and Fersht, A. R. (2003). Is there a unifying mechanism for protein folding? *TIBS*, **28**, 18–25.

DeFelice, L. J. (1981). *Introduction to Membrane Noise*. New York: Plenum Press. (2004). Transporter structure and mechanism. *TINS*, **27**, 352–359.

De Gennes, P. G. (1972). Exponents for the excluded volume problem as derived by the Wilson method. *Phys. Letts*, **38A**, 339–340.

De Lean, A., Stadel, J. M. and Lefkowitz, R. J. (1980). A ternary complex model explains the agonist-specific binding properties of the adenylate cyclase-coupled b-adrenergic receptor. *J. Biol. Chem.*, **255**, 7108–7117.

de Schutter, E. (1986). Alternative equations for the molluscan ion currents described by Connor and Stevens. *Brain Res.*, **382**, 134–138.

Dill, K. A. (1990). Dominant forces in protein folding. *Biochemistry*, **29**, 7133–7155. [Cited in Chapter 2. This is an excellent review with an emphasis on the hydrophobic effect.]

Dill, D. A., Bromberg, S., Yue, K. *et al.* (1995). Principles of protein folding – a perspective from simple exact models. *Protein Sci.*, **4**, 561–602.

Dinner, A. R. and Karplus, M. (2001). Comment on the communication "The key to solving the protein-folding problem lies in an accurate description of the denatured state" by van Gunsteren *et al. Angew. Chem. Int. Ed.*, **40**, 4615–4616.

Doyle, D. A., Cabral, J. M., Pfuetzner, R. A. *et al.* (1998). The structure of the potassium channel: molecular basis of K^+ conduction and selectivity. *Science*, **280**, 69–77. [Cited in Chapter 14. The first crystal structure of a selective channel and a major advance.]

Dunitz, J. D. (1994). The entropic cost of bound water in crystals and biomolecules. *Science*, **264**, 670. [Cited in Chapter 4. This is a lucid account of the thermodynamics of water association with proteins.]

Eaton, W. A., Henry, E. R. and Hofrichter, J. (1991). Application of linear free energy relations to protein conformational changes: the quaternary structural change of hemoglobin. *Proc. Natl Acad. Sci.*, **88**, 4472–4475.

Edwards, S., Corry, B., Kuyucak, S. and Chung, S.-H. (2002). Continuum electrostatics fails to describe ion permeation in the gramicidin channel. *Biophys. J.*, **83**, 1348–1360.

Ehrenstein, G., Blumenthal, R., Latorre, R. and Lecar, H. (1974). Kinetics of the opening and closing of individual excitability-inducing material channels in a lipid bilayer. *J. Gen. Physiol.*, **63**, 707–721.

Eigen, M. and Hammes, G. G. (1963). Elementary steps in enzyme reactions. *Adv. Enyzmol.*, **25**, 1–38.

Einstein, A. (1956). *Investigations on the Theory of Brownian Movement*. New York: Dover. [Cited in Chapter 6. This small book contains five papers published by Einstein from 1905 to 1908. This work is timeless and still worthy of careful study by students of biophysics.]

Eisenman, G. and Horn, R. (1983). Ionic selectivity revisited: the role of kinetic and equilibrium processes in ion permeation through channels. *J. Membrane Biol.*, **76**, 197–225.

Elson, E. L. and Magde, D. (1974). Fluorescence correlation spectroscopy: I. Conceptual basis. *Biopolymers*, **13**, 1–27.

Falke, J. J., Drake, S. K., Hazard, A. L. and Peersen, O. B. (1994). Molecular tuning of ion binding to calcium signaling proteins. *Q. Rev. Biophys.*, **27**, 219–290.

Fersht, A. R. (1987). The hydrogen bond in molecular recognition. *Trends Biochem. Sci.*, **12**, 301–304. (1998). *Structure and Mechanism in Protein Science*. New York: Freeman. [Cited in Chapters 7 and 10. This is an excellent resource for kinetics and enzyme mechanisms.]

Fersht, A. R., Shi, J.-P., Knill-Jones, J. *et al.* (1985). Hydrogen bonding and biological specificity analysed by protein engineering. *Nature*, **314**, 235–238.

Fersht, A. R., Itzhaki, L. S., El Masry, N. F., Matthews, J. M. and Otzen, D. E. (1994). Single versus parallel pathways of protein folding and fractional formation of structure in the transition state. *Proc. Natl Acad. Sci.*, **91**, 10 426–10 429. [Cited in Chapter 7. This provides a very nice example of how to use Φ plots to address the mechanism of protein unfolding.]

Fielder, E. M., Roberts, P. B., Bray, R. C. *et al.* (1974). The mechanism of action of superoxide dismutase from pulse radiolysis and electron paramagnetic resonance. *Biochem. J.*, **139**, 49–60.

Finkel, A. S. and Redman, S. J. (1983). The synaptic current evoked in cat spinal motoneurons by impulse in single group Ia axons. *J. Physiol.*, **342**, 615–632.

Finkelstein, A. (1987). *Water Movement Through Lipid Bilayers, Pores, and Plasma Membranes*. New York: Wiley-Interscience.

Finkelstein, A. and Andersen, O. S. (1981). The gramicidin A channel: A review of its permeability characteristics with special reference to the single-file aspect of transport. *J. Membrane Biol.*, **59**, 155–171.

Finkelstein, A. V. and Janin, J. (1989). The price of lost freedom: entropy of bimolecular complex formation. *Protein Engineering*, **3**, 1–3.

Flory, P. J. (1969). *Statistical Mechanics of Chain Molecules*. New York: Interscience Publishers.

Fredkin, D. R., Montal, M. and Rice, J. A. (1985). Identification of aggregated Markovian models: application to the nicotinic acetylcholine receptor. In *Proceedings of the Berkeley Conference in Honor of Jerzy Neyman and Jack Kiefer*, ed. L. M. Le Carn and R. A. Olshen. Wadsworth, Vol. 1, pp. 269–289.

Gallivan, J. P. and Dougherty, D. A. (1999). Cation-π interactions in structural biology. *Proc. Natl Acad. Sci.*, **96**, 9459–9464.

(2000). A computational study of cation-π interactions vs salt bridges in aqueous media: Implications for protein engineering. *J. Amer. Chem. Soc.*, **122**, 870–874.

Garofoli, S. and Jordan, P. C. (2003). Modeling permeation energetics in the KcsA potassium channel. *Biophys. J.*, **84**, 2814–2830.

Gavish, B. and Werber, M. M. (1979). Viscosity-dependent structural fluctuations in enzyme catalysis. *Biochemistry*, **18**, 1269–1275.

Gentet, L. J., Stuart, G. J. and Clements, J. D. (2000). Direct measurement of specific membrane capacitance in neurons. *Biophys. J.*, **79**, 314–320.

Gilson, M. K., Given, J. A., Bush, B. L. and McCammon, J. A. (1997). The statistical-thermodynamic basis for computation of binding affinities: a critical review. *Biophys. J.*, **72**, 1047–1069. [Cited in Chapter 4. This is an excellent review of the statistical mechanics of molecular associations.]

Goldman, L. and Albus, J. S. (1968). Computation of impulse conduction in myelinated fibers; theoretical basis of the velocity-diameter relation. *Biophys. J.*, **8**, 596–607.

Goldstein, S. S. and Rall, W. (1974). Changes of action potential shape and velocity for changing core conductor geometry. *Biophys. J.*, **14**, 731–757.

Goychuk, I. and Hänggi, P. (2002). Ion channel gating: a first-passage time analysis of the Kramers type. *Proc. Natl Acad. Sci.*, **99**, 3552–3556.

Green, W. N., Weiss, L. B. and Andersen, O. S. (1987). Batrachotoxin-modified sodium channels in planar lipid bilayers. *J. Gen. Physiol.*, **89**, 841–872.

Griko, Y. V. and Privalov, P. L. (1992). Calorimetric study of the heat and cold denaturation of beta-lactoglobulin. *Biochemistry*, **31**, 8810–8815.

Grosman, C. (2003). Free-energy landscapes of ion-channel gating are malleable: changes in the number of bound ligands are accompanied by changes in the location of the transition state of the acetylcholine-receptor channels. *Biochemistry*, **42** (50), 14 977–14 987.

Grosman, C., Zhou, M. and Auerbach, A. (2000). Mapping the conformational wave of acetylcholine receptor channel gating. *Nature*, **403**, 773–776. [Cited in Chapter 7. This is an important advance mapping the gating transition of an ion channel and an excellent illustration of the use of Φ plots in biophysics.]

Gurney, R. W. (1953). *Ionic Processes in Solution*. New York: McGraw-Hill.

Gutman, M. and Nachliel, E. (1997). Time-resolved dynamics of proton transfer in proteinous systems. *Ann. Rev. Phys. Chem.*, **48**, 329–356.

Hagen, S. J. and Eaton, W. A. (1996). Nonexponential structural relaxations in proteins. *J. Chem. Phys.*, **104**, 3395–3398.

Hall, J. E., Mead, C. A. and Szabo, G. (1973). A barrier model for current flow in lipid bilayer membranes. *J. Membr. Biol.*, **11**, 75–97. [Cited in Chapter 14. This is an early example of how barriers shape the current–voltage behavior of a membrane.]

Hammes, G. G. (1978). *Principles of Chemical Kinetics*. New York: Academic Press.

Han, W.-G., Jalkanen, K. J., Elstner, M. and Suhai, S. (1998). Theoretical study of aqueous N-acetyl-L-alanine N'-methylamine: structures and Raman, VCD, and ROA spectra. *J. Phys. Chem. B*, **102**, 2487–2602.

Hänggi, P., Talkner, P. and Borkovec, M. (1990). Reaction-rate theory: fifty years after Kramers. *Rev. Modern Phys.*, **62**, 251–341. [Cited in Chapter 7. This is an excellent review of modern approaches to rate theory.]

Hardy, L. W. and Kirsch, J. F. (1984). Diffusion-limited components of reactions catalyzed by bacillus cereus b-lactanase I. *Biochemistry*, **23**, 1275–1282.

Hecht, S., Schlaer, S. and Pirenne, M. (1942). Energy, quanta, and vision. *J. Gen. Physiol.*, **25**, 819–840.

Hedstrom, L., Perona, J. J. and Rutter, W. J. (1994). Converting trypsin to chymotrypsin: Residue 172 is a substrate specificity determinant. *Biochemistry*, **33**, 8757–8763.

Hess, P. and Tsien, R. W. (1984). Mechanism of ion permeation through calcium channels. *Nature*, **309**, 453–456.

Hestrin, S., Nicoll, R. A., Perkel, D. J. and Sah, P. (1990). Analysis of excitatory synaptic action in pyramidal cells using whole-cell recording from rat hippocampal slices. *J. Physiol.*, **422**, 203–225.

Hill, T. L. (1960). *An Introduction to Statistical Thermodynamics*. Reading: Addison-Wesley. [This is highly recommended as a resource for statistical mechanics.]

Hille, B. (1977). Ionic basis of resting and action potentials. In *Handbook of Physiology. The Nervous System. Cellular Biology of Neurons*, ed. J. M. Brookhart and V. B. Mountcastle. Bethesda: American Physiological Society, pp. 99–136.

(1991). *Ion Channels of Excitable Membranes*. Sunderland: Sinauer Associates.

Hille, B. and Schwarz, W. (1978). Potassium channels as multi-ion single-file pores. *J. Gen. Physiol.*, **72**, 409–442. [Cited in Chapter 14. A full treatment of single-file models and an excellent summary of their basic properties.]

Hines, M. L. and Carnevale, N. T. (1997). The NEURON simulation environment. *Neural Computation*, **9**, 1179–1209.

Hodgkin, A. L. (1964). *The Conduction of Nervous Impulse*. Springfield, IL: Charles Thomas.

Hodgkin, A. L. and Horowicz, P. (1959). The influence of potassium and chloride ions on the membrane potential of single muscle fibres. *J. Physiol.*, **148**, 127–160.

Hodgkin, A. L. and Huxley, A. F. (1952). A quantitative description of membrane current and its application to conduction and excitation in nerve. *J. Physiol.*, **117**, 500–544. [Cited in Chapter 16. This is the culminating paper in the seminal series on membrane excitability.]

Hodgkin, A. L. and Keynes, R. D. (1955). The potassium permeability of a giant nerve fibre. *J. Physiol.*, **128**, 61–88. [Cited in Chapter 14. This is the classical development of the single-file theory of permeation.]

Hodgkin, A. L. and Rushton, W. A. H. (1946). The electrical constants of a crustacean nerve fiber. *Proc. R. Soc. Lond.*, B**133**, 444–479.

Hodgkin, A. L., Huxley, A. F. and Katz, B. (1952). Measurement of current voltage relations in the membrane of the giant axon of *Loligo*. *J. Physiol.*, **116**, 424–448.

Hol, W. G. J., van Duijnen, P. T. and Berendsen, H. J. C. (1978). The α-helix dipole and the properties of proteins. *Nature*, **273**, 443–446.

Horn, R. and Lange, K. (1983). Estimating kinetic constants from single channel data. *Biophys. J.*, **43**, 207–223.

Horn, R. and Vandenberg, C. A. (1984). Statistical properties of single sodium channels. *J. Gen. Physiol.*, **84**, 505–534.

Huxley, A. F. and Stämpfli, R. (1949). Evidence for saltatory conduction in peripheral myelinated nerve fibres. *J. Physiol.*, **108**, 315–339.

Imoto, K., Busch, C., Sakmann, B. *et al.* (1988). Rings of negatively charged amino acids determine the acetylcholine receptor channel conductance. *Nature*, **335**, 645–648.

Itzhaki, L. S., Otzen, D. E. and Fersht, A. R. (1995). The structure of the transition state for protein folding of chymotrypsin inhibitor 2 analysed by protein engineering methods: evidence for a nucleation condensation mechanism for protein folding. *J. Mol. Biol.*, **254**, 260–288.

Jack, J. J. B. and Redman, S. J. (1971). An electrical description of the motoneuron and its application to the analysis of synaptic potentials. *J. Physiol.*, **215**, 321–352.

Jack, J. J. B., Noble, D. and Tsien, R. W. (1983). *Electric Current Flow in Excitable Cells*. Oxford: Oxford University Press. [Cited in Chapters 15 and 16. This is a comprehensive textbook that covers cable theory and excitability very well.]

Jackson, J. D. (1975). *Classical Electrodynamics*. New York: John Wiley & Sons.

Jackson, M. B. (1985). The stochastic behavior of a many channel membrane system. *Biophys. J.*, **47**, 129–137.

(1992). Cable analysis with the whole-cell patch clamp: theory and experiment. *Biophys. J.*, **61**, 756–766.

(1993a). Passive current flow and morphology in the terminal arborizations of the posterior pituitary. *J. Neurophysiol.*, **69**, 692–702.

(1993b). Binding specificity of receptor chimeras revisited. *Biophys. J.*, **63**, 1443–1444.

(1993c). On the time scale and time course of protein conformational changes. *J. Chem. Phys.*, **99**, 7253–7259.

(1994). Single channel currents in the nicotinic receptor: A direct demonstration of allosteric transitions. *TIBS*, **19**, 396–399.

(1997a). Adding up the energies in the acetylcholine receptor channel: relevance to allosteric theory. In *The Nicotinic Acetylcholine Receptor: Current Views and Future Trends*, ed. F. Barrantes. Austin: Landes Bioscience, pp. 61–84.

(1997b). Inversion of Markov processes to determine rate constants from single channel data. *Biophys. J.*, **73**, 1382–1394.

(1998). Allosteric mechanisms in the activation of ligand-gated channels. *Biophysics Textbook of the Biophysical Society*. Rockville: Biophysical Society. [Cited in Chapter 5. A discussion of allosteric mechanisms using ligand-gated channels to illustrate important concepts.]

Jackson, M. B. and Zhang, S. J. (1995). Action potential propagation and propagation block by GABA in rat posterior pituitary nerve terminals. *J. Physiol.*, **483**(3), 597–611.

Jackson, M. B., Wong, B. M., Morris, C. E., Lecar, H. and Christian, C. N. (1983). Successive openings of the same acetylcholine receptor channel are correlated in open time. *Biophys. J.*, **42**, 109–114.

Jackson, M. B., Konnerth, A. and Augustine, G. J. (1991). Action potential broadening and frequency-dependent facilitation of calcium signals in pituitary nerve terminals. *Proc. Natl Acad. Sci.*, **88**, 380–384.

Jacobson, K., Ishihara, A. and Inman, R. (1987). Lateral diffusion of proteins in membranes. *Ann. Rev. Physiol.*, **49**, 163–175.

Jeffrey, G. A. and Saenger, W. (1991). *Hydrogen Bonding in Biological Structures*. Berlin: Springer–Verlag.

Jencks, W. P. (1975). Binding energy, specificity, and enzyme catalysis: the Circe effect. *Adv. Enzymol.*, **43**, 219–410.

Jencks, W. P. and Carriuolo, J. (1961). General base catalysis of ester hydrolysis. *Journal of the American Chemical Society*, **83**, 1743–1750.

Jentsch, T. J., Stein, V., Weinreich, F. and Zdebik, A. A. (2002). Molecular structure and physiological function of chloride channels. *Physiol. Rev.*, **82**, 503–568.

Jones, S. W. (1989). On the resting potential of isolated frog sympathetic neurons. *Neuron*, **3**, 153–161.

Jordan, P. C. (1990). Ion-water and ion-polypeptide interactions in a gramicidin-like channel. A molecular dynamics study. *Biophys. J.*, **58**, 1133–1156.

Jordan, P. C., Bacquet, R. J., McCammon, A. J. and Tran, P. (1989). How electrolyte shielding influences the electrical potential in transmembrane ion channels. *Biophys. J.*, **55**, 1041–1052.

Kallenbach, N. (2001). Breaking open a protein barrel. *Proc. Natl Acad. Sci.*, **98**, 2958–2960. [Cited in Chapter 2. This presents a clear presentation of the oil-droplet versus jigsaw puzzle pictures of a protein interior.]

Kao, J. P. Y. and Tsien, R. Y. (1988). Ca^{2+} binding kinetics of fura-2 and azo-1 from temperature jump relaxation measurements. *Biophys. J.*, **53**, 635–639.

Karplus, M. (2002). Molecular dynamics simulations of biomolecules. *Acc. Chem. Res.*, **35**, 321–323. [Cited in Chapter 2. This is the editorial for a special issue that covers a wide range of applications.]

Katz, B. (1966). *Nerve, Muscle, and Synapse*. New York: McGraw-Hill.

Kaya, H. and Chan, H. S. (2000). Polymer principles of protein calorimetric two-state cooperativity. *Proteins: Structure, Function, and Genetics*, **40**, 637–661.

Keizer, J. (1987). Diffusion effects on rapid bimolecular chemical reactions. *Chem. Rev.*, **87**, 167–180.

Kell, M. J. and DeFelice, L. J. (1988). Surface charge near the cardiac inward-rectifier channel measured from single-channel conductance. *J. Membrane Biol.*, **102**, 1–10.

Kellermayer, M. S. Z., Smith, S. B., Granzier, H. L. and Bustamante, C. (1997). Folding–unfolding transitions in single titin molecules characterized with laser tweezers. *Science*, **276**, 1112–1116.

Khanin, R., Parnas, H. and Segel, L. (1994). Diffusion cannot govern the discharge of neurotransmitter in fast synapses. *Biophys. J.*, **67**, 966–972.

Kijima, S. and Kijima, H. (1987). Statistical analysis of channel current from a membrane patch I. Some stochastic properties of ion channels or molecular systems at equilibrium. *J. Theor. Biol.*, **128**, 423–434.

Kittel, C. (1958). *Elementary Statistical Physics*. New York: John Wiley & Sons.

Kolinski, A., Godzik, A. and Skolnick, J. (1993). A general method for the prediction of the three dimensional structure and folding pathway of globular proteins: Application to designed helical proteins. *J. Chem. Phys.*, **98**, 7420–7433.

Kolinski, A., Galazka, W. and Skolnick, J. (1996). On the origin of the co-operativity of protein folding: implications from model simulations. *Proteins: Structure, Function, and Genetics*, **26**, 271–287.

Koshland, D. E., Nemethy, G. and Filmer, D. (1966). Comparison of experimental binding data and theoretical models in proteins containing subunits. *Biochemistry*, **5**, 365–384.

Kramers, H. A. (1940). Brownian motion in a field of force. *Physica*, **7**, 284–304.

Kuyucak, S., Andersen, O. S. and Chung, S. -H. (2001). Models of permeation in ion channels. *Reports of Progress in Physics*, **64**, 1427–1472.

Latorre, R., Labarca, P. and Naranjo, D. (1992). Surface charge effects on ion conduction in ion channels. In *Ion Channels (Methods in Enzymology)*, ed. L. Iverson and B. Rudy, vol. 207, pp. 471–501. [Cited in Chapter 11. A very clear overview of surface charge effects in ion channels.]

Läuger, P. (1973). Ion transport through pores: a rate-theory analysis. *Biophysica and Biochimica Acta*, **311**, 423–441. [Cited in Chapter 14. A thorough development of barrier models for permeation.]

Lecar, H. and Sachs, F. (1981). Membrane noise analysis. In *Excitable Cells in Tissue Culture*, ed. P. G. Nelson and M. Lieberman. New York: Plenum, pp. 137–172.

Lee, A. W., Karplus, M., Poyart, C. and Bursaux, E. (1988). Analysis of proton release in oxygen binding by hemoglobin: implications for cooperative mechanism. *Biochemistry*, **27**, 1285–1301.

Lesk, A. M., Lo Conte, L. and Hubbard, T. J. P. (2001). Assessment of novel fold targets in CASP4: Predictions of three-dimensional structures, secondary structures, and interresidue contacts. *Proteins: Structure, Function, and Genetics*, **45**(S5), 98–118.

Levinthal, C. (1968). Are there pathways for protein folding. *J. Chem. Phys.*, **65**, 44–45.

Levitt, D. G (1978a). Electrostatic calculations for an ion channel. I. Energy and potential profiles and interactions between ions. *Biophys. J.*, **22**, 209–219. (1978b). Electrostatic calculations for an ion channel II. Kinetic behavior of the gramicidin A channel. *Biophys. J.*, **22**, 221–248.

Levitt, M., Sander, C. and Stern, P. S. (1985). Protein normal-mode dynamics: trypsin inhibitor, crambin, ribonuclease and lysozyme. *J. Mol. Biol.*, **181**, 423–447.

Levitt, M., Hirshberg, M., Sharon, R. and Daggett, V. (1995). Potential energy function and parameters for simulations of the molecular dynamics of proteins and nucleic acids in solution. *Comput. Phys. Communs*, **91**, 215–231.

Lewis, C. A. (1979). Ion-concentration dependence of the reversal potential and the single channel conductance of ion channels at the frog neuromuscular junction. *J. Physiol.*, **286**, 417–445.

Lifson, S. and Roig, A. (1963). On the theory of helix–coil transition in polypeptides. *J. Chem. Phys.*, **34**, 1961–1974. [Cited in Chapter 3. This is an elegent and clear development of the mathematical theory of helix–coil transitions.]

Lifson, S. and Warshel, A. (1968). Consistent force field for calculations of conformations, vibrational spectra, and enthalpies of cycloalkane and n-alkane molecules. *J. Chem. Phys.*, **49**, 5119–5129.

Lin, J., Cassidy, C. S. and Frey, P. A. (1998). Correlations of the basicity of His 57 with transition state analogue binding, substrate reactivity, and the strength of the low-barrier hydrogen bond in chymotrypsin. *Biochemistry*, **37**, 11 940–11 948.

Linderstrøm-Lang, K. (1924). On the ionization of proteins. *Compt. rend. trav. Carlsberg*, **17**, 1–29. (See also Linderstrøm-Lang, K. (1962) *Selected Papers*. New York: Academic Press.)

Ma, J. C. and Dougherty, D. A. (1997). The cation–π interaction. *Chem. Rev.*, **97**, 1303–1324.

MacInnes, D. A. (1961). *The Principles of Electrochemistry*. New York: Dover.

MacKinnon, R., Latorre, R. and Miller, C. (1989). Role of surface electrostatics in the operation of a high-conductance Ca^{2+}-activated K^+ channel. *Biochemistry*, **28**, 8092–8099.

Magee, J. C. and Cook, E. P. (2000). Somatic EPSP amplitude is independent of synapse location in hippocampal pyramidal neurons. *Nature Neurosci.*, **3**, 895–903.

Mainen, Z. F., Joerges, J., Huguenard, J. R. and Sejnowski, T. J. (1995). A model of spike initiation in neocortical pyramidal neurons. *Neuron*, **15**, 1427–1439.

Major, G., Evan, J. D. and Jack, J. B. (1993). Solutions for transients in arbitrarily branching cables: I. Voltage recording with a somatic shunt. *Biophys. J.*, **65**, 423–449. [Cited in Chapter 15. This paper explores the cable equation in complicated dendritic arbors.]

Manning, G. S. (1969). Limiting laws and counterion condensation in polyelectrolyte solutions. I. Colligative properties. *J. Chem. Phys.*, **51**, 924–933.

(1978). The molecular theory of polyelectrolyte solutions with applications to the electrostatic properties. *Q. Rev. Biophys.*, **11**, 179–246.

Marcus, R. A. (1964). Chemical and electrochemical electron-transfer theory. *Ann. Rev. Phys. Chem.*, **15**, 155–196.

(1968). Theoretical relations among rate constants, barriers, and Brønsted slopes of chemical reactions. *J. Phys. Chem.*, **72**, 891–899.

Martinez, M. B., Flickinger, M. C. and Nelsestuen, G. L. (1996). Accurate kinetic modeling of alkaline phosphatase in the *Escherichia coli* periplasm: implications for enzyme properties and substrate diffusion. *Biochemistry*, **35**, 1179–1186.

Marty, A, and Neher, E. (1995). Tight-seal whole-cell recording. In *Single-Channel Recording*, ed. B. Sakmann and E. Neher. New York: Plenum, pp. 31–51.

McCammon, A. J., Wolynes, P. G. and Karplus, M. (1979). Picosecond dynamics of tyrosine side chains in proteins. *Biochemistry*, **18**, 927–942. [Cited in Chapter 10. This is a seminal paper on the internal dynamics of proteins.]

McLaughlin, S. (1989). The electrostatic properties of membranes. *Ann. Rev. Biophys. Biophys. Chem.*, **18**, 113–136.

McQuarrie, D. A., (1976). *Statistical Mechanics*. New York: Harper & Row. [This is highly recommended as a resource for statistical mechanics.]

Meyer, E., Müller, C. O. and Fromherz, P. (1997). Cable properties of dendrites in hippocampal neurons of the rat mapped by a voltage-sensitive dye. *Eur. J. Neurosci.*, **9**, 778–785.

Miedema, H. (2002). Surface potentials and the calculated selectivity of ion channels. *Biophys. J.*, **82**, 156–159.

Migliore, M., Hoffman, D. A., Magee, J. C. and Johnston, D. (1999). Role of an A-type K^+ conductance in the back-propagation of action potentials in the dendrites of hippocampal pyramidal neurons. *J. Comp. Neurosci.*, **7**, 5–15.

Millar, J. A., Barrett, L., Southan, A. P., Page, K. M., Fyffe, R. E. W. and Robertson, B. (2000). A functional role for the two-pore domain potassium channel TASK-1 in cerebellar granule neurons. *Proc. Natl Acad. Sci.*, **97**, 3614–3618.

Miller, B. G. and Wolfenden, R. (2002). Catalytic proficiency: The unusual case of OMP decarboxylase. *Ann. Rev. Biochem.*, **71**, 847–885.

Moczydlowski, E., Alvarez, O., Vergara, C. and Latorre, R. (1985). Effect of phospholipid surface charge on the conductance and gating of a Ca^{2+}-activated K^+ channel in planar lipid bilayers. *J. Membrane Biol.*, **83**, 273–282.

Monod, J., Wyman, J. and Changeux, J.-P. (1965). On the nature of allosteric transitions: A plausible model. *J. Mol. Biol.*, **12**, 88–118. [Cited in Chapter 5. This is the seminal paper on allosteric regulation of proteins. A remarkable conceptual advance and still well worth reading.]

Moore, W. J. (1972). *Physical Chemistry*. Englewood Cliffs: Prentice-Hall. [This is highly recommended as a resource for statistical mechanics.]

Morais-Cabral, J. H., Zhou, Y. and MacKinnon, R. (2001). Energetic optimization of ion conduction rate by the K^+ selectivity filter. *Nature*, **414**, 37–42.

Morris, C. E. and Lecar, H. (1981). Voltage oscillations in the barnacle giant muscle fiber. *Biophys. J.*, **35**, 193–213. [Cited in Chapter 16. A clear and illuminating account of membrane oscillations.]

Moy, G., Corry, B., Kuyucak, S. and Chung, S.-H. (2000). Tests of continuum theories as models of ion channels. I. Poisson–Boltzmann theory versus Brownian dynamics. *Biophys. J.*, **78**, 2349–2363.

Myers, J. K. and Pace, C. N. (1996). Hydrogen bonding stabilizes globular proteins. *Biophys. J.*, **71**, 2033–2039.

Nakajima, Y., Nakajima, S. and Inoue, M. (1988). Pertussis toxin-insensitive G protein mediates substance P-induced inhibition of potassium channels in brain neurons. *Proc. Natl Acad. Sci.*, **85**, 3643–3647.

Nakatani, H. and Dunford, H. B. (1979). Meaning of diffusion-controlled association rate constants in enzymology. *J. Phys. Chem.*, **83**, 2662–2665.

Neher, E. and Steinbach, J. H. (1978). Local anesthetics transiently block currents through single acetylcholine-receptor channels. *J. Physiol.*, **277**, 153–176. [Cited in Chapter 9. This is a striking example of the power of single-channel kinetics in the analysis of mechanisms of drug action.]

Nelsestuen, G. L. and Martinez, M. B. (1997). Steady state enzyme velocities that are independent of [enzyme]: an important behavior in many membrane and particle-bound states. *Biochemistry*, **36**, 9081–9086.

Nolte, H.-J., Rosenberry, T. L. and Neumann, E. (1980). Effective charge on acetylcholinesterase active sites determined from the ionic strength dependence of association rate constants with cationic ligands. *Biochemistry*, **19**, 3705–3711.

Noskov, S. Y., Berniche, S. and Roux, B. (2004). Control of ion selectivity in potassium channels by electrostatic and dynamic properties of carbonyl ligands. *Nature*, **431**, 830–834.

Obaid, A. L. and Salzberg, B. M. (1996). Micromolar 4-aminopyridine enhances invasion of a vertebrate neurosecretory terminal arborization. *J. Gen. Physiol.*, **107**, 353–368.

Oberhauser, A. F. and Fernandez, J. M. (1995). Hydrophobic ions amplify the capacitance currents used to measure exocytotic fusion. *Biophys. J.*, **69**, 451–459.

O'Mara, M., Barry, P. H. and Chung, S.-H. (2003). A model of the glycine receptor deduced from Brownian dynamics studies. *Proc. Natl Acad. Sci.*, **100**, 4310–4315.

O'Neil, K. T. and DeGrado, W. F. (1990). A thermodynamic scale for the helix-forming tendencies of the commonly occurring amino acids. *Science*, **250**, 646–651.

Oosawa, F. (1971). *Polyelectrolytes*. New York: Marcel Dekker, Inc.

Overbeek, J. Th. G. (1952). The electrochemistry of the double layer. In *Colloid Science*, Vol. 1, ed. H. R. Kruyt. Amsterdam: Elsevier Publishing Company, pp. 115–193.

Overbeek, J. T. G. and Wiersema, P. H. (1967). The interpretation of electrophoretic mobilities. In *Electrophoresis*, ed. M. Bier. New York: Academic Press, pp. 1–52.

Pace, C. N. (1992). Contributions of the hydrophobic effect to globular protein stability. *J. Mol. Biol.*, **226**, 29–35.

Papazian, D. M., Timpe, L. C., Jan, Y. N. and Jan, L. Y. (1991). Alteration of voltage-dependence of *Shaker* potassium channel by mutations in the S4 sequence. *Nature*, **349**, 305–310.

Parsegian, V. A. (1969). Energy of an ion crossing a low dielectric membrane: Solutions to four relevant electrostatics problems. *Nature*, **221**, 844–846. [Cited in Chapters 2 and 14. This is an early study that defined the basic energetic parameters of permeation.]

(1973). Long-range physical forces in the biological milieu. *Ann. Rev. Biophys. Bioeng.*, **2**, 221–255.

Patlak, C. S. (1960). Derivation of an equation for the diffusion potential. *Nature*, **188**, 944–945.

Perkel, D. H., Mulloney, B. and Budelli, R. W. (1981). Quantitative methods for predicting neuronal behavior. *Neuroscience*, **5**, 823–837.

Perutz, M. F., Wilkenson, A. J., Paoli, M. and Dodson, D. D. (1998). The stereochemical mechanism of the cooperative effects in hemoglobin revisited. *Ann. Rev. Biophys. Biomol. Structure*, **27**, 1–34. [Cited in Chapter 5. This contains a clear and thorough discussion of the evidence favoring a two-state description of hemoglobin.]

Peters, R. and Cherry, R. J. (1982). Lateral and rotational diffusion of bacteriorhodopsin in lipid bilayers: experimental test of the Saffman–Delbrück equations. *Proc. Natl Acad. Sci.*, **79**, 4317–4321.

Piek, T. (1975). Ionic and electrical properties. In *Insect Muscle*, ed. P. N. R. Usherwood. New York: Academic Press, pp. 275–336.

Plowman, K. M. (1972). *Enzyme Kinetics*. New York: McGraw Hill.

Pokarowski, P., Kolinski, A. and Skolnick, J. (2003). A minimal physically realistic protein-like lattice model: designing an energy landscape that ensures all-or-none folding to a unique native state. *Biophys. J.*, **84**, 1518–1526.

Poland, D. and Scheraga, H. A. (1970). *Theory of Helix–Coil Transitions*. New York: Academic Press.

Privalov, P. L. (1979). Stability of proteins. *Adv. Protein Chem.*, **33**, 167–241.

(1982). Stability of proteins: Proteins which do not present a single cooperative system. *Adv. Protein Chem.*, **35**, 1–104. [Cited in Chapters 2 and 3. This reference presents thorough reviews of the thermodynamics of thermal transitions in proteins.]

Prod'hom, B., Peitrobon, D. and Hess, P. (1987). Direct measurement of proton transfer rates to a group controlling the dihydropyridine-sensitive Ca^{2+} channel. *Nature*, **329**, 243.

Pumphrey, R. J. and Young, J. Z. (1938). The rates of conduction of nerve fibres of various diameters in cephalopods. *J. Exp. Biol.*, **14**, 453–466.

Putnam, S. J., Coulson, A. F., Farley, I. R., Ridd Leston, B. and Knowles, J. R. (1972). Specificity and kinetics of triose phosphate isomerase from chicken muscle. *Biochem. J.*, **129**, 301–310.

Raleigh, D. P. and DeGrado, W. F. (1992). A de novo designed protein shows a thermally induced transition from a native to a molten globule-like state. *J. Amer. Chem. Soc.*, **114**, 10 079–10 081.

Rall, W. (1959) Branching dendritic trees and motoneuron membrane resistivity. *Exp. Neurol.*, **1**, 491–527. [Cited in Chapter 15. This provides a remarkable insight into how to simplify the cable analysis of dendrites.]
(1967). Distinguishing theoretical synaptic potentials computed for different soma-dendritic distributions of synaptic input. *J. Neurophysiol.*, **30**, 1138–1168.
(1969). Time constants and electrotonic length of membrane cylinders and neurons. *Biophys. J.*, **9**, 1483–1508. [Cited in Chapter 15. A thorough and clear exposition of practical aspects of cable analysis.]
(1977). Core conductor theory and cable properties of neurons. In *Handbook of Physiology. The Nervous System. Cellular Biology of Neurons*, ed. J. M. Brookhart and V. B. Mountcastle. Bethesda: American Physiological Society, pp. 39–97.

Rall, W., Burke, R. E., Smith, T. G., Nelson, P. G. and Frank, K. (1967). Dendritic location of synapses and possible mechanisms for the monosynaptic EPSPs in motoneurons. *J. Neurophysiol.*, **30**, 1169–1193.

Ramachandran, G. N. and Sasisekharan, V. (1968). Conformation of polypeptides and proteins. *Adv. Protein Chem.*, **23**, 284–437.

Ramachandran, G. N., Venkatachalam, C. M. and Krimm, S. (1966). Stereochemical criteria for polypeptide and protein chain conformations. 3. Helical and hydrogen-bonded polypeptide chains. *Biophys. J.*, **6**(6), 849–872.

Rand, R. P. (1981). Interacting phospholipid bilayers: measured forces and induced structural changes. *Ann. Rev. Biophys. Bioengin.*, **10**, 288–314.

Rapp, M., Yarom, Y. and Segev, I. (1996). Modeling back propagating action potentials in weakly excitable dendrites of neocortical pyramidal cells. *Proc. Natl Acad. Sci.*, **93**, 11 985–11 990.

Rashin, A. A. and Honig, B. (1985). Reevaluation of the Born model of ion hydration. *J. Phys. Chem.*, **89**, 5588–5593.

Record, M. T. (1975). Effects of Na^+ and Mg^{++} ions on the helix–coil transition of DNA. *Biopolymers*, **14**, 2137–2158.

Record, M. T., Mazur, S. J., Melancon, P., Roe, J.-H., Shaner, S. L. and Unger, L. (1981). Double helical DNA: conformations, physical properties, and interactions with ligands. *Ann. Rev. Biochem.*, **50**, 997–1024.

Redman, S. and Walmsley, B. (1983). The time course of synaptic potentials evoked in cat spinal motoneurons at identified group 1a synapses. *J. Physiol.*, **343**, 117–133.

Reed, A. E. and Weinhold, F. (1991). Natural bond orbital analysis of internal rotation barriers and related phenomena. *Isr. J. Chem.*, **31**, 277–285.

Rees, D. C., DeAntonio, L. and Eisenberg, D. (1989). Hydrophobic organization of membrane proteins. *Science*, **245**, 510–513.

Richard, J. P. (1998). The enhancement of enzymatic rate accelerations by Brønsted acid–base catalysis. *Biochemistry*, **37**, 4305–4309.

Rigler, R. and Elson, E. L. (eds.) (2001). *Fluorescence Correlaton Spectroscopy: Theory and Applications*. Berlin: Springer.

Rogawski, M. A. (1985). The A-current: how ubiquitous a feature of excitable cells is it? *Trends Neurosci.*, **8**, 214–219.

Rose, G. D., Gesolowitz, A. R., Lesser, G. J., Lee, R. H. and Zehfus, M. H. (1985). Hydrophobicity of amino acid residues in globular proteins. *Science*, **229** (4716), 834–838.

Roseman, M. A. (1988). Hydrophobicity of the peptide $C = O \cdots H–N$ hydrogen-bonded group. *J. Mol. Biol.*, **201**, 621–623.

Rosenthal, L., Rabolt, J. F. and Hummel, J. (1982). An investigation of the conformational equilibrium of *n*-butane in a solvent using Raman spectroscopy. *J. Chem. Phys.*, **76**, 817–820.

Rothberg, B. S. and Magleby, K. L. (2001). Testing for detailed balance (microscopic reversibility) in ion channel gating. *Biophys. J.*, **80**, 3025–3026.

Roux, B. and MacKinnon, R. (1999). The cavity and pore helices in the KcsA K^+ channel: electrostatic stabilization of monovalent cations. *Science*, **285**, 100–102.

Roux, B., Berniche, S. and Im, W. (2000). Ion channels, permeation, and electrostatics: insight into the function of KcsA. *Biochemistry*, **39**, 13 295–13 306.

Rushton, W. A. H. (1951). A theory of the effects of fibre size in medullated nerve. *J. Physiol.*, **115**, 101–122.

Saffman, P. G. and Delbrück, M. (1975). Brownian motion in biological membranes. *Proc. Natl Acad. Sci.*, **72**, 3111–3113.

Sanchez, I. C. (1979). Phase transition behavior of the isolated polymer chain. *Macromolecules*, **12**, 980–988.

Saxton, M. J. and Jacobson, K. (1997). Single-particle tracking: applications to membrane dynamics. *Ann. Rev. Biophys. Biomol. Struct.*, **26**, 373–399.

Scatchard, G. (1949). The attractions of proteins for small molecules and ions. *Ann. N. Y. Acad. Sci.*, **51**, 660–671.

Schafmeister, C. E., LaPorte, S. L., Miercke, L. J. W. and Stroud, R. M. (1997). A designed four helix bundle protein with native-like structure. *Nature Struct. Biol.*, **4**, 1039–1046.

Scheer, A., Fanelli, F., Costa, T., De Benedetti, P. G. and Cotecchia, S. (1997). The activation process of the α_{1B}-adrenergic receptor: potential role of protonation and hydrophobicity of a highly conserved aspartate. *Proc. Natl Acad. Sci.*, **94**, 808–818.

Schirmer, T. and Evans, P. R. (1990). Structural basis of the allosteric behavior of phosphofructokinase. *Nature*, **343**, 140–145.

Schoppa, N. E., McCormack, K., Tanouye, M. A. and Sigworth, F. J. (1992). The size of the gating charge in wild-type and mutant *Shaker* potassium channels. *Science*, **255**, 1712–1715. [Cited in Chapter 1. This is an excellent synthesis of the charge and steepness in a voltage-induced protein transition.]

Schultz, P. G. and Lerner R. A. (1995). From molecular diversity to catalysis: Lessons from the immune system. *Science*, **269**: 1835–1842.

Schumaker, M. F. and MacKinnon, R. (1990). A simple model for multi-ion permeation: single vacancy conduction in a simple pore model. *Biophys. J.*, **58**, 975–984.

Segev, I. (1990). Computer study of presynaptic inhibition controlling the spread of action potentials into nerve terminals. *J. Neurophys.*, **63**, 987–997.

Segev, I., Fleshman, J. W. and Burke, R. E. (1989). Compartmental models of complex neurons. In *Methods in Neural Modeling*, ed. C. Koch and I. Segev. Cambridge: MIT Press, pp. 63–96.

Serrano, L., Matouschek, A. and Fersht, A. R. (1992). The folding of an enzyme III. Structure of the transition state of barnase analysed by a protein engineering procedure. *J. Mol. Biol.*, **224**, 805–818.

Setlow, R. B., and Pollard, E. C. (1962). *Molecular Biophysics*, chapter 6. Palo Alto: Addison-Wesley Publishing Co. Inc. [Cited in Chapter 2. This forgotten text contains a lucid presentation of molecular forces.]

Shi, Z., Krantz, B. A., Kallenbach, N. and Sosnick, T. R. (2002a). Contribution of hydrogen bonding to protein stability estimated from isotope effects. *Biochemistry*, **41**, 2120–2129.

Shi, Z., Olson, C. A. and Kallenbach, N. R. (2002b). Cation-π interaction in model α-helical peptides. *J. Amer. Chem. Soc.*, **124**, 3284–3291.

Shi, Z., Olson C. A., Rose, G. D., Baldwin, R. L. and Kallenbach, N. R. (2002c). Polyproline II structure in a sequence of seven alanine residues. *Proc. Natl Acad. Sci.*, **99**, 9190–9195.

Shoup, D., Lipari, G. and Szabo, A. (1981). Diffusion-controlled bimolecular reaction rates. *Biophys. J.*, **36**, 697–714.

Sigworth, F. J. (1994). Voltage gating of ion channels. *Q. Rev. Biophys.*, **27**, 1–27.

Silverman, D. N. (2000). Marcus rate theory applied to enzymatic proton transfer. *Biochim. Biophys. Acta*, **1458**, 88–103.

Silverman, D. N., Tu, C., Chen, X., Tanhauser, S. M., Kresge, A. J. and Laipis, P. J. (1993). Rate-equilibria relationships in intramolecular proton transfer in human carbonic anhydrase III. *Biochemistry*, **32**, 10 757–10 761.

Silverman, J. A., Balakrishnan, R. and Harbury, P. B. (2001). Reverse engineering the $(\beta/\alpha)_8$ barrel fold. *Proc. Natl Acad Sci.*, **98**, 3092–3097.

Sine, S. M., Claudio, T. and Sigworth, F. (1990). Activation of *Torpedo* acetylcholine receptors expressed in mouse fibroblasts. *J. Gen. Physiol.*, **96**, 395–437.

Spolar, R. S. and Record, M. T. (1994). Coupling of local folding to site-specific binding of proteins to DNA. *Science*, **263**, 777–784. [Cited in Chapter 4. This is a study that considers many different contributions to the free energy of association.]

Spolar, R. S., Livingstone, J. R. and Record, M. T. (1992). Use of liquid hydrocarbon and amide transfer data to estimate contributions to thermodynamic functions of protein folding from the removal of nonpolar and polar surfaces from water. *Biochemistry*, **31**, 3947–3955.

Steinberg, I. Z. (1987). Relationship between statistical properties of single ion channel recordings and thermodynamic state of the channels. *J. Theor. Biol.*, **124**, 71–87.

Steinberg, I. Z. and Scheraga, H. A. (1963). Entropy changes accompanying association reactions of proteins. *J. Biol. Chem.*, **238**, 172–181.

Stigter, D. and Dill, K. A. (1990). Charge effects on folded and unfolded proteins. *Biochemistry*, **29**, 1262–1271.

Stillinger, F. H. (1980). Water revisited. *Science*, **209**, 451–457.

Stockbridge, N. (1988). Etiology of the supernormal period. *Biophys. J.*, **54**, 777–780.

Stuart, G. J. and Sakmann, B. (1994). Active propagation of somatic action potentials into neocortical pyramidal cell dendrites. *Nature*, **367**, 69–72.

Stuart, G., Spruston, N. and Hausser, M. (2000). *Dendrites*. Oxford: Oxford University Press.

Stuehmer, W., Conti, F., Suzuki, H. *et al.* (1989). Structural parts involved in activation and inactivation of the sodium channel. *Nature*, **339**, 597–603.

Sukharev, S., Blount, P., Martinac, B. and Kung, C. (1997). Mechanosensitive channels of *Escherichia coli*: the MscL gene, protein, and activities. *Ann. Rev. Physiol.*, **59**, 633–657.

Sukharev, S., Durell, S. R. and Guy, H. R. (2001). Structural models of the MscL gating mechanism. *Biophys. J.*, **81** (2), 917–936.

Sussman, J. L., Harel, M., Frolow, F. *et al.* (1991). Atomic structure of acetylcholinesterase from *Torpedo californica*: a prototypic acetylcholine-binding protein. *Science*, **253**, 872–879.

Swadlow, H. A., Kocsis, J. D. and Waxman, S. G. (1980). Modulation of impulse conduction along the axonal tree. *Ann. Rev. Biophys. Bioeng.*, **9**, 143–179.

Szabo, A. and Karplus, M. (1972). A mathematical model for structure-function relations in hemoglobin. *J. Mol. Biol.*, **72**, 163–197. [Cited in Chapter 5. This is an important theoretical effort to relate ideas of allosteric regulation to a more detailed picture of protein structure.]

Tainer, J. A., Getzoff, E. D., Richardson, J. S. and Richardson, D. C. (1983). Structure and mechanism of copper, zinc superoxide dismutase. *Nature*, **306**, 284–287.

Tanford, C. (1955). Hydrogen ion titration curves of proteins. In *Electrochemistry in Biology and Medicine* (ed. T. Shedlovsky). New York: John Wiley & Sons, pp. 248–265.

(1961). *Physical Chemistry of Macromolecules*. New York: John Wiley & Sons.

(1968). Protein denaturation. *Adv. Protein Chem.*, **23**, 121–282.

(1970). Protein denaturation Part C. Theoretical models for the mechanisms of denaturation. *Adv. Protein Chem.*, **24**, 1–95.

Tauc, L. (1962). Site of origin and propagation of spike in the giant neuron of Aplysia. *J. Gen. Physiol.*, **45**, 1077–1097.

Terada, S., Kinjo, M. and Hirokawa, N. (2000). Oligomeric tubulin in large transporting complex is transported via kinesin in squid giant axons. *Cell*, **103**, 141–155.

Tian, F. and Cross, T. A. (1999). Cation transport: an example of structural based selectivity. *J. Mol. Biol.*, **285**, 1993–2003.

Tidor, B. and Karplus, M. (1994). The contribution of vibrational entropy to molecular association: the dimerization of insulin. *J. Mol. Biol.*, **238**, 405–414.

Tucek, S. (1997). Is the R and R* dichotomy real? *TIPS*, **18**, 414–416.

Tytgat, J. and Hess, P. (1992). Evidence for cooperative interactions in potassium channel gating. *Nature*, **359**, 420–423.

Ussing, H. H. (1949). The distinction by means of tracers between active transport and diffusion. *Acta Physiologica Scand.*, **19**, 43–56.

Van Kampen, N. (1981). *Stochastic Processes in Physics and Chemistry*. New York: North Holland.

Wallace, B. A. (1990). Gramicidin channels and pores. *Ann. Rev. Biophys. Biophys. Chem.*, **19**, 127–157.

Wang, W., Donini, O., Reyes, C. M. and Kollman, P. A. (2001). Biomolecular simulations: Recent developments in force fields, simulations of enzyme catalysis, protein–ligand, protein–protein, and protein–nucleic acid noncovalent interactions. *Ann. Rev. Biophys. Biomol. Struct.*, **30**, 211–243.

Waxman, S. G. and Swadlow, H. A. (1977). The conduction properties of axons in central white matter. *Progress Neurobiol.*, **8**, 297–324.

Warshel, A. and Levitt, M. (1976). Theoretical studies of enzymatic reactions: dielectric, electrostatic and steric stabilization of the carbonium ion in the reaction of lysozyme. *J. Mol. Biol.*, **103**, 227–249. [Cited in Chapter 10. A seminal paper using computational methods to investigate the energetics of enzyme catalysis.]

Weiner, M. C. and White, S. H. (1992). Structure of a fluid dioleoylphosphatidylcholine bilayer determined by joint refinement of x-ray and neutron diffraction data III. Complete structure. *Biophys. J.*, **61**, 434–447.

Weiner, S. J., Kollman, P. A., Nguyen, D. T. and Case, D. A. (1986). An all atom force field for simulations of proteins and nucleic acids. *J. Comput. Chem.*, **7**, 230–252.

Weinhold, F. (1997). Nature of H-bonding in clusters, liquids, and enzymes: an ab initio, natural bond perspective. *J. Mol. Struct. (Theochem).*, **398–399**, 181–197.

Wells, T. N. C. and Fersht, A. R. (1986). Use of binding energy in catalysis analyzed by mutagenesis of the tyrosyl-tRNA synthetase. *Biochemistry*, **25**, 1881.

Wess, J., Gdula, D. and Brann, M. R. (1990). Site-directed mutagenesis of the m3 muscarinic receptor: identification of a series of threonine and tyrosine residues involved in agonist but not antagonist binding. *EMBO J.*, **10**, 3729–3734.

Wilson, E. B., Decius, J. C. and Cross, P. C. (1955). *Molecular Vibrations*, chapter 8. New York: Dover Publications, Inc.

Wu, N., Mo, Y., Gao, J. and Pai, E. F. (2000). Electrostatic stress in catalysis: structure and mechanism of the enzyme orotidine monophosphate decarboxylase. *Proc. Natl Acad. Sci.*, **97** (5), 2017–2022.

Yue, K., Fiebig, K. M., Thomas, P. D., Chan, H. S., Shakhnovich, E. I. and Dill, K. A. (1995). A test of lattice protein folding algorithms. *Proc. Natl Acad. Sci.*, **92**, 325–329.

Zeltwanger, S., Wang, F., Wang, G.-T., Gilles, K. D. and Hwang, T.-C. (1999). Gating of the cystic fibrosis transmembrane conductance regulator chloride channels by adenosine triphosphate hydrolysis: quantitative analysis of a cyclic gating scheme. *J. Gen. Physiol.*, **113**, 541–554.

Zerangue, N. and Kavanaugh, M. P. (1996) Flux coupling in a neuronal glutamate transporter. *Nature*, **383**, 634–637.

Zhang, S. J. and Jackson, M. B. (1995). Properties of the GABAa receptor of rat posterior pituitary nerve terminals. *J. Neurophysiol.*, **73**, 1135–1144.

Zhong, W., Gallivan, J. P., Zhang, Y., Li, L., Lester, H. A. and Dougherty, D. A. (1998). From *ab initio* quantum mechanics to molecular neurobiology: a cation-π binding site in the nicotinic receptor. *Proc. Natl Acad. Sci.*, **95**, 12 088–12 093.

Zimm, B. H. and Bragg, J. K. (1959). Theory of the phase transition between helix and random coil in polypeptide chains. *J. Chem. Phys.*, **31**, 526–533.

Index

Page numbers in italics refer to figures. Page numbers in bold denotes entries in tables.